BIOZONE

Student Edition

Textbook Lite | Activities | Study Guide

AP®
BIOLOGY 2

AP BIOLOGY 2

Student Edition

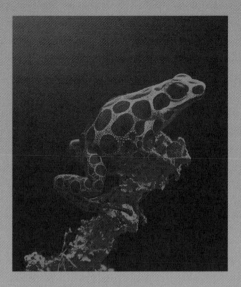

Meet the writing team

Tracey
Senior Author

Tracey Greenwood
I have been writing resources for students since 1993. I have a Ph.D in biology, specialising in lake ecology and I have taught both graduate and undergraduate biology.

Kent
Author

Kent Pryor
I have a BSc from Massey University majoring in zoology and ecology and taught secondary school biology and chemistry for 9 years before joining BIOZONE as an author in 2009.

Lissa
Author

Lissa Bainbridge-Smith
I worked in industry in a research and development capacity for 8 years before joining BIOZONE in 2006. I have a M.Sc from Waikato University.

Richard
Founder & CEO

Richard Allan
I have had 11 years experience teaching senior secondary school biology. I have a Masters degree in biology and founded BIOZONE in the 1980s after developing resources for my own students.

Cover photograph

The splash-back poison frog (*Ranitomeya variabilis*) is a small species of poison dart frog widely distributed throughout Peru, Ecuador, and adjacent lowlands. It is semi-arboreal, occupying the understory and canopy and breeding in bromeliads. Like other poison frogs, it contains alkaloid poisons. The well known mimic poison frog is a Müllerian mimic of this species.

PHOTO: Dirk Ercken
fineartamerica.com

Thanks to:

The staff at BIOZONE, including Mike Campbell and Holly Coon for design and graphics support, Paolo Curray and Malaki Toleafoa for IT support, Allan Young and Arahi Hippolite for office handling and logistics, and the BIOZONE sales team.

Special thanks to **Jason Crean** and **Cherylann Hollinger** for their advice and lively discussion throughout.

Second edition 2017
Second printing with corrections

ISBN: 978-1-927309-65-0

Copyright © 2017 Richard Allan
Published by BIOZONE International Ltd

Printed by REPLIKA PRESS PVT LTD using paper produced from renewable and waste materials

Purchases of this workbook may be made direct from the publisher:

BIOZONE Corporation
USA and Canada
FREE phone: 1-855-246-4555
FREE fax: 1-855-935-3555
Email: sales@thebiozone.com
Web: www.thebiozone.com

Contents

CODES: **Activity** is marked: ☐ to be done   ☑ when completed

Contents

CODES: **Activity** is marked: • to be done   ✓ when completed

Contents

CODES: **Activity** is marked: ⬚● to be done   ✓ when completed

Using The Student Edition

Activities make up most of this book. These are usually presented as short instructional sequences allowing you to build a deeper understanding of core concepts and content as you progress through each chapter. Each activity is accompanied by questions or specific tasks for you to complete. Throughout each chapter, there is a varied and interesting mix of knowledge-based, data-driven, and inquiry based activities.

Book structure

▶ The outline of the book structure below will help you to navigate through the material in each chapter.

Chapter introduction
A checklist of essential knowledge and skills and their associated activities.

Activity pages
These make up the bulk of the book. Each covers one concept and leads in with a key idea capturing the main focus of the page.

Did you get it?
Tests your knowledge and understanding of key terms and ideas in the chapter.

Synoptic questions
These conclude each major section of work and can be used as a formal assessment of the content in the preceding chapters.

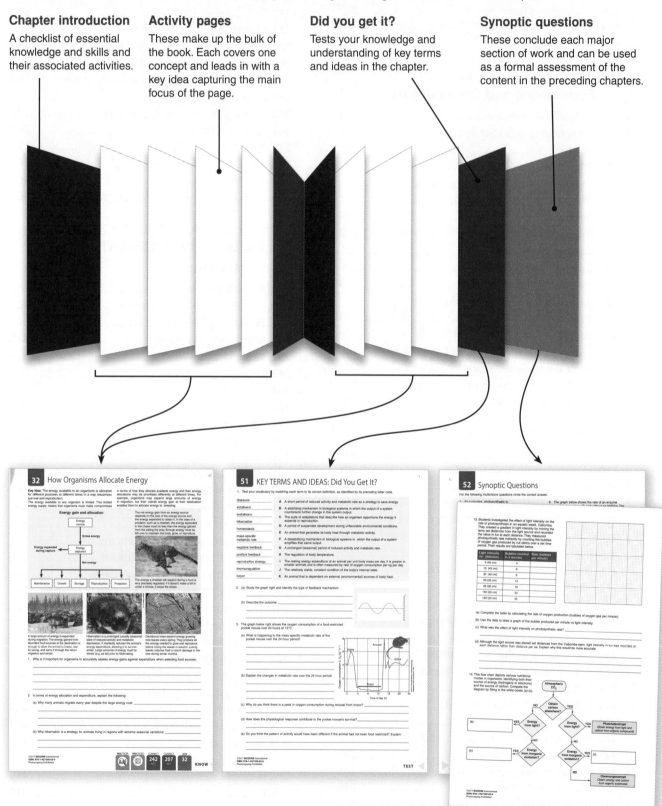

▶ Understanding the activity coding system and making use of the online material identified will enable you to get the most out of this book. The chapter content is structured to build knowledge and skills but this structure does not necessarily represent a strict order of treatment. Be guided by your teacher, who will assign activities as part of a wider program of independent and group-based work.

Look out for these features and know how to use them:

The chapter introduction provides a summary of the **essential knowledge** and **skills** required for the topic, phrased as a set of learning tasks. Use the check boxes to identify and mark off the points as you complete them. Activities to support required AP investigations are identified by a blue flag. A list of key terms for the chapter is also provided, from which you can construct your own glossary.

The activities form most of this workbook. They are numbered sequentially and each has a task code identifying the skill emphasized. Each activity has a short introduction with a key idea identifying the main message of the page. Most of the information is associated with illustrations, photographs, and diagrams, and your understanding of the content is reviewed through the questions. Some of the activities involve modeling and group work.

Free response questions allow you to use the information provided to answer questions about the content of the activity, either directly or by applying the same principles to a new situation. In some cases, an activity will assume understanding of prior content.

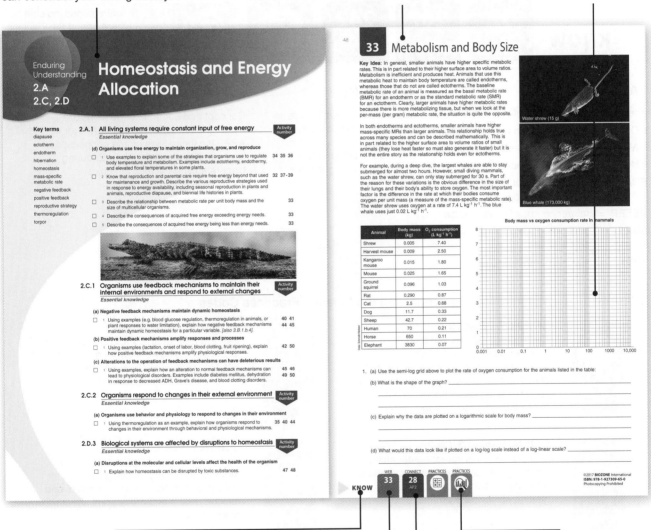

A **TASK CODE** identifies the type of activity. For example, is it primarily information-based (**KNOW**) or does the content prepare for or support one of the 13 AP Investigations (**PRAC**)? A full list of codes is given on the following page but the codes are relatively self explanatory.

WEB tabs alert the reader to the Weblinks resource, which provides external, online support for the activity, usually in the form of an animation, video clip, photo library, 3-D model, or quiz. Bookmark the Weblinks page (see page ix) and visit it often as you progress through the book.

CONNECT tabs identify related concepts and content across the entire AP program. The tab identifies the relevant activity number in either AP1 or AP2.

PRACTICES tabs identify where a specific science practice is emphasized. The codes are explained on the following page.

Using the Tab System

The tab system is a useful way to quickly identify science practices and connected ideas across the AP program. It also indicates the type of task involved and shows whether or not the activity is supported online.

▶ The **CONNECT** tabs indicate concept and content connections across Big Ideas and their Enduring Understandings. The connections made are some of the important ones, but you can make your own using the concept map provided on pages x-xi.

▶ The **PRACTICES** picture codes indicate that a particular Science Practice is emphasized in the activity. A guide to the picture codes and the science practices they represent is provided below. There may be more than one or none.

▶ The **WEBLINKS** code is always the same as the activity number on which it is cited. On visiting the WEBLINKS page (opposite), find the activity number and it will link you directly to one or more external websites providing supporting content or one of BIOZONE's spreadsheet activities. From this page you can also access a variety of categorized 3D models to support your work.

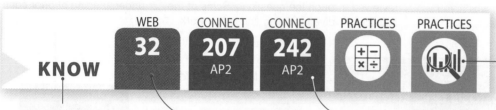

PRACTICES

Picture codes indicate where one of the 7 science practices is emphasized. See guide below.

TASK CODES

These identify the nature of the activity

COMP = comprehension of text

DATA = data handling and interpretation

KNOW = content you need to know

PRAC = supports one of the 13 AP Biology investigations

REFER = reference - use for information

SKILL = supporting a mathematical or practical skill

TEST = test your understanding

WEBLINKS

Bookmark the weblinks page: www.thebiozone.com/weblink/AP2-9650
Access the external URL for the activity by clicking the link.

CONNECT

Concept and content connections are made to other activities within AP1 or in AP2. Use these to reinforce connections between Big Ideas and their Enduring Understandings across all topics.

Recognizing the Science Practices Codes

SCIENCE PRACTICE 1
Use representations and models to communicate scientific phenomena and solve scientific problems. Includes creating, describing, refining, and using representations and models of natural or man-made phenomena and systems.

SCIENCE PRACTICE 2
Use mathematics appropriately, including justifying the use of mathematical routines, applying mathematical routines, and making numerical estimates.

SCIENCE PRACTICE 3
Engage in scientific questioning to extend thinking or to guide investigations, including posing, refining, and evaluating scientific questions

SCIENCE PRACTICE 4
Plan and implement data collection strategies appropriate to a particular scientific question. Includes posing, refining, and evaluating scientific questions as well as drawing conclusions from the experimental results of other scientists.

SCIENCE PRACTICE 5
Perform data analysis and evaluation of evidence, including analyzing data to identify patterns or relationships and evaluating evidence provided by data in relation to a particular question.

SCIENCE PRACTICE 6
Work with scientific explanations and theories, including justifying claims with evidence, constructing explanations and making claims and predictions about natural phenomena.

SCIENCE PRACTICE 7
Connect and relate knowledge across various scales, concepts, and representations in and across domains. Includes connecting phenomena and models across scales such as time, size, and complexity, and describing how enduring understandings and/or big ideas are connected.

BIOZONE's Online Resources

WEBLINKS is an online resource compiled by BIOZONE to enhance or extend the content provided in the activities, largely though explanatory animations and short videos. All external websites have been selected for their suitability and accuracy and regularly checked. From this page, you can also access a wide range of annotated 3D models provided by BIOZONE and check for any errata or clarifications to the book or model answers since printing.

www.thebiozone.com/weblink/AP2-9650

▶ This WEBLINKS page provides links to external websites and **3D models** supporting the activities.

▶ The external websites are, for the most part, narrowly focused animations and video clips directly relevant to some aspect of the activity on which they are cited. They provide great support to help your understanding.

▶ The comprehensive collection of annotated 3D models provides a different way to visualize and understand theoretical content. Choose those models relevant to your program or interests.

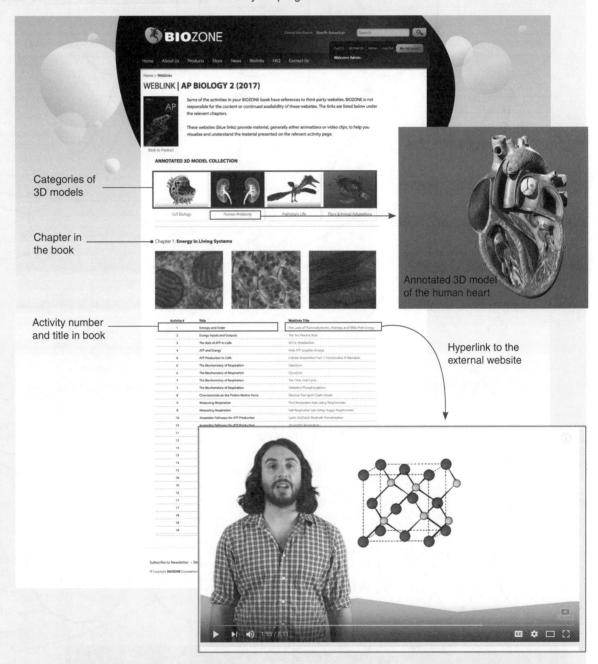

Categories of 3D models

Chapter in the book

Activity number and title in book

Annotated 3D model of the human heart

Hyperlink to the external website

Bookmark weblinks by typing in the address: it is not accessible directly from BIOZONE's website
Corrections and clarifications to current editions are always posted on the weblinks page

AP Biology: Concepts and Connections

This map shows the structure of the AP Biology program, showing the four **Big Ideas**, their **Enduring Understandings** in blue ovals and some of the connections between them. You can draw in your own connections and add labels with keys words to show how different parts of the program relate to each other. The white clouds with blue text indicate content in AP Biology 1. The gray clouds with gray text indicate content covered in AP Biology 2.

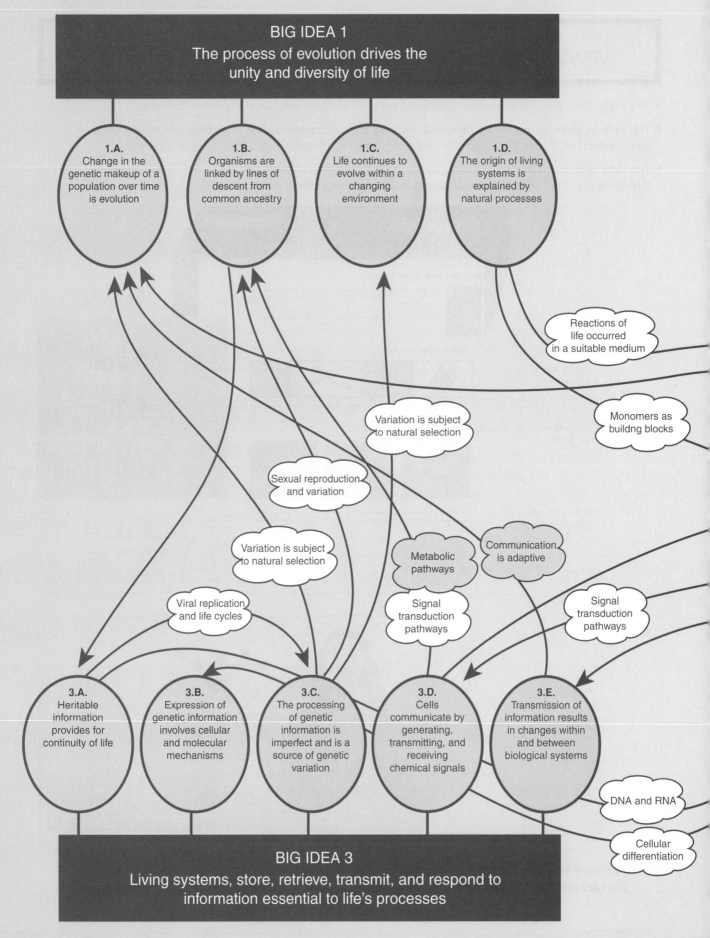

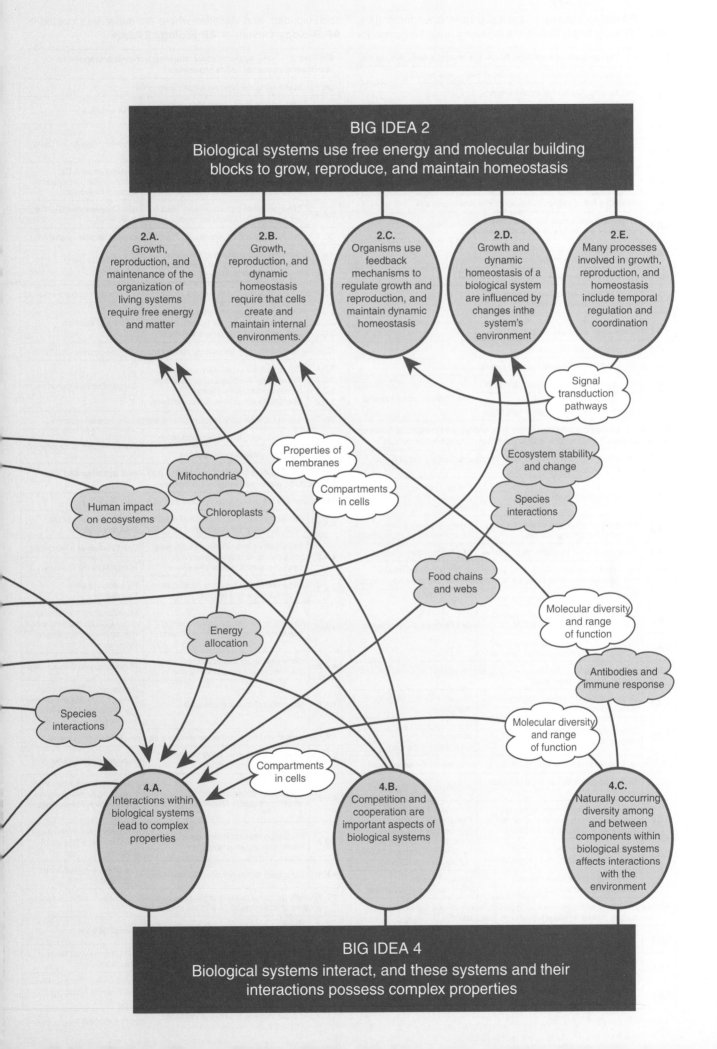

AP Biology Guide

The AP biology program is organized into four underlying big ideas. The guide below lists the enduring understandings for each big idea, and identifies where the material is located in **AP Biology 1** (blue) or **AP Biology 2** (black).

Big Idea 1: The process of evolution drives the diversity and unity of life

1A: Change in the genetic makeup of a population over time is evolution

1.A.1	Natural selection is a major mechanism of evolution	Genetic Change in Populations
1.A.2	Natural selection acts on phenotypic variations in populations	
1.A.3	Evolutionary change is also driven by random processes	
1.A.4	Biological evolution is supported by scientific evidence from many disciplines	Evidence for Biological Evolution

1B: Organisms are linked by lines of descent from common ancestry

1.B.1	Organisms share many conserved core processes and features that have evolved	The Relatedness of Organisms
1.B.2	Phylogenetic trees and cladograms are graphical models of evolutionary history	

1C: Life continues to evolve within a changing environment

1.C.1	Speciation and extinction have occurred throughout the Earth's history	Speciation and Extinction
1.C.2	Speciation may occur when populations become reproductively isolated	
1.C.3	Populations continue to evolve	

1D: The origin of living systems is explained by natural processes

1.D.1	Hypotheses about the natural origin of life	The Origin of Living Systems
1.D.2	Scientific evidence from different disciplines supports models of life's origin	

Big Idea 2: Biological systems utilize free energy and molecular building blocks to grow, to reproduce and to maintain dynamic homeostasis

2A: Growth, reproduction and maintenance of the organization of living systems require free energy and matter

2.A.1	All living systems require energy	Energy in Living Systems, Homeostasis & Energy Allocation
2.A.2	Organisms capture and store free energy for use in biological processes	Energy in Living Systems, Energy Flow & Nutrient Cycles
2.A.3	Energy exchange maintains life processes	The Biochemistry of Life

2B: Growth, reproduction and dynamic homeostasis require that cells create and maintain internal environments that are different from their external environments

2.B.1	Cell membranes are selectively permeable	Cell Structure and Processes
2.B.2	Movement of molecules across membranes maintains growth and homeostasis	
2.B.3	Internal membranes in eukaryotic cells partition the cell into specialized regions	

2C: Organisms use feedback mechanisms to regulate growth and reproduction, and to maintain dynamic homeostasis

2.C.1	Organisms used feedback mechanisms to maintain internal environments	Homeostasis & Energy Allocation
2.C.2	Organisms respond to change in their external environments	Homeostasis & Energy Allocation, Timing & Coordination

2D: Growth & dynamic homeostasis are influenced by changes in the environment

2.D.1	Biotic and abiotic factors affect biological systems	Populations & Communities
2.D.2	Homeostatic mechanisms reflect both common ancestry and divergence due to adaptation in different environments	Homeostasis & Energy Allocation, Plant Structure & Adaptation, Comparing Animal Systems, Interactions in Physiological Systems, The Diversity & Stability of Ecosystems (2.D.3 only)
2.D.3	Biological systems are affected by disruptions to their dynamic homeostasis*	
2.D.4	Plants and animals have chemical defenses against infections	Internal Defense, Plant Structure & Adaptation

2E: Many biological processes involved in growth, reproduction & dynamic homeostasis include temporal regulation & coordination

2.E.1	Timing and coordination of events are regulated and necessary for development	Regulation of Gene Expression
2.E.2	Multiple mechanisms regulate timing & coordination of physiological events	Timing & Coordination
2.E.3	Timing and coordination are regulated and are important in natural selection	

Big Idea 3: Living systems store, retrieve, transmit & respond to information essential to life processes

3A: Heritable information provides for continuity of life

3.A.1	DNA, and in some cases RNA, is the primary source of heritable information	DNA and RNA
3.A.2	In eukaryotes, heritable information is passed on via the cell cycle and mitosis or meiosis plus fertilization	Chromosomes & Cell Division
3.A.3	The chromosomal basis of inheritance gives an understanding of transmission of genes from parent to offspring	Chromosomes & Cell Division, The Chromosomal Basis of Inheritance
3.A.4	The inheritance pattern of many traits is not explained by Mendelian genetics	The Chromosomal Basis of Inheritance

3B: Expression of genetic information involves cellular and molecular mechanisms

3.B.1	Gene regulation results in differential gene expression and cell specialization	Regulation of Gene Expression
3.B.2	Signals mediate gene expression	

3C: Processing of genetic information is imperfect and a source of genetic variation

3.C.1	Genotype changes can alter phenotype	Sources of Variation
3.C.2	Processes that increase genetic variation	
3.C.3	Viral replication and genetic variation	

3D: Cells communicate by generating, transmitting and receiving chemical signals

3.D.1	Commonalities in cell communication	Cellular Communication
3.D.2	Signaling by direct contact or chemicals	
3.D.3	Signal transduction pathways	
3.D.4	Changes to signal transduction pathways	

3E: Transmission of information results in changes within and between systems

3.E.1	Communicating information with others	Communicating & Responding
3.E.2	Nervous systems and responses	

Big Idea 4: Biological systems interact, and these systems and their interactions possess complex properties

4A: Interactions within biological systems lead to complex properties

4.A.1	Properties of a molecule are determined by its molecular construction	The Biochemistry of Life, DNA and RNA
4.A.2	The structure and function of subcellular components, and their interactions, provide essential cellular processes	Cell Structure and Processes / Energy in Living Systems
4.A.3	Gene expression results in specialization of cells, tissues and organs	Regulation of Gene Expression
4.A.4	Organisms exhibit complex properties due to interactions between their parts	Plant Structure & Adaptation, Comparing Animal Systems, Interactions in Physiological Systems
4.A.5	Communities are composed of populations that interact in complex ways	Populations and Communities
4.A.6	Movement of matter and energy	Populations & Communities, Energy Flow & Nutrient Cycles, The Diversity and Stability of Ecosystems

4B: Competition and cooperation are important aspects of biological systems

4.B.1	Interactions between molecules affect their structure and function	Enzymes & Metabolism
4.B.2	Cooperative interactions within organisms promote efficiency	Plant Structure & Adaptation, Comparing Animal Systems, Interactions in Physiological Systems
4.B.3	Population interactions influence species distribution and abundance	Populations & Communities, The Diversity & Stability of Ecosystems
4.B.4	Ecosystem distribution changes over time	The Diversity & Stability of Ecosystems

4C: Naturally occurring diversity among and between components within biological systems affects interactions with the environment

4.C.1*	Variation in molecular units provides cells with a wider range of functions	Internal Defense
4.C.2*	Environmental factors influence the expression of the genotype	The Chromosomal Basis of Inheritance
4.C.3	Variation in populations affects dynamics	Populations & Communities, The Diversity & Stability of Ecosystems
4.C.4	Diversity may influence ecosystem stability	

* 4.C.1 and 4.C.2 also see Sources of Variation

©2017 **BIOZONE** International

Energy in Living Systems

Key terms

ATP/ADP
alcoholic fermentation
autotroph
Calvin cycle
cellular respiration
chemiosmosis
chemoautotroph
chloroplast
cristae
electron carrier
electron transport chain
endergonic
entropy
exergonic
FAD/FADH$_2$
fermentation
glycolysis
grana
heterotroph
Krebs cycle
lactic acid fermentation
light dependent reactions
light independent reactions
link reaction
matrix
mitochondrion
NAD$^+$/NADH
NADP$^+$/NADPH
oxidative phosphorylation
photoautotroph
photosynthesis
photosystem
stroma
substrate-level phosphorylation
thylakoid membranes

2.A.1 All living systems require a constant input of free energy

Essential knowledge

Activity number

(a) Life requires a highly ordered system

☐ 1 Explain how biological order is maintained by constant free energy input into the system, be that cell, organism, or ecosystem. **1 2**

☐ 2 Understand that the loss of order or free energy flow results in death. **1 2**

☐ 3 Using examples, explain how increased disorder and entropy are offset by biological processes that maintain or increase order. **1 2**

(b) Living systems do not violate the second law of thermodynamics

☐ 1 State the second law of thermodynamics. Explain how order is maintained by coupling cellular processes that increase entropy (negative ΔG) with those that decrease entropy (positive ΔG). **1**

☐ 2 Explain why energy input must exceed free energy lost to the environment to maintain order and power cellular processes. **1**

☐ 3 Using examples, explain why exergonic reactions can be used to maintain or increase order in a system by being coupled with endergonic reactions. **1 5**

Dartmouth College   Kristian Peters   Dartmouth College

(c) Energy-related pathways are sequential with multiple entry points

☐ 1 Use examples to describe how energy-related pathways in biological systems are sequential and may be entered at multiple points. Examples include: **5**
   • Krebs cycle in which intermediates are transformed sequentially. **6 7**
   • Glycolysis in which glucose is phosphorylated and split into 2 x 3C molecules. **7**
   • Calvin cycle in which three sequential steps are linked by intermediates. **17**
   • Fermentation in which the end product of glycolysis, pyruvate, is broken down into either acetaldehyde (then ethanol) or lactic acid. **10**

2.A.2 Organisms capture and store free energy

Essential knowledge

Activity number

(a) Autotrophs capture free energy from physical sources in the environment

☐ 1 Describe the role of photosynthetic organisms in capturing the free energy in sunlight as organic molecules (sugars). Explain what these sugars are used for. **2 5 14**

☐ 2 Understand the basis by which chemosynthetic organisms manufacture food and know that the process may be anaerobic or aerobic. **2**

(b) Heterotrophs capture free energy present in carbon compounds produced by other organisms

☐ 1 Explain how heterotrophs metabolize carbon compounds (e.g. carbohydrates, proteins, and lipids) produced by other organisms. **2 7**

☐ 2 Outline fermentation in yeasts and mammalian muscle and their end products. **10**

(c) Different energy-capturing processes use different type of electron acceptors

☐ 1 Identify the electron acceptors in photosynthesis and cellular respiration. **7 16**

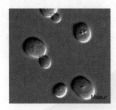

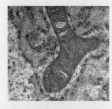

(d) The light-dependent reactions of photosynthesis capture the free energy in light

☐ 1 Describe the role of chlorophylls in photosynthesis. 12 15

☐ 2 Describe the location and organization of the photosystems and explain how they are connected. 12 16

☐ 3 Explain how transfer of electrons between molecules in the ETC establishes an electrochemical gradient of protons across the thylakoid membrane. 8 16

☐ 4 Explain how the proton-motive force is used to drive the synthesis of ATP via ATP synthase. 8 16

☐ 5 Identify the products of the light dependent reactions and explain their role in the Calvin cycle in the stroma of the chloroplast. 16 17

PR-5 ▶ Investigate factors affecting photosynthesis in leaf disks. 18

(e) Photosynthesis first evolved in prokaryotes

☐ 1 Describe the scientific evidence supporting the hypothesis that: 13
i prokaryotic photosynthesis was responsible for an oxygenated atmosphere,
ii eukaryotic photosynthesis evolved from prokaryotic photosynthetic pathways.

(f) Eukaryotic cellular respiration involves a series of coordinated enzyme-catalyzed reactions that harvest free energy from sunlight

☐ 1 Describe the events in glycolysis, identifying the inputs and outputs. 6 7

☐ 2 Describe the fate of the pyruvate in aerobic cellular respiration. 6 7

☐ 3 Outline the events in the Krebs cycle, identifying inputs and outputs. 7

☐ 4 Describe what happens to the electrons carried from the Krebs cycle by NADH and $FADH_2$ 7

(g) The electron transport chain captures free energy from electrons

☐ 1 Identify the location of electron transport chain reactions in prokaryotes, and in chloroplasts and mitochondria. 7 8

☐ 2 Outline the events in the electron transport chain in cellular respiration, identifying the terminal electron acceptor. Recall that the terminal electron acceptor in photosynthesis is $NADP^+$. 7 8

☐ 3 Describe how and where a proton gradient is established in mitochondria and chloroplasts, and in prokaryotes. 7 8 16

☐ 4 Explain how the flow of electrons down their concentration gradient va ATP synthase is used to generate ATP by chemiosmosis. Understand that this process also regenerates electron carriers. 7 8

☐ 5 Describe how decoupling oxidative phosphorylation from electron transport is involved in thermoregulation. 8

PR-6 ▶ Use a simple respirometer to measure the rate of cellular respiration in living organisms. 9

(h) The conversion of ATP to ADP provides free energy for metabolism

☐ 1 Describe how the energy provided by the hydrolysis of ATP is used in metabolism. 4

4.A.2 The structure an function of subcellular components and their interactions provide essential cellular processes

Activity number

Essential knowledge

(d) Mitochondria specialize in energy capture and transformation

☐ 1 Describe how the compartments in mitochondria relate to their function. 3 6 7

☐ 2 Describe the features of the outer and inner mitochondrial membranes. 3 7 8

☐ 3 Explain how the structure and features of the cristae relate to their functional role. 3 7 8

(g) Chloroplasts are specialized organelles that capture energy through photosynthesis

☐ 1 Describe how the relationship between structure and function in a chloroplast is related to its role in photosynthesis. 12

☐ 2 Describe the location and role of chlorophylls in chloroplasts. 12 15

☐ 3 Describe the compartmentalized structure of chloroplasts and relate this to the two stages of photosynthesis (energy capture and carbon fixation). 8 12 15 16 17

1 Entropy and Order

Key Idea: Organisms follow the laws of thermodynamics by using energy to maintain order.

Thermodynamics lays out the fundamental laws of energy that govern the universe. Firstly, the energy in an isolated system is constant. Secondly, disorder (**entropy**) increases over time. In other words, to maintain order within a system, there must be an input of energy. Living organisms follow these laws by using energy in the form of light (in plants) or food (in animals) to drive the chemical reactions that maintain order in their bodies. Without the input of energy, order in a living system is quickly lost and organisms die. The disorder in their constituent parts then continues to increase until it reaches an equilibrium with the environment.

Entropy

Entropy is a measure of the disorder in a system, or the amount of energy not available to do work, and increases with time. The greater the entropy, the greater the disorder in the system.

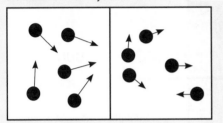

Hot   Cold

The system shown above has order. Hot, fast-moving molecules (high energy) are separated from cold, slow moving molecules (low energy).

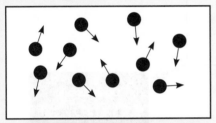

If the separation in the system is removed, the high energy molecules move into the area of low energy molecules. Collisions of the high energy molecules with the low energy molecules decrease the energy of the high energy molecules and increase the energy of the low energy molecules. Energy is also lost as heat. Eventually all molecules will have the same energy (equilibrium is reached). Entropy in the system has increased to its maximum level.

Order and the cell

Cells maintain order by using energy, either directly or indirectly. ATP can be used to directly drive a chemical reaction or it can used to produce a gradient, which is then used to move molecules into or out of the cell.

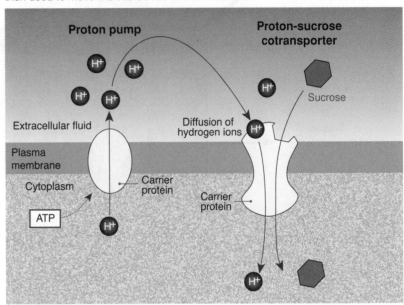

The proton-sucrose co-transporter shown above is common in the plasma membranes of plant cells. Energy in ATP is used to pump protons out of the cell, decreasing entropy with respect to the protons and producing a gradient. Sucrose is coupled to the flow of protons back down the gradient (increasing entropy with respect to the protons). Thus as the entropy with respect to the protons increases, the entropy with respect to the sucrose decreases.

The arrow of time

Entropy provides an answer as to why time moves forward (never backwards) and why we age. The universe is (we assume) an isolated system and its entropy can never decrease. Thus time is a result of the entropy of the universe moving towards its maximum state. Aging can be viewed as an increase in the body's entropy. Death is effectively the highest and therefore most thermodynamically favored level of entropy. So favorable, in fact, that no amount of energy input can prevent it.

1. (a) Define entropy: _____

_____

(b) Why has the entropy in the hot-cold system in the blue panel increased in the second image?_____

_____

_____

2. (a) Explain how living organisms maintain order in their cells: _____

_____

(b) How does the proton-sucrose co-transporter decrease entropy with respect to the sucrose inside and outside a cell.

_____

_____

_____

©2017 **BIOZONE** International
ISBN: 978-1-927309-65-0
Photocopying Prohibited

2 Energy Inputs and Outputs

Key Idea: Autotrophs manufacture their food using the energy from sunlight or chemicals. Heterotrophs consume other organisms to obtain their energy and carbon.

Living things obtain their energy for metabolism in two main ways. **Autotrophs** (producers) use the energy in sunlight or inorganic molecules to make their own food. **Heterotrophs** (consumers) rely on other organisms as a source of energy and carbon. All other organisms depend on producers, even if they do not consume them directly. The energy flow into and out of each trophic (feeding) level can be represented on a diagram using arrows of different sizes to represent relative amounts of energy lost from different trophic levels.

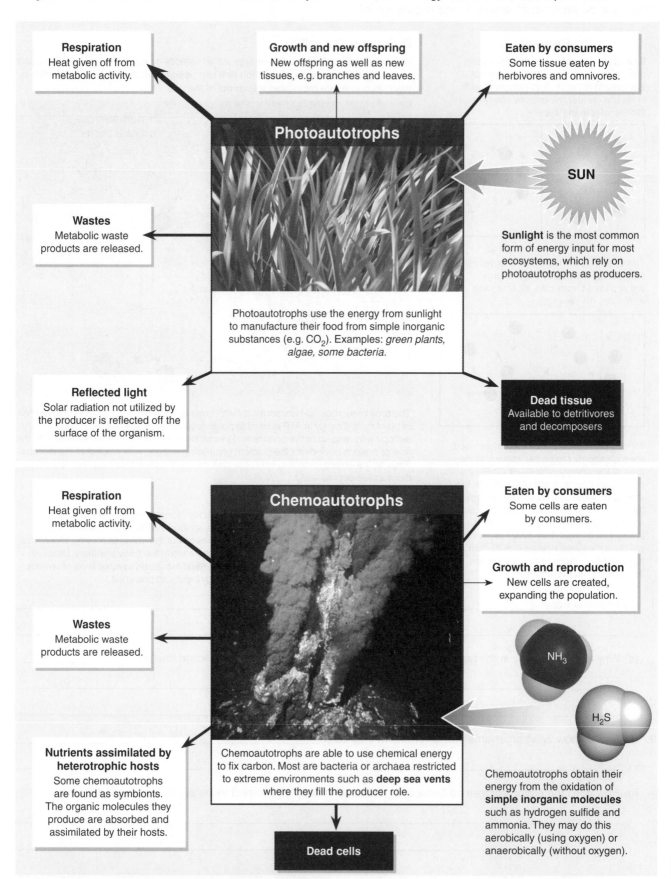

Respiration
Heat given off from metabolic activity.

Growth and new offspring
New offspring as well as new tissues, e.g. branches and leaves.

Eaten by consumers
Some tissue eaten by herbivores and omnivores.

Photoautotrophs

SUN

Sunlight is the most common form of energy input for most ecosystems, which rely on photoautotrophs as producers.

Wastes
Metabolic waste products are released.

Photoautotrophs use the energy from sunlight to manufacture their food from simple inorganic substances (e.g. CO_2). Examples: *green plants, algae, some bacteria*.

Reflected light
Solar radiation not utilized by the producer is reflected off the surface of the organism.

Dead tissue
Available to detritivores and decomposers

Respiration
Heat given off from metabolic activity.

Chemoautotrophs

Eaten by consumers
Some cells are eaten by consumers.

Growth and reproduction
New cells are created, expanding the population.

Wastes
Metabolic waste products are released.

NH_3

H_2S

Nutrients assimilated by heterotrophic hosts
Some chemoautotrophs are found as symbionts. The organic molecules they produce are absorbed and assimilated by their hosts.

Chemoautotrophs are able to use chemical energy to fix carbon. Most are bacteria or archaea restricted to extreme environments such as **deep sea vents** where they fill the producer role.

Chemoautotrophs obtain their energy from the oxidation of **simple inorganic molecules** such as hydrogen sulfide and ammonia. They may do this aerobically (using oxygen) or anaerobically (without oxygen).

Dead cells

WEB **2**

CONNECT **232** AP2

CONNECT **233** AP2

CONNECT **237** AP2

PRACTICES

KNOW

©2017 **BIOZONE** International
ISBN: 978-1-927309-65-0
Photocopying Prohibited

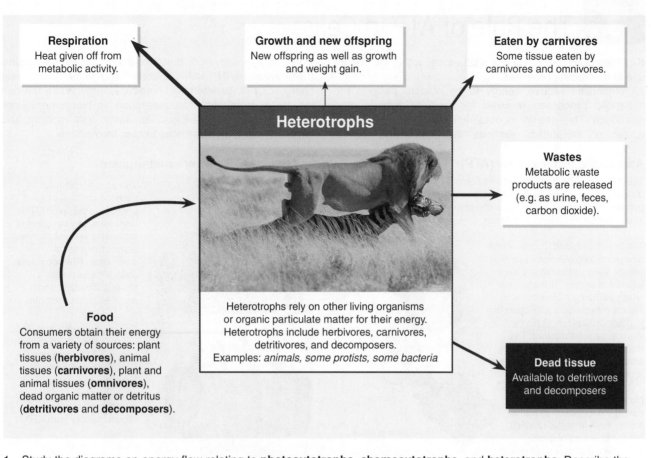

Respiration
Heat given off from metabolic activity.

Growth and new offspring
New offspring as well as growth and weight gain.

Eaten by carnivores
Some tissue eaten by carnivores and omnivores.

Heterotrophs

Wastes
Metabolic waste products are released (e.g. as urine, feces, carbon dioxide).

Food
Consumers obtain their energy from a variety of sources: plant tissues (**herbivores**), animal tissues (**carnivores**), plant and animal tissues (**omnivores**), dead organic matter or detritus (**detritivores** and **decomposers**).

Heterotrophs rely on other living organisms or organic particulate matter for their energy. Heterotrophs include herbivores, carnivores, detritivores, and decomposers.
Examples: *animals, some protists, some bacteria*

Dead tissue
Available to detritivores and decomposers

1. Study the diagrams on energy flow relating to **photoautotrophs**, **chemoautotrophs**, and **heterotrophs**. Describe the differences between these categories of organisms with respect to how they obtain their energy:

_____

_____

_____

_____

_____

2. Describe how energy may be lost from organisms in the form of:

(a) Wastes: _____

(b) Respiration: _____

3. Explain why so little energy is available for growth and reproduction, regardless of trophic group: _____

_____

_____

_____

4. Describe the ecological importance of chemoautotrophic organisms in deep sea environments: _____

_____

_____

_____

5. In what way is the chemoautotrophic system of the deep sea thermal vent linked to other ecological systems?

_____

_____

_____

©2017 **BIOZONE** International
ISBN: 978-1-927309-65-0
Photocopying Prohibited

3 The Role of ATP in Cells

Key Idea: ATP transports chemical energy within the cell for use in metabolic processes.

All organisms require energy to be able to perform the metabolic processes required for them to function and reproduce. This energy is obtained by cellular respiration, a set of metabolic reactions that ultimately convert biochemical energy from 'food' into the nucleotide **adenosine triphosphate** (ATP). ATP is considered to be a universal energy carrier, transferring chemical energy within the cell for use in metabolic processes such as biosynthesis, cell division, cell signaling, thermoregulation, cell mobility, and active transport of substances across membranes.

Adenosine triphosphate (ATP)

The ATP molecule consists of three components; a purine base (**adenine**), a pentose sugar (**ribose**), and **three phosphate groups** which attach to the 5' carbon of the pentose sugar. The structure of ATP is described below.

The bonds between the phosphate groups contain electrons in a high energy state which store a large amount of energy. The energy is released during ATP hydrolysis. Typically, hydrolysis is coupled to another cellular reaction to which the energy is transferred. The end products of the reaction are adenosine diphosphate (ADP) and an inorganic phosphate (Pi).

Note that energy is released during the formation of bonds during the hydrolysis reaction, not the breaking of bonds between the phosphates (which requires energy input).

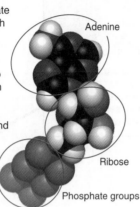

Adenine

Ribose

Phosphate groups

The mitochondrion

Cellular respiration and ATP production occur in mitochondria. A mitochondrion is bounded by a double membrane. The inner and outer membranes are separated by an intermembrane space, compartmentalizing the regions where the different reactions of cellular respiration take place.

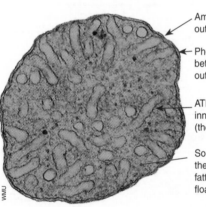

Amine oxidases on the outer membrane surface

Phosphorylases between the inner and outer membranes

ATPases on the inner membranes (the cristae)

Soluble enzymes for the Krebs cycle and fatty acid degradation floating in the matrix

ATP powers metabolism

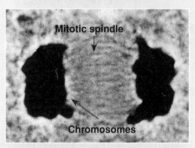

Solid particle

The energy released from the removal of a phosphate group of ATP is used for active transport of molecules and substances across the plasma membrane. **Phagocytosis** (left), which involves the engulfment of solid particles, is an example.

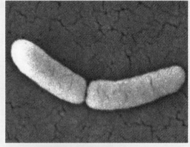

Mitotic spindle

Chromosomes

Cell division (**mitosis**), as observed in this onion cell, requires ATP to proceed. Formation of the mitotic spindle and chromosome separation are two aspects of cell division which require energy from ATP hydrolysis.

ATP is required when bacteria divide by binary fission (left). For example, ATP is required in DNA replication and to synthesize components of the peptidoglycan cell wall.

Not all of the energy released in the oxidation of glucose is captured in ATP. The rest is lost as heat. This heat energy is used to maintain body temperature in some animals. Thermoregulatory mechanisms such as shivering and sweating also involve energy expenditure.

1. Why do organisms need to respire? _____

_____

2. (a) Describe the general role of mitochondria in cell respiration: _____

_____

(b) Explain the importance of compartmentalization in the mitochondrion: _____

_____

3. Explain why thermoregulation is associated with energy expenditure: _____

_____

_____

WEB **3**   CONNECT **35** AP2   CONNECT **44** AP2   CONNECT **134** AP1   PRACTICES

KNOW

©2017 **BIOZONE** International
ISBN: 978-1-927309-65-0
Photocopying Prohibited

4 ATP and Energy

Key Idea: ATP is the universal energy carrier in cells. Energy is stored in the covalent bonds between phosphate groups. The molecule ATP (adenosine triphosphate) is the universal energy carrier for the cell. ATP can release its energy quickly by hydrolysis of the terminal phosphate. This reaction is catalyzed by the enzyme ATPase. Once ATP has released its energy, it becomes ADP (adenosine diphosphate), a low energy molecule that can be recharged by adding a phosphate. The energy to do this is supplied by the controlled breakdown of respiratory substrates in cellular respiration.

How does ATP provide energy?

ATP releases its energy during hydrolysis. Water is split and added to the terminal phosphate group resulting in ADP and Pi. For every mole of ATP hydrolyzed **30.7 kJ** of energy is released. Note that energy is released during the formation of chemical bonds not from the breaking of chemical bonds.

The reaction of A + B is endergonic. It requires energy to proceed and will not occur spontaneously.

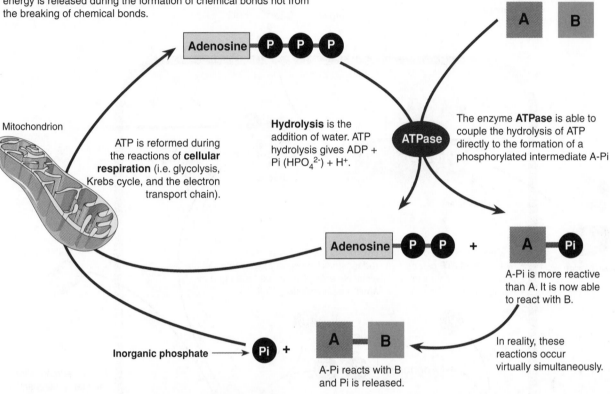

Mitochondrion

ATP is reformed during the reactions of **cellular respiration** (i.e. glycolysis, Krebs cycle, and the electron transport chain).

Hydrolysis is the addition of water. ATP hydrolysis gives ADP + Pi (HPO_4^{2-}) + H^+.

The enzyme **ATPase** is able to couple the hydrolysis of ATP directly to the formation of a phosphorylated intermediate A-Pi

A-Pi is more reactive than A. It is now able to react with B.

In reality, these reactions occur virtually simultaneously.

Inorganic phosphate → Pi +

A-Pi reacts with B and Pi is released.

Note! The phosphate bonds in ATP are often referred to as high energy bonds. This can be misleading. The bonds contain *electrons in a high energy state* (making the bonds themselves relatively weak). A small amount of energy is required to break the bonds, but when the intermediates recombine and form new chemical bonds a large amount of energy is released. The final product is less reactive than the original reactants.

In many textbooks the reaction series above is simplified and the intermediates are left out:

1. (a) How does ATP supply energy to power metabolism? _____

_____

(b) In what way is the ADP/ATP system like a rechargeable battery? _____

_____

2. What is the immediate source of energy for reforming ATP from ADP? _____

3. Explain the purpose of the folded inner membrane in mitochondria: _____

_____

_____

4. Explain why highly active cells (e.g. sperm cells) have large numbers of mitochondria: _____

_____

_____

©2017 **BIOZONE** International
ISBN: 978-1-927309-65-0
Photocopying Prohibited

PRACTICES   WEB

4

KNOW

5 Energy Transformations in Cells

Key Idea: Photosynthesis uses energy from the sun to produce glucose. Glucose breakdown produces ATP, which is used by all cells to provide the energy for metabolism.
A summary of the flow of energy within a plant cell is illustrated below. Heterotrophic cells (animals and fungi) have a similar flow except the glucose is supplied by ingestion or absorption of food molecules rather than by photosynthesis. The energy not immediately stored in chemical bonds is lost as heat. Note that ATP provides the energy for most metabolic reactions, including photosynthesis.

Summary of energy transformations in a photosynthetic plant cell

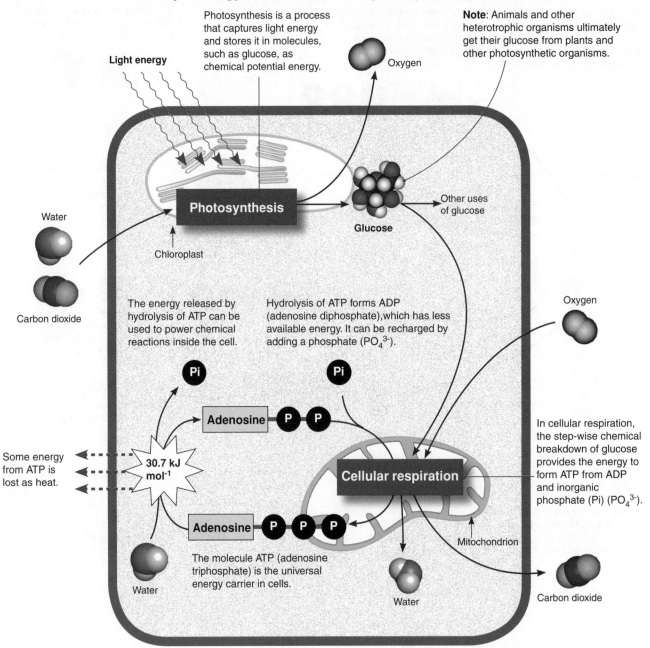

Photosynthesis is a process that captures light energy and stores it in molecules, such as glucose, as chemical potential energy.

Light energy

Oxygen

Note: Animals and other heterotrophic organisms ultimately get their glucose from plants and other photosynthetic organisms.

Water

Carbon dioxide

Photosynthesis

Chloroplast

Other uses of glucose

Glucose

The energy released by hydrolysis of ATP can be used to power chemical reactions inside the cell.

Hydrolysis of ATP forms ADP (adenosine diphosphate), which has less available energy. It can be recharged by adding a phosphate (PO_4^{3-}).

Oxygen

Pi

Pi

Adenosine P P

Some energy from ATP is lost as heat.

30.7 kJ mol^{-1}

Cellular respiration

In cellular respiration, the step-wise chemical breakdown of glucose provides the energy to form ATP from ADP and inorganic phosphate (Pi) (PO_4^{3-}).

Adenosine P P P

The molecule ATP (adenosine triphosphate) is the universal energy carrier in cells.

Water

Mitochondrion

Water

Carbon dioxide

1. (a) What are the raw materials for photosynthesis? _____

_____

(b) What are the raw materials for respiration? _____

_____

2. What is the immediate source of energy for reforming ATP from ADP? _____

3. What is the ultimate source of energy for plants? _____

4. What is the ultimate source of energy for animals? _____

8
AP2

KNOW

©2017 **BIOZONE** International
ISBN: 978-1-927309-65-0
Photocopying Prohibited

6 ATP Production in Cells

Key Idea: Cellular respiration is the process by which the energy in glucose is transferred to ATP.

Cellular respiration can be **aerobic** (requires oxygen) or **anaerobic** (does not require oxygen). Plants and animals respire aerobically, although they may generate ATP anaerobically for short periods using only the first step in cellular respiration, glycolysis. Anaerobic bacteria use only anaerobic respiration and live in oxygen-free environments. Cellular respiration occurs in the cytoplasm and mitochondria and involves a series of linked reactions.

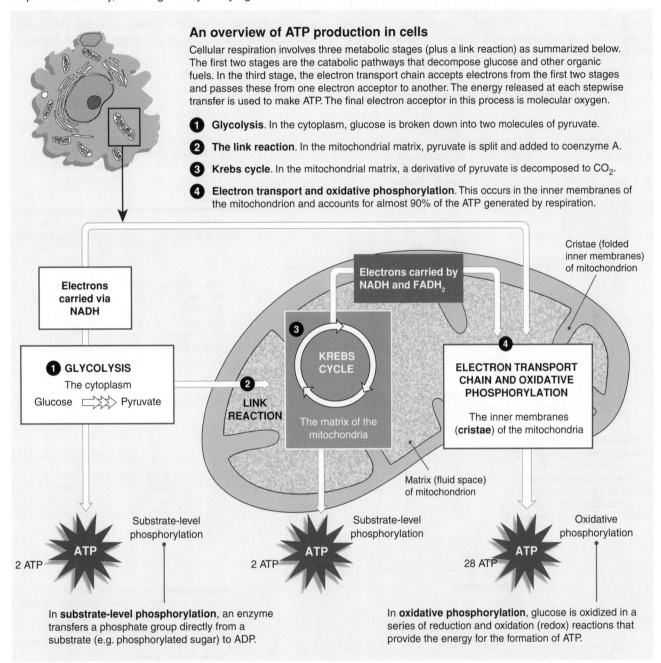

An overview of ATP production in cells

Cellular respiration involves three metabolic stages (plus a link reaction) as summarized below. The first two stages are the catabolic pathways that decompose glucose and other organic fuels. In the third stage, the electron transport chain accepts electrons from the first two stages and passes these from one electron acceptor to another. The energy released at each stepwise transfer is used to make ATP. The final electron acceptor in this process is molecular oxygen.

1. **Glycolysis**. In the cytoplasm, glucose is broken down into two molecules of pyruvate.

2. **The link reaction**. In the mitochondrial matrix, pyruvate is split and added to coenzyme A.

3. **Krebs cycle**. In the mitochondrial matrix, a derivative of pyruvate is decomposed to CO_2.

4. **Electron transport and oxidative phosphorylation**. This occurs in the inner membranes of the mitochondrion and accounts for almost 90% of the ATP generated by respiration.

Cristae (folded inner membranes) of mitochondrion

Electrons carried by NADH and FADH$_2$

Electrons carried via NADH

1 **GLYCOLYSIS**
The cytoplasm
Glucose ⟹ Pyruvate

2 **LINK REACTION**

3 **KREBS CYCLE**
The matrix of the mitochondria

Matrix (fluid space) of mitochondrion

4 **ELECTRON TRANSPORT CHAIN AND OXIDATIVE PHOSPHORYLATION**
The inner membranes (**cristae**) of the mitochondria

Substrate-level phosphorylation
ATP
2 ATP

Substrate-level phosphorylation
ATP
2 ATP

Oxidative phosphorylation
ATP
28 ATP

In **substrate-level phosphorylation**, an enzyme transfers a phosphate group directly from a substrate (e.g. phosphorylated sugar) to ADP.

In **oxidative phosphorylation**, glucose is oxidized in a series of reduction and oxidation (redox) reactions that provide the energy for the formation of ATP.

---

1. Describe precisely in which part of the cell the following take place:

   (a) Glycolysis: _____

   (b) The link reaction: _____

   (c) Krebs cycle reactions: _____

   (d) Electron transport chain: _____

2. How does ATP generation in glycolysis and the Krebs cycle differ from ATP generation via the electron transport chain?

   _____

   _____

   _____

PRACTICES   PRACTICES   WEB

KNOW

7 The Biochemistry of Respiration

Key Idea: During cellular respiration, the energy in glucose is transferred to ATP in a series of enzyme controlled steps. The oxidation of glucose is a catabolic, energy yielding pathway. The breakdown of glucose and other organic fuels (such as fats and proteins) to simpler molecules releases energy for ATP synthesis. Glycolysis and the Krebs cycle supply electrons to the electron transport chain, which drives oxidative phosphorylation. Glycolysis nets two ATP. The conversion of pyruvate (the end product of glycolysis) to acetyl CoA links glycolysis to the Krebs cycle. One "turn" of the cycle releases carbon dioxide, forms one ATP, and passes electrons to three NAD^+ and one FAD. Most of the ATP generated in cellular respiration is produced by oxidative phosphorylation when $NADH + H^+$ and $FADH_2$ donate electrons to the series of electron carriers in the electron transport chain. At the end of the chain, electrons are passed to molecular oxygen, reducing it to water. Electron transport is coupled to ATP synthesis.

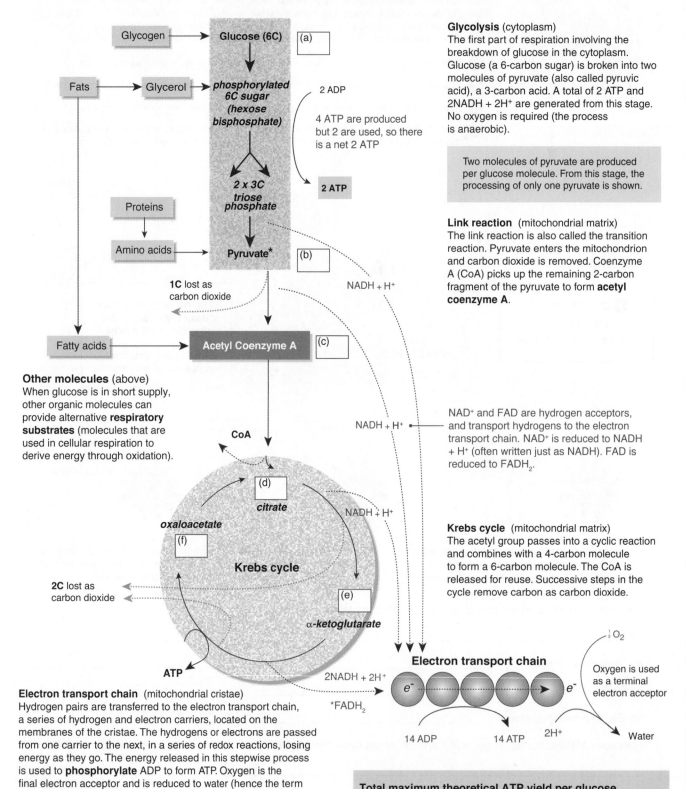

Glycolysis (cytoplasm)
The first part of respiration involving the breakdown of glucose in the cytoplasm. Glucose (a 6-carbon sugar) is broken into two molecules of pyruvate (also called pyruvic acid), a 3-carbon acid. A total of 2 ATP and $2NADH + 2H^+$ are generated from this stage. No oxygen is required (the process is anaerobic).

Two molecules of pyruvate are produced per glucose molecule. From this stage, the processing of only one pyruvate is shown.

Link reaction (mitochondrial matrix)
The link reaction is also called the transition reaction. Pyruvate enters the mitochondrion and carbon dioxide is removed. Coenzyme A (CoA) picks up the remaining 2-carbon fragment of the pyruvate to form **acetyl coenzyme A**.

NAD+ and FAD are hydrogen acceptors, and transport hydrogens to the electron transport chain. NAD+ is reduced to NADH + H+ (often written just as NADH). FAD is reduced to $FADH_2$.

Krebs cycle (mitochondrial matrix)
The acetyl group passes into a cyclic reaction and combines with a 4-carbon molecule to form a 6-carbon molecule. The CoA is released for reuse. Successive steps in the cycle remove carbon as carbon dioxide.

Other molecules (above)
When glucose is in short supply, other organic molecules can provide alternative **respiratory substrates** (molecules that are used in cellular respiration to derive energy through oxidation).

Oxygen is used as a terminal electron acceptor

Total maximum theoretical ATP yield per glucose
Glycolysis: 2 ATP, *Krebs cycle*: 2 ATP, *Electron transport*: 28 ATP

Electron transport chain (mitochondrial cristae)
Hydrogen pairs are transferred to the electron transport chain, a series of hydrogen and electron carriers, located on the membranes of the cristae. The hydrogens or electrons are passed from one carrier to the next, in a series of redox reactions, losing energy as they go. The energy released in this stepwise process is used to **phosphorylate** ADP to form ATP. Oxygen is the final electron acceptor and is reduced to water (hence the term **oxidative phosphorylation**).
Note FAD enters the electron transport chain at a lower energy level than NAD, and only 2ATP are generated per $FADH_2$.

©2017 **BIOZONE** International
ISBN: 978-1-927309-65-0
Photocopying Prohibited

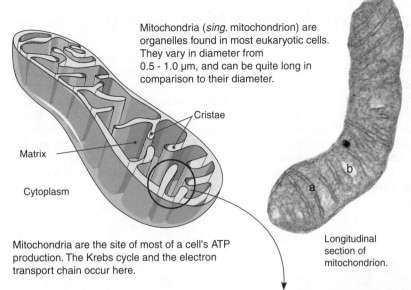

Mitochondria (*sing.* mitochondrion) are organelles found in most eukaryotic cells. They vary in diameter from 0.5 - 1.0 μm, and can be quite long in comparison to their diameter.

Cristae

Matrix

Cytoplasm

Mitochondria are the site of most of a cell's ATP production. The Krebs cycle and the electron transport chain occur here.

Longitudinal section of mitochondrion.

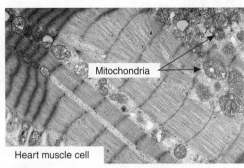

Mitochondria

Heart muscle cell

Cells that require a lot of ATP for cellular processes have a lot of mitochondria. Sperm cells contain a large number of mitochondria near the base of the tail. Liver cells have around 2000 mitochondria per cell, taking up 25% of the cytoplasmic space. Heart muscle cells (above) may have 40% of the cytoplasmic space taken up by mitochondria.

Location of cellular respiration

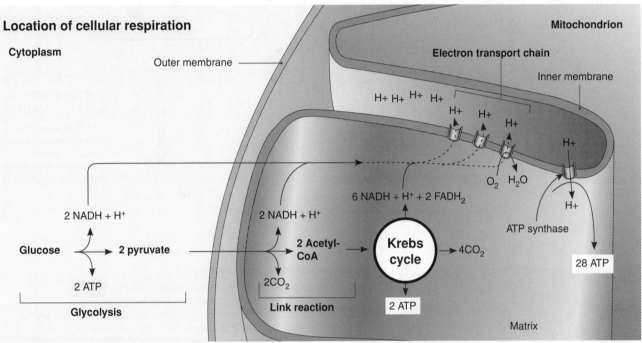

1. In the section of a mitochondrion (above), identify the regions (a) _____ (b) _____

2. Explain the purpose of the link reaction: _____

_____

_____

3. On the diagram of cell respiration (previous page), state the number of carbon atoms in each of the molecules (a)-(f):

4. How many ATP molecules **per molecule of glucose** are generated during the following stages of respiration?

    (a) Glycolysis: _____ (b) Krebs cycle: _____ (c) Electron transport chain: _____ (d) Total: _____

5. Explain what happens to the carbon atoms lost during respiration: _____

6. Explain what happens during oxidative phosphorylation: _____

_____

_____

_____

_____

_____

©2017 **BIOZONE** International
ISBN: 978-1-927309-65-0
Photocopying Prohibited

8 Chemiosmosis and the Proton Motive Force

Key Idea: Chemiosmosis is the process in which electron transport is coupled to ATP synthesis.

Chemiosmosis occurs in the membranes of mitochondria, the chloroplasts of plants, and across the plasma membrane of bacteria. Chemiosmosis involves the establishment of a proton (hydrogen) gradient across a membrane. The concentration gradient is used to drive ATP synthesis. Chemiosmosis has two key components: an **electron transport chain** sets up a proton gradient as electrons pass along it to a final electron acceptor, and an enzyme called **ATP synthase**

uses the proton gradient to catalyze ATP synthesis. In cellular respiration, electron carriers on the inner membrane of the mitochondrion oxidize $NADH + H^+$ and $FADH_2$. The energy released from this process is used to move protons against their concentration gradient, from the mitochondrial matrix into the space between the two membranes. The return of protons to the matrix via ATP synthase is coupled to ATP synthesis. Similarly, in the chloroplasts of green plants, ATP is produced when protons pass from the thylakoid lumen to the chloroplast stroma via ATP synthase.

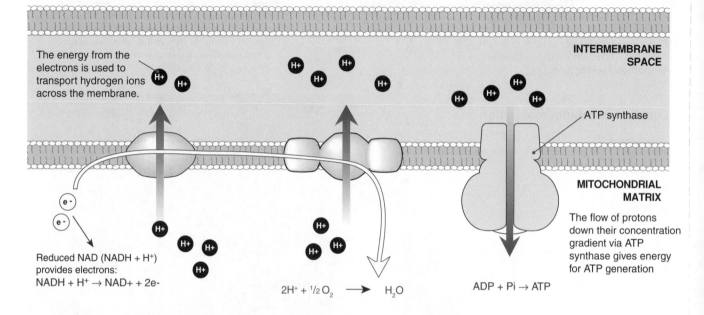

The energy from the electrons is used to transport hydrogen ions across the membrane.

INTERMEMBRANE SPACE

ATP synthase

MITOCHONDRIAL MATRIX

The flow of protons down their concentration gradient via ATP synthase gives energy for ATP generation

Reduced NAD (NADH + H⁺) provides electrons:
$NADH + H^+ \rightarrow NAD+ + 2e-$

$2H^+ + \frac{1}{2}O_2 \longrightarrow H_2O$

$ADP + Pi \rightarrow ATP$

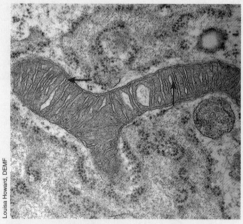

The intermembrane spaces can be seen (arrows) in this transverse section of mitochondria.

Louisa Howard, DEMF

The evidence for chemiosmosis

The British biochemist Peter Mitchell proposed the chemiosmotic hypothesis in 1961. He proposed that, because living cells have membrane potential, electrochemical gradients could be used to do work, i.e. provide the energy for ATP synthesis. Scientists at the time were skeptical, but the evidence for chemiosmosis was extensive and came from studies of isolated mitochondria and chloroplasts. Evidence included:

▶ The outer membranes of mitochondria were removed leaving the inner membranes intact. Adding protons to the treated mitochondria increased ATP synthesis.

▶ When isolated chloroplasts were illuminated, the medium in which they were suspended became alkaline.

▶ Isolated chloroplasts were kept in the dark and transferred first to a low pH medium (to acidify the thylakoid interior) and then to an alkaline medium (low protons). They then spontaneously synthesized ATP (no light was needed).

1. Summarize the process of chemiosmosis in a mitochondrion: _____

_____

_____

_____

_____

_____

_____

©2017 **BIOZONE** International
ISBN: 978-1-927309-65-0
Photocopying Prohibited

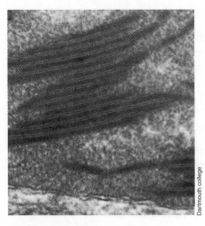

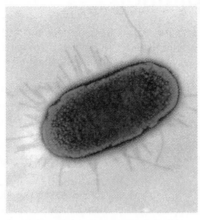

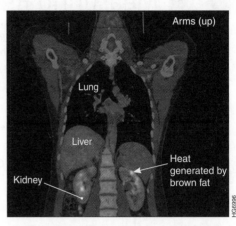

Arms (up)

Lung

Liver

Heat generated by brown fat

Kidney

Chemiosmosis in plants

The light reactions that occur in the chloroplasts of plants exploit the same chemiosmotic process as mitochondria. In this case, electrons are excited by the energy of photons and flow down the ETC causing protons to move across the thylakoid membrane into the stroma.

Chemiosmosis in prokaryotes

Prokaryotes also use chemiosmosis to produce ATP. Their electron transport chains are located in the plasma membrane and as electrons move down the ETC, H^+ are pumped out of the cell. H^+ flow back into the cell through the ATP synthase complex generating ATP.

Uncoupling H^+ flow and ATP generation

In some cells, such as the brown fat cells of mammals, 'uncoupling' proteins in the inner mitochondrial membrane act as channels, allowing protons to pass directly to the matrix without traveling through ATP synthase. This alternate route for H^+ flow allows the energy of the gradient to be dissipated, generating body heat (bright spots above).

2. Why did the addition of protons to the treated mitochondria increase ATP synthesis? _____

_____

_____

3. Why did the suspension of isolated chloroplasts become alkaline when illuminated? _____

_____

4 (a) What was the purpose of transferring the chloroplasts first to an acid then to an alkaline medium? _____

_____

_____

(b) Why did ATP synthesis occur spontaneously in these treated chloroplasts? _____

_____

_____

5. Compare and contrast how and where chemiosmosis occurs in animals, plants, and prokaryotes: _____

_____

_____

_____

_____

_____

6. (a) In mammals, what happens to energy stored in the proton gradient if it is not used to generate ATP? _____

_____

(b) Suggest how this apparently wasteful process might benefit a hibernating mammal: _____

_____

_____

_____

©2017 **BIOZONE** International
ISBN: 978-1-927309-65-0
Photocopying Prohibited

9 Measuring Respiration

Key Idea: Oxygen consumption and carbon dioxide production in respiring organisms can be measured with a respirometer. A respirometer measures the amount of oxygen consumed and the amount of carbon dioxide produced during cellular respiration. Respirometers are quite simple pieces of apparatus but can give accurate results if set up carefully.

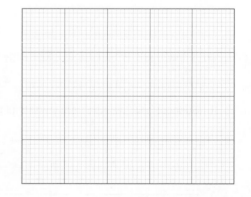

Screw clip

Scale

Capillary tube

Colored bubble

Perforated metal cage

Germinating seeds

Soda lime (or KOH) pellets (CO_2 absorbent)

⚠ Caution is required when handling KOH as it is caustic. Wear protective eyewear and gloves.

Clamp stand

Measuring respiration with a simple respirometer

The diagram on the left shows a **simple respirometer**. It measures the change in gases as respiration occurs.

▶ Respiring organisms, in this case germinating seeds, are placed into the bottom of the chamber.

▶ Soda lime or potassium hydroxide is added to absorb any carbon dioxide produced during respiration. Therefore the respirometer measures oxygen consumption.

▶ Once the organisms have been placed into the chamber the screw clip is closed. The start position of the colored bubble is measured (this is the time zero reading).

▶ The colored bubble in the capillary tube moves in response to the change in oxygen consumption. Measuring the movement of the liquid (e.g. with a ruler) allows the change in volume of gas to be estimated.

▶ Care needs to be taken when using a simple respirometer because changes in temperature or atmospheric pressure may change the readings and give a false measure of respiration.

▶ Differential respirometers (not shown) use two chambers (a control chamber with no organisms and a test chamber) connected by a U-tube. Changes in temperature or atmospheric pressure act equally on both chambers. Observed changes are only due to the activities of the respiring organism.

1. Why does the bubble in the capillary tube move?

_____

_____

_____

2. A student used a simple respirometer (like the one above) to measure respiration in maggots. Their results are presented in the table (right). The maggots were left to acclimatize for 10 minutes before the experiment was started.

(a) Calculate the rate of respiration and record this in the table. The first two calculations have been done for you.

(b) Plot the rate of respiration on the grid, below right.

(c) Describe the results in your plot: _____

_____

_____

_____

Time (minutes)	Distance bubble moved (mm)	Rate (mm min^{-1})
0	0	–
5	25	5
10	65	
15	95	
20	130	
25	160	

(d) Why was there an acclimatization period before the experiment began?

_____

_____

3. Why would it have been better to use a differential respirometer? _____

_____

_____

_____

_____

PRACTICES PRACTICES PRACTICES

PRAC

©2017 **BIOZONE** International
ISBN: 978-1-927309-65-0
Photocopying Prohibited

10 Anaerobic Pathways for ATP Production

Key Idea: Glucose can be metabolized aerobically and anaerobically to produce ATP. The ATP yield from aerobic processes is higher than from anaerobic processes.

Aerobic respiration occurs in the presence of oxygen. Organisms can also generate ATP anaerobically (without oxygen) by using a molecule other than oxygen as the terminal electron acceptor for the pathway. In alcoholic fermentation, the electron acceptor is ethanal. In lactic acid fermentation, which occurs in mammalian muscle even when oxygen is present, the electron acceptor is pyruvate itself.

Alcoholic fermentation

In alcoholic fermentation, the H^+ acceptor is ethanal which is reduced to ethanol with the release of carbon dioxide (CO_2). Yeasts respire aerobically when oxygen is available but can use alcoholic fermentation when it is not. At ethanol levels above 12-15%, the ethanol produced by alcoholic fermentation is toxic and this limits their ability to use this pathway indefinitely. The root cells of plants also use fermentation as a pathway when oxygen is unavailable but the ethanol must be converted back to respiratory intermediates and respired aerobically.

Lactic acid fermentation

Skeletal muscles produce ATP in the absence of oxygen using lactic acid fermentation. In this pathway, pyruvate is reduced to lactic acid, which dissociates to form lactate and H^+. The conversion of pyruvate to lactate is reversible and this pathway operates alongside the aerobic system all the time to enable greater intensity and duration of activity. Lactate can be metabolized in the muscle itself or it can enter the circulation and be taken up by the liver to replenish carbohydrate stores. This 'lactate shuttle' is an important mechanism for balancing the distribution of substrates and waste products.

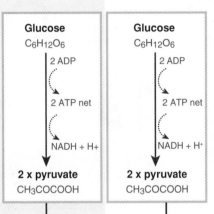

Glucose
$C_6H_{12}O_6$

2 ADP

2 ATP net

NADH + H+

2 x pyruvate
$CH_3COCOOH$

Glucose
$C_6H_{12}O_6$

2 ADP

2 ATP net

NADH + H$^+$

2 x pyruvate
$CH_3COCOOH$

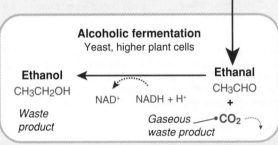

Alcoholic fermentation
Yeast, higher plant cells

Ethanol ← **Ethanal**
CH_3CH_2OH CH_3CHO
NAD$^+$ NADH + H$^+$ +
Waste product   Gaseous → •CO_2
                         waste product

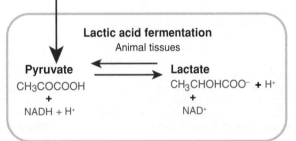

Lactic acid fermentation
Animal tissues

Pyruvate ⇄ **Lactate**
$CH_3COCOOH$ $CH_3CHOHCOO^-$ + H^+
        +   +
NADH + H$^+$ NAD$^+$

The alcohol and CO_2 produced from alcoholic fermentation form the basis of the brewing and baking industries. In baking, the dough is left to ferment and the yeast metabolizes sugars to produce ethanol and CO_2. The CO_2 causes the dough to rise.

Yeasts are used to produce almost all alcoholic beverages (e.g. wine and beers). The yeast used in the process breaks down the sugars into ethanol (alcohol) and CO_2. The alcohol produced is a metabolic by-product of fermentation by the yeast.

The lactate shuttle in vertebrate skeletal muscle works alongside the aerobic system to enable maximal muscle activity. Lactate moves from its site of production to regions within and outside the muscle (e.g. liver) where it can be respired aerobically.

1. Describe the key difference between aerobic respiration and fermentation: _____

_____

_____

2. (a) Refer to page 10 and determine the efficiency of fermentation compared to aerobic respiration: _____ %

   (b) Why is the efficiency of these anaerobic pathways so low? _____

   _____

3. Why can't alcoholic fermentation go on indefinitely? _____

_____

CONNECT
237
AP2

WEB
10

KNOW

11 Investigating Fermentation in Yeast

Key Idea: Brewer's yeast preferentially uses alcoholic fermentation when there is excess sugar, releasing CO_2, which can be collected as a measure of fermentation rate. Brewer's yeast is a facultative anaerobe (meaning it can respire aerobically or use fermentation). It will preferentially use alcoholic fermentation when sugars are in excess. One would expect glucose to be the preferred substrate, as it is the starting molecule in cellular respiration, but brewer's yeast is capable of utilizing a variety of sugars, including disaccharides, which can be broken down into single units.

The aim

To investigate the suitability of different mono- and disaccharide sugars as substrates for alcoholic fermentation in yeast.

The hypothesis

If glucose is the preferred substrate for fermentation in yeast, then the rate of fermentation will be highest when the yeast is grown on glucose rather than on other sugars.

Background

The rate at which brewer's or baker's yeast (*Saccharomyces cerevisiae*) metabolizes carbohydrate substrates is influenced by factors such as temperature, solution pH, and type of carbohydrate available.
The literature describes yeast metabolism as optimal in warm, acid (pH 4-6) environments. High levels of sugars suppress aerobic respiration in yeast, so yeast will preferentially use the fermentation pathway in the presence of excess substrate.

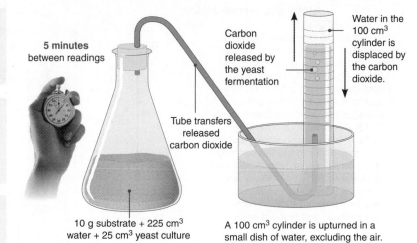

5 minutes between readings

Carbon dioxide released by the yeast fermentation

Tube transfers released carbon dioxide

Water in the 100 cm³ cylinder is displaced by the carbon dioxide.

10 g substrate + 225 cm³ water + 25 cm³ yeast culture

A 100 cm³ cylinder is upturned in a small dish of water, excluding the air.

The apparatus

In this experiment, all substrates tested used the same source culture of 30 g active yeast dissolved in 150 cm³ of room temperature (24°C) tap water. 25 g of each substrate to be tested was added to 225 cm³ room temperature (24°C) tap water buffered to pH 4.5. Then 25 cm³ of source culture was added to the test solution. The control contained yeast solution but no substrate.

The substrates

Glucose is a monosaccharide. Maltose (glucose-glucose), sucrose (glucose-fructose), and lactose (glucose-galactose) are disaccharides.

Time (min)	Volume of carbon dioxide collected (cm³)				
	None	Glucose	Maltose	Sucrose	Lactose
0	0	0	0	0	0
5	0	0	0.8	0	0
10	0	0	0.8	0	0
15	0	0	0.8	0.1	0
20	0	0.5	2.0	0.8	0
25	0	1.2	3.0	1.8	0
30	0	2.8	3.6	3.0	0
35	0	4.2	5.4	4.8	0
40	0	4.6	5.6	4.8	0
45	0	7.4	8.0	7.2	0
50	0	10.8	8.9	7.6	0
55	0	13.6	9.6	7.7	0
60	0	16.1	10.4	9.6	0
65	0	22.0	12.1	10.2	0
70	0	23.8	14.4	12.0	0
75	0	26.7	15.2	12.6	0
80	0	32.5	17.3	14.3	0
85	0	37.0	18.7	14.9	0
90	0	39.9	21.6	17.2	0

1. Write the equation for the fermentation of glucose by yeast:

   _____

2. The results are presented on the table left. Using the final values, calculate the rate of CO_2 production per minute for each substrate:

   (a) None: _____

   (b) Glucose: _____

   (c) Maltose: _____

   (d) Sucrose: _____

   (e) Lactose: _____

3. What assumptions are being made in this experimental design and do you think they were reasonable?

   _____

   _____

   _____

   _____

   _____

   _____

Experimental design and results adapted from Tom Schuster, Rosalie Van Zyl, & Harold Coller, California State University Northridge 2005

WEB
11

CONNECT
5
AP1

PRACTICES

PRACTICES

DATA

©2017 **BIOZONE** International
ISBN: 978-1-927309-65-0
Photocopying Prohibited

4. Use the tabulated data to plot an appropriate graph of the results on the grid provided:

5. (a) Identify the independent variable: _____

   (b) State the range of values for the independent variable: _____

   (c) Name the unit for the independent variable: _____

6. (a) Identify the dependent variable: _____

   (b) Name the unit for the dependent variable: _____

7. (a) Summarize the results of the fermentation experiment: _____

   _____

   _____

   _____

   (b) Why do you think CO_2 production was highest when glucose was the substrate? _____

   _____

   _____

   (c) Suggest why fermentation rates were lower on maltose and sucrose than on glucose: _____

   _____

   _____

   (d) Suggest why there may have been no fermentation on the lactose substrate: _____

   _____

8. Predict what would happen to CO_2 production rates if the yeast cells were respiring aerobically: _____

   _____

   _____

12 Chloroplasts

Key Idea: Chloroplasts have a complicated internal membrane structure that provides the sites for the light dependent reactions of photosynthesis.

Chloroplasts are the specialized plastids in which photosynthesis occurs. A mesophyll leaf cell contains between 50-100 chloroplasts. The chloroplasts are generally aligned so that their broad surface runs parallel to the cell wall

to maximize the surface area available for light absorption. Chloroplasts have an internal structure characterized by a system of membranous structures called **thylakoids** arranged into stacks called **grana**. Special pigments, called **chlorophylls** and **carotenoids**, are bound to the membranes as part of light-capturing photosystems. They absorb light of specific wavelengths and thereby capture the light energy.

The structure of a chloroplast

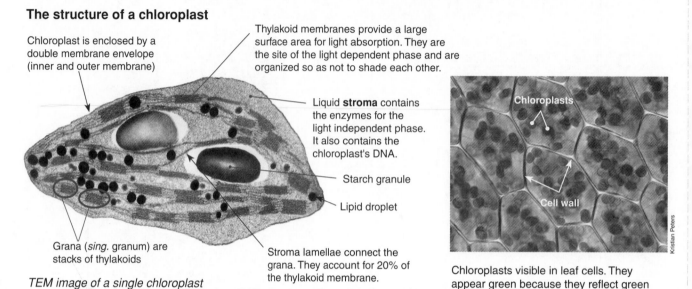

Chloroplast is enclosed by a double membrane envelope (inner and outer membrane)

Thylakoid membranes provide a large surface area for light absorption. They are the site of the light dependent phase and are organized so as not to shade each other.

Liquid **stroma** contains the enzymes for the light independent phase. It also contains the chloroplast's DNA.

Starch granule

Lipid droplet

Grana (*sing.* granum) are stacks of thylakoids

Stroma lamellae connect the grana. They account for 20% of the thylakoid membrane.

TEM image of a single chloroplast

Chloroplasts

Cell wall

Chloroplasts visible in leaf cells. They appear green because they reflect green light, absorbing blue and red light.

Kristian Peters

1. Label the transmission electron microscope image of a chloroplast below:

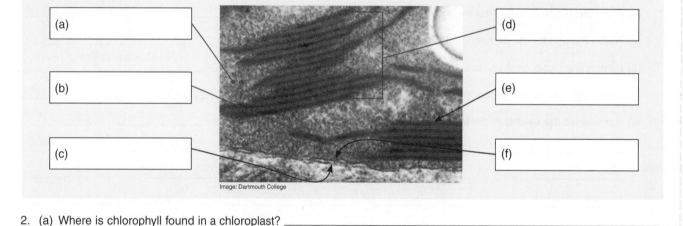

(a) _____

(b) _____

(c) _____

(d) _____

(e) _____

(f) _____

Image: Dartmouth College

2. (a) Where is chlorophyll found in a chloroplast? _____

_____

(b) Why is chlorophyll found there? _____

_____

_____

3. Explain how the internal structure of chloroplasts helps absorb the maximum amount of light: _____

_____

_____

_____

4. Explain why plant leaves appear green: _____

_____

©2017 **BIOZONE** International
ISBN: **978-1-927309-65-0**
Photocopying Prohibited

13 Origin of Eukaryotic Photosynthesis

Key Idea: Chloroplasts arose as an endosymbiosis between an early pre-eukaryotic cell and a cyanobacterium.

Oxygenic photosynthesis is thought to have evolved in cyanobacteria about 2.3 billion years ago (bya) and eukaryotes evolved about 1 bya. It is thought that eukaryotic cells evolved from pre-eukaryotic (bacterial) cells that ingested other free-floating bacteria. They formed a symbiotic relationship with the cells they engulfed (**endosymbiosis**). The two organelles that evolved in eukaryotic cells as a result of bacterial endosymbiosis were **mitochondria**, for aerobic respiration, and **chloroplasts**, for photosynthesis in aerobic conditions. Primitive eukaryotes probably acquired mitochondria by engulfing purple bacteria. Similarly, chloroplasts may have been acquired by engulfing photosynthetic cyanobacteria.

Evolution of a plant cell

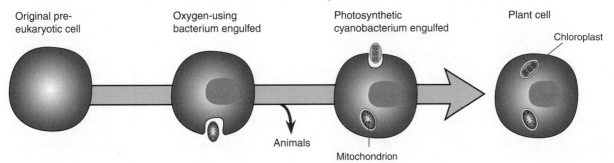

Original pre-eukaryotic cell — Oxygen-using bacterium engulfed — Photosynthetic cyanobacterium engulfed — Plant cell — Chloroplast — Animals — Mitochondrion

Evidence for the bacterial origin of chloroplasts

Evidence for the bacterial origin of chloroplasts includes:

▶ Chloroplasts have a similar morphology to bacteria.

▶ Chloroplasts divide by binary fission, they split in half to form new chloroplasts, just like bacteria. Thus new chloroplasts arise from preexisting chloroplasts, they are not manufactured by the cell.

▶ They have a chemically distinct inner membrane. The outer membrane is similar to the plasma membrane (as if a vesicle formed around the engulfed bacteria) but the inner membrane is similar to the membrane of bacteria.

▶ Bacterial DNA is a single circular molecule and chloroplasts also have their own single circular DNA. Like bacterial DNA, the DNA of chloroplasts has no introns or histones. Also the chloroplast DNA evolves or mutates at a different rate to the nuclear DNA.

▶ Chloroplasts contain ribosomes that are more similar in size to bacterial ribosomes than ribosomes in the cytoplasm.

▶ Antibiotics that inhibit protein synthesis in bacteria also inhibit protein synthesis in chloroplasts. Conversely, bacterial toxins that inhibit protein synthesis in eukaryotes do not affect chloroplasts.

▶ Analysis of chloroplast DNA shows that they are related to cyanobacteria.

Examples of engulfment

Paramecium bursaria

Paramecium bursaria (right) is a single celled protozoan. It engulfs cells of *Zoochlorella*, a photosynthetic green alga. It houses the algae and carries them to light areas in a pond where they can photosynthesize. In return, it uses the food made by the algae.

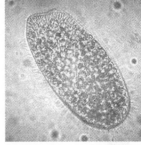

Bob Blaylock

Amoeba proteus

From 1972 microbiologist Kwang Jeon studied the infection of *Amoeba proteus* by *Legionella*-like bacteria. He found that most infected amoebae died. The few that survived were cultured over many generations. Eventually, the amoebae became dependent on the bacteria for nuclear function. Experiments showed that when the nucleus of an infected cell was placed in an uninfected cell which had also had its nucleus removed the new cell quickly died.

1. Which endosymbiosis occurred first in the evolution of eukaryotic cells? Explain your reasoning: _____

_____

_____

2. Outline four pieces of evidence for the bacterial origin of chloroplasts:

(a) _____

(b) _____

(c) _____

(d) _____

3. How do the examples of *Paramecium bursaria* and *Amoeba proteus* support the endosymbiotic origin of chloroplasts?

_____

_____

_____

©2017 **BIOZONE** International
ISBN: 978-1-927309-65-0
Photocopying Prohibited

PRACTICES   PRACTICES

CONNECT **251** AP1   CONNECT **54** AP1   CONNECT **51** AP1   WEB **13**

KNOW

14 Photosynthesis

Key Idea: Photosynthesis is the process by which light energy is used to convert CO_2 and water into glucose and oxygen. **Photosynthesis** is of fundamental importance to living things because it transforms sunlight energy into chemical energy stored in molecules, releases free oxygen gas, and absorbs carbon dioxide (a waste product of cellular metabolism).

Photosynthetic organisms use special pigments, called **chlorophylls**, to absorb light of specific wavelengths and capture the light energy. Photosynthesis involves reduction and oxidation (redox) reactions. In photosynthesis, water is split and electrons are transferred together with hydrogen ions from water to CO_2, reducing it to sugar.

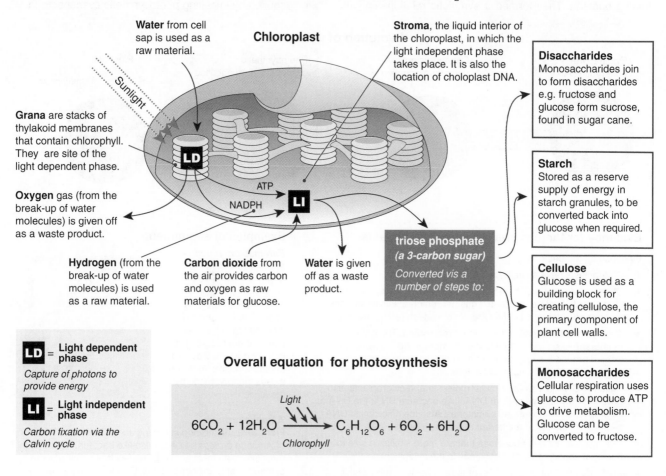

Water from cell sap is used as a raw material.

Chloroplast

Stroma, the liquid interior of the chloroplast, in which the light independent phase takes place. It is also the location of choloplast DNA.

Grana are stacks of thylakoid membranes that contain chlorophyll. They are site of the light dependent phase.

Oxygen gas (from the break-up of water molecules) is given off as a waste product.

Hydrogen (from the break-up of water molecules) is used as a raw material.

Carbon dioxide from the air provides carbon and oxygen as raw materials for glucose.

Water is given off as a waste product.

Disaccharides
Monosaccharides join to form disaccharides e.g. fructose and glucose form sucrose, found in sugar cane.

Starch
Stored as a reserve supply of energy in starch granules, to be converted back into glucose when required.

triose phosphate (a 3-carbon sugar)
Converted vis a number of steps to:

Cellulose
Glucose is used as a building block for creating cellulose, the primary component of plant cell walls.

LD = **Light dependent phase**
Capture of photons to provide energy

LI = **Light independent phase**
Carbon fixation via the Calvin cycle

Monosaccharides
Cellular respiration uses glucose to produce ATP to drive metabolism. Glucose can be converted to fructose.

Overall equation for photosynthesis

$$6CO_2 + 12H_2O \xrightarrow[\text{Chlorophyll}]{\text{Light}} C_6H_{12}O_6 + 6O_2 + 6H_2O$$

1. Distinguish between the two different regions of a chloroplast and describe the biochemical processes that occur in each:

(a) _____

_____

_____

(b) _____

_____

_____

2. State the origin and fate of the following molecules involved in photosynthesis:

(a) Carbon dioxide: _____

(b) Oxygen: _____

(c) Hydrogen: _____

3. Discuss the potential uses for the end products of photosynthesis: _____

_____

_____

_____

©2017 **BIOZONE** International
ISBN: 978-1-927309-65-0
Photocopying Prohibited

15 Pigments and Light Absorption

Key Idea: Chlorophyll pigments absorb light of specific wavelengths and capture light energy for photosynthesis. Substances that absorb visible light are called **pigments**, and different pigments absorb light of different wavelengths. The ability of a pigment to absorb particular wavelengths of light can be measured with a spectrophotometer. The light absorption vs the wavelength is called the **absorption**

spectrum of that pigment. The absorption spectrum of different photosynthetic pigments provides clues to their role in photosynthesis, since light can only perform work if it is absorbed. An **action spectrum** profiles the effectiveness of different wavelengths of light in fueling photosynthesis. It is obtained by plotting wavelength against a measure of photosynthetic rate (e.g. O_2 production).

The electromagnetic spectrum

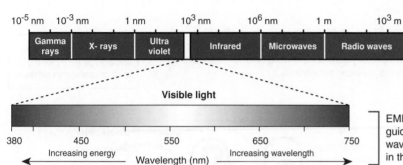

Light is a form of energy known as electromagnetic radiation (EMR). The segment of the electromagnetic spectrum most important to life is the narrow band between about 380 nm and 750 nm. This radiation is known as visible light because it is detected as colors by the human eye. It is visible light that drives photosynthesis.

EMR travels in waves, where wavelength provides a guide to the energy of the photons. The greater the wavelength of EMR, the lower the energy of the photons in that radiation.

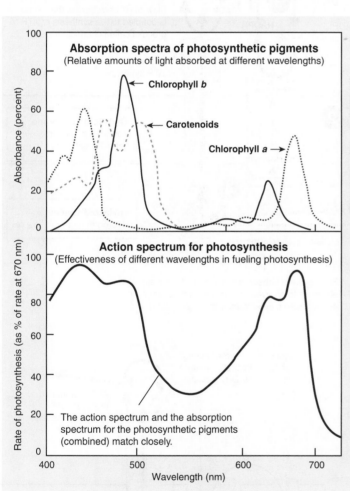

Absorption spectra of photosynthetic pigments
(Relative amounts of light absorbed at different wavelengths)

Chlorophyll *b*

Carotenoids

Chlorophyll *a* →

Absorbance (percent)

Action spectrum for photosynthesis
(Effectiveness of different wavelengths in fueling photosynthesis)

The action spectrum and the absorption spectrum for the photosynthetic pigments (combined) match closely.

Rate of photosynthesis (as % of rate at 670 nm)

Wavelength (nm)

The photosynthetic pigments of plants

The photosynthetic pigments of plants fall into two categories: **chlorophylls** (which absorb red and blue-violet light) and **carotenoids** (which absorb strongly in the blue-violet range and appear orange, yellow, or red). The pigments are located on the chloroplast membranes (the thylakoids) and are associated with membrane transport systems.

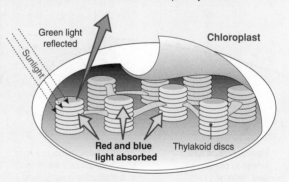

The pigments of chloroplasts in higher plants (above) absorb blue and red light, and the leaves therefore appear green (which is reflected). Each photosynthetic pigment has its own characteristic absorption spectrum (top left). Only chlorophyll *a* participates directly in the light reactions of photosynthesis, but the accessory pigments (chlorophyll *b* and carotenoids) can absorb wavelengths of light that chlorophyll *a* cannot and pass the energy (photons) to chlorophyll *a*, thus broadening the spectrum that can effectively drive photosynthesis.

Left: Graphs comparing absorption spectra of photosynthetic pigments compared with the action spectrum for photosynthesis.

1. What is meant by the absorption spectrum of a pigment? _____

_____

2. Why doesn't the action spectrum for photosynthesis exactly match the absorption spectrum of chlorophyll *a*? _____

_____

_____

©2017 **BIOZONE** International
ISBN: 978-1-927309-65-0
Photocopying Prohibited

WEB

15

KNOW

16 Light Dependent Reactions

Key Idea: In light dependent reactions of photosynthesis, the energy from photons of light is used to drive the reduction of $NADP^+$ and the production of ATP.

Like cellular respiration, photosynthesis is a redox process, but in photosynthesis, water is split, and electrons and hydrogen ions, are transferred from water to CO_2, reducing it to sugar. The electrons increase in potential energy as they move from water to sugar. The energy to do this is provided by light. Photosynthesis has two phases. In the **light dependent** **reactions**, light energy is converted to chemical energy (ATP and NADPH). In the **light independent reactions**, the chemical energy is used to synthesize carbohydrate. The light dependent reactions most commonly involve **non-cyclic phosphorylation**, which produces ATP and NADPH in roughly equal quantities. The electrons lost are replaced from water. In **cyclic phosphorylation**, the electrons lost from photosystem II are replaced by those from photosystem I. ATP is generated, but not NADPH.

Non-cyclic phosphorylation

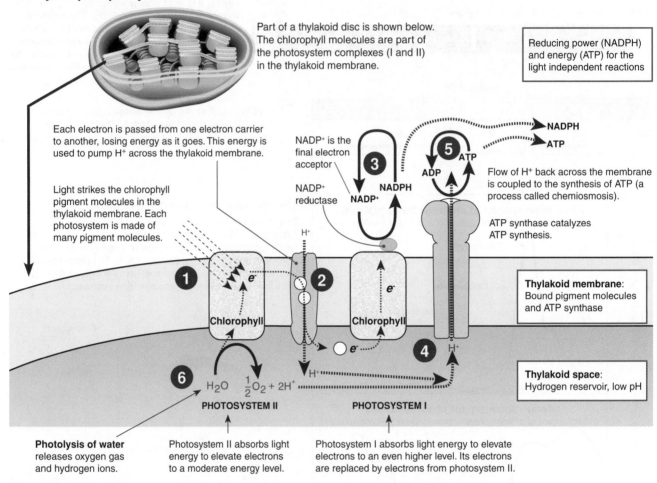

Part of a thylakoid disc is shown below. The chlorophyll molecules are part of the photosystem complexes (I and II) in the thylakoid membrane.

Reducing power (NADPH) and energy (ATP) for the light independent reactions

Each electron is passed from one electron carrier to another, losing energy as it goes. This energy is used to pump H^+ across the thylakoid membrane.

Light strikes the chlorophyll pigment molecules in the thylakoid membrane. Each photosystem is made of many pigment molecules.

$NADP^+$ is the final electron acceptor

$NADP^+$ reductase

3 NADP⁺ → NADPH

5 ATP
ADP

Flow of H^+ back across the membrane is coupled to the synthesis of ATP (a process called chemiosmosis).

ATP synthase catalyzes ATP synthesis.

1 Chlorophyll e^-

H^+

2

Chlorophyll e^-

e^-

4 H^+

Thylakoid membrane: Bound pigment molecules and ATP synthase

6 H_2O → $\frac{1}{2}O_2 + 2H^+$

H^+

Thylakoid space: Hydrogen reservoir, low pH

PHOTOSYSTEM II

PHOTOSYSTEM I

Photolysis of water releases oxygen gas and hydrogen ions.

Photosystem II absorbs light energy to elevate electrons to a moderate energy level.

Photosystem I absorbs light energy to elevate electrons to an even higher level. Its electrons are replaced by electrons from photosystem II.

Cyclic phosphorylation

Cyclic phosphorylation involves only photosystem I and no NADPH is generated. Electrons from photosystem I are shunted back to the electron carriers in the membrane so this pathway produces ATP only. The Calvin cycle uses more ATP than NADPH, so cyclic phosphorylation makes up the difference. It is activated when NADPH levels build up, and remains active until enough ATP is made to meet demand.

Electrons are cycled through a pathway that takes them away from $NADP^+$ reductase.

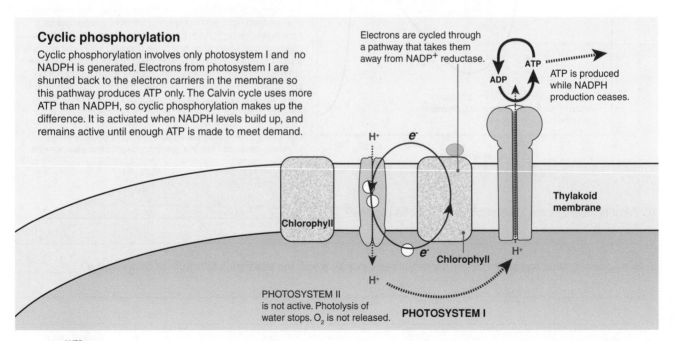

ATP
ADP

ATP is produced while NADPH production ceases.

H^+ e^-

Chlorophyll

Thylakoid membrane

e^- **Chlorophyll**

H^+

H^+

PHOTOSYSTEM II is not active. Photolysis of water stops. O_2 is not released.

PHOTOSYSTEM I

©2017 **BIOZONE** International
ISBN: 978-1-927309-65-0
Photocopying Prohibited

1. Describe the role of the carrier molecule NADP in photosynthesis: _____

_____

2. Explain the role of chlorophyll molecules in photosynthesis: _____

_____

_____

_____

3. Summarize the events of the light dependent reactions and identify where they occur: _____

_____

_____

_____

_____

4. Describe how ATP is produced as a result of light striking chlorophyll molecules during the light dependent phase:

_____

_____

_____

_____

_____

5. (a) Explain what you understand by the term non-cyclic phosphorylation: _____

_____

_____

_____

(b) Suggest why this process is also known as non-cyclic photophosphorylation: _____

_____

_____

6. (a) Describe how cyclic photophosphorylation differs from non-cyclic photophosphorylation: _____

_____

_____

(b) Both cyclic and non-cyclic pathways operate to varying degrees during photosynthesis. Since the non-cyclic pathway produces both ATP and NAPH, explain the purpose of the cyclic pathway of electron flow:

_____

_____

7. Complete the summary table of the light dependent reactions of photosynthesis

	Non-cyclic phosphorylation	Cyclic phosphorylation
Photosystem involved		
Energy carrier(s) produced		
Photolysis of water (yes / no)		
Production of oxygen (yes / no)		

©2017 **BIOZONE** International
ISBN: 978-1-927309-65-0
Photocopying Prohibited

17 Light Independent Reactions

Key Idea: The light independent reactions of photosynthesis take place in the stroma of the chloroplast and do not require light to proceed.

In the **light independent reactions** (the **Calvin cycle**) hydrogen (H⁺) is added to CO_2 and a 5C intermediate to

make carbohydrate. The H⁺ and ATP are supplied by the light dependent reactions. The Calvin cycle uses more ATP than NADPH, but the cell uses cyclic phosphorylation (which does not produce NADPH) when it runs low on ATP to make up the difference.

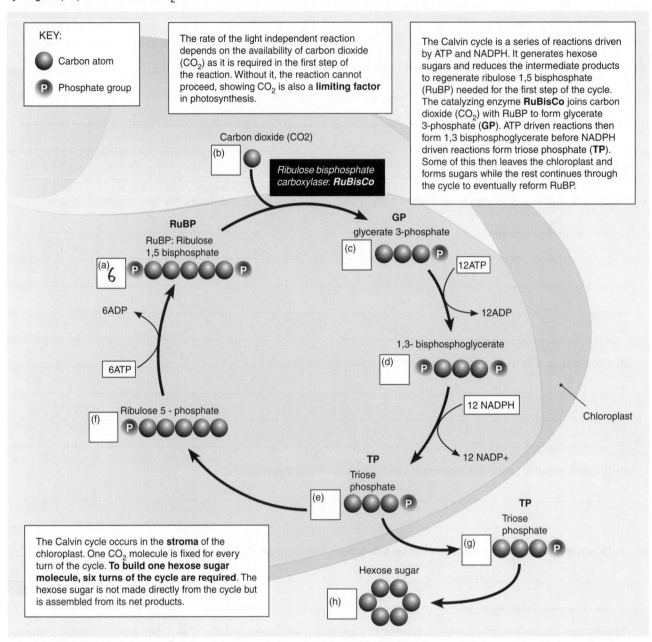

KEY:

● Carbon atom

Ⓟ Phosphate group

The rate of the light independent reaction depends on the availability of carbon dioxide (CO_2) as it is required in the first step of the reaction. Without it, the reaction cannot proceed, showing CO_2 is also a **limiting factor** in photosynthesis.

The Calvin cycle is a series of reactions driven by ATP and NADPH. It generates hexose sugars and reduces the intermediate products to regenerate ribulose 1,5 bisphosphate (RuBP) needed for the first step of the cycle. The catalyzing enzyme **RuBisCo** joins carbon dioxide (CO_2) with RuBP to form glycerate 3-phosphate (**GP**). ATP driven reactions then form 1,3 bisphosphoglycerate before NADPH driven reactions form triose phosphate (**TP**). Some of this then leaves the chloroplast and forms sugars while the rest continues through the cycle to eventually reform RuBP.

Carbon dioxide (CO2)

(b)

Ribulose bisphosphate carboxylase: RuBisCo

RuBP
RuBP: Ribulose 1,5 bisphosphate

(a) 6

GP
glycerate 3-phosphate

(c)

12ATP

12ADP

6ADP

1,3- bisphosphoglycerate

(d)

6ATP

12 NADPH

Chloroplast

Ribulose 5 - phosphate

(f)

12 NADP+

TP
Triose phosphate

(e)

TP
Triose phosphate

(g)

The Calvin cycle occurs in the **stroma** of the chloroplast. One CO_2 molecule is fixed for every turn of the cycle. **To build one hexose sugar molecule, six turns of the cycle are required**. The hexose sugar is not made directly from the cycle but is assembled from its net products.

Hexose sugar

(h)

1. In the boxes on the diagram above, write the number of molecules formed at each step during the formation of **one hexose sugar molecule**. The first one has been done for you:

2. Explain the importance of RuBisCo in the Calvin cycle: _____

_____

3. Identify the actual end product on the Calvin cycle: _____

4. Write the equation for the production of one hexose sugar molecule from carbon dioxide: _____

_____

5. Explain why the Calvin cycle is likely to cease in the dark for most plants, even though it is independent of light:

_____

_____

©2017 **BIOZONE** International
ISBN: 978-1-927309-65-0
Photocopying Prohibited

18 Investigating Photosynthesis

Key Idea: The rate of photosynthesis varies with different wavelengths of visible light.
Photosynthetic pigments absorb specific wavelengths of light and capture the energy within it to drive photosynthesis.

However, some wavelengths are absorbed more strongly than others. The experiment described below investigates the effect of different wavelengths on the photosynthetic rate of a green plant.

Aim

To investigate the effect of wavelength on the photosynthetic rate of a green plant.

Method

▶ Select several green leaves of the same type. Avoiding areas with major leaf veins, use a hole punch to cut out 40 discs of a uniform size. Place the discs into a large syringe containing a 0.2% bicarbonate solution. Place a finger tightly over the tip of the syringe and slowly pull back on the plunger. Repeat until all the discs sink. Do not use any that remain floating. Keep the syringe containing the discs in a dark place until required.

▶ Label four 150 mL glass beakers as red, blue, green, and clear. To each beaker add 100 mL of 0.2% bicarbonate solution and 5 mL of detergent. Color the solutions by adding 10 drops of the appropriate color food coloring to the bicarbonate solution. No food coloring is added to the clear container.

▶ Place 10 leaf discs into the beaker, and place it 15 cm from a 100 watt light bulb. Start a timer immediately and record the time taken for all 10 leaf discs to float. Repeat with the remaining colors.

Background

Leaf disc assays are commonly used to investigate photosynthesis in the classroom because they are simple to perform and do not require any specialized equipment. The bicarbonate solution under pressure removes any oxygen in the leaf by replacing the air in the leaf air spaces and it also serves as a source of CO_2 during the experiment. As photosynthesis occurs, O_2 is produced and the leaf disks become buoyant and eventually float. The rate of flotation is an indirect measure of the rate of photosynthesis. The detergent is added to break down the water-repellent barrier on the leaf surface, allowing sodium bicarbonate to enter the leaf more easily.

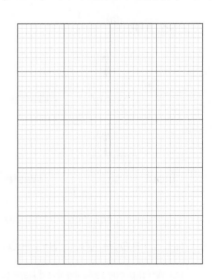

1. Generate a brief hypothesis for this experiment: _____

   _____

   _____

   _____

   _____

Results

The results from the experiment are shown below.

Light color	Time taken for 10 discs to float (s)
Blue	162
Red	558
Green	998
White	694

2. Why do the leaf discs float? _____

   _____

   _____

3. (a) Graph the results on the grid provided (right):

   (b) Describe how photosynthesis was affected by light color:

   _____

   _____

   _____

   _____

   _____

4. Did the results support your hypothesis? Explain: _____

   _____

   _____

   _____

PRACTICES   PRACTICES   PRACTICES   CONNECT **58** AP1 CONNECT **54** AP1 CONNECT **238** AP2 CONNECT **60** AP2 WEB **18**

PRAC

19 KEY TERMS AND IDEAS: Did You Get It?

1. Test your vocabulary by matching each term to its correct definition, as identified by its preceding letter code.

cellular respiration

chemiosmosis

electron transport chain

glycolysis

Krebs cycle

mitochondrion

oxidative phosphorylation

photosynthesis

stroma

A The biochemical process that uses light energy to convert carbon dioxide and water into glucose molecules and oxygen.

B The anaerobic breakdown of glucose into two molecules of pyruvate.

C A series of electron transport proteins in the membrane of mitochondria which generate a protein gradient which can be used to generate ATP.

D Organelle responsible for producing the cell's ATP. They appear oval in shape with an outer double membrane and a convoluted interior membrane.

E The catabolic process in which the chemical energy in complex organic molecules is coupled to ATP production.

F The stage in cellular respiration in which a derivative of pyruvate is broken down to carbon dioxide with the release of hydrogen ions.

G The liquid interior of the chloroplast where the light independent phase takes place.

H The process by which the synthesis of ATP is coupled to electron transport and the movement of protons.

I The process by which electrons are transferred in a series of redox reactions to a final electron acceptor such as oxygen. The energy released is used to form ATP.

2. Complete the schematic diagram of photosynthesis below:

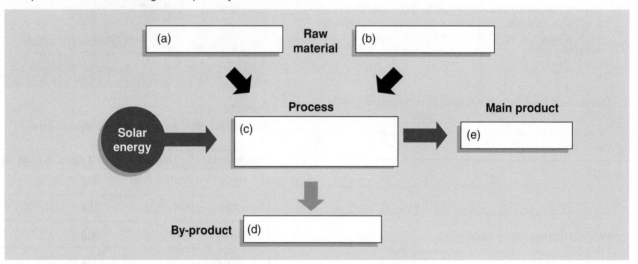

3. Complete the diagram of cellular respiration below by filling in the boxes below:

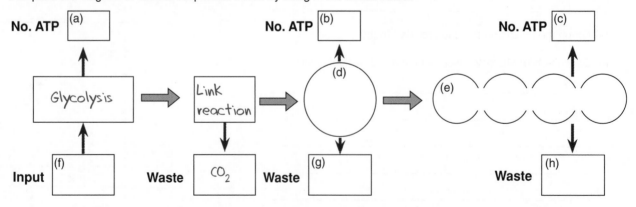

4. The following equation summarizes a process in the bacterium *Nitrosomonas*, an ammonia-oxidizing bacterium:

$$NH_3 + 1.5\ O_2 \rightarrow NO_2^- + H^+ + H_2O\ (274.91\ kJ\,mol^{-1})$$

   (a) What process is involved here (circle correct answer)? photosynthesis / cellular respiration / chemosynthesis

   (b) *Nitrosomonas* is a photoautotroph / chemoautotroph / heterotroph (circle correct answer).

©2017 **BIOZONE** International
ISBN: 978-1-927309-65-0
Photocopying Prohibited

TEST

Enzymes and Metabolism

4.B.1 Interactions between molecules affect their structure and function

Essential knowledge

(a) Change in the structure of a molecular system may result in a change of function

☐ 1 Use examples to illustrate how changes in a molecule's structure can change its function. Examples include enzymes and receptor molecules. *[also 3.D.3.a]*   24 25

(b) An enzyme's shape, active site and interactions with the substrate are essential to its function

☐ 1 Explain how enzyme-catalyzed reactions rely on the substrate being complementary to the shape and charge of the enzyme's active site.   20 21 22

☐ 2 Explain the role of cofactors and coenzymes in enzyme function.   24

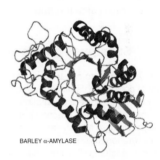

PEPSIN   LACTASE   BARLEY α-AMYLASE

(c) Enzyme activity can be enhanced or inhibited by other molecules and the environment

☐ 1 Explain how enzyme activity is affected by the reversible or irreversible binding of molecules to the active or allosteric sites of an enzyme.   25

(d) Change in enzyme activity can be interpreted from information about the concentrations of products or substrate as a function of time

☐ 1 Use graphical representations of enzyme-catalyzed reactions over time to describe the relationship between enzyme activity, disappearance of substrate, and/or presence of a competitive inhibitor.   23

☐ Develop a method for measuring baseline enzyme activity in plant material. Investigate the effect of pH on enzyme activity.   26

4.B.2 Cooperative interactions within organisms promote efficiency in the use of energy and matter

Essential knowledge

(a) Organisms have areas or compartments that perform a subset of functions related to energy and matter and these parts contribute to the whole

☐ 1 Explain how compartmentalization within the cell contributes to overall specialization and functioning of the cell.   28 29

☐ 2 Using examples, explain how specialization of organs within multicellular organisms contributes to overall functioning. Examples include specialization for exchange of gases, digestion, circulation of fluids, and excretion of wastes.   27

☐ 3 Describe how interactions among cells of a population of unicellular organisms can be similar to those of multicellular organisms and can lead to increases efficiency and use of energy and matter.   30

20 Enzymes

Key Idea: Enzymes are biological catalysts. The active site is critical to this functional role.

Most enzymes are globular proteins. Enzymes are biological catalysts because they speed up biochemical reactions, but the enzyme itself remains unchanged. The substrate in a reaction binds to a region of the enzyme called the active site, which is formed by the precise folding of the enzyme's amino acid chain(s). Enzymes control metabolic pathways. One enzyme will act on a substance to produce the next reactant in a pathway, which will be acted on by a different enzyme.

The active site

Enzymes have an **active site** to which specific substrates bind. The shape and chemistry of the active site is specific to an enzyme, and is a function of the polypeptide's complex tertiary structure.

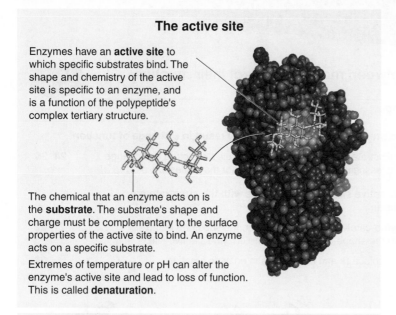

The chemical that an enzyme acts on is the **substrate**. The substrate's shape and charge must be complementary to the surface properties of the active site to bind. An enzyme acts on a specific substrate.

Extremes of temperature or pH can alter the enzyme's active site and lead to loss of function. This is called **denaturation**.

Substrates collide with an enzyme's active site

For a reaction to occur reactants must collide with sufficient speed and with the correct orientation. Enzymes enhance reaction rates by providing a site for reactants to come together in such a way that a reaction will occur. They do this by orientating the reactants so that the reactive regions are brought together. They may also destabilize the bonds within the reactants making it easier for a reaction to occur.

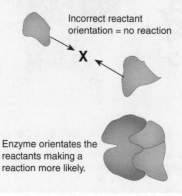

Incorrect reactant orientation = no reaction

X

Enzyme orientates the reactants making a reaction more likely.

Enzymes can be intracellular or extracellular

Enzymes can be defined based on where they are produced relative to where they are active.

An **intracellular enzyme** is an enzyme that performs its function within the cell that produces it. Most enzymes are intracellular enzymes, e.g. respiratory enzymes. **Example**: Catalase.

Many metabolic processes produce hydrogen peroxide, which is harmful to cells. Catalase converts hydrogen peroxide into water and oxygen (below) to prevent damage to cells and tissues.

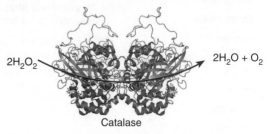

$2H_2O_2$ → $2H_2O + O_2$

Catalase

An **extracellular enzyme** is an enzyme that functions outside the cell from which it originates (i.e. it is produced in one location but active in another). **Examples**: Amylase and trypsin.

Amylase is a digestive enzyme produced in the salivary glands and pancreas in humans. However, it acts in the mouth and small intestine respectively to hydrolyze starch into sugars.

Trypsin is a protein-digesting enzyme and hydrolyzes the peptide bond immediately after a basic residue (e.g. arginine). It is produced in an inactive form (called trypsinogen) and secreted into the small intestine by the pancreas. It is activated in the intestine by the enzyme enteropeptidase to form trypsin. Active trypsin can convert more trypsinogen to trypsin.

1. (a) Describe what is meant by the active site of an enzyme and relate it to the enzyme's tertiary structure: _____

_____

_____

_____

(b) Describe how the chemistry and shape of the active site and an enzyme's substrate are complementary: _____

_____

_____

(c) How does this complementary relationship determine enzyme specificity? _____

_____

_____

2. How do substrate molecules come into contact with an enzyme's active site? _____

_____

_____

©2017 **BIOZONE** International
ISBN: 978-1-927309-65-0
Photocopying Prohibited

21 Models of Enzyme Activity

Key Idea: Enzymes catalyze reactions by providing a reaction site for a substrate. The model that describes the behavior of enzymes the best is the induced fit model.

The initial model of enzyme activity was the lock and key model proposed by Emil Fischer in the 1890s. Fischer proposed enzymes were rigid structures, similar to a lock, and the substrate was the key. While some aspects of

Fischer's model were correct, for example, substrates align with enzymes in a way that is likely to make a reaction more likely, the model has been adapted as techniques to study molecular structures have developed. The current 'induced-fit' model of enzyme function is supported by studies of enzyme inhibitors, which show that enzymes are flexible and change shape when interacting with the substrate.

The lock and key model of enzyme function

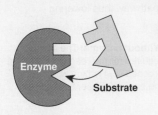

1 The substrate molecule is drawn into the active site of the enzyme. The enzyme's active site does not change shape.

2 The enzyme-substrate (ES) complex is formed.

3 The enzyme reaction takes place to form the enzyme-product (EP) complex.

4 The products are released from the enzyme. Note there has been no change in the shape of the active site throughout the reaction.

The **lock and key** model proposed in 1894 suggested that the (perfectly fitting) substrate was simply drawn into a matching site on the enzyme molecule. If the substrate did not perfectly fit the active site, the reaction did not proceed. This model was supported by early X-ray crystallography studies but has since been modified to recognize the flexibility of enzymes (the induced fit model).

The current induced fit model

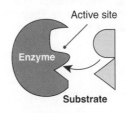

1 A substrate molecule is drawn into the enzyme's active site, which is like a cleft into which the substrate molecule(s) fit.

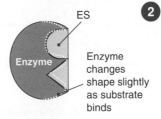

2 The enzyme changes shape as the substrate binds an enzyme-substrate (ES) complex. The shape change makes the substrate more amenable to alteration. In this way, the enzyme's interaction with its substrate is best regarded as an induced fit.

3 The ES interaction results in an intermediate enzyme-product (EP) complex. The substrate becomes bound to the enzyme by weak chemical bonds, straining bonds in the substrate and allowing the reaction to proceed more readily.

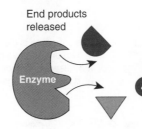

4 The end products are released and the enzyme returns to its previous shape.

Once the substrate enters the active site, the shape of the active site changes to form an active enzyme-substrate (ES) complex. This strains substrate bonds and lowers the energy required to reach the transition state. The **induced-fit model** is supported by X-ray crystallography, chemical analysis, and studies of enzyme inhibitors, which show that enzymes are flexible and change shape when interacting with the substrate.

1. Describe the key features of the 'lock and key' model of enzyme action and explain its deficiencies as a working model:

_____

_____

_____

_____

2. How does the current 'induced fit' model of enzyme action differ from the lock and key model? _____

_____

_____

_____

©2017 **BIOZONE** International
ISBN: 978-1-927309-65-0
Photocopying Prohibited

PRACTICES CONNECT **265** AP2 WEB **21** **KNOW**

22 How Enzymes Work

Key Idea: Enzymes increase the rate of biological reactions by lowering the reaction's activation energy.

Chemical reactions in cells are accompanied by energy changes. The amount of energy released or taken up is directly related to the tendency of a reaction to run to completion (for all the reactants to form products). Any reaction needs to raise the energy of the substrate to an unstable transition state before the reaction will proceed (below). The amount of energy needed to do this is the **activation energy** (*Ea*). Enzymes lower the *Ea* by destabilizing bonds in the substrate so that it is more reactive. Enzyme reactions can break down a single substrate molecule into simpler substances (catabolic reactions) or join two or more substrate molecules together (anabolic reactions).

Lowering the activation energy

The presence of an enzyme simply makes it easier for a reaction to take place. All catalysts speed up reactions by influencing the stability of bonds in the reactants. They may also provide an alternative reaction pathway, thus lowering the activation energy (*Ea*) needed for a reaction to take place (see the graph below).

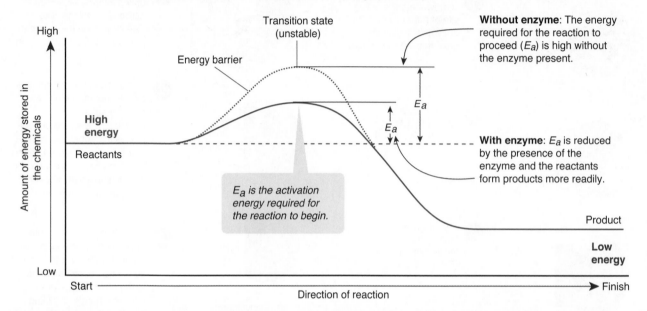

Without enzyme: The energy required for the reaction to proceed (*Ea*) is high without the enzyme present.

With enzyme: *Ea* is reduced by the presence of the enzyme and the reactants form products more readily.

Ea is the activation energy required for the reaction to begin.

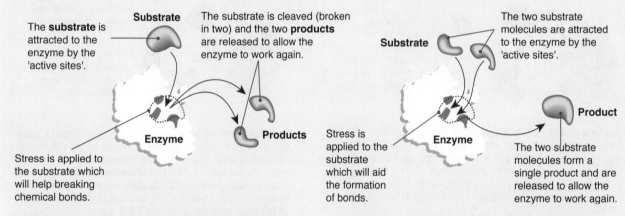

The **substrate** is attracted to the enzyme by the 'active sites'.

The substrate is cleaved (broken in two) and the two **products** are released to allow the enzyme to work again.

Stress is applied to the substrate which will help breaking chemical bonds.

The two substrate molecules are attracted to the enzyme by the 'active sites'.

Stress is applied to the substrate which will aid the formation of bonds.

The two substrate molecules form a single product and are released to allow the enzyme to work again.

Catabolic reactions produce smaller molecules

Some enzymes can cause a single substrate molecule to be drawn into the active site. Chemical bonds are broken, causing the substrate molecule to break apart to become two separate molecules. Catabolic reactions break down complex molecules into simpler ones and involve a net release of energy, so they are called exergonic.
Examples: *hydrolysis, cellular respiration*.

Anabolic reactions produce larger molecules

Some enzymes can cause two substrate molecules to be drawn into the active site. Chemical bonds are formed, causing the two substrate molecules to form bonds and become a single molecule. Anabolic reactions involve a net use of energy (they are endergonic) and build more complex molecules and structures from simpler ones.
Examples: *protein synthesis, photosynthesis*.

1. How do enzymes lower the activation energy for a reaction? _____

_____

_____

2. Describe the difference between a catabolic and anabolic reaction: _____

_____

©2017 **BIOZONE** International
ISBN: 978-1-927309-65-0
Photocopying Prohibited

23 Enzyme Kinetics

Key Idea: Enzymes operate most effectively within a narrow range of conditions. The rate of enzyme-catalyzed reactions is influenced by both enzyme and substrate concentration. Enzymes usually have an optimum set of conditions (e.g. of pH and temperature) under which their activity is greatest. Many plant and animal enzymes show little activity at low temperatures. Enzyme activity increases with increasing temperature, but falls off after the optimum temperature is exceeded and the enzyme is denatured. Extremes in pH can also cause denaturation. Within their normal operating conditions, enzyme reaction rates are influenced by enzyme and substrate concentration in a predictable way.

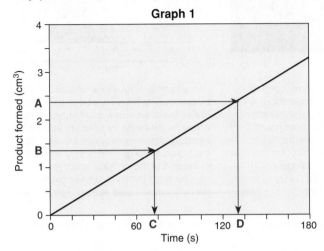

Graph 1

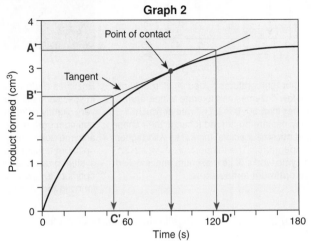

Graph 2

The rate of a reaction can be calculated from the amount of product produced during a given time period. For a reaction in which the rate does not vary (graph 1) the reaction rate calculated at any one point in time will be the same. For example: $B/C = A/D = A-B/D-C = (\Delta p/\Delta t)$ (the change in product divided by the change in time).

In a reaction in which the rate varies (graph 2) a reaction rate can be calculated for any instantaneous moment in time by using a tangent. The tangent must touch the curve at only one point. The gradient of the tangent can then be used to calculate the rate of reaction at that point in time ($A'-B'/D'-C'$).

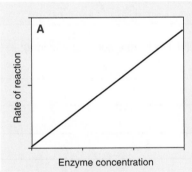

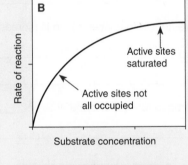

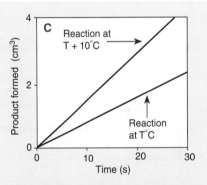

Given an unlimited amount of substrate, the rate of reaction will continue to increase as enzyme concentration increases. More enzyme means more reactions between substrates can be catalyzed in any given time (graph A).

If there is unlimited substrate but the enzyme is limited, the reaction rate will increase until the enzyme is saturated, at which point the rate will remain static (graph B).

The effect of temperature on a reaction rate is expressed as the temperature coefficient, usually given as the Q_{10}. Q_{10} expresses the increase in the rate of reaction for every rise of 10°C. **Q_{10} = rate of reaction at (T + 10°C)/ rate of reaction at T**, where T is the temperature in °C (graph C).

1. Calculate the reaction rate in graph 1: _____

2. For graph 2:

   (a) The reaction rate at 90 seconds: _____

   (b) The reaction rate at 30 seconds: _____

3. (a) What must be happening to the reaction mix in graph 1 to produce the straight line (constant reaction rate)?

   _____

   _____

   (b) Explain why the reaction rate in graph 2 changes over time: _____

   _____

©2017 **BIOZONE** International
ISBN: 978-1-927309-65-0
Photocopying Prohibited

PRACTICES PRACTICES CONNECT CONNECT WEB

    14 AP1 **265** AP2 **23**

KNOW

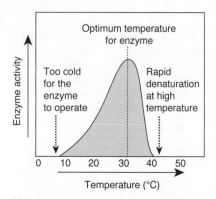

Optimum temperature for enzyme

Too cold for the enzyme to operate

Rapid denaturation at high temperature

Professor Dr. habil. Uwe Kils CC3.0

Antarctic icefish

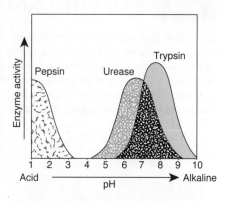

Higher temperatures speed up all reactions, but few enzymes can tolerate temperatures higher than 50–60°C. The rate at which enzymes are denatured (change their shape and become inactive) increases with higher temperatures. The temperature at which an enzyme works at its maximum rate is called the **optimum temperature**.

Enzymes performing the same function in species in different environments are very slightly different in order to maintain optimum performance. For example, the enzyme acetylcholinesterase has an optimum temperature of -2°C in the nervous system of an Antarctic icefish but an optimum temperature of 25°C in grey mullet found in the Mediterranean.

Like all proteins, enzymes are denatured by extremes of pH (very acid or alkaline). Within these extremes, most enzymes have a specific pH range for optimum activity. For example, digestive enzymes are specific to the region of the gut where they act: pepsin in the acid of the stomach and trypsin in the alkaline small intestine. Urease catalyzes the hydrolysis of urea at a pH near neutral.

4. (a) Describe the change in reaction rate when the enzyme concentration is increased and the substrate is not limiting:

_____

_____

(b) Suggest how a cell may vary the amount of enzyme present: _____

_____

_____

5. Describe the change in reaction rate when the substrate concentration is increased (with a fixed amount of enzyme):

_____

_____

6. (a) Describe what is meant by an optimum temperature for enzyme activity: _____

_____

_____

(b) Why do most enzymes perform poorly at low temperatures? _____

_____

_____

(c) For graph C on the previous page, calculate the Q_{10} for the reaction: _____

_____

_____

_____

7. (a) State the optimum pH for each of the enzymes:

Pepsin: _____ Trypsin: _____ Urease: _____

(b) Explain how the pH optima of each of these enzymes is suited to its working environment: _____

_____

_____

_____

©2017 **BIOZONE** International
ISBN: 978-1-927309-65-0
Photocopying Prohibited

24 Enzyme Cofactors

Key Idea: Some enzymes act as catalysts on their own, but others need another molecule, called a cofactor, to function. Nearly all enzymes are made of protein. Some enzymes are functional protein-only molecules, but others require additional non-protein components, called **cofactors**, to have catalytic activity. Cofactors may be **inorganic ions** (e.g. Ca^{2+} or Zn^{2+}) or non-protein organic molecules, called **coenzymes**. Many vitamins or vitamin derivatives are coenzymes. Cofactors may be tightly or loosely bound to the enzyme. Permanently bound cofactors are called **prosthetic groups**. Many enzymes require several cofactors, often both organic coenzymes and inorganic ions, to function.

Conjugated protein enzymes

Where extra non-protein components are required for enzyme function, the enzyme (protein) is called the **apoenzyme** and the additional chemical component is called a **cofactor**. Neither the apoenzyme nor the cofactor has catalytic activity on its own.

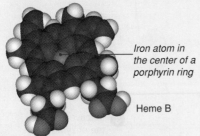

Iron atom in the center of a porphyrin ring

Heme B

Enzyme with prosthetic group

Prosthetic groups are cofactors that are permanently bound to the enzyme. The heme prosthetic groups, which consist of an iron atom in the center of a porphyrin ring, are an example of a prosthetic group. Heme groups are also cofactors in catalase and some respiratory enzymes.

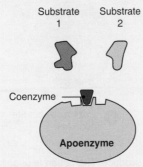

Substrate 1   Substrate 2

One substrate combines with the coenzyme to form a coenzyme/substrate complex

The coenzyme/substrate complex combines with the second substrate to form the product (P).

Coenzyme

Apoenzyme

Enzyme with coenzyme loosely attached

Other coenzymes are loosely attached to the enzyme and detach after the reaction to participate in other reactions.
Example: dehydrogenases + NAD. NAD is the coenzyme form of the vitamin niacin (B_3). Many coenzymes are vitamin derivatives.

The enzyme α amylase is present in saliva where it starts the hydrolysis of starch into the simple sugars maltose and glucose. To work correctly, it needs the ions Ca^{2+} and Cl^-. Cl^- increases the binding of Ca^{2+} by 100 times. It also shifts the optimum pH for amylase from 6 to 6.8.

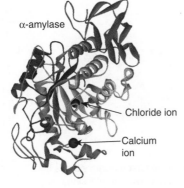

α-amylase

Chloride ion

Calcium ion

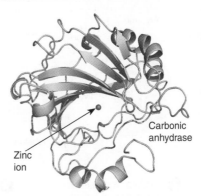

Zinc ion

Carbonic anhydrase

Carbonic anhydrase is an important molecule in the transport of CO_2 into and out of the cell. It contains a central Zn^{2+} ion as a prosthetic group.

---

1. What is a cofactor? _____
   _____

2. Describe the difference between the two different broad categories of cofactors: _____
   _____
   _____
   _____

3. How do cofactors enable an enzyme's catalytic activity? _____
   _____
   _____

4. Distinguish between the apoenzyme and the cofactor: _____
   _____
   _____

25 Enzyme Inhibitors

Key Idea: Enzyme activity can be reduced or stopped by inhibitors. These may be competitive or non-competitive. Enzyme activity can be stopped, temporarily or permanently, by chemicals called enzyme inhibitors. **Irreversible inhibitors** bind tightly to the enzyme and are not easily displaced. **Reversible inhibitors** can be displaced from the enzyme and have a role as enzyme regulators in metabolic pathways.

Competitive inhibitors compete directly with the substrate for the active site and their effect can be overcome by increasing the concentration of available substrate. A **non-competitive inhibitor** does not occupy the active site, but distorts it so that the substrate and enzyme can no longer interact. Both competitive and non-competitive inhibition may be irreversible, in which case the inhibitors act as poisons.

Allosteric enzyme regulation

Allosteric site: The place on an enzyme where a molecule that is not a substrate may bind. The allosteric binding site is never the active site.

Active site

Substrate molecules

1 Active form of the enzyme

Enzyme catalyzes the reaction between the substrates producing a new molecule.

2 Enzyme-substrate complex

The new molecule attaches to the allosteric site of the enzyme, inhibiting the enzyme's activity.

3 Inactive form of the enzyme

Metabolic pathways can be regulated by the products they produce. The action is usually by **feedback inhibition** (negative feedback). This can be achieved by **allosteric regulation**. When the concentration of the end product is high, the end product will bind to the allosteric site of the first enzyme in the pathway, inhibiting the enzyme and shutting down the pathway. When the concentration of the end product is reduced, the allosteric site is freed and the pathway is activated again.

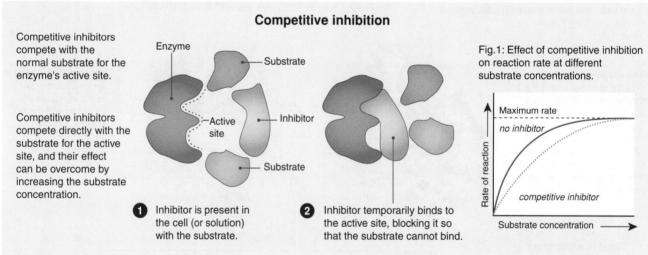

Competitive inhibition

Competitive inhibitors compete with the normal substrate for the enzyme's active site.

Competitive inhibitors compete directly with the substrate for the active site, and their effect can be overcome by increasing the substrate concentration.

Enzyme

Substrate

Active site

Inhibitor

Substrate

1 Inhibitor is present in the cell (or solution) with the substrate.

2 Inhibitor temporarily binds to the active site, blocking it so that the substrate cannot bind.

Fig.1: Effect of competitive inhibition on reaction rate at different substrate concentrations.

Maximum rate

no inhibitor

competitive inhibitor

Rate of reaction

Substrate concentration ⟶

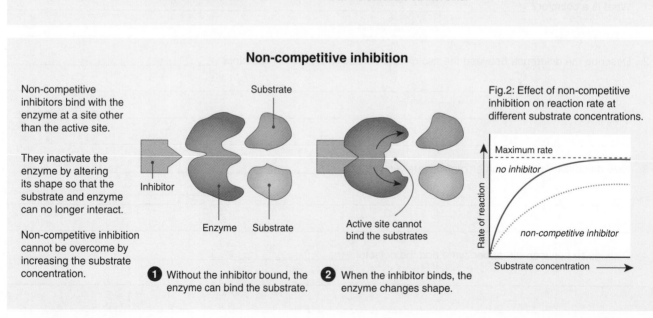

Non-competitive inhibition

Non-competitive inhibitors bind with the enzyme at a site other than the active site.

They inactivate the enzyme by altering its shape so that the substrate and enzyme can no longer interact.

Non-competitive inhibition cannot be overcome by increasing the substrate concentration.

Substrate

Inhibitor

Enzyme   Substrate

Active site cannot bind the substrates

1 Without the inhibitor bound, the enzyme can bind the substrate.

2 When the inhibitor binds, the enzyme changes shape.

Fig.2: Effect of non-competitive inhibition on reaction rate at different substrate concentrations.

Maximum rate

no inhibitor

non-competitive inhibitor

Rate of reaction

Substrate concentration ⟶

KNOW

©2017 **BIOZONE** International
ISBN: 978-1-927309-65-0
Photocopying Prohibited

Poisons are irreversible inhibitors

Some enzyme inhibitors are poisons because the enzyme-inhibitor binding is irreversible. Irreversible inhibitors form strong covalent bonds with an enzyme. These inhibitors may act at, near, or remotely from the active site and modify the enzyme's structure to such an extent that it ceases to work. For example, the poison cyanide is an irreversible enzyme inhibitor that combines with the copper and iron in the active site of cytochrome c oxidase and blocks cellular respiration.

Since many enzymes contain sulfhydryl (-SH), alcohol, or acidic groups as part of their active sites, any chemical that can react with them may act as an irreversible inhibitor. Heavy metals, Ag^+, Hg^{2+}, or Pb^{2+}, have strong affinities for -SH groups and destroy catalytic activity. Most heavy metals are non-competitive inhibitors.

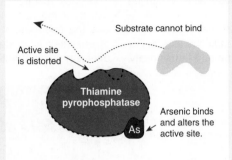

Substrate cannot bind

Active site is distorted

Thiamine pyrophosphatase

Arsenic binds and alters the active site.

As

Arsenic and phosphorus share some structural similarities so arsenic will often substitute for phosphorus in biological systems. It therefore targets a wide variety of enzyme reactions. Arsenic can act as either a competitive or a non-competitive inhibitor (as above) depending on the enzyme.

Drugs

Many drugs work by irreversible inhibition of a pathogen's enzymes. Penicillin (below) and related antibiotics inhibit transpeptidase, a bacterial enzyme which forms some of the linkages in the bacterial cell wall. Susceptible bacteria cannot complete cell wall synthesis and cannot divide. Human cells are unaffected by the drug.

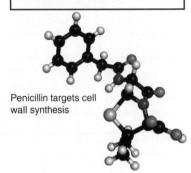

Penicillin targets cell wall synthesis

1. Distinguish between competitive and non-competitive inhibition: _____

2. (a) Compare and contrast the effect of competitive and non-competitive inhibition on the relationship between the substrate concentration and the rate of an enzyme controlled reaction (figures 1 and 2 on the previous page): _____

   (b) Suggest how you could distinguish between competitive and non-competitive inhibition in an isolated system: _____

3. Describe how an allosteric regulator can regulate enzyme activity: _____

4. Explain why heavy metals, such as lead and arsenic, are poisonous: _____

5. (a) In the context of enzymes, explain how penicillin is exploited to control human diseases: _____

   (b) Explain why the drug is poisonous to the target organism, but not to humans: _____

26 Investigating Enzyme Activity

Key Idea: The factors affecting peroxidase activity can be measured using the indicator guaiacol.

Enzymes control all the metabolic activities required to sustain life. Changes to environmental conditions (e.g. pH or temperature) may alter an enzyme's shape and functionality.

This may result in decreased activity or complete loss of activity if the enzyme is denatured. In this activity you will use the information provided and your own understanding of enzymes to design an experiment to investigate factors affecting enzyme activity.

Background

Peroxidase breaks down hydrogen peroxide (H_2O_2), a toxic metabolic by-product of respiration, into water and oxygen.

$$\text{Hydrogen peroxide} \xrightarrow{\textit{Peroxidase}} \text{Water + Oxygen}$$

Like all enzymes, peroxidase activity is highest within specific ranges of pH and temperature, and activity drops off or is halted altogether when the conditions fall outside of the optimal range. The conversion of H_2O_2 is also influenced by other factors such as the levels of substrate and enzyme.

The effect of turnip peroxidase on H_2O_2 breakdown can be studied using the indicator guaiacol. **Guaiacol** has a high affinity for oxygen. In solution, guaiacol binds oxygen and forms tetraguaiacol, which is a brown color. The greater the amount of oxygen produced, the darker brown the solution becomes (right). The color palette provides a standard way to measure relative oxygen production (and therefore peroxidase activity).

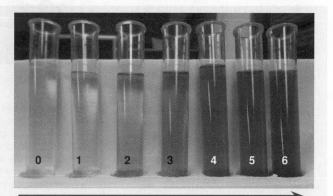

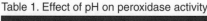

Increasing levels of oxygen production over time (minutes)

The students were provided with a reference color palette (above) against which to compare their results. The palette was produced by adding a set amount of peroxidase to a solution containing hydrogen peroxidase and water. The color change was measured after set time points (0-6 minutes).

Determining the effect of pH on peroxidase activity

Students examined the effect of pH on peroxidase activity using the following procedure:

▶ **Substrate tubes** were prepared by adding 7 mL of distilled water, 0.3 mL of 0.1% H_2O_2 solution, and 0.2 mL of prepared guaiacol solution into 6 clean test tubes. The tubes were covered with parafilm and mixed.

▶ **Enzyme tubes** were prepared by adding 6.0 mL of prepared buffered pH solution (pH 3, 5, 6, 7, 8, 10) and 1.5 mL of prepared turnip peroxidase solution into 6 clean test tubes. The tubes were covered with parafilm and mixed.

▶ The substrate and enzyme tubes were combined, covered in parafilm, mixed and placed back into a test tube rack at room temperature. Timing began immediately. Students took photos with their phones to record the color change (relative to the reference color palette) every minute from time 0-6 minutes. Results are provided in Table 1.

Table 1. Effect of pH on peroxidase activity

	Color reference number					
	0 min	1 min	2 min	3 min	4 min	5 min
pH 3	0	2	2	3	3	3
pH 5	0	2	4	5	6	6
pH 6	0	3	3	3	3	3
pH 7	0	3	4	4	4	4
pH 8	0	3	3	3	3	3
pH 10	0	0	0	0	0	0

1. Graph the students' results on the grid (right).

2. (a) Describe the effect of pH on peroxidase activity:

_____

_____

_____

_____

_____

_____

_____

©2017 **BIOZONE** International
ISBN: 978-1-927309-65-0
Photocopying Prohibited

(b) No color change was recorded at pH 10. Explain why and relate this finding to the enzyme's structure and the way it interacts with its substrate:

_____

_____

_____

3. The color palette (opposite) shows the relative amounts of tetraguaiacol formed when oxygen binds to guaiacol. How can this be used to determine enzyme activity?

_____

_____

4. In the pH experiment, the students measured the rate of enzyme activity by comparing their results against a color palette. How could they have measured the results quantitatively?

_____

_____

5. How might the results be affected if the students did not begin timing immediately after mixing the enzyme and substrate tubes together?

_____

_____

_____

6. Why is peroxidase written above the arrow in the equation for enzymatic breakdown of H_2O_2? _____

_____

7. Using the information provided, design an experiment to test the effect of concentration of turnip peroxidase on oxygen production. In the space below, summarize your method as step by step instructions. Note how you will record and display the data as well as any limitations or sources of potential error with your design:

_____

_____

_____

_____

_____

_____

_____

_____

_____

_____

_____

_____

_____

_____

_____

_____

27 Levels of Organization

Key Idea: Structural organization in multicellular organisms is hierarchical, with new properties arising at each level.

Organization and the emergence of novel properties in complex systems are two of the defining features of living organisms. Organisms are organized according to a hierarchy of structural levels, each level building on the one before it. At each level, novel properties emerge that were not present at the simpler level. Hierarchical organization allows specialized cells to group together into tissues and organs to perform a particular function. This improves efficiency of function in the organism.

In the spaces provided for each question below, assign each of the examples listed to one of the levels of organization as indicated.

1. **Animals**: *epinephrine, blood, bone, brain, cardiac muscle, cartilage, collagen, DNA, heart, leukocyte, lysosome, mast cell, nervous system, neuron, phospholipid, reproductive system, ribosomes, Schwann cell, spleen, squamous epithelium.*

   (a) Molecular level: _____

   _____

   (b) Organelles: _____

   _____

   (c) Cells: _____

   _____

   (d) Tissues: _____

   _____

   (e) Organs: _____

   _____

   (f) Organ system: _____

   _____

   _____

2. **Plants**: *cellulose, chloroplasts, collenchyma, companion cells, DNA, epidermal cell, fibers, flowers, leaf, mesophyll, parenchyma, pectin, phloem, phospholipid, ribosomes, roots, sclerenchyma, tracheid.*

   (a) Molecular level: _____

   _____

   (b) Organelles: _____

   _____

   (c) Cells: _____

   _____

   (d) Tissues: _____

   _____

   (e) Organs: _____

   _____

   _____

Molecular level
Atoms and molecules form the most basic level of organization. This level includes all the chemicals essential for maintaining life, e.g. water, ions, fats, carbohydrates, amino acids, proteins, and nucleic acids.

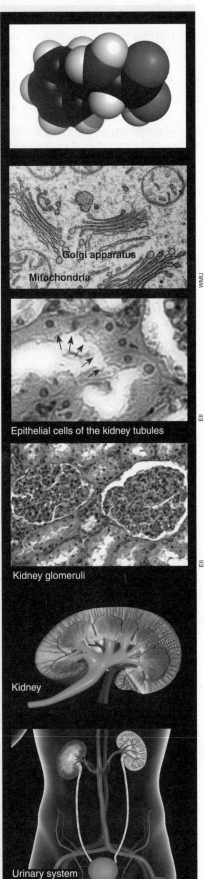

Organelle level
Many diverse molecules may associate together to form complex, specialized cellular organelles, where metabolic reactions may be compartmentalized, e.g. mitochondria, Golgi apparatus, endoplasmic reticulum, chloroplasts.

Golgi apparatus

Mitochondria

Cellular level
Cells are the basic structural and functional units of an organism. Each specialized cell type has a different structure and role as a result of cellular differentiation during development.
Animal examples: *epithelial cells, osteoblasts, muscle fibers.*
Plant examples: *sclereids, xylem vessels, sieve tubes.*

Epithelial cells of the kidney tubules

Tissue level
Tissues are collections of specialized cells of the same origin that together carry out a specific function.
Animal examples: *epithelial tissue, bone, muscle.*
Plant examples: *phloem, chlorenchyma, endodermis, xylem.*

Kidney glomeruli

Organ level
Organs are formed by the functional grouping together of multiple tissues. They have a definite form and structure.
Animal examples: *stomach, heart, lungs, brain, kidney.*
Plant examples: *leaves, roots, storage organs, ovary.*

Kidney

Organ system level
In animals, organs form parts of larger units called organ systems. An organ system is an association of organs with a common function, e.g. digestive system, cardiovascular system, urinary system. In all, eleven organ systems make up a mammalian organism.

Urinary system

WEB
CONNECT
CONNECT

27
52
AP1
54
AP1

KNOW

©2017 **BIOZONE** International
ISBN: 978-1-927309-65-0
Photocopying Prohibited

28 Achieving Metabolic Efficiency

Key Idea: Cells and organelles have compartments. These separate regions help achieve metabolic efficiency.

Metabolic reactions often occur as a linked series in which each step in the pathway relies on the completion of a previous step and each step is controlled by specific enzymes. The end product of one enzyme-controlled step provides the substrate for the next step, so failure of one step causes failure of all later steps. Metabolic pathways are tightly controlled to prevent energy being wasted. This energy conservation is termed metabolic efficiency. Metabolic reactions are often localized within specific organelles so that all the components of the pathway are kept together.

Achieving efficiency by compartmentalization

To increase metabolic efficiency, regions within a cell or an organelle are **compartmentalized** (separated) by membranes. Particular metabolic reactions are restricted to certain regions where all the necessary metabolic components are located. Compartmentalization prevents interference between different reaction pathways and enables radically different reaction environments to be accommodated within different organelles.

Example: Cellular respiration in the mitochondrion
The membrane system of the mitochondrion divides it into several regions. Glycolysis takes place outside of the mitochondrion, in the cell's cytoplasm, but the remaining steps take place in different specialized regions of the mitochondrion. This helps to regulate movement of substrates and end-products and therefore reaction rates, increasing efficiency of the process (below).

1 Cytoplasm (outside the mitochondrion): Glycolysis

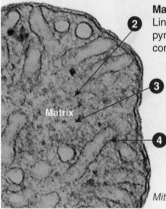

2 **Matrix**: Link reaction. Link reaction enzymes (e.g. pyruvate dehydrogenase complex) are in the matrix.

3 **Matrix**: Krebs cycle. Krebs cycle enzymes (e.g. fumarase) are in the matrix.

Matrix

4 **Cristae**: Electron transport chain. Membrane-bound enzymes include ATP synthase

Mitochondrion

Achieving efficiency by inhibition

Many metabolic pathways are controlled by **feedback inhibition** (negative feedback loop). The pathway is stopped when there is a build-up of end product (or certain intermediate products). The build-up stops the enzymes in the pathway from working and allows the cell to shut down a pathway when it is not needed. This conserves the cell's energy, so it is not manufacturing products it does not need.

Both linear pathways (e.g. glycolysis), and cyclic pathways (e.g. the Krebs cycle) and can be regulated this way (below).

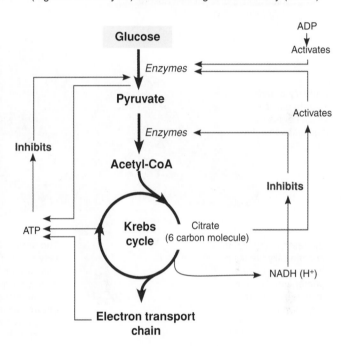

1. What does metabolic efficiency mean? _____

_____

_____

2. Describe how cells achieve metabolic efficiency through:

(a) Compartmentalization: _____

_____

_____

(b) Feedback inhibition: _____

_____

_____

3. What would happen if cells could not regulate their metabolic pathways? _____

_____

_____

©2017 **BIOZONE** International
ISBN: 978-1-927309-65-0
Photocopying Prohibited

PRACTICES

CONNECT **54** AP1

CONNECT **52** AP1

KNOW

29 Regional Specialization and Functional Efficiency

Key Idea: The specialization of organs to carry out specific roles and regional specialization within an organ system allow for functional efficiency in multicellular organisms.
Functional efficiency is the ability to carry out a process in the most efficient manner, with minimal energy or resource cost. In multicellular organisms, functional efficiency is enhanced by having specialized organs and organ systems that perform

a specific function. Examples include the digestive system and the immune system. Organs that perform a specialized role can work more efficiently than if they were required to carry out a number of different, unrelated tasks. Within an organ system, further specialization occurs. Specific organs or regions perform certain tasks and this increases the functional efficiency of the system as a whole.

Functional efficiency in the mammalian digestive system

Efficiency of digestion in mammals is enhanced by the one-way movement of food through the digestive system. It allows the gut to become regionally specialized for processing food. Each region of the digestive system has a different role.

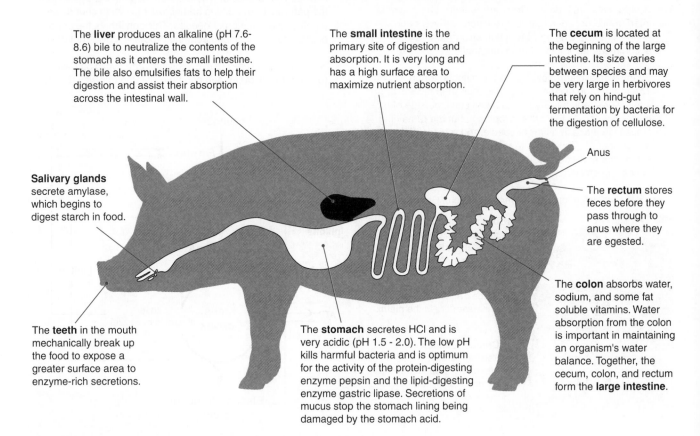

The **liver** produces an alkaline (pH 7.6-8.6) bile to neutralize the contents of the stomach as it enters the small intestine. The bile also emulsifies fats to help their digestion and assist their absorption across the intestinal wall.

The **small intestine** is the primary site of digestion and absorption. It is very long and has a high surface area to maximize nutrient absorption.

The **cecum** is located at the beginning of the large intestine. Its size varies between species and may be very large in herbivores that rely on hind-gut fermentation by bacteria for the digestion of cellulose.

Anus

The **rectum** stores feces before they pass through to anus where they are egested.

Salivary glands secrete amylase, which begins to digest starch in food.

The **teeth** in the mouth mechanically break up the food to expose a greater surface area to enzyme-rich secretions.

The **stomach** secretes HCl and is very acidic (pH 1.5 - 2.0). The low pH kills harmful bacteria and is optimum for the activity of the protein-digesting enzyme pepsin and the lipid-digesting enzyme gastric lipase. Secretions of mucus stop the stomach lining being damaged by the stomach acid.

The **colon** absorbs water, sodium, and some fat soluble vitamins. Water absorption from the colon is important in maintaining an organism's water balance. Together, the cecum, colon, and rectum form the **large intestine**.

Regional specialization and functional efficiency

The chemical environment within each region of the mammalian digestive tract (gut) varies. For example, the stomach is a very acidic environment, whereas the small intestine has a more neutral pH. Each region provides the optimal chemical environment for the enzymes operating there. This increases digestive efficiency, reducing the time it takes for food to pass through the gut and enabling the animal to maximize food intake.

For example, the protein-digesting enzymes pepsin and trypsin both degrade proteins to peptides. However, each enzyme recognizes a different hydrolysis site, so the protein is degraded at different places. Pepsin is found in the stomach, operates at pH 2.0, and is inactive in the neutral pH of the small intestine. Trypsin works optimally in the more neutral pH of the small intestine. Having several different enzymes performing a similar task maximizes the digestion of food.

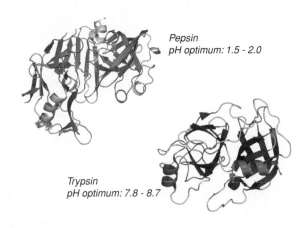

Pepsin
pH optimum: 1.5 - 2.0

Trypsin
pH optimum: 7.8 - 8.7

1. What is the advantage of specialization of organs within organ systems? _____

_____

_____

_____

©2017 **BIOZONE** International
ISBN: **978-1-927309-65-0**
Photocopying Prohibited

Organ systems work together

Although multicellular organisms have specialized organ systems that perform specific roles, the various organ systems must interact and work together to contribute to the overall functioning of the organism.
For example:

▶ The skeletal and muscular systems protect the heart and lungs from damage and enable ventilation of the lungs.

▶ The gas exchange system provides the environment for the exchange of respiratory gases, O_2 and CO_2, between the blood and the air.

▶ The circulatory system transports O_2 from the gas exchange surfaces to the tissues and transports CO_2 in the opposite direction (right).

Circulatory system

Respiratory system

In mammals, the **respiratory system** and **cardiovascular system** interact to supply oxygen and remove carbon dioxide from the body.

Function

Delivers oxygen (O_2) and nutrients to all cells and tissues. Removes carbon dioxide (CO_2) and other waste products of metabolism. CO_2 is transported to the lungs.

Function

Provides a surface for gas exchange. Moves air high in O_2 into the body and air high in CO_2 out of the body.

2. How do multicellular organisms achieve functional efficiency? _____

_____

_____

_____

3. (a) Describe the movement of food through the mammalian digestive system: _____

_____

(b) How does this flow through the digestive system help to achieve functional efficiency? _____

_____

_____

_____

_____

_____

4. Several different digestive enzymes are present in the mammalian digestive system.

(a) Predict what would happen to trypsin if it was subjected to the acidic pH of the stomach: _____

_____

(b) What is the functional advantage of pepsin and trypsin hydrolyzing (breaking down) proteins at different sites?

_____

_____

_____

5. Briefly outline the nature of the interaction between the circulatory and respiratory systems: _____

_____

_____

_____

©2017 **BIOZONE** International
ISBN: 978-1-927309-65-0
Photocopying Prohibited

30 Metabolism in Bacterial Communities

Key Idea: Bacterial cells can interact to increase the efficiency and utilization of energy and matter.

Until relatively recently, bacteria were viewed as unicellular organisms that showed only individual (not collective) behaviors. There are now many documented examples of bacterial communities showing behavior that is closer to multicellularity than unicellularity. By behaving somewhat like a multicellular organism, bacterial communities can access nutrient and energy sources that would otherwise be unavailable to them.

Under certain conditions, microbial communities can attach to surfaces, forming organized structures called **biofilms**. During the formation of biofilms, cells undergo numerous changes in gene expression, leading to changes in the metabolism of the community as a whole. The changes offer new properties to the biofilm including enhanced production of extracellular polymeric substances (for anchoring the colony), antimicrobial resistance, quorum sensing, and gene transfer. The biofilm also offers protection against dehydration, osmotic stress, and nutrient limitations. Biofilms allow individual cells to interact in a way that optimizes use of available resources. Biofilms are problematic in both medicine and industry. In medicine they are responsible for producing plaques (e.g. dental plaque) and infections that are extremely difficult to eliminate. In industry, biofilms can cause numerous problems, including reducing fuel efficiency in ships, fouling heat exchangers, and contaminating food processing machinery.

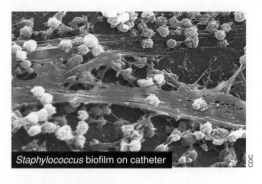
Staphylococcus biofilm on catheter

Myxococcus fruiting bodies

Myxococcus xanthus is well known for the cooperative ability of individual cells. When starved, cells come together to form fruiting bodies, but when food is plentiful, cells show cooperative predatory behavior. When in an aqueous environment with cyanobacteria, *M. xanthus* will secrete lytic enzymes and feed off the cyanobacteria. However, the lytic enzymes are diluted in water. *M. xanthus* deals with this by forming spherical colonies with cyanobacteria trapped inside. The colony is then able to feed effectively *en masse*, thus utilizing the resource by multicellular behavior.

Cyanobacteria (e.g. *Anabaena*) are a taxon of bacteria that are able to photosynthesize. The bacteria often form long filaments of individual cells joined together. Under low-nitrogen conditions, some of these cells will differentiate into **heterocysts**. These cells are able to fix nitrogen from molecular N_2. They show quite different gene expression from neighboring undifferentiated cells, most importantly in the production of the enzyme **nitrogenase** and their inability to photosynthesize. Moreover, the heterocysts share the nitrogen they fix with neighboring cells, while receiving other nutrients from them.

Heterocysts

Anabaena

In marine bacterial communities, the bacterium *Thiovulum* forms meshes of up to a million cells per cm^2. These are formed when *Thiovulum* cells swarm to a transition zone between oxygen rich water above and sulfide rich water below (e.g. seeping from a vent). The population produces slime threads that hold the cells together. The mesh separates the two layers of water and enables *Thiovulum* to regulate the flow of each. By swimming as a group, the mesh community is able to move the mesh up or down to maintain its position on the boundary and control the amount of oxygen or sulfide they receive.

1. (a) How do biofilms form? _____

_____

(b) How does a bacterial community benefit by forming a biofilm? _____

_____

(c) Why is understanding biofilm formation and metabolism important for medicine and industry?_____

_____

_____

2. Using an example from above, explain how unicellular organisms benefit by forming communities that behave like multicellular organisms:

_____

_____

_____

_____

WEB
30

CONNECT
136
AP2

CONNECT
79
AP1

PRACTICES

KNOW

©2017 **BIOZONE** International
ISBN: 978-1-927309-65-0
Photocopying Prohibited

31 KEY TERMS AND IDEAS: Did You Get It?

1. Test your vocabulary by matching each term to its correct definition, as identified by its preceding letter code.

active site _____

allosteric site _____

biological catalyst _____

coenzyme _____

cofactor _____

enzyme _____

inhibitor _____

organs _____

substrate _____

tissue _____

A An organic molecule that acts as a cofactor in an enzyme reaction but is only loosely bound to the enzyme.

B Structures formed by the functional grouping together of multiple tissues. They have a definite form and structure

C The name given to a molecule that can stop or limit enzyme activity, either permanently or temporarily.

D The chemical the enzyme acts upon.

E A site other than the active site where a molecule can bind to an enzyme and alter its activity.

F Collections of specialized cells of the same origin together with their supporting extracellular substances that together carry out a specific function, e.g. muscle.

G The region of an enzyme responsible for substrate binding and reaction catalysis.

H A biological molecule (e.g. protein) that lowers the activation energy of a reaction but is not itself used up during the reaction.

I A globular protein that acts as catalyst to speed up a specific biological reaction.

J An additional non-protein substance essential for the operation of some enzymes.

2. Study the enzymatic word equation below and answer the following questions:

$$\text{Sucrose + Water} \xrightarrow{\text{Sucrase}} \text{Glucose + Fructose}$$

(a) Identify the substrate: _____

(b) Identify the products: _____

(c) Identify the enzyme: _____

3. With reference to enzyme activity, describe the importance of adequate vitamin and mineral intake in the diet: _____

_____

_____

_____

4. (a) Label the graph right with axes and the following labels: Reactants, products, activation energy, transition state.

(b) Assume the reaction has had no enzyme added. Draw the shape of the graph when an enzyme is added to the reaction mix.

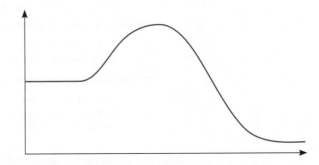

5. The graph (right) shows the effect of an enzyme inhibitor in enzyme reaction rate.

(a) It shows competitive inhibition/non-competitive inhibition (delete one).

(b) Identify the diagram below that illustrates your choice in (a): _____

 A

 B

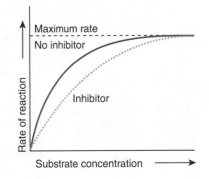

TEST

<table>
<tr><td rowspan="11">

Enduring Understanding
2.A
2.C, 2.D
</td><td colspan="2">

Homeostasis and Energy Allocation
</td></tr>
</table>

Key terms

diapause

ectotherm

endotherm

hibernation

homeostasis

mass-specific metabolic rate

negative feedback

positive feedback

reproductive strategy

thermoregulation

torpor

2.A.1 All living systems require constant input of free energy
Essential knowledge

Activity number

(d) Organisms use free energy to maintain organization, grow, and reproduce

☐ 1 Use examples to explain some of the strategies that organisms use to regulate body temperature and metabolism. Examples include ectothermy, endothermy, and elevated floral temperatures in some plants.   34 35 36

☐ 2 Know that reproduction and parental care require free energy beyond that used for maintenance and growth. Describe the various reproductive strategies used in response to energy availability, including seasonal reproduction in plants and animals, reproductive diapause, and biennial life histories in plants.   32 37-39

☐ 3 Describe the relationship between metabolic rate per unit body mass and the size of multicellular organisms.   33

☐ 4 Describe the consequences of acquired free energy exceeding energy needs.   33

☐ 5 Describe the consequences of acquired free energy being less than energy needs.   33

2.C.1 Organisms use feedback mechanisms to maintain their internal environments and respond to external changes
Essential knowledge

Activity number

(a) Negative feedback mechanisms maintain dynamic homeostasis

☐ 1 Using examples (e.g. blood glucose regulation, thermoregulation in animals, or plant responses to water limitation), explain how negative feedback mechanisms maintain dynamic homeostasis for a particular variable. *[also 3.B.1.b.4]*.   40 41 44 45

(b) Positive feedback mechanisms amplify responses and processes

☐ 1 Using examples (lactation, onset of labor, blood clotting, fruit ripening), explain how positive feedback mechanisms amplify physiological responses.   42 50

(c) Alterations to the operation of feedback mechanisms can have deleterious results

☐ 1 Using examples, explain how an alteration to normal feedback mechanisms can lead to physiological disorders. Examples include diabetes mellitus, dehydration in response to decreased ADH, Grave's disease, and blood clotting disorders.   45 46 49 50

2.C.2 Organisms respond to changes in their external environment
Essential knowledge

Activity number

(a) Organisms use behavior and physiology to respond to changes in their environment

☐ 1 Using thermoregulation as an example, explain how organisms respond to changes in their environment through behavioral and physiological mechanisms.   35 40 44

2.D.3 Biological systems are affected by disruptions to homeostasis
Essential knowledge

Activity number

(a) Disruptions at the molecular and cellular levels affect the health of the organism

☐ 1 Explain how homeostasis can be disrupted by toxic substances.   47 48

32 How Organisms Allocate Energy

Key Idea: The energy available to an organism is allocated for different purposes at different times in a way that maximizes survival and reproduction.

The energy available to any organism is limited. This limited energy supply means that organisms must make compromises in terms of how they allocate available energy and their energy allocations may be prioritized differently at different times. For example, organisms may expend large amounts of energy in migration, but their overall energy gain at their destination enables them to allocate energy to breeding.

Energy gain and allocation

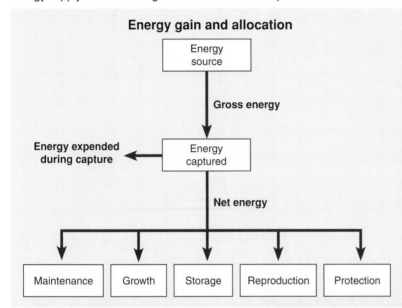

The net energy gain from an energy source depends on the size of the energy source and the energy expended to obtain it. In the case of a predator, such as a cheetah, the energy expended in the chase must be less than the energy gained from the eating the prey. Enough energy must be left over to maintain the body, grow, or reproduce.

The energy a cheetah will expend during a hunt is very precisely regulated. If it doesn't make a kill in under a minute, it stops the chase.

A large amount of energy is expended during migration. The energy gained from abundant food sources at the destination is enough to allow the animal to breed, rear its young, and carry it through the return migration and winter.

Hibernating Northern bat, Norway

Magne Flåten cc 3.0

Hibernation is a prolonged (usually seasonal) state of reduced activity and metabolic depression. It markedly reduces the animal's energy expenditure, allowing it to survive winter. Large amounts of energy must be stored (e.g. as fat) prior to hibernating.

Deciduous trees expend energy growing new leaves every spring. They produce all the energy needed to grow and reproduce before losing the leaves in fall. Losing leaves reduces frost or storm damage to the tree during winter months.

1. Why is it important for organisms to accurately assess energy gains against expenditure when selecting food sources:

_____

_____

_____

_____

2. In terms of energy allocation and expenditure, explain the following:

(a) Why many animals migrate every year despite the large energy cost: _____

_____

_____

_____

(b) Why hibernation is a strategy for animals living in regions with extreme seasonal variations: _____

_____

_____

PRACTICES   PRACTICES   PRACTICES CONNECT **242** AP2 CONNECT **207** AP2 CONNECT **148** AP2 WEB **32**

DATA

Daily energy requirements for infants 0-12 months

Age (months)	Energy expenditure (E) (MJ day^{-1})	Energy deposition (D) (MJ day^{-1})	Total energy requirement (T = E + D) (MJ day^{-1})	Percent D of T (D ÷ T) x100
0-1	2.5	1.63		
1-2	3.1	1.43		
2-3	3.6	1.14		
3-4	4.1	0.51		
4-5	4.5	0.43		
5-6	4.8	0.35		
6-7	5.0	0.15		
7-8	5.3	0.13		
8-9	5.5	0.12		
9-10	5.7	0.16		
10-11	5.9	0.15		
11-12	6.1	0.15		

Data: Combined data FAO

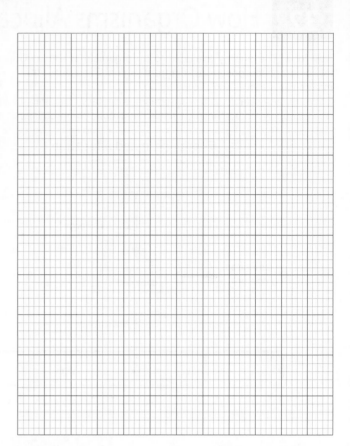

To grow, an organism must obtain more energy than it expends in activity and maintenance. As the body grows, new tissue also requires maintaining and so the daily energy needs of a growing organism increase. As growth slows, the daily energy requirement above that needed for maintenance is reduced. The table above lists daily total energy expenditure and energy deposition (used for growth) for infants from birth to 12 months.

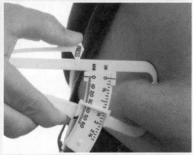

When people obtain more energy than they need for daily activities or growth, the extra energy is stored as fat. This is a growing problem many countries where high energy food is easily obtainable.

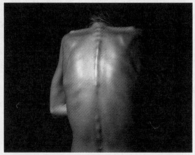

When the body does not get enough energy to carry out its daily activities, it starts to obtain energy by breaking down tissues in the body. This begins with fat tissue, then proteins (normally muscle).

Balancing energy input and energy output is important for managing weight or being able to carry out high energy physical tasks like building muscle or playing sport. High energy tasks require greater energy input.

3. (a) Complete the missing columns in the table above.

(b) Plot the total energy expenditure, energy deposition, and energy requirement for infants aged 0 to 12 months.

(c) What happens to the total energy expenditure of infants as they grow? _____

_____

(d) What happens to the amount of energy allocated to deposition as infants grow? _____

_____

(e) How does the percentage of the total energy requirement allocated to energy deposition change as the infants grow?

_____

4. How does the body deal with an excess energy intake in someone with a sedentary (inactive) lifestyle? _____

_____

33 Metabolism and Body Size

Key Idea: In general, smaller animals have higher specific metabolic rates. This is in part related to their higher surface area to volume ratios. Metabolism is inefficient and produces heat. Animals that use this metabolic heat to maintain body temperature are called endotherms, whereas those that do not are called ectotherms. The baseline metabolic rate of an animal is measured as the basal metabolic rate (BMR) for an endotherm or as the standard metabolic rate (SMR) for an ectotherm. Clearly, larger animals have higher metabolic rates (total kJ used per day) because there is more metabolizing tissue, but when we look at the per-mass (per gram) metabolic rate, the situation is quite the opposite.

In both endotherms and ectotherms, smaller animals have higher mass-specific MRs than larger animals. This relationship holds true across many species and can be described mathematically. This is in part related to the higher surface area to volume ratios of small animals (they lose heat faster so must also generate it faster) but it is not the entire story as the relationship holds even for ectotherms.

For example, during a deep dive, the largest whales are able to stay submerged for almost two hours. However, small diving mammals, such as the water shrew, can only stay submerged for 30 s. Part of the reason for these variations is the obvious difference in the size of their lungs and their body's ability to store oxygen. The most important factor is the difference in the rate at which their bodies consume oxygen per unit mass (a measure of the mass-specific metabolic rate). The water shrew uses oxygen at a rate of 7.4 L kg^{-1} h^{-1}. The blue whale uses just 0.02 L kg^{-1} h^{-1}.

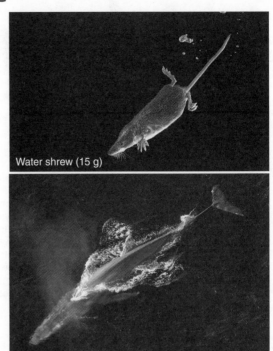
Water shrew (15 g)

Blue whale (173,000 kg)

Animal	Body mass (kg)	O₂ consumption (L kg^{-1} h^{-1})
Shrew	0.005	7.40
Harvest mouse	0.009	2.50
Kangaroo mouse	0.015	1.80
Mouse	0.025	1.65
Ground squirrel	0.096	1.03
Rat	0.290	0.87
Cat	2.5	0.68
Dog	11.7	0.33
Sheep	42.7	0.22
Human	70	0.21
Horse	650	0.11
Elephant	3830	0.07

Data: Schmidt-Nielsen

Body mass vs oxygen consumption rate in mammals

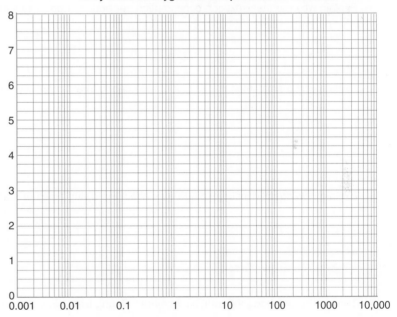

1. (a) Use the semi-log grid above to plot the rate of oxygen consumption for the animals listed in the table:

   (b) What is the shape of the graph? _____

   _____

   _____

   (c) Explain why the data are plotted on a logarithmic scale for body mass? _____

   _____

   _____

   (d) What would this data look like if plotted on a log-log scale instead of a log-linear scale? _____

   _____

©2017 **BIOZONE** International
ISBN: 978-1-927309-65-0
Photocopying Prohibited

PRACTICES   PRACTICES   CONNECT   WEB

   28 AP2   33

DATA

The relationship between specific metabolic rate and body mass for mammals is expressed as B ~ 70 M$^{-0.25}$. That is, the specific metabolic rate (B) of any mammal is proportional to its mass (M (kg)) to the power of -0.25. Thus a 2 kg cat has a specific metabolic rate 3.1 times less than a 0.02 kg mouse. However, there appear to be many exceptions so there is some debate over the validity of the function.

The relationship of specific metabolic rate and size also appears in organisms other than mammals. Birds follow a relationship very close to that of mammals. Invertebrates and unicellular organisms all follow similar patterns in that the bigger they are, the slower their mass-specific metabolic rate. Trees follow a relationship close to 1 (i.e. a tree half the size of another will have a specific metabolic rate twice as great).

Exactly why metabolic rates are slower in large animals is uncertain. Certainly the surface area to volume ratio of organisms appears to be involved (small animals radiate more heat per volume than large ones). However, this does not explain why ectotherms follow a similar relationship to mammals. It may be because smaller animals have more structural mass per volume, which has a higher cost to maintain.

2. Calculate the following:

    (a) How many times greater the specific metabolic rate of a shrew is than a horse: _____

    (b) How many times less the specific metabolic rate of a human is than a rat: _____

3. Explain why per mass oxygen consumption per hour is a good indicator of mass specific metabolic rate: _____

   _____

   _____

4. Explain why a small mammal (e.g. a shrew) needs to eat proportionally more than a large mammal (e.g. an elephant):

   _____

   _____

   _____

5. Explain how specific metabolic rate may affect the diving times of mammals: _____

   _____

   _____

   _____

6. The plot below shows the total surface area of lung tissue in various mammals against the rate of oxygen consumption:

    (a) Describe the relationship between a mammal's lung surface area and their oxygen consumption:

    _____

    _____

    (b) Is this what you would expect? Explain: _____

    _____

    _____

    _____

    _____

    _____

©2017 **BIOZONE** International
ISBN: 978-1-927309-65-0
Photocopying Prohibited

34 Endothermy vs Ectothermy

Key Idea: Ectotherms depend on heat from the environment whereas endotherms generate heat through metabolic activity. Endotherms and many ectotherms thermoregulate to maintain an optimum temperature for functioning.

Animals are classified into two broad groups based on the source of their body heat. **Ectotherms** depend on the environment for their heat energy (e.g. heat from the sun) and their metabolic demands are relatively low, whereas **endotherms** generate their body heat from metabolism. All endotherms and many ectotherms **thermoregulate** (control body temperature) in order to maintain an optimum temperature for functioning. Ectotherms rely on behavioral mechanisms to do this, whereas in endotherms both behavioral and physiological responses are involved.

Most fish and all amphibians are ectothermic (they rely on environmental sources of heat energy). Unlike many reptiles, they do not thermoregulate, so their body temperature fluctuates with the environment (they are poikilothermic) and they are usually restricted to thermally stable environments.

Reptiles, such as snakes, lizards, and crocodiles, are also ectothermic but regulate body temperature using behavior. They bask and use body positioning to raise their body temperature for activity and seek shade when temperatures are too high. Some larger reptiles are able to maintain a relatively elevated body temperature for much of the time.

Birds and mammals are endotherms and achieve a high body temperature through metabolic activity and reduction of heat exchanges. They can function independently of the environmental temperature (within the species-specific tolerance range) and maintain high metabolic rates. Their body temperature remains stable (they are homeothermic) and their energy costs are high.

Daily temperature variations in ectotherms and endotherms

Ectotherm: Diurnal lizard
Body temperature is regulated by behavior so that it does not rise above 40°C. Basking increases heat uptake from the sun. Activity occurs when body temperature is high. Underground burrows are used for retreat.

Endotherm: Human
Body temperature fluctuates within narrow limits over a 24 hour period. Exercise and eating increase body temperature for a short time. Body temperature falls during rest and is partly controlled by an internal rhythm.

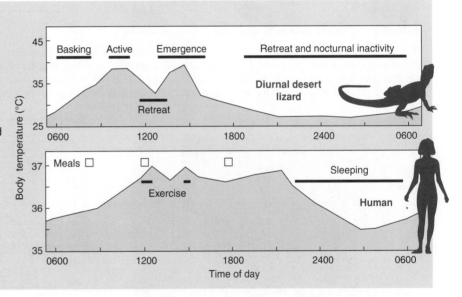

1. Distinguish between ectotherms and endotherms in terms of their sources of body heat: _____

_____

_____

_____

2. Explain why ectotherms that do not thermoregulate are restricted to environments with relatively stable temperatures:

_____

_____

3. The diagrams above show daily temperature variations in an ectotherm and an endotherm.

   (a) Which animal has the largest temperature variation? _____

   (b) Explain why this is the case: _____

   _____

©2017 **BIOZONE** International
ISBN: 978-1-927309-65-0
Photocopying Prohibited

PRACTICES CONNECT **44** AP2 WEB **34**

KNOW

4. Again, with reference to the diagram of daily temperature fluctuations on the previous page:

(a) Describe the effect of eating and exercise on body temperature in humans: _____

_____

(b) What effect does sleeping have on human body temperature? _____

_____

(c) What do you think is happening to the metabolic rate during these times? _____

_____

_____

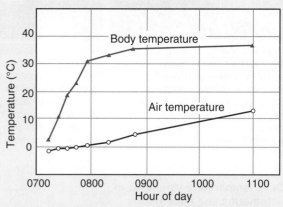

The Peruvian mountain lizard (*Liolaemus*) lives at altitudes of ~4000 m in Peru, where the air temperature is low, even in summer. It emerges in the morning when the air temperature is below freezing. By exposing itself to the sun, it rapidly heats up to a body temperature that enables it to be fully active. Once warm, the lizard maintains its preferred body temperature of around 35°C by changing posture and orientation to the sun and thereby controlling the amount of heat energy absorbed.

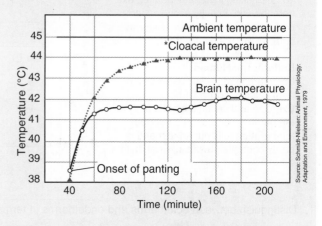

The chuckwalla (*Sauromalus*) is a widespread lizard species in the deserts of the southwestern United States and northern Mexico and is active in the temperature range 26-39°C (higher for basking). If moved from 15°C to 45°C, cloacal and brain temperatures increase rapidly. At ~41°C, these temperatures diverge and the brain stays at ~2°C below the cloacal temperature and 3°C below air temperature. The chuckwalla achieves this by panting. Its carotid arteries supplying the brain run close to the surface of the pharynx and heat is lost there by evaporative cooling. *Cloacal temperature measures deep body temperature through the cloaca. It is equivalent to rectal temperature in mammals.

5. (a) In the examples above, the increase in body temperature is very rapid. Why is this important for an ectotherm?

_____

_____

(b) What is the purpose of 'panting' in the chuckwalla? _____

_____

_____

©2017 **BIOZONE** International
ISBN: 978-1-927309-65-0
Photocopying Prohibited

6. Compare and contrast the thermoregulatory strategies of the Peruvian mountain lizard and the desert chuckwalla, relating any differences to the difference in their respective environments:

_____

_____

_____

_____

_____

7. As illustrated in the examples opposite, ectotherms are capable of achieving and maintaining high, relatively constant body temperatures for relatively long periods in spite of environmental fluctuations. However, they also tolerate marked declines in body temperature to levels lower than are tolerated by most endotherms.

   (a) What could be the advantage of allowing body temperature to fall when ambient temperature drops? _____

   _____

   _____

   _____

   (b) Why might ectothermy be regarded as an adaptation to low or variable food supplies?_____

   _____

   _____

8. (a) In the generalized graph right, identify the optimum temperature range for an endotherm:

   _____

   (b) Describe the energetic costs of thermoregulation (as measured by oxygen consumption) in an endotherm:

   _____

   _____

   _____

   (c) Explain why this is the case: _____

   _____

   _____

Body temperature and oxygen consumption in an endotherm at different ambient temperatures

9. The graph right shows the body temperatures of two species of North American bats. When the animals are torpid, their body temperature falls with that of the environment:

   (a) What two behaviors do the bats show at lower temperatures (10°-15°C)?

   _____

   (b) What is the most common behavior at 10°C and below:

   _____

   (c) The oxygen consumption of active bats increases with decreasing air temperature. Explain this observation:

   _____

   _____

   _____

Source: Schmidt-Nielsen: Animal Physiology: Adaptation and Environment, 1979

35 Mechanisms of Thermoregulation in Endotherms

Key Idea: Endotherms regulate their body temperature by controlling heat exchanges with the environment and generating heat from metabolism.

Heat exchanges with the environment occur via **conduction** (direct heat transfer), **radiation** (indirect heat transfer), and **evaporation**. To maintain a relatively constant body temperature, endotherms must balance heat losses and gains. They use a range of structural, behavioral, and physiological mechanisms to maintain a body temperature that is optimum for functioning.

Water has a much greater capacity than air to transfer heat away from organisms, so aquatic mammals have heavily insulated surfaces of vascularized fat called blubber (up to 60% of body thickness). Blood is diverted to the outside of the blubber if heat needs to be lost.

Mammals generate their body heat through metabolism.

Heat loss from flippers and tail flukes is minimized by the use of countercurrent heat exchangers in which heat is transferred between arterial and venous blood flows.

Temperature regulation mechanisms in water

▸ Heat generation from metabolic activity
▸ Insulation layer of blubber
▸ Changes in circulation patterns when swimming
▸ Large body size
▸ Heat exchange systems in limbs or high activity muscle

Aquatic birds have heat exchangers in their webbed feet that transfer heat from the outgoing arterial blood to the incoming blood in the veins. Their feet are therefore close to the ambient temperature of the water and they reduce heat loss to the environment.

Temperature regulation mechanisms in air

▸ Behavior or habitat choice
▸ Heat generation from metabolic activity, including shivering.
▸ Insulation (fat, fur, feathers)
▸ Circulatory changes including constriction and dilation of blood vessels
▸ Large body size
▸ Sweating and panting
▸ Tolerance of fluctuation in body temperature

Large body size reduces heat loss by lowering the surface area to volume ratio.

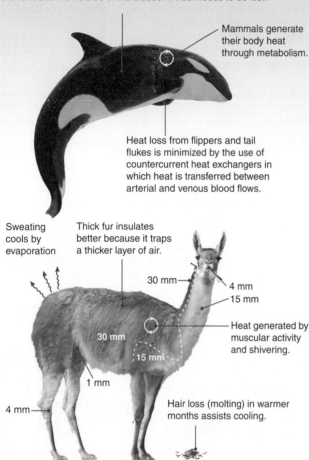

Sweating cools by evaporation

Thick fur insulates better because it traps a thicker layer of air.

30 mm — 4 mm — 15 mm

30 mm

15 mm

1 mm

4 mm

Heat generated by muscular activity and shivering.

Hair loss (molting) in warmer months assists cooling.

Water is lost in evaporative cooling

The greater the temperature gradient between the air and the body, the greater the rate of heat gain (or loss).

For most mammals, the thickness of the fur or hair varies around the body (as indicated above). Thermoregulation is assisted by adopting body positions that expose or cover areas of thin fur (the figures above are for the llama-like guanaco).

Animals adapted to temperature extremes (hot or cold) often tolerate large fluctuations in their body temperature. In well watered camels, body temperature fluctuates less than 2°C, but when they are deprived of water, the body temperature may fluctuate up to 7°C (34°C to 41°C) over a 24 hour period. By allowing their body temperature to rise, heat gain is reduced and the animal conserves water and energy.

1. (a) What is the thermoregulatory role of thick hair or fur? _____

_____

(b) Why is fur/hair thickness variable over different regions of the body? _____

_____

(c) How would you expect fur thickness to vary between related mammal species at high and low altitude?

_____

_____

(d) How do marine mammals compensate for lack of thick hair or fur? _____

2. Explain the adaptive advantage of heat exchangers in the feet of a duck: _____

_____

_____

©2017 **BIOZONE** International
ISBN: 978-1-927309-65-0
Photocopying Prohibited

36 Responses to Temperature in Plants

Key Idea: Low environmental temperatures trigger a range of responses in plants, including (in some) heat generation. Plants are capable of marked physiological responses to a wide range of environmental variables, including temperature, which has a critical role in plant growth and development. A prolonged period of low temperature is an important cue to indicate winter. Woody plants are able to survive freezing temperatures because of metabolic changes that occur in the plant between summer and winter. This process of acclimation involves producing thicker cell walls and accumulating growth inhibitors in the tissues. A small number of plant species are also able to regulate the temperature of their flowers.

Alternating periods of growth with periods of **dormancy** allows plants to survive water shortages and temperature extremes. When dormant, growth will not resume until the right combination of environmental cues are met, e.g. cold exposure and suitable photoperiod.

Low temperature stimulation of seed germination (stratification) is common in many species. The seeds of many cold-climate plants will not germinate until exposed to a period of wet, cold (5°C) conditions. This is called **cold stratification**.

In many plants, including bulbs, bud burst and flowering often follow exposure to the prolonged cold of winter. This process is called **vernalization**. Like stratification, the period of cold exposure must be long enough to reliably indicate winter has passed.

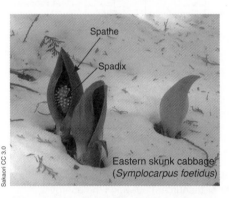

Sakaori CC 3.0

Eastern skunk cabbage (*Symplocarpus foetidus*)

Although a number of plants are endothermic (produce internal warmth) only a few can thermoregulate. Two of these are the sacred lotus and the eastern skunk cabbage (above). Both of these are able to maintain steady floral temperatures well above the air temperature even when the air temperature changes.

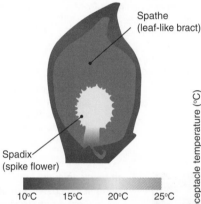

Spathe (leaf-like bract)

Spadix (spike flower)

10°C   15°C   20°C   25°C

High floral temperatures are maintained by uncoupling electron transport from ATP generation in the mitochondria, so energy is lost as heat. The skunk cabbage produces enough heat to melt the snow around it.

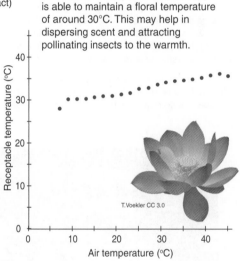

The sacred lotus (*Nelumbo nucifera*) is able to maintain a floral temperature of around 30°C. This may help in dispersing scent and attracting pollinating insects to the warmth.

T. Voekler CC 3.0

1. Describe three physiological responses of plants to changes in temperature:

(a) _____

(b) _____

(c) _____

2. What is the adaptive advantage of cold stratification in seeds? _____

3. (a) Describe how some plants are able to produce heat internally: _____

(b) Give two reasons why regulating floral temperature could be an advantage to plants: _____

©2017 **BIOZONE** International
ISBN: 978-1-927309-65-0
Photocopying Prohibited

CONNECT **8** AP2   WEB **36**   **KNOW**

37 Energy and Seasonal Breeding

Key Idea: Reproduction represents a major energy investment, so animals have various strategies to ensure success. Reproduction is an energetically expensive process, with energy expended in finding a mate, producing gametes, and raising young. The breeding cycles of organisms are normally timed so that young are produced when the food available (energy) is most abundant and conditions are favorable for survival of the young (spring and summer).

The reproductive cycle for animals that breed seasonally often begins in late autumn, with a winter gestation and birth of the young in spring. This maximizes the time available for raising young before the next winter and coincides with the time of maximum food availability.

For Antarctic penguins, the environment dictates a slightly different strategy. Emperor penguins spend their summer feeding and begin their reproductive cycle in March, at the beginning of the Antarctic winter. They move to breeding colonies where the chicks are raised through the winter, becoming independent by the start of the Antarctic summer in November-December. Raising chicks through the harsh Antarctic winter allows them to be independent by summer and so be better able to survive the following winter.

Emperor penguins must also have enough time after raising their chick to feed before their annual molt, during which they cannot enter the sea and so must again rely on their fat stores.

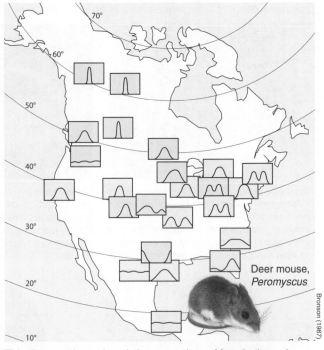

Deer mouse, *Peromyscus*

Bronson (1987)

This diagram shows the relative proportions of female deer mice pregnant at any one time of the year (January on the left side of each graph, December on the right). At higher latitudes, females breed during summer months only, whereas at lower latitudes they breed throughout the year (the blue boxes show one single species).

▶ In many animals, male-male competition develops where the benefits (access to females or resources) outweigh the energy costs (fighting or territory defense). The amount of energy expended by males in attracting or monopolizing several females needs to be weighed against the assistance the females might need to raise the young and how far they range.

▶ It is energetically uneconomic for males to defend a large number of females if they require male assistance to rear the young or travel a great distance to find food. Generally, the most energy expensive reproductive activities for males involve courtship and mating. In females, it is in pregnancy and rearing young.

1. Explain why many animals and plants have seasonal reproductive cycles: _____

   _____

2. Suggest why deer mice at 50°-60° latitude have a strict summer breeding period, unlike the deer mice at 20° latitude:

   _____

   _____

   _____

3. (a) In which part of the reproductive process do males tend to spend the most energy? Explain: _____

   _____

   (b) In which part of the reproductive process do females tend to spend the most energy? _____

   _____

CONNECT **144** AP2   CONNECT **170** AP2   CONNECT **171** AP2   CONNECT **208** AP1

©2017 **BIOZONE** International
ISBN: 978-1-927309-65-0
Photocopying Prohibited

KNOW

38 Reproductive Allocation and Parental Care

Key Idea: The way in which an animal allocates its reproductive effort is part of its reproductive strategy. Effort can be expended in producing offspring or caring for them or both. Different strategies carry different costs and benefits.

The **reproductive effort** is the amount of energy allocated to reproduction (production and care of young). Of the total reproductive effort, the amount remaining after production of the offspring can be allocated to parental care. At one extreme, most invertebrates expend their total reproductive effort in producing eggs and sperm and there is no parental care. At the other extreme, mammals invest heavily in a small number of offspring and the parental care cost is substantial. Between this is a continuum, with some animals adopting alternative strategies, such as brood parasitism. No strategy is necessarily 'better' than any other. They are different solutions to the problem of successful reproduction.

Many eggs or young produced	Moderate number of eggs produced	Few eggs or young produced
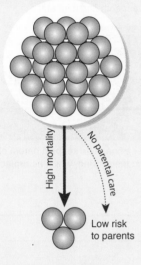	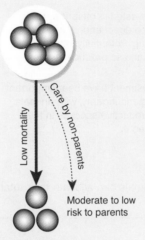	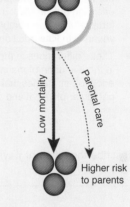
High mortality / No parental care → Low risk to parents	Low mortality / Care by non-parents → Moderate to low risk to parents	Low mortality / Parental care → Higher risk to parents
Low percentage of offspring survive to reproductive age	High percentage of offspring survive to reproductive age	High percentage of offspring survive to reproductive age

Little or no parental care

▶ Large number of offspring produced.
▶ Reproductive effort per offspring is low.
▶ Little or no parental care of offspring.
▶ Reproductive effort is put into producing the offspring, not parental care.
▶ Examples: most fish, amphibians, reptiles, and invertebrates.

Brood parasites

▶ Moderate number of offspring.
▶ Reproductive effort per offspring is moderate to low.
▶ Rely on others to raise offspring.
▶ Risk of egg loss is mitigated by distributing eggs amongst a number of hosts.
▶ Examples: some birds, insects, and fish.

Parental care

▶ Few offspring.
▶ Reproductive effort per offspring is high.
▶ Moderate to substantial care of offspring.
▶ Large reproductive effort put into raising offspring to a less vulnerable stage.
▶ Examples: most birds and mammals, some fish.

Broadcast spawning involves no parental investment after the gametes are released.

A shiny cowbird chick is fed by its host parent, a smaller rufous collared sparrow.

Mammals have a high level of investment in offspring before and after birth.

Many invertebrates allocate all their reproductive effort to producing offspring and there is no parental care. Broadcast spawners such as clams and corals (above), release millions of gametes into the water. Very few of the planktonic larvae will survive. This is the most common reproductive strategy in the oceans and is typical of most marine invertebrates and many fish. Many amphibians are also broadcast spawners although there are many exceptions, including New Zealand's native frogs (*Leiopelma*) in which the males carry the offspring on their backs.

Brood parasitism is a strategy adopted by some birds, notably cuckoos and cowbirds. The brood parasite removes an egg from the nest of a host species and lays one of its own in its place. To reduce the risk of eggs being discovered and destroyed, the eggs are spread around a large number of hosts. Most avian brood parasites have short incubation times, so the egg hatches before those of the host and the imposter will eliminate all or most of the host's eggs/nestlings. The host then raises the parasite chick as if it were its own, even when the chick is larger and differs in appearance. The strategy is not without risk - only about half of the parasite's young survive.

Both mammals and birds are well known for their high levels of parental care and mammals also have a high level of prenatal investment. Other vertebrates, such as some amphibians, fish, and reptiles also provide care until the offspring are capable of fending for themselves. Bird parents are required to incubate their eggs in a nest and then feed the chicks until they are independent. Although most mammals give birth to well developed offspring, they are dependent on their mother for nourishment via suckling milk, as well as learning behaviors essential to their survival.

CONNECT **160** AP2 WEB **38**

KNOW

The significance of reproductive investment

▶ Producing offspring demands enormous amount of energy and risk. In many vertebrate species, reproduction is almost entirely up to the female (males contributing only sperm) but in other species the male also provides support (e.g. by defending a territory or providing food for the female).

▶ Because the female is most heavily invested in the offspring, it is important that she has the best possible reproductive outcome each time she mates, and mate choice is critical. In general, females have a limited reproductive outcome and can only produce so many eggs or offspring in a lifetime. For example, a human female produces one egg a month for about 40 years, a maximum of ~480 eggs in a lifetime. Given that gestation and breast feeding (which suppresses ovulation) may take two years, only about 20 children can be raised in the average lifetime (the record is reportedly 69).

▶ Males, on the other hand, have less invested in offspring. They produce sperm continuously and put no direct energy into the offspring until at least birth or egg laying. Potentially, males could fertilize unlimited numbers of females and so produce far more young with little additional effort.

▶ These differences in reproductive investment have been important in the evolution of mating systems, e.g. monogamy, with animals adopting strategies that maximize reproductive success in their particular physical and social environment.

In ospreys, both parents are needed to successfully rear the chicks. Monogamy is a common mating system when biparental care is needed for offspring survival. Ospreys usually mate for life and raising the young requires a five month commitment.

A South American sea lion male keeps a harem of up to 18 females and their young, which he protects.

1. Describe the different ways in which animals can allocate their total reproductive effort: _____

_____

_____

_____

2. Animals with parental care protect the investment they have already made in offspring. Explain how factors in the environment (e.g. food resources and risks to young) might influence how much care is provided by each parent:

_____

_____

_____

3. (a) What might be the benefits of a brood parasitism strategy to the brood parasite? _____

_____

(b) What adaptations of the brood parasite help to maximize the success of its strategy? _____

_____

_____

_____

4. The common cuckoo (*Cuculus canorus*) is a widespread brood parasite. It parasitizes a range of species including dunnocks, meadow pipits, and reed warblers but females specialize in parasitizing a single species, whose eggs it mimics. This characteristic is heritable, although males fertilize females of all lines, which maintains gene flow in the population.

(a) Explain how a female cuckoo's strategy of parasitizing only a single species contributes to the success of their reproductive strategy:

_____

_____

(b) What is the genetic effect of males showing no such preference for females?

_____

Reed warbler feeds common cuckoo

©2017 **BIOZONE** International
ISBN: 978-1-927309-65-0
Photocopying Prohibited

39 Diapause as a Reproductive Strategy

Key Idea: The red kangaroo can adjust its reproductive cycle to suit the prevailing environmental conditions.

The red kangaroo has a reproductive strategy that enables it to make the most of favorable conditions without compromising survival when conditions are difficult (Australia is an arid continent and availability of food and water are often low). In favorable conditions, it produces young in rapid succession. The female may have a joey at heel, one in the pouch, and an embryo in **diapause** (suspended development) ready to replace the pouch offspring when it leaves (below).

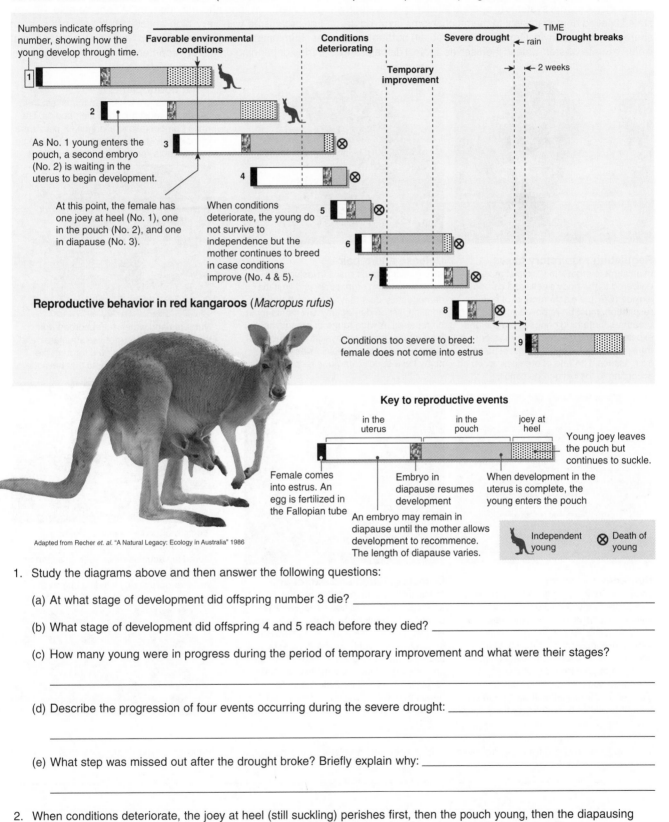

Numbers indicate offspring number, showing how the young develop through time.

As No. 1 young enters the pouch, a second embryo (No. 2) is waiting in the uterus to begin development.

At this point, the female has one joey at heel (No. 1), one in the pouch (No. 2), and one in diapause (No. 3).

When conditions deteriorate, the young do not survive to independence but the mother continues to breed in case conditions improve (No. 4 & 5).

Reproductive behavior in red kangaroos (*Macropus rufus*)

Conditions too severe to breed: female does not come into estrus

Key to reproductive events

Young joey leaves the pouch but continues to suckle.

Female comes into estrus. An egg is fertilized in the Fallopian tube

Embryo in diapause resumes development

When development in the uterus is complete, the young enters the pouch

An embryo may remain in diapause until the mother allows development to recommence. The length of diapause varies.

Independent young   Death of young

Adapted from Recher *et. al.* "A Natural Legacy: Ecology in Australia" 1986

1. Study the diagrams above and then answer the following questions:

   (a) At what stage of development did offspring number 3 die? _____

   (b) What stage of development did offspring 4 and 5 reach before they died? _____

   (c) How many young were in progress during the period of temporary improvement and what were their stages?

   _____

   (d) Describe the progression of four events occurring during the severe drought: _____

   _____

   (e) What step was missed out after the drought broke? Briefly explain why: _____

   _____

2. When conditions deteriorate, the joey at heel (still suckling) perishes first, then the pouch young, then the diapausing embryo. Why does the mother withdraw nourishment from the young in this order when times are hard?

   _____

   _____

PRACTICES

DATA

40 Maintaining Homeostasis

Key Idea: Maintaining homeostasis involves the interaction of the body's organ systems. However, the strategies an organism employs to maintain its steady state may vary according to environmental conditions if energy is limited.

All organisms must maintain steady internal conditions that enable their enzyme systems to operate efficiently. This steady state is called homeostasis and the mechanisms involved are shared across all phyla. Organisms must obtain nutrients and water, excrete wastes, defend themselves against disease and damage, and coordinate the responses required for survival. For unicellular eukaryotes, homeostasis is the job of the cell's organelles. For more complex multicellular organisms, the job of homeostatic regulation falls to the various organ systems that make up the organism. Maintaining homeostasis, particularly in endotherms, demands considerable energy. For organisms with very high specific metabolic rates, such as hummingbirds, this may mean having a strategy that varies depending on the environmental conditions at the time.

Regulating respiratory gases
In all aerobic organisms, oxygen must be delivered to the body's cells and carbon dioxide (CO_2), a waste product of cellular respiration, must be removed. In air breathers, breathing brings in oxygen and expels CO_2, and the circulatory system transports these respiratory gases to and from tissues. The rate of breathing is varied according to the oxygen demands.

Defense and repair
All organisms must resist invasion from disease-causing organisms, and limit the damage they do if they enter the body. In addition, tissue damage by any cause must be repaired. A wide range of organisms have similar types of physical and chemical defenses against microbial invasion and many have specific immune responses also.

Nutrients and water
Food and drink are taken in to maintain supply of energy and fluids. In most animals, a digestive system makes these nutrients available and the circulatory system distributes them throughout the body. In humans, food intake is regulated largely through nervous mechanisms, while hormones control the cellular uptake of glucose. The liver metabolizes proteins to form urea, which is excreted by the kidneys.

Osmoregulation and excretion
In animals, excretion of nitrogenous wastes and regulation of the body's fluid and ion content are major homeostatic challenges. Nitrogenous wastes from the metabolism of proteins are highly toxic and must be excreted. In mammals, the kidneys are the primary organs for excreting wastes and regulating fluid and ion balance. In fish however, osmoregulation and excretion are primarily the job of the gills, not the kidneys.

What about plants?
From the smallest moss to the tallest tree, plants too must maintain their bodies. The main homeostatic challenges for plants are in maintaining water balance and obtaining the nutrients required for photosynthesis.

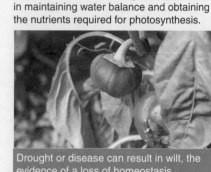
Drought or disease can result in wilt, the evidence of a loss of homeostasis

Nutrients and water are absorbed from the soil by roots or root-like structures and water losses are regulated through the activity of stomata in plants that have them.

Stoma on tomato leaf

Plants produce very few nitrogenous wastes and any waste products tend to be stored in vacuoles or wood (e.g. as tannins or resins) or utilized within the plant body for other purposes, including chemical defense against browsers and pathogens.

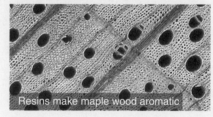
Resins make maple wood aromatic

1. Compare and contrast the approaches of plants and animals to the homeostatic challenge of eliminating wastes:

_____

_____

_____

_____

_____

WEB	CONNECT	CONNECT	CONNECT	PRACTICES
40	60 AP2	94 AP2	122 AP2	

KNOW

©2017 **BIOZONE** International
ISBN: 978-1-927309-65-0
Photocopying Prohibited

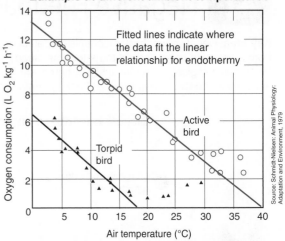

Oxygen consumption and activity of *Eulampis* at different ambient temperatures

Eulampis enters torpor each night when not feeding

cc 3.0 Charlesjsharp

Source: Schmidt-Nielsen: Animal Physiology: Adaptation and Environment, 1979

Fitted lines indicate where the data fit the linear relationship for endothermy

Active bird

Torpid bird

Homeostasis involves accurate physiological control

▶ Homeostasis in all organisms requires precise coordination of the body's regulatory systems. For many very small endotherms, including small bats, pocket mice, and hummingbirds, the cost of maintaining their high specific metabolic rate when environmental temperatures are low may be too high when food is scarce or unavailable. Such animals save energy (and are able to survive) by entering into a torpid state (during which the metabolic rate is lowered), eliminating the increased cost of keeping warm.

▶ The tropical purple throated carib (hummingbird), *Eulampis*, is unusual as it enters a torpid state each night even when the environment is not particularly cold. Torpor refers to decreased physiological activity in an animal, usually achieved by a reduced body temperature and metabolic rate. During both torpor and activity the metabolic rate is tightly regulated showing that torpor is not a failure to thermoregulate.

2. (a) From what you know of the energetic costs of metabolic activity, explain why maintaining homeostasis carries a considerable energetic cost:

_____

_____

(b) From what you have learned about thermoregulation so far, explain why these costs are higher in an endotherm than in an animal that allows its body temperature to fluctuate with the ambient temperature:

_____

_____

Study the plot of the oxygen consumption and activity of *Eulampis* above right and answer the following.

3. (a) Describe the relationship between specific metabolic rate (as measured by oxygen consumption) and air temperature in active birds:

_____

(b) Explain this relationship: _____

_____

4. Compare the relative energetic costs of activity vs torpor over a range of ambient air temperatures:

_____

_____

_____

5. (a) What is the approximate range in per mass rate of oxygen consumption in torpid birds at 18-30°C?

_____

(b) What do think would be happening to the body temperature of the birds in torpor over this temperature range?

_____

(c) What is different about the per mass rate of oxygen consumption in torpid birds below 18°C vs those over 18°C?

_____

(d) What do you think would be happening to the body temperature of the birds in torpor below 18°C? _____

_____

41 Negative Feedback

Key Idea: Negative feedback mechanisms detect departures from a set point norm and act to restore the steady state. Most physiological systems achieve homeostasis through negative feedback. In negative feedback systems, movement away from a steady state is detected and triggers a mechanism to counteract that change. **Negative feedback** has a stabilizing effect, dampening variations from a set point and returning internal conditions to a steady state.

Negative feedback and control systems

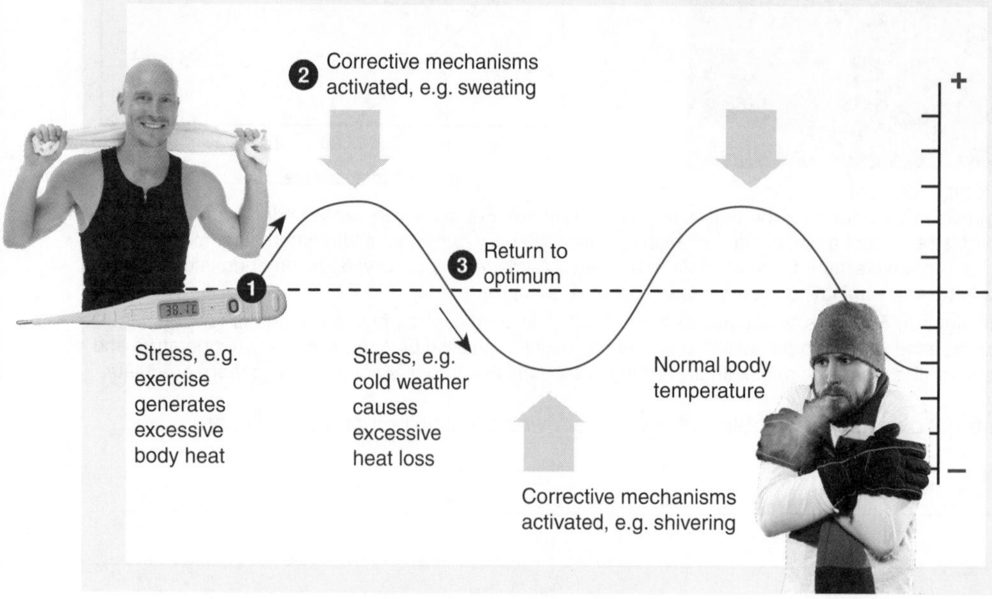

2 Corrective mechanisms activated, e.g. sweating

3 Return to optimum

Stress, e.g. exercise generates excessive body heat

Stress, e.g. cold weather causes excessive heat loss

Normal body temperature

Corrective mechanisms activated, e.g. shivering

1 A **stressor**, e.g. exercise, takes the internal environment away from optimum.

2 Stress is detected by receptors and corrective mechanisms (e.g. sweating or shivering) are activated.

3 Corrective mechanisms act to restore optimum conditions.

Negative feedback acts to counteract departures from steady state. The diagram shows how stress is counteracted in the case of body temperature.

Negative feedback in calcium homeostasis

Blood calcium is regulated by several hormones, including parathyroid hormone (PTH). Low blood Ca^{2+} stimulates release of PTH. When blood Ca^{2+} is restored, PTH secretion stops.

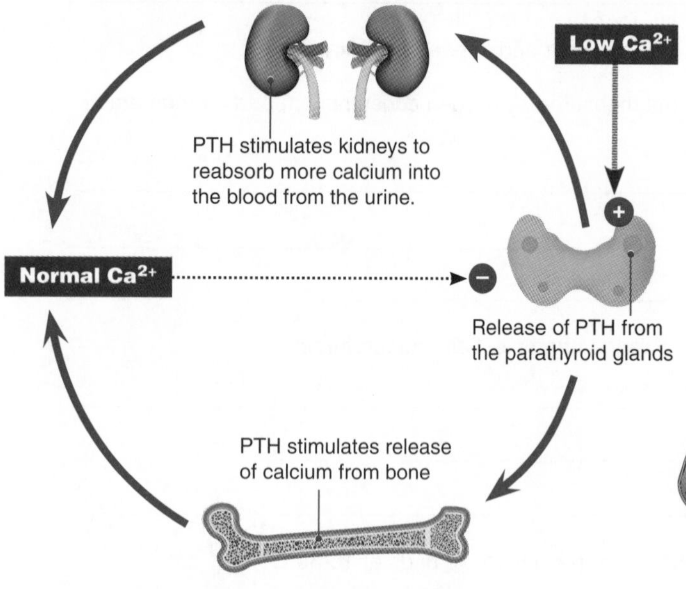

Low Ca^{2+}

PTH stimulates kidneys to reabsorb more calcium into the blood from the urine.

Normal Ca^{2+}

Release of PTH from the parathyroid glands

PTH stimulates release of calcium from bone

Negative feedback in stomach emptying

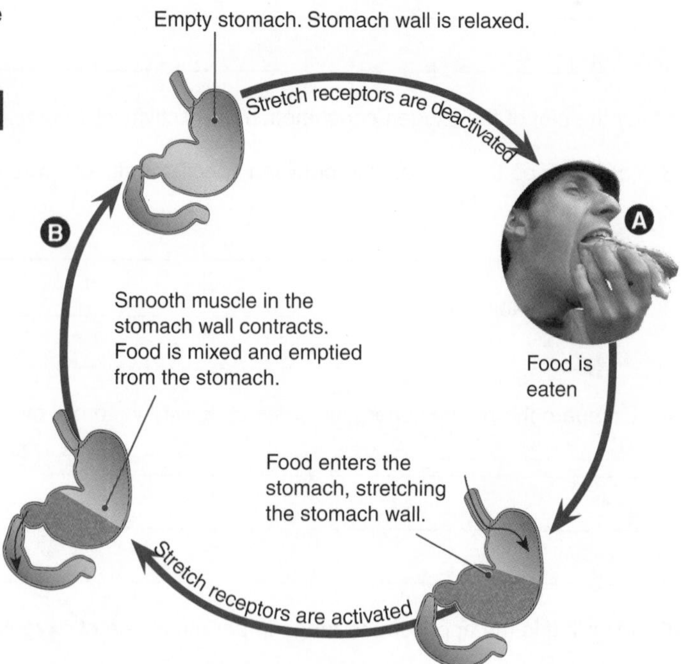

Empty stomach. Stomach wall is relaxed.

Stretch receptors are deactivated

B

A

Food is eaten

Smooth muscle in the stomach wall contracts. Food is mixed and emptied from the stomach.

Food enters the stomach, stretching the stomach wall.

Stretch receptors are activated

1. How do negative feedback mechanisms maintain homeostasis in a variable environment?_____

_____

_____

2. On the diagram of stomach emptying:

(a) State the stimulus at A: _____   State the response at B: _____

(b) Name the effector in this system: _____

(c) What is the steady state for this example?_____

©2017 **BIOZONE** International
ISBN: 978-1-927309-65-0
Photocopying Prohibited

42 Positive Feedback

Key Idea: Positive feedback results in the escalation of a response to a stimulus. It causes system instability and is used where a particular outcome or resolution is required.

Positive feedback mechanisms amplify a physiological response. Their usual purpose is to achieve a particular outcome. Fever, aspects of female reproduction, and blood clotting all involve positive feedback. Normally, a positive feedback loop is ended when a natural resolution is reached (e.g. baby is born, pathogen is destroyed, blood clot forms). Very few physiological processes involve positive feedback because such mechanisms are unstable. If left unchecked, they can be dangerous or even fatal.

Fever, positive feedback, and response escalation

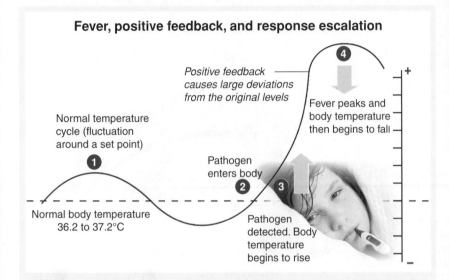

Positive feedback causes large deviations from the original levels

Normal temperature cycle (fluctuation around a set point)

1

Normal body temperature 36.2 to 37.2°C

Pathogen enters body

2

3

Pathogen detected. Body temperature begins to rise

Fever peaks and body temperature then begins to fall

4

+

-

Positive feedback also occurs in blood clotting. A wound releases signal chemicals that activate platelets in the blood. Activated platelets release chemicals that activate more platelets, so a blood clot is rapidly formed.

1 Body temperature fluctuates regularly around a narrow set point.

2 Pathogen enters the body.

3 The body detects the pathogen and macrophages (white blood cells) attack it. Macrophages release interleukins which stimulate the hypothalamus to increase prostaglandin production and reset the body's thermostat to a higher 'fever' level by shivering (the **chill phase**).

4 The fever breaks when the infection subsides. Levels of circulating interleukins (and other fever-associated chemicals) fall, and the body's thermostat is reset to normal. This ends the positive feedback escalation and normal controls resume. If the infection persists, the escalation may continue, and the fever may intensify. Body temperatures in excess of 43°C are often fatal or result in brain damage.

Ethylene is a gaseous plant hormone involved in fruit ripening. It accelerates the ripening of fruit in its vicinity so nearby fruit also ripens, releasing more ethylene. Over-exposure to ethylene causes fruit to over-ripen (rot).

1. (a) What is the biological role of positive feedback? Describe an example: _____

_____

_____

_____

(b) Why is positive feedback inherently unstable (compare with negative feedback)? _____

_____

_____

_____

(c) How is a positive feedback loop normally stopped? _____

_____

(d) Describe a situation in which this might not happen. What would be the result? _____

_____

_____

PRACTICES   PRACTICES   CONNECT   WEB

261   **42**

AP2

KNOW

43 Feedback Systems Can Interact

Key Idea: Positive and negative feedback loops can operate side by side to control a regular sequence of events.

We have seen how positive feedback acts to amplify the original response instead of to dampen it. The purpose of positive feedback is to achieve a specific physiological outcome and the positive feedback loop normally stops when the natural resolution is reached. Positive feedback is involved in certain aspects of reproduction in humans and other mammals, specifically ovulation, labor, and lactation. It is important in these instances because the usual physiological response must be amplified in order to bring about a certain event, e.g. release of an egg, birth of a baby, or let down of milk. In the female reproductive cycle, positive and negative feedback work together to closely regulate a regular cycle.

Feedback and ovulation

The human female reproductive or menstrual cycle is the regular cycle of changes in the reproductive system that results in release of oocytes (eggs) and preparation of the uterus for pregnancy. It involves an interplay of hormones and both positive and negative feedback regulation.

Before ovulation

▶ Before ovulation, the ovary releases the hormone **estrogen**, which stimulates the hypothalamus to release **gonadotropin releasing hormone** (GnRH) and the anterior pituitary to release **luteinizing hormone** (LH).

▶ GnRH also promotes release of LH from the anterior pituitary. The rapid rise in LH stimulates release of more estrogen from the ovary in a positive feedback loop. A peak in LH levels stimulates ovulation (release of an egg from an ovarian follicle) and the positive feedback loop ends.

After ovulation

▶ After ovulation, higher LH causes the ruptured follicle to form a corpus luteum. The corpus luteum secretes both estrogen and **progesterone**.

▶ When progesterone is present, the release of GnRH from the hypothalamus and anterior pituitary is inhibited (negative feedback) and the uterine lining is prepared to receive a fertilized egg.

Ovulation: positive feedback

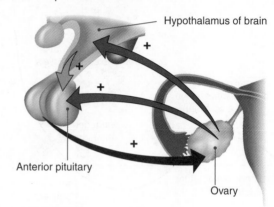

Hypothalamus of brain

Anterior pituitary

Ovary

Post ovulation: negative feedback

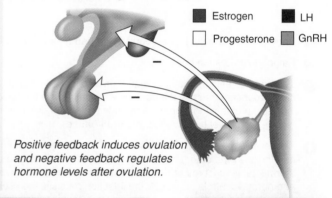

■ Estrogen ■ LH
□ Progesterone ■ GnRH

Positive feedback induces ovulation and negative feedback regulates hormone levels after ovulation.

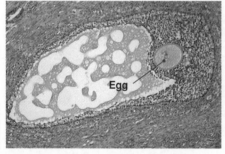

Egg

Ovulation: Positive feedback brings about ovulation, the release of an oocyte from a mature ovarian follicle. In the photograph above, the soon-to-be-released egg is visible within the mature (or Graafian) follicle. When the egg is released, the positive feedback loop ends and negative feedback regulation of hormone levels resumes.

Labor: During childbirth, receptors in the cervix detect stretching and signal the hypothalamus to release the hormone oxytocin. Oxytocin intensifies uterine contractions, moving the baby further into the birth canal and causing further stretching of the cervix, which in turn causes the release of more oxytocin. The birth itself (above) restores the system by removing the initiating stimulus.

Lactation: After birth, levels of the milk-production hormone prolactin increase. Suckling maintains prolactin secretion and causes the release of oxytocin, resulting in milk release (let down). The more an infant suckles, the more these hormones are produced.

1. Summarize how positive and negative feedback are involved in preparing for reproduction: _____

_____

_____

2. Describe how the positive feedback loop is ended in the case of milk production and let down: _____

_____

©2017 **BIOZONE** International
ISBN: 978-1-927309-65-0
Photocopying Prohibited

44 | Thermoregulation in Humans

Key Idea: The temperature regulation center in humans is in the hypothalamus. Thermoregulation relies on negative feedback mechanisms and involves several body systems. The temperature regulation center of the human body is in the hypothalamus of the brain which has a 'set-point' temperature of 36.7°C. The hypothalamus acts like a thermostat, regulating body temperature through negative feedback. Changes in core body temperature or in skin temperature are registered by the hypothalamus, which then coordinates the nervous and hormonal responses to counteract the changes and restore normal body temperature. When normal temperature is restored, the corrective mechanisms are switched off.

Counteracting heat loss

The hypothalamus detects a fall in skin or core temperature below 35.8°C and coordinates responses that generate and conserve heat.

Increased metabolic rate produces heat.

Body hairs become raised (goosebumps) and increase the insulating air layer around the body.

In extreme cold, two hormones (adrenaline and thyroxine) increase the energy-releasing activity of the liver.

The flow of blood to the skin decreases, keeping warm blood near the core (where the vital organs are).

Shivering (fast contraction and relaxation of muscles) produces internal heat.

Factors causing heat loss
► Cold external temperature
► Wind
► Insufficient insulation
► Being wet or in cold water
► Dehydration or circulatory shock

Counteracting heat gain

The hypothalamus monitors any rise in skin or core temperature above 37.5°C and coordinates responses that increase heat loss.

Sweating occurs. This cools the body by evaporation.

Decreased metabolic rate. This reduces the amount of heat generated by the body.

Body hairs become flattened against the skin. This reduces the insulating air layer around the body and helps heat loss.

The flow of blood to the skin increases. Warm blood from the body core is transported to the skin and the heat is lost from the skin surface.

Factors causing heat gain
► Warm external temperature
► High humidity
► Excessive fat deposits
► Wearing too much clothing
► Intense physical activity

1. (a) Where is the temperature regulation center in humans located? _____

   (b) What is its role in thermoregulation: _____

2. State two mechanisms by which body temperature could be reduced after intense activity: _____

CONNECT **100** AP2   WEB **44**   KNOW

Skin section

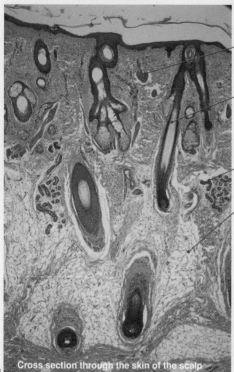

Cross section through the skin of the scalp

Blood vessels in the dermis dilate or constrict to promote or restrict heat loss.

Hairs raised or lowered to increase or decrease the thickness of the insulating air layer between the skin and the environment.

Sweat glands produce sweat, which cools through evaporation.

Fat in the sub-dermal layers insulates the organs against heat loss.

Thermoreceptors in the dermis are free nerve endings, which respond to changes in skin temperature and send that information to the hypothalamus. Hot thermoreceptors detect a rise in skin temperature above 37.5°C and cold thermoreceptors detect a fall below 35.8°C.

Regulating blood flow to the skin

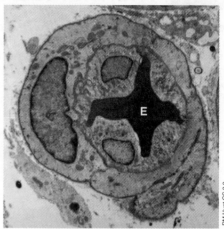

Constriction of a small blood vessel. An erythrocyte (E) (red blood cell) is in the center of the vessel.

To regulate heat loss or gain from the skin, the blood vessels beneath the skin's surface constrict (**vasoconstriction**) to reduce blood flow or dilate (**vasodilation**) to increase blood flow. When blood vessels are fully constricted there may be as much as a 10°C temperature gradient from the outer to inner layers of the skin. Extremities such the hands and feet have additional vascular controls which can reduce blood flow to them in times of severe cooling.

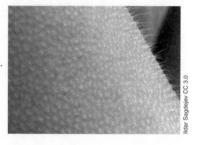

The hair erector muscles, sweat glands, and blood vessels are the effectors for mediating a response to information from thermoreceptors. Temperature regulation by the skin involves **negative feedback** because the output is fed back to the skin receptors and becomes part of a new stimulus-response cycle.

Left photograph shows vasodilation and sweating in response high temperature or exertion.
Right photograph shows vasoconstriction and goosebumps in response low temperature or inactivity.

3. (a) What is the purpose of sweating and how does it achieve its effect?_____

_____

_____

(b) Why does a dab of methanol or ethanol on the skin feels cold, even if the liquid is at room temperature? _____

_____

4. Describe the feedback system that regulates body temperature: _____

_____

_____

_____

_____

5. How do the blood vessels help to regulate the amount of heat lost from the skin and body? _____

_____

_____

6. (a) What is the role of subcutaneous fat in temperature regulation in humans: _____

_____

(b) Why do excessive deposits of fat tend to lead to overheating during exercise?_____

_____

©2017 **BIOZONE** International
ISBN: 978-1-927309-65-0
Photocopying Prohibited

45 | Control of Blood Glucose

Key Idea: The endocrine portion of the pancreas produces two hormones, insulin and glucagon, which maintain blood glucose at a steady state through negative feedback.

Blood glucose levels are controlled by negative feedback involving two hormones, insulin and glucagon. These hormones are produced by the islet cells of the pancreas, and act in opposition to control blood glucose levels. **Insulin** lowers blood glucose by promoting the uptake of glucose by the body's cells and the conversion of glucose into the storage molecule glycogen in the liver. **Glucagon** increases blood glucose by stimulating the breakdown of stored glycogen and the synthesis of glucose from amino acids. Negative feedback stops hormone secretion when normal blood glucose levels are restored. Blood glucose homeostasis allows energy to be available to cells as required. The liver has a central role in these carbohydrate conversions.

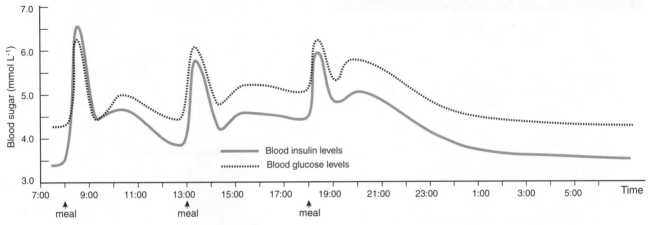

Negative feedback in blood glucose regulation

Blood glucose can be tested using a finger prick test. The glucose in the blood reacts with an enzyme electrode, generating an electric charge proportional to the glucose concentration. This is displayed as a digital readout.

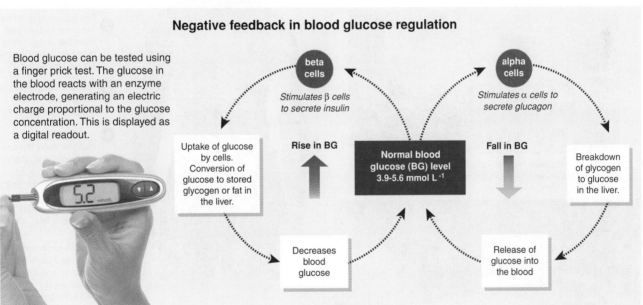

1. (a) Identify the stimulus for the release of insulin: _____

   (b) Identify the stimulus for the release of glucagon: _____

   (c) How does glucagon increase blood glucose level? _____

   _____

   _____

   (d) How does insulin decrease blood glucose level? _____

   _____

2. Explain the pattern of fluctuations in blood glucose and blood insulin levels in the graph above: _____

   _____

   _____

   _____

©2017 **BIOZONE** International
ISBN: 978-1-927309-65-0
Photocopying Prohibited

PRACTICES PRACTICES CONNECT **84** AP1 CONNECT **83** AP1 WEB **45**

KNOW

46 Type 1 Diabetes Mellitus

Key Idea: Diabetes mellitus is a condition in which blood glucose levels are too high and glucose appears in the urine. In type 1 diabetes, the insulin-producing cells of the pancreas are destroyed and insulin is not produced.

Diabetes mellitus (often called diabetes) is a condition in which blood glucose is too high because the body's cells cannot take up glucose in the normal way. Diabetes mellitus is characterized by large volumes (diabetes) of sweet (mellitus) urine and extreme thirst. In **type 1 diabetes**, the insulin producing cells of the pancreas are destroyed and no insulin is produced. Patients must have regular insulin injections to stabilize blood glucose levels.

Type 1 diabetes mellitus

The body does not produce insulin. Type 1 diabetes is also called insulin dependent diabetes.

Age at onset: Early in life (often in childhood).

Symptoms: The cells cannot take up glucose so blood glucose is high. The glucose in the blood exceeds the reabsorption capacity of the kidney and glucose spills over in the urine resulting in large volumes of sweet (high glucose) urine, thirst, hunger, weight loss, fatigue, and infections.

Cause: The beta cells are destroyed by the body's own immune system. A genetic predisposition and environmental factors (e.g. a viral infection) may be the trigger.

Treatment: There is no cure. Blood glucose is monitored regularly. Insulin injections combined with dietary management keep blood sugar levels stable.

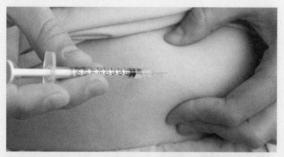

Type 1 diabetics must regularly check their blood glucose levels and administer insulin injections if it is too high.

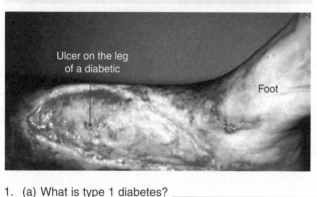

Ulcer on the leg of a diabetic

Foot

The effects of diabetes mellitus

Insulin promotes the uptake of glucose from the blood by the cells of the body, where it is used as an energy source. However, in type 1 diabetes no insulin is produced, so blood glucose levels remain high. High blood glucose irritates and damages blood vessels and nerves. This can cause a number of problems including numbness, loss of vision as cells in the eye are damaged, gangrene, and failure of wounds to heal.

When glucose is not available, the body's cells metabolize fat as an alternative energy source. This can also cause problems such as a fall in blood pH (which can be fatal), and damage to blood vessels and the development of cardiovascular disease.

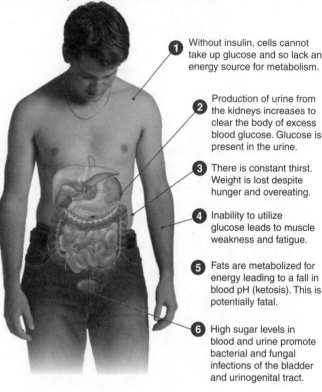

1. Without insulin, cells cannot take up glucose and so lack an energy source for metabolism.

2. Production of urine from the kidneys increases to clear the body of excess blood glucose. Glucose is present in the urine.

3. There is constant thirst. Weight is lost despite hunger and overeating.

4. Inability to utilize glucose leads to muscle weakness and fatigue.

5. Fats are metabolized for energy leading to a fall in blood pH (ketosis). This is potentially fatal.

6. High sugar levels in blood and urine promote bacterial and fungal infections of the bladder and urinogenital tract.

Poor blood flow to the limbs, changes in metabolic function, and damage to nerves can result in ulcers. These can be very difficult to treat in diabetics.

1. (a) What is type 1 diabetes? _____

_____

(b) Explain how the usual negative feedback mechanisms for blood glucose homeostasis are disrupted in a person with type 1 diabetes:

_____

_____

_____

2. How do regular insulin injections help a person with type 1 diabetes to maintain their blood glucose homeostasis?

_____

_____

©2017 **BIOZONE** International
ISBN: 978-1-927309-65-0
Photocopying Prohibited

47 Alcohol and Blood Glucose

Key Idea: The ability of the body to regulate blood glucose is impaired by the consumption of alcohol.

Alcohol affects blood glucose levels and its regulation. The alcohol in alcoholic beverages acts as a preservative and adds to its flavor. Alcohol is toxic and its metabolism reduces the body's capacity to regulate blood glucose levels. The liver metabolizes alcohol, so while there is alcohol in the blood, it has less capacity to regulate blood glucose (by converting glycogen into glucose). Alcohol lowers blood glucose levels by stimulating insulin production. Low to moderate alcohol intakes don't affect the body's response to insulin. However, long-term alcohol consumption damages the liver and interferes with many aspects of metabolism, including metabolism of glucose.

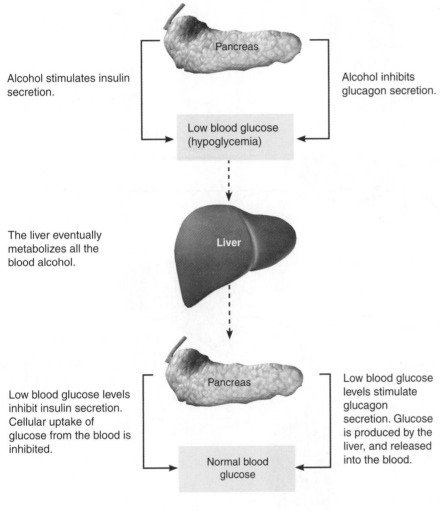

Alcohol stimulates insulin secretion.

Alcohol inhibits glucagon secretion.

Low blood glucose (hypoglycemia)

The liver eventually metabolizes all the blood alcohol.

Low blood glucose levels inhibit insulin secretion. Cellular uptake of glucose from the blood is inhibited.

Low blood glucose levels stimulate glucagon secretion. Glucose is produced by the liver, and released into the blood.

Normal blood glucose

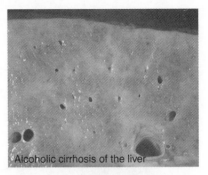

Alcoholic cirrhosis of the liver

Long term consumption of large amounts of alcohol can lead to liver disease, due to the release of inflammatory proteins by liver cells.

Alcohol not only reduces blood glucose but also acts as a diuretic. Drinking alcohol can reduce performance in sports or physical activity due to dehydration and reduced ability to access glucose for energy (not to mention impairing decision making).

1. (a) How does drinking alcohol result in low blood glucose? _____

(b) The liver prioritizes the metabolism of alcohol before restoring blood glucose. Suggest why this is the case:

2. Why does drinking alcohol reduce sports performance? _____

48 Nicotine and Blood Glucose

Key Idea: Regulation of blood glucose levels is affected by nicotine, which acts indirectly to increase blood glucose.
Nicotine also affects blood glucose levels and its regulation. Nicotine is the highly addictive component of tobacco, and a potent carcinogen (cancer causing agent). It is also responsible for depressing appetite in smokers, partly through its indirect effect on the liver and the regulation of blood glucose levels.

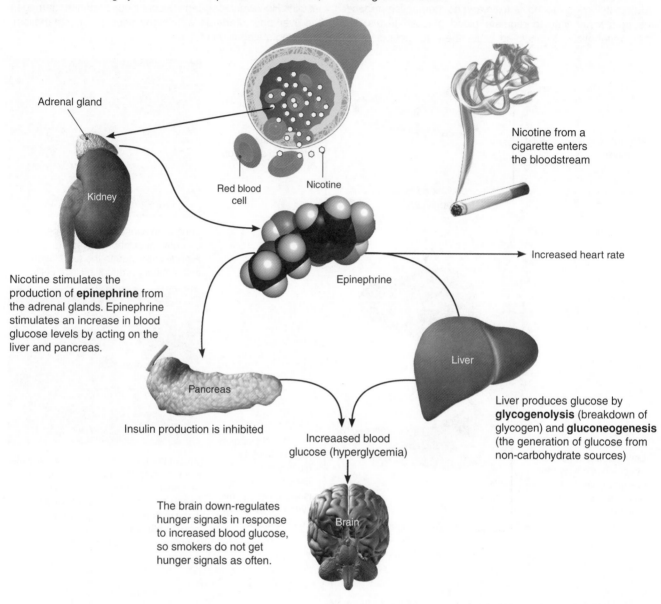

Adrenal gland

Kidney

Red blood cell

Nicotine

Nicotine from a cigarette enters the bloodstream

Increased heart rate

Epinephrine

Nicotine stimulates the production of **epinephrine** from the adrenal glands. Epinephrine stimulates an increase in blood glucose levels by acting on the liver and pancreas.

Pancreas

Liver

Insulin production is inhibited

Increaased blood glucose (hyperglycemia)

Liver produces glucose by **glycogenolysis** (breakdown of glycogen) and **gluconeogenesis** (the generation of glucose from non-carbohydrate sources)

The brain down-regulates hunger signals in response to increased blood glucose, so smokers do not get hunger signals as often.

Brain

1. (a) How does nicotine reduce appetite? _____

_____

_____

_____

(b) Why do people often put on weight when they stop smoking? _____

_____

_____

_____

2. Explain why nicotine acts as a stimulant: _____

_____

_____

©2017 **BIOZONE** International
ISBN: 978-1-927309-65-0
Photocopying Prohibited

49 Hyperthyroidism and Thermoregulation

Key Idea: Hyperthyroidism refers to over-production of the hormone thyroxine by the thyroid gland. This in turn can affect the body's ability to thermoregulate properly.

The **thyroid gland** is a large endocrine gland found in the neck. The thyroid gland produces a number of hormones involved in regulating many aspects of metabolism, including temperature regulation. **Hyperthyroidism** (overactive thyroid) is a medical condition in which the thyroid gland produces too much of the hormone thyroxine. One of the effects of too much thyroxine is that it increases body temperature.

The thyroid gland

The thyroid gland is a butterfly shaped endocrine gland located just below the Adam's apple at the front of the wind pipe (trachea). The thyroid secretes several hormones, collectively called thyroid hormones, but the main hormone produced is thyroxine. Thyroxine is also called T4. Thyroid hormones have many functions including regulating metabolism, growth and development, and body temperature.

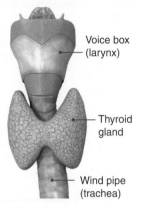

Voice box
(larynx)

Thyroid
gland

Wind pipe
(trachea)

Hyperthyroidism and temperature regulation

One of the effects of thyroxine is that it speeds up metabolic activity in cells. The increase in metabolic activity also results in the production of heat and, under normal conditions, this is one of the mechanisms by which the body raises body temperature.

The negative feedback mechanism for thyroxine production can be disrupted by hyperthyroidism, a condition where too much thyroxine is produced by the thyroid gland. This can disrupt temperature regulation.

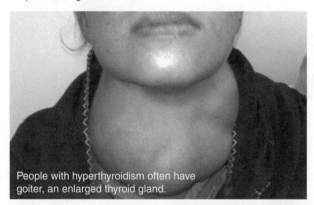

People with hyperthyroidism often have goiter, an enlarged thyroid gland.

The most common cause of hyperthyroidism is Graves' disease. In Graves' disease, the negative feedback loop is bypassed because a protein called thyroid stimulating immunoglobulin (TSI) binds directly to the thyroid and stimulates thyroxine production. In this instance, thyroxine production is independent of TSH production, so the usual regulatory mechanisms are ineffective.

Negative feedback regulates thyroxine production

▶ Thyroxine (T4) production is controlled by negative feedback. This mechanism involves two parts of the brain, the hypothalamus and the pituitary gland.

▶ Low body temperature stimulates the hypothalamus to secrete thyroid releasing hormone (TRH), which in turn stimulates cells in the anterior pituitary to secrete thyroid stimulating hormone (TSH).

▶ TSH acts on the thyroid gland, causing it to produce thyroid hormones, including thyroxine. Thyroxine binds to target cells, increasing their metabolic activity, resulting in the production of heat.

▶ High levels of circulating thyroid hormones inhibit production of TRH and TSH. As a result, thyroid secretion is reduced. When the level of thyroid hormones drops below a certain threshold, TRH and TSH production begins again.

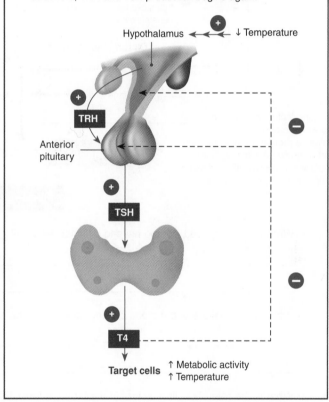

Hypothalamus ← + ← ↓ Temperature

TRH

Anterior pituitary

+

TSH

+

T4

Target cells ↑ Metabolic activity ↑ Temperature

1. Briefly outline how negative feedback of thyroxine production is involved in temperature regulation: _____

_____

_____

_____

_____

2. Why do high levels of thyroxine not inhibit its production from the thyroid gland in a person with Graves' disease?

_____

_____

_____

©2017 **BIOZONE** International
ISBN: 978-1-927309-65-0
Photocopying Prohibited

CONNECT
84
AP1

WEB
49

KNOW

50 Blood Clotting

Key Idea: Blood clotting is a homeostatic process involving positive feedback. It restricts blood loss limits pathogen entry. Blood has a role in the body's defense against infection. Tearing or puncturing of a blood vessel initiates **blood clotting** through a positive feedback loop involving platelets, clotting factors, and plasma proteins (below). Clotting quickly seals off the tear, preventing blood loss and infection. Clot formation is triggered by the release of clotting factors from the damaged cells. A hardened clot forms a scab, which acts to prevent further blood loss and acts as a mechanical barrier to the entry of pathogens. A failure of the blood clotting mechanism represents a severe threat to homeostasis.

Blood clotting pathway

1 Injury to the lining of a blood vessels exposes collagen fibers to the blood. Platelets stick to the collagen fibers.

3 As the platelets clump together, more chemicals are released, accelerating the clot formation (positive feedback). The platelet plug forms immediate protection against blood loss.

When blood clotting fails
Failure of the blood to clot leads to bleeding disorders, in which bleeding (including into joints) is heavy or prolonged. Bleeding disorders include von Willebrand's disease, which is the most common, and hemophilia, which is rarer but more severe. Both are inherited diseases resulting from a lack of clotting factors. Bleeding disorders related to platelets (e.g. abnormally low platelet counts) are usually acquired later in life.

4 A fibrin clot reinforces the seal. The clot traps blood cells and the positive feedback loop ends. The clot eventually dries to form a **scab**.

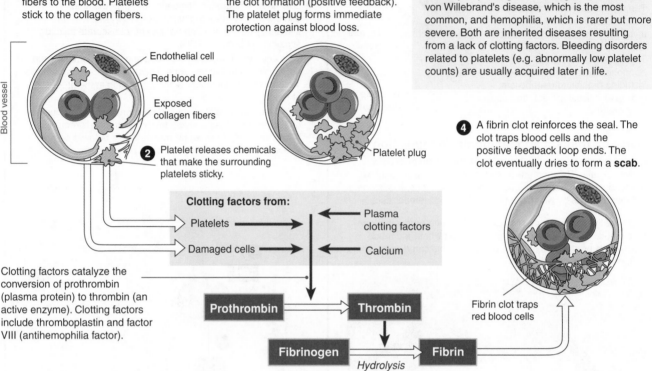

Blood vessel

Endothelial cell
Red blood cell
Exposed collagen fibers

2 Platelet releases chemicals that make the surrounding platelets sticky.

Platelet plug

Clotting factors catalyze the conversion of prothrombin (plasma protein) to thrombin (an active enzyme). Clotting factors include thromboplastin and factor VIII (antihemophilia factor).

Clotting factors from:
Platelets
Damaged cells
Plasma clotting factors
Calcium

Prothrombin → Thrombin

Fibrinogen → Fibrin
Hydrolysis

Fibrin clot traps red blood cells

1. (a) What role does blood clotting have in internal defense? _____

_____

(b) Why would a failure of the blood clotting mechanism be so damaging to homeostasis? _____

_____

2. (a) What is the role of clotting factors in the blood in formation of the clot? _____

_____

(b) Why are these clotting factors not normally present in the plasma?_____

_____

_____

3. (a) Explain why a deficiency in one or more clotting factors leads to a failure of the blood to clot: _____

_____

(b) Name an inherited disorder caused by this: _____

©2017 **BIOZONE** International
ISBN: 978-1-927309-65-0
Photocopying Prohibited

51 KEY TERMS AND IDEAS: Did You Get It?

1. Test your vocabulary by matching each term to its correct definition, as identified by its preceding letter code.

diapause ....

ectotherm ....

endotherm ....

hibernation ....

homeostasis ....

mass-specific metabolic rate ....

negative feedback ....

positive feedback ....

reproductive strategy ....

thermoregulation ....

torpor ....

A A short period of reduced activity and metabolic rate as a strategy to save energy.

B A stabilizing mechanism in biological systems in which the output of a system counteracts further change in the system output.

C The suite of adaptations that describe how an organism apportions the energy it expends in reproduction.

D A period of suspended development during unfavorable environmental conditions.

E An animal that generates its body heat through metabolic activity.

F A destabilizing mechanism in biological systems in which the output of a system amplifies that same output.

G A prolonged (seasonal) period of reduced activity and metabolic rate.

H The regulation of body temperature.

I The resting energy expenditure of an animal per unit body mass per day. It is greater in smaller animals and is often measured by rate of oxygen consumption per kg per day.

J The relatively stable, constant condition of the body's internal state.

K An animal that is dependent on external (environmental) sources of body heat.

2. (a) Study the graph right and identify the type of feedback mechanism:

_____

(b) Describe the outcome: _____

_____

_____

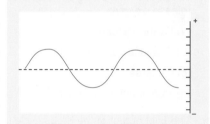

3. The graph below right shows the oxygen consumption of a food-restricted pocket mouse over 24 hours at 15°C.

(a) What is happening to the mass specific metabolic rate of the pocket mouse over the 24 hour period?

_____

_____

_____

_____

(b) Explain the changes in metabolic rate over the 24 hour period:

_____

_____

_____

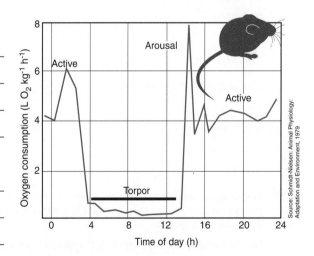

Source: Schmidt-Nielsen: Animal Physiology: Adaptation and Environment, 1979

(c) Why do you think there is a peak in oxygen consumption during arousal from torpor? _____

_____

(d) How does this physiological response contribute to the pocket mouse's survival?_____

_____

(e) Do you think the pattern of activity would have been different if the animal had not been food restricted? Explain:

_____

_____

TEST

52 Synoptic Questions

For the following multiple choice questions circle the correct answer.

1. As a process, photosynthesis is:

   (a) Reductive, exergonic, and catabolic

   (b) Reductive, endergonic, and catabolic

   (c) Reductive, endergonic, and anabolic

   (d) Reductive, exergonic, and anabolic

2. The site of the Calvin cycle in chloroplasts is the:

   (a) Stroma

   (b) Stroma lamellae

   (c) Thylakoids

   (d) Both (a) and (c)

3. Which of the following is least effective in photosynthesis:

   (a) Blue light

   (b) Red light

   (c) Green light

   (d) Sunlight

4. Which of the following photosynthetic organisms do not release oxygen:

   (a) Cyanobacteria

   (b) Green algae

   (c) Oak tree

   (d) Purple sulfur bacterium

5. The final output of the Krebs cycle includes all of the following except:

   (a) NADP

   (b) $FADH_2$

   (c) ATP

   (d) CO_2

6. The end product of glycolysis is:

   (a) CO_2

   (b) Pyruvate

   (c) Lactate

   (d) Acetyl-CoA

7. Under what cellular conditions would you expect the proton gradient in the mitochondria to be highest and ATP synthesis to proceed?

   (a) Pyruvate (present)-oxygen (present)-ATP levels (low)

   (b) Pyruvate (present)-oxygen (absent)-ATP levels (high)

   (c) Pyruvate (absent)-oxygen (present)-ATP levels (low)

   (d) Pyruvate (present)-oxygen (present)-ATP levels (high)

8. Some enzymes require a non-protein substance if they are to catalyze a reaction. Such a substance is called a:

   (a) Cofactor

   (b) Coenzyme

   (c) Prosthetic group

   (d) Activator

9. The graph below shows the rate of an enzyme catalyzed reaction with and without an inhibitor. The inhibition is:

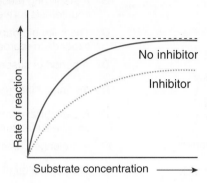

   (a) Competitive

   (b) Non-competitive

   (c) In concentrations too low to work properly

   (d) Overcome by increase in substrate concentration

Questions 10 - 12 refer to the following:
The rate of respiration was measured in mice and crickets at two different temperatures using a respirometer. The results are summarized below:

Organism	Temperature (°C)	Mean respiration rate (mL O_2 g^{-1} min^{-1})
Mouse	10	0.052
Mouse	25	0.032
Cricket	10	0.0013
Cricket	25	0.0038

10. According to the data, the mice at 10°C had a greater rate of oxygen consumption than mice at 25°C. The most likely explanation for this:

    (a) The mice at 10°C had a higher rate of ATP production than the mice at 25°C

    (b) The mice at 25°C weighed less than the mice at 10°C.

    (c) The mice at 25°C were more active than the mice at 10°C

    (d) The mice at 10°C had a lower metabolic rate than the mice at 25°C

11. According to the data, the crickets at 10°C had a lower rate of oxygen consumption than crickets at 25°C. This was the opposite to what was found for the mice. The most likely explanation for this is:

    (a) Relative size of crickets compared to mice

    (b) Crickets and mice have different nutritional modes

    (c) Crickets and mice have different modes of temperature regulation

    (d) Crickets and mice generate ATP in different ways

12. (a) This data set indicates that mice are endothermic / ectothermic (circle correct answer)

    (b) This data set indicates that crickets are endothermic / ectothermic (circle correct answer)

©2017 **BIOZONE** International
ISBN: 978-1-927309-65-0
Photocopying Prohibited

13. Students investigated the effect of light intensity on the rate of photosynthesis in an aquatic weed, *Cabomba*. They created a gradient in light intensity by moving the lamp set distances from the light source and recorded the value in lux at each distance. They measured photosynthetic rate indirectly by counting the bubbles of oxygen gas produced by cut stems over a set time period. Their results are tabulated below.

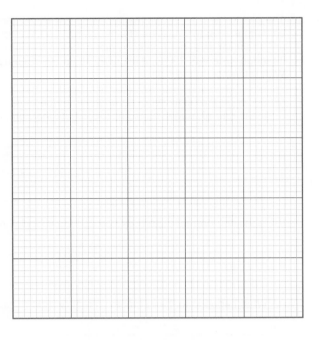

Light intensity (lx) (distance)	Bubbles counted in 3 minutes	Rate (bubbles per minute)
5 (50 cm)	0	
13 (45 cm)	6	
30 (40 cm)	9	
60 (35 cm)	12	
95 (30 cm)	18	
150 (25 cm)	33	
190 (20 cm)	35	

(a) Complete the table by calculating the rate of oxygen production (bubbles of oxygen gas per minute):

(b) Use the data to draw a graph of the bubble produced per minute vs light intensity:

(c) What was the effect of light intensity on photosynthetic rate? _____

_____

(d) Although the light source was placed set distances from the *Cabomba* stem, light intensity in lux was recorded at each distance rather than distance *per se*. Explain why this would be more accurate:

_____

_____

_____

14. This flow chart depicts various nutritional modes in organisms, identifying both their source of energy (hydrogens or electrons) and the source of carbon. Compete the diagram by filling in the white boxes (a)-(c).

Atmospheric CO₂

Obtain carbon elsewhere?
NO — YES

(a)

YES — Energy from light?

Energy from light? — YES

Photoheterotroph
Obtain energy from light and carbon from organic compounds

NO

NO

(b)

YES — Energy from inorganic oxidation?

Energy from inorganic oxidation? — YES (c)

NO

Chemoorganotroph
Obtain energy and carbon from organic substrates

15. The plot to the right shows the per mass rate of oxygen consumption in two species of small North American bats. The bats are of similar size, but *Eptesicus* hibernates in the northern part of its range, whereas *Tadarida* migrates south but does not hibernate. With reference to the plot:

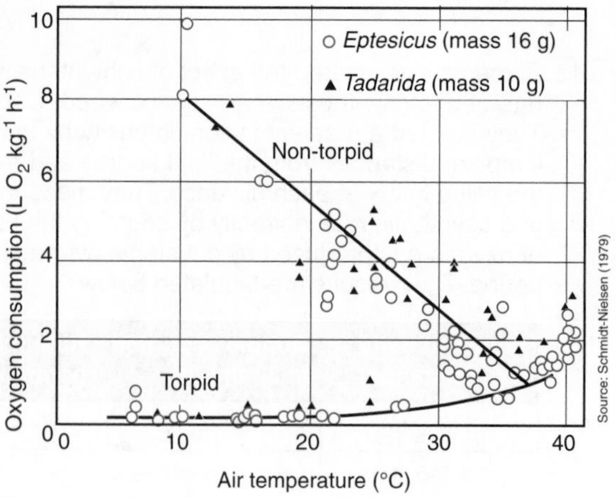

(a) Describe the relationship between air temperature and per mass rate of oxygen consumption in bats that remained active:

_____

_____

_____

(b) Is this the normal pattern for homeothermic endotherms? Explain: _____

_____

_____

(c) What happens to the per mass rate of oxygen consumption in bats that enter torpor (relative to that when active)?

_____

_____

(d) Describe the survival advantage of this behavior to the bats: _____

_____

_____

(e) These bats have different strategies for surviving the winter period (hibernation vs migration). Describe the energetic costs associated with each of these survival strategies:

_____

_____

_____

_____

16. Study the graph of oxygen consumption of lobsters, right.

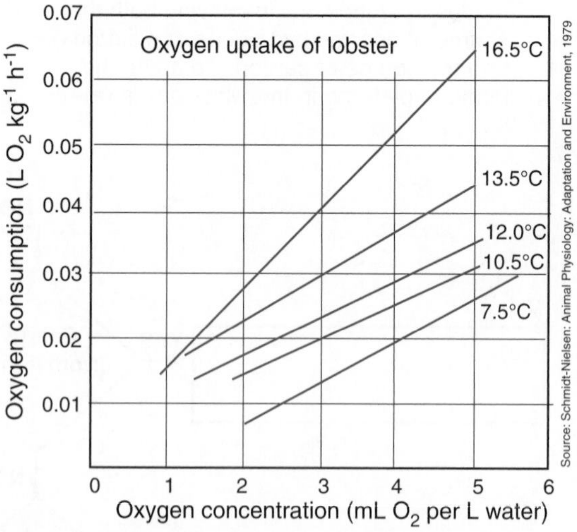

(a) What happens to the rate of oxygen consumption as oxygen concentration increases?

_____

(b) What happens to the rate of oxygen consumption as temperature increases?

_____

(c) Give a reason for the relationship between rate of oxygen consumption and water temperature in lobsters:

_____

_____

(d) The relationship between rate of oxygen consumption and oxygen concentration is less easy to explain. Suggest why an ectothermic animal, such as a lobster, which does not thermoregulate, might show dependency between rate of oxygen consumption and oxygen concentration in its environment:

_____

_____

_____

_____

Plant Structure and Adaptation

2.D.2 Homeostatic mechanisms reflect common ancestry and divergence due to adaptation in different environments

Activity number

Essential knowledge

(a) Continuity of homeostatic mechanisms reflects common ancestry, while changes may occur in response to different environments

☐ 1 Use examples to show that homeostatic mechanisms in organisms reflect common ancestry but there is adaptation to different environments.   60

(b) Organisms have various mechanisms for obtaining nutrients and eliminating wastes

☐ 1 Describe the relationship between gas exchange, nutrition, and elimination of wastes in green plants. Use examples to show that plants have various mechanisms for gas exchange in different environments.   59 60 62 63

(c) Homeostatic control systems across phyla support common ancestry

☐ 1 Describe examples of osmoregulation in aquatic and terrestrial plants that illustrate the common ancestry of homeostatic control systems.   61

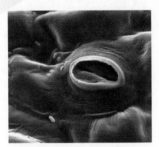

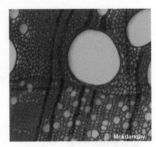

4.A.4 Organisms exhibit complex properties due to interactions between their component parts

Activity number

Essential knowledge

(a) Interactions and coordination between organs provide essential biological activities

☐ 1 Describe how interactions between the roots, stems, and leaves in plants provide essential activities such as nutrient and water uptake, gas exchange, photosynthesis, and transport of sugars.   53-63 65

(b) Interactions and coordination between systems provide essential biological activities

☐ 1 Describe how interactions between the plant vascular system and the leaves in plants provide essential activities such as nutrient and water uptake, gas exchange, photosynthesis, and transport of sugars. *[also 2.A.1.d.1]*   53-63 65

☐ **PR-11** Calculate leaf surface area and use a potometer to investigate factors affecting transpiration in plant shoots.   64

4.B.2 Cooperative interactions within organisms promote efficiency in the use of energy and matter

Activity number

Essential knowledge

(a) Organisms have areas or compartments that perform a subset of functions related to energy and matter and these parts contribute to the whole

☐ 1 Explain how compartmentalization within the cell, e.g. location of the reactions of photosynthesis within chloroplasts, contributes to overall specialization and functioning of the cell.   58 61

☐ 2 Using the example of gas exchange in plants, explain how specialization of the leaf contributes to overall functioning of the plant.   58-60

53 The Plant Body

Key Idea: The plant body comprises connected shoot and root systems. The shoot system collects carbon dioxide, oxygen, and light and produces sugars. The root system collects water and nutrients from the soil.

As terrestrial organisms, plants have two interdependent systems to take advantage of and to solve the problems of living on land. The **shoot system**, consisting of **stems**, **leaves** and **reproductive structures**, has evolved to collect carbon dioxide, oxygen and light, and to disperse pollen and seeds. The **root system** has evolved to collect water

and nutrients from the soil and to provide anchorage to the ground or substrate. These systems are integrated to form the closely linked support and transport systems. If a plant is to grow to any size, it must have ways to hold itself up against gravity and to move materials around its body. Vascular tissues (xylem and phloem) link all plant parts. Water and minerals are transported in the xylem, while manufactured food is transported in the phloem. All plants rely on fluid pressure within their cells (turgor) to give some support to their less rigid structures e.g. leaves and flowers.

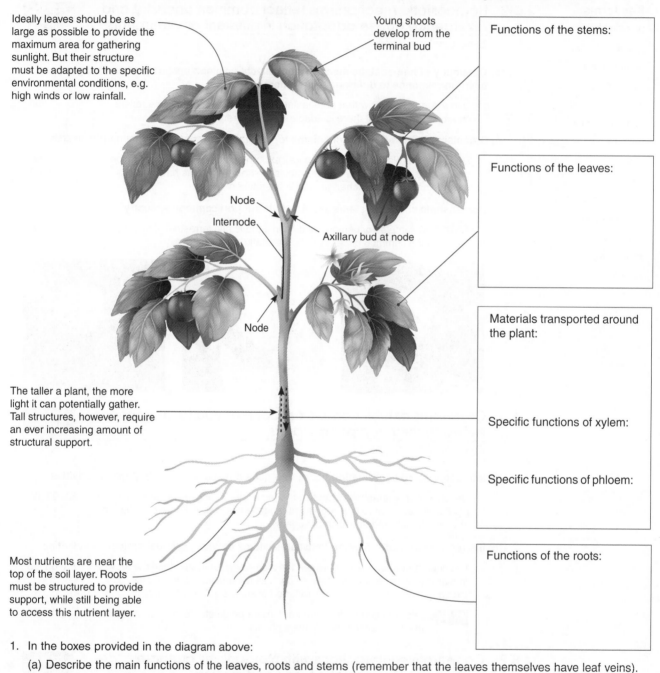

Ideally leaves should be as large as possible to provide the maximum area for gathering sunlight. But their structure must be adapted to the specific environmental conditions, e.g. high winds or low rainfall.

Young shoots develop from the terminal bud

Node

Internode

Axillary bud at node

Node

The taller a plant, the more light it can potentially gather. Tall structures, however, require an ever increasing amount of structural support.

Most nutrients are near the top of the soil layer. Roots must be structured to provide support, while still being able to access this nutrient layer.

Functions of the stems:

Functions of the leaves:

Materials transported around the plant:

Specific functions of xylem:

Specific functions of phloem:

Functions of the roots:

1. In the boxes provided in the diagram above:

   (a) Describe the main functions of the leaves, roots and stems (remember that the leaves themselves have leaf veins).

   (b) List the materials that are transported around the plant body.

   (c) Describe the functions of the transport tissues: xylem and phloem.

2. Name the solvent for all the materials that are transported around the plant: _____

3. What factors are involved in determining how tall a plant could potentially grow? _____

   _____

   _____

©2017 **BIOZONE** International
ISBN: 978-1-927309-65-0
Photocopying Prohibited

KNOW

54 Xylem

Key Idea: The xylem is involved in water and mineral transport in vascular plants.

Xylem is the principal water conducting tissue in vascular plants. It is also involved in conducting dissolved minerals, in food storage, and in supporting the plant body. As in animals, tissues in plants are groupings of different cell types that work together for a common function. In angiosperms, it is composed of five cell types: tracheids, vessels, xylem parenchyma, sclereids (short sclerenchyma cells), and fibers. The tracheids and vessel elements form the bulk of the tissue. They are heavily strengthened and are the conducting cells of the xylem. Parenchyma cells are involved in storage, while fibers and sclereids provide support. When mature, xylem is dead.

1. (a) What cells conduct the water in xylem?

   _____

   _____

   (b) What other cells are present in xylem tissue and what are their roles?

   _____

   _____

   _____

   _____

   _____

2. (a) How does water pass between vessels?

   _____

   _____

   _____

   (b) How does water pass between tracheids:

   _____

   _____

   (c) Which cell type do you think provides the most rapid transport of water and why?

   _____

   _____

   _____

   (d) Why do you think the tracheids and vessel elements have/need secondary thickening?

   _____

   _____

   _____

   _____

3. How can xylem vessels and tracheids be dead when mature and functional?

   _____

   _____

   _____

   _____

Water moves through the continuous tubes made by the vessel elements of the xylem.

Smaller tracheids are connected by pits in the walls but do not have end wall perforations

SEM

Vessels

Xylem is dead when mature. Note how the cells have lost their cytoplasm.

LM

McKDandy cc 2.5

As shown in these SEM and light micrographs of xylem, the **tracheids** and **vessel elements** form the bulk of the xylem tissue. They are heavily strengthened and are involved in moving water through the plant. The transporting elements are supported by parenchyma (packing and storage cells) and sclerenchyma cells (fibers and sclereids), which provide mechanical support to the xylem.

The xylem cells form continuous tubes through which water is conducted.

Spiral thickening of **lignin** around the walls of the vessel elements give extra strength and rigidity.

RCN

Vessel element
Diameter up to 500 μm

Secondary walls of cellulose are laid down after the cell has elongated or enlarged and lignin is deposited to add strength. This thickening is a feature of tracheids and vessels.

Vessels connect end to end. The end walls of the vessels are perforated to allow rapid water transport.

Tip of tracheid
Diameter ~80 μm

Pits and bordered pits allow transfer of water between cells but there are no end wall perforations.

No cytoplasm or nucleus in mature cell.

Tracheids are longer and thinner than vessels.

Vessel elements and tracheids are the two water conducting cell types in the xylem of flowering plants. Tracheids are long, tapering hollow cells. Water passes from one tracheid to another through thin regions in the wall called pits. Vessel elements are much larger cells with secondary thickening in different patterns (e.g. spirals). Vessel end walls are perforated to allow efficient conduction of water.

CONNECT **54** AP1 CONNECT **63** AP2 WEB **54**

KNOW

55 Phloem

Key Idea: Phloem is the principal food (sugar) conducting tissue in vascular plants, transporting dissolved sugars around the plant.

Like xylem, **phloem** is a complex tissue, comprising a variable number of cell types. The bulk of phloem tissue is made up of the **sieve tubes** (sieve tube elements and sieve cells) and their companion cells. The sieve tubes are the principal conducting cells in phloem and are closely associated with the **companion cells** (modified parenchyma cells) with which they share a mutually dependent relationship. Other parenchyma cells, concerned with storage, occur in phloem, and strengthening fibers and sclereids (short sclerenchyma cells) may also be present. Unlike xylem, phloem is alive when mature.

LS through a sieve tube end plate

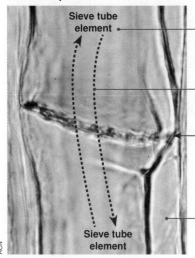

Sieve tube element

The sieve tube elements (also called sieve tube members) lose most of their organelles but are still alive when mature.

Sugar solution flows in both directions

Sieve tube end plate
Tiny holes (arrowed in the photograph below) perforate the sieve tube elements allowing the sugar solution to pass through.

Companion cell
A cell adjacent to the sieve tube member, responsible for keeping it alive.

Sieve tube element

TS through a sieve tube end plate

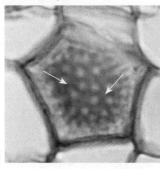

Adjacent sieve tube elements are connected through **sieve plates** through which phloem sap flows.

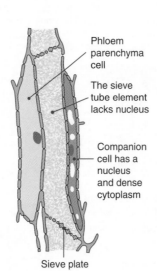

Phloem parenchyma cell

The sieve tube element lacks nucleus

Companion cell has a nucleus and dense cytoplasm

Sieve plate

The structure of phloem tissue

Phloem is alive at maturity and functions in the transport of sugars and minerals around the plant. Like xylem, it forms part of the structural vascular tissue of plants.

Fibers are associated with phloem as they are in xylem. Here they are seen in cross section where you can see the extremely thick cell walls and the way the fibers are clustered in groups. See the previous page for a view of fibers in longitudinal section.

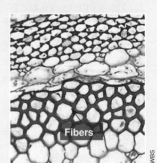
Fibers

In this cross section through a buttercup root, the smaller companion cells can be seen lying alongside the sieve tube members. It is the sieve tube elements that, end on end, produce the **sieve tubes**. They are the conducting tissue of phloem.

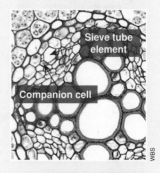
Sieve tube element
Companion cell

In this longitudinal section of a buttercup root, each sieve tube element has a thin **companion cell** associated with it. Companion cells retain their nucleus and control the metabolism of the sieve tube member next to them. They also have a role in the loading and unloading of sugar into the phloem.

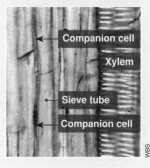
Companion cell
Xylem
Sieve tube
Companion cell

1. (a) What is the conducting cell type in phloem? _____

   (b) What other cell type is associated with these conducting cells? _____

   (c) Describe two roles of these associated cells: _____

   _____

2. Mature phloem is a live tissue, whereas xylem (the water transporting tissue) is dead when mature. Why is it necessary for phloem to be alive to be functional, whereas xylem can function as a dead tissue?

   _____

3. What is the role of fibers and sclereids in phloem?_____

   _____

4. What are the large open cells next to the phloem in the centre photo above right? _____

©2017 **BIOZONE** International
ISBN: 978-1-927309-65-0
Photocopying Prohibited

56 Stems

Key Idea: The vascular tissue in dicots can be identified by its appearance in sections viewed with a light microscope. The vascular tissue (xylem and phloem) in the stems of dicotyledonous plants (dicots) is organized in bundles, which are distributed in a regular fashion around the outer edge of the stem. Each vascular bundle contains xylem tissue to the inside and phloem to the outside, separated by a ring of vascular cambium (actively dividing cells).

Dicot stem structure

In dicots, the vascular bundles are arranged in an orderly fashion around the periphery of the stem. Each vascular bundle contains **xylem** (to the inside) and **phloem** (to the outside). Between the phloem and the xylem is the **vascular cambium**. The cambium is a layer of cells that divide to produce the thickening of the stem, secondary xylem to the inside and secondary phloem to the outside.

Between the vascular bundles and the epidermis is the cortex of thick-walled collenchyma cells, which provide support and structure to the stem. The center of the stem, called the **pith**, is filled with thin-walled parenchyma cells.

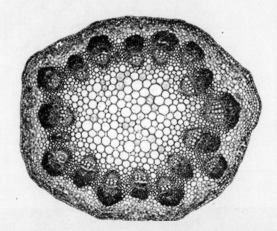

WBS

The image above shows a transverse section through the stem of a typical dicot stem (sunflower) as viewed with a light microscope. A plan diagram, illustrating the distribution and arrangement of the tissues, is shown below.

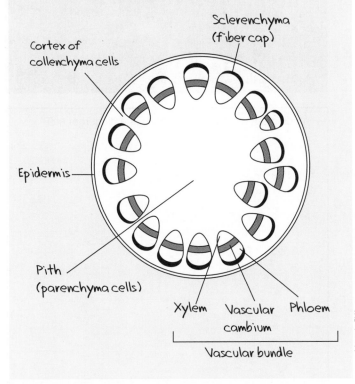

- Cortex of collenchyma cells
- Sclerenchyma (fiber cap)
- Epidermis
- Pith (parenchyma cells)
- Xylem
- Vascular cambium
- Phloem
- Vascular bundle

John main, PLU

1. Describe how the vascular tissue in the stem of a dicot is arranged:

_____

_____

2. Identify the phloem (P) and xylem (X) tissue In the micrograph of a dicot stem below:

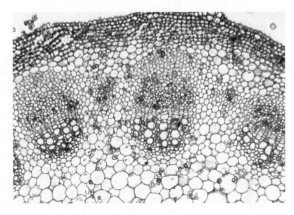

3. The diagram below shows a cross section of a typical dicot stem. identify the labels A - F:

A. _____

B. _____

C. _____

D. _____

E. _____

F. _____

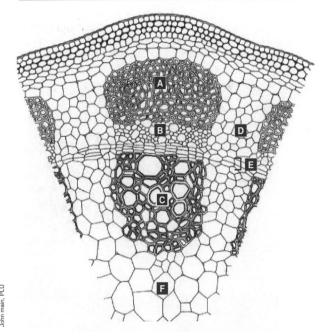

Cross section through a typical dicot stem

CONNECT
50
AP1

WEB
56

KNOW

57 Roots

Key Idea: The xylem and phloem in dicot roots are arranged together forming a central cylinder called the stele.

The vascular tissue in dicot roots forms a central cylinder called the stele. Outside the stele is a cortex of large packing cells, which store starch and other substances. Roots absorb water and minerals from the soil so they can be transported to the rest of the plant. The root epidermis has a thin cuticle that presents no barrier to water entry. The surface area for water absorption is increased by the presence of root hairs, single cell extensions of the epidermal cells.

Dicot root structure

The primary tissues of a dicot root are structurally simple. The vascular tissue (xylem and phloem) forms a central cylinder through the root called the **stele**. The vascular tissue is surrounded by the pericycle, a ring of cells from which lateral roots arise. The large cortex is made up of parenchyma (packing) cells, which store starch and other substances. The air spaces between the cells are essential for aeration of the root tissue, which is non-photosynthetic.

The primary xylem (X) of dicot roots forms a star shape in the centre of the vascular cylinder with usually 3 or 4 points (below). The phloem (P) is located between the regions of xylem tissue.

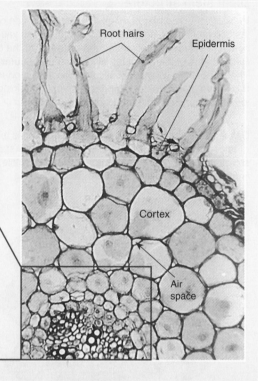

Pericycle: outermost part of the stele

Endodermis: innermost layer of the cortex. The cells of the endodermis have a waterproof band, called the Casparian strip, which forces water to enter the cells.

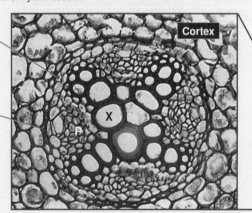

1. The image on the right is an electron micrograph showing xylem vessels in a toothpick. Calculate the magnification of the image:

_____

_____

_____

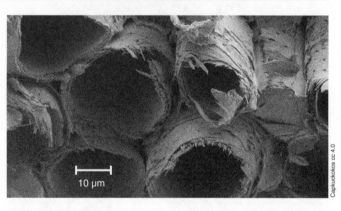

2. The image below is a photomicrograph (x50) showing a partial section through a dicot root. Use this image and the information above to construct a plan diagram:

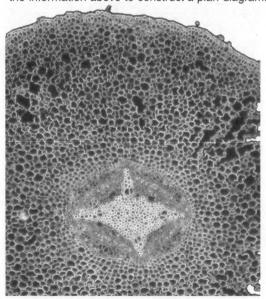

Plan diagram, partial section through a dicot root (x50)

©2017 **BIOZONE** International
ISBN: 978-1-927309-65-0
Photocopying Prohibited

58 | Leaves

Key Idea: The role of leaves is to capture sunlight and fix carbon as carbohydrate for use elsewhere in the plant. To do this, leaves also enable gas exchange with the environment. The vascular tissue in leaves (xylem uppermost and phloem below) forms the **leaf veins**, which are located in the photosynthetic spongy mesophyll tissue of the leaf. Each vein is surrounded by a bundle sheath, layers of parenchyma cells that form a protective layer around the vein. The leaf is the site of photosynthesis in the plant, so the phloem transfers sugars from the source (where the sugars are made) to where they are used (the sink). Experiments show the sources and the sink are always on the same side of the plant.

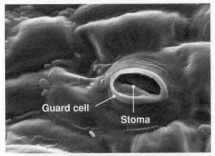

Guard cells on each side of a stoma (pl. stomata) regulate the entry and exit of gases (including water vapor). Stomata permit gas exchange but are also the major routes for water loss.

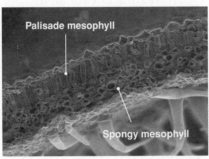

Column-like palisade cells lie beneath the upper epidermis, forming the palisade mesophyll. Below these are the more irregularly shaped cells of the spongy mesophyll. Both types contain chloroplasts.

Alternating the placement of leaves is an adaptation to gather the most amount of light without shading leaves on the same plant. Leaves are most often found only on the outer edges of a plant or tree.

1. The image below shows a cross section through a dicot leaf (light microscope x 30). Use the word list to label the photograph and circle the leaf vein. Construct a plan diagram to show the arrangement of the tissues:

Word list:
xylem, phloem,
air space,
guard cells,
upper epidermis,
palisade mesophyll,
spongy mesophyll,
bundle sheath,
lower epidermis

(a) _ _ _ _ _ _ _ _ _ _ _ _ _ _ _

(b) _ _ _ _ _ _ _ _ _ _ _ _ _ _

(c) _ _ _ _ _ _ _ _ _ _ _ _ _ _ _

(d) _ _ _ _ _ _ _ _ _ _ _ _

(e) _ _ _ _ _ _ _ _ _ _ _ _

(f) _ _ _ _ _ _ _ _ _ _ _ _ _

(g) _ _ _ _ _ _ _ _ _ _ _ _ _ _ _ _ _

(h) _ _ _ _ _ _ _ _ _ _ _ _ _ _ _ _ _

(i) _ _ _ _ _ _ _ _ _ _ _ _ _ _ _ _ _ _ _

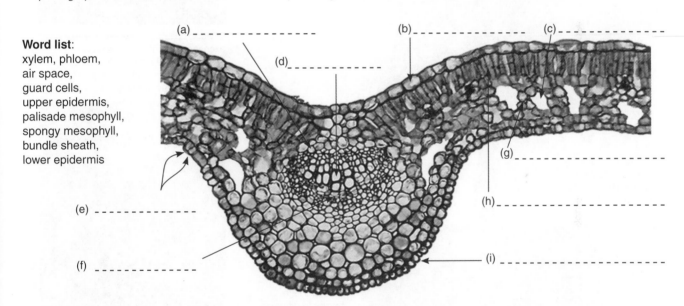

2. Aphids feed on phloem sap. Suggest why aphids are most abundant on the underside of a dicot leaf: _____

_____

3. What are the primary functions of the leaf? _____

_____

4. What is the function of the stomata? _____

_____

5. What is the function of the guard cells? _____

_____

6. Leaves developing in shaded parts of a tree generally grow larger than in sunlit parts. Explain why this might happen:

_____

_____

©2017 **BIOZONE** International
ISBN: 978-1-927309-65-0
Photocopying Prohibited

CONNECT **61** AP2   CONNECT **60** AP2   WEB **58**   **KNOW**

59 Gas Exchange in Plants

Key Idea: Gas exchange in plants occurs through stomata. Respiring tissues require oxygen, and the photosynthetic tissues of plants also require carbon dioxide in order to produce the sugars for growth and maintenance. The main gas exchange organs in plants are the leaves, and sometimes the stems. In most plants, gases cannot diffuse directly across the leaf surface because of its waxy cuticle, so gases enter and leave the leaf via **stomata** (pores) in the leaf surface. The plant has to balance its need for CO_2 (stomata open) against its need to reduce water loss (stomata closed).

Terrestrial environment

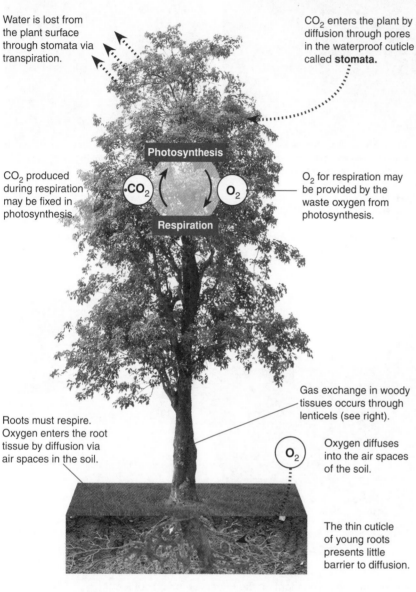

Water is lost from the plant surface through stomata via transpiration.

CO_2 enters the plant by diffusion through pores in the waterproof cuticle called **stomata.**

Photosynthesis

CO_2 produced during respiration may be fixed in photosynthesis.

O_2 for respiration may be provided by the waste oxygen from photosynthesis.

Respiration

Roots must respire. Oxygen enters the root tissue by diffusion via air spaces in the soil.

Gas exchange in woody tissues occurs through lenticels (see right).

O_2

Oxygen diffuses into the air spaces of the soil.

The thin cuticle of young roots presents little barrier to diffusion.

Most gas exchange in plants occurs through the leaves, but some also occurs through the stems and roots. Leaves are very thin with a high surface area and this assists gas exchange by diffusion.

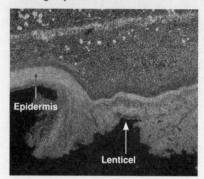

Epidermis

Lenticel

In woody plants, the wood prevents gas exchange. A lenticel is a small area in the bark where the loosely arranged cells allow entry and exit of gases into the stem tissue underneath.

Lenticels

Dave Powell USDA CC 3.0

1. By which process does oxygen enter the plant tissues? _____

2. Where does most gas exchange occur in plants? _____

3. What is the role of lenticels in plant gas exchange? _____
_____
_____

4. Identify two properties of leaves that assist gas exchange: _____

5. With respect to gas exchange and water balance, describe the most important considerations for:

   (a) Terrestrial plants: _____

   (b) Aquatic plants: _____

6. Identify two ways in which plants obtain CO_2: _____

©2017 **BIOZONE** International
ISBN: 978-1-927309-65-0
Photocopying Prohibited

Aquatic environment

Water loss in an aquatic environment is not a problem for plants, but CO_2 availability is often limited because most of the dissolved CO_2 is present as bicarbonate ions, which are not directly available to plants. Maximizing uptake of gaseous CO_2 by reducing barriers to diffusion is therefore important.

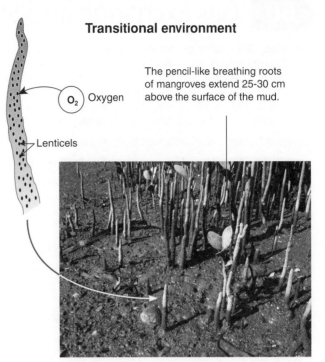

Transitional environment

The pencil-like breathing roots of mangroves extend 25-30 cm above the surface of the mud.

O_2 Oxygen

Lenticels

Absorption of CO_2 by direct diffusion.

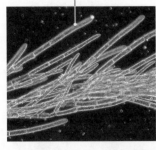

Gas exchange through stomata on the upper surface.

Algae lack stomata but achieve adequate gas exchange through simple diffusion into the cells.

Floating leaves, such as the water lilies above, generally lack stomata on their lower surface.

Most plants have stomata to provide for gas exchange. CO_2 uptake is aided in submerged plants because they have little or no cuticle to form a barrier to diffusion of gases. The few submerged aquatics that lack stomata altogether rely only on diffusion through the epidermis. Most aquatic plants also have air spaces in their spongy tissues (which also assist buoyancy).

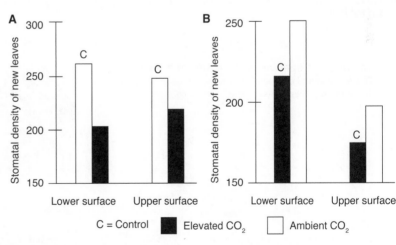

In waterlogged soils, there is little oxygen available for respiring roots and many plants have aerial roots. In mangroves, these are called pneumatophores. The inside of the root is composed of spongy tissue filled with air entering via lenticels in the bark.

7. Describe an adaptation for gas exchange in the following plants:

(a) A submerged aquatic angiosperm: _____

_____

(b) A mangrove in a salty mudflat: _____

_____

8. Study the graphs right (C = control):

The graphs show the effect of CO_2 concentration on stomatal density.

Graph A: Mature leaves were subjected to 720 ppm of CO_2. The effect of this on the stomatal density of new leaves is shown compared to plants in ambient CO_2 of 360 ppm (C).

Graph B: Mature leaves were subjected to ambient (360 ppm) CO_2 while the young leaves were subjected to elevated (720 ppm) CO_2 (the reverse experiment). The effect of this on the stomatal density of new leaves is shown compared to plants left in elevated (720 ppm) CO_2 (C).

A

Stomatal density of new leaves

300 — 250 — 200 — 150 —

C   C

Lower surface   Upper surface

B

Stomatal density of new leaves

250 — 200 — 150 —

C   C

Lower surface   Upper surface

C = Control   ■ Elevated CO_2 □ Ambient CO_2

(a) What is the effect of increasing CO_2 concentration on stomatal density? _____

_____

(b) What is the effect of decreasing CO_2 concentration on stomatal density? _____

_____

(c) Explain these results: _____

_____

_____

©2017 **BIOZONE** International
ISBN: 978-1-927309-65-0
Photocopying Prohibited

60 Gas Exchange and Stomata

Key Idea: Gas exchange through stomata is associated with water losses. Plants have adaptations to limit this loss.

The leaf epidermis of angiosperms is covered with tiny pores, called **stomata**. Angiosperms have many air spaces between the cells of the stems, leaves, and roots. These air spaces are continuous and gases are able to move freely through them and into the plant's cells via the stomata. Each stoma is bounded by two **guard cells**, which together regulate the entry and exit of gases and water vapor. Although stomata permit gas exchange between the air and the photosynthetic cells inside the leaf, they are also the major routes for water loss through transpiration.

Gas exchanges and the function of stomata

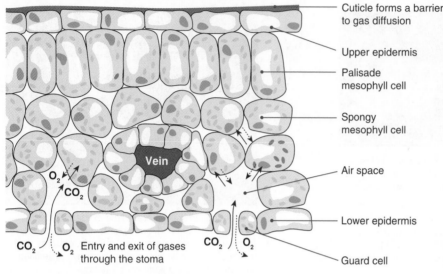

Net gas exchanges in a photosynthesizing dicot leaf

Labels: Cuticle forms a barrier to gas diffusion; Upper epidermis; Palisade mesophyll cell; Spongy mesophyll cell; Air space; Lower epidermis; Guard cell; Vein; Entry and exit of gases through the stoma; O_2; CO_2

The number of stomata is influenced by the environment

Stomatal index*

Lower surface   Upper surface

■ Shade   □ Light

*Stomatal index is the percentage number of stomata compared to all the epidermal cells in a unit area of leaf.

An increase in light intensity on mature leaves increases the number of stomata developing on young leaves.

▶ Gases enter and leave the leaf through stomata. Inside the leaf (as illustrated for a dicot, above), the large air spaces and loose arrangement of the spongy mesophyll facilitate the diffusion of gases and provide a large surface area for gas exchanges.

▶ Respiring plant cells use oxygen (O_2) and produce carbon dioxide (CO_2). These gases move in and out of the plant and through the air spaces by diffusion.

▶ When the plant is photosynthesizing, the situation is more complex. Overall there is a net consumption of CO_2 and a net production of oxygen. The fixation of CO_2 maintains a gradient in CO_2 concentration between the atmosphere (high) and the leaf tissue (low).

▶ Oxygen is produced in excess of respiratory needs and diffuses out of the leaf. These **net** exchanges are indicated by the arrows on the diagram.

The cycle of opening and closing of stomata

The opening and closing of stomata shows a daily cycle that is largely determined by the hours of light and dark.

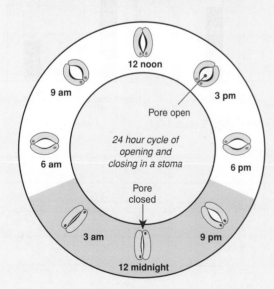

12 noon · 3 pm · 6 pm · 9 pm · 12 midnight · 3 am · 6 am · 9 am

Pore open

24 hour cycle of opening and closing in a stoma

Pore closed

The image left shows a scanning electron micrograph (SEM) of a single stoma from the leaf epidermis of a dicot. Note the guard cells (G), which are swollen tight and open the pore (S) to allow gas exchange between the leaf tissue and the environment.

Factors influencing stomatal opening

Stomata	Guard cells	Daylight	CO_2	Soil water
Open	Turgid	Light	Low	High
Closed	Flaccid	Dark	High	Low

The opening and closing of stomata depends on environmental factors, the most important being light, CO_2 concentration in the leaf tissue, and water supply. Stomata tend to open during daylight in response to light, and close at night (left and above). Low CO_2 levels also promote stomatal opening. Conditions that induce water stress cause the stomata to close, regardless of light or CO_2 level.

WEB 60 · CONNECT 137 AP2 · PRACTICES

©2017 **BIOZONE** International
ISBN: 978-1-927309-65-0
Photocopying Prohibited

KNOW

The guard cells on each side of a stoma control the diameter of the pore by changing shape. When the guard cells take up water by osmosis they swell and become turgid, opening the pore. When the guard cells lose water, they become flaccid and the pore closes. By this mechanism a plant can control the amount of gas entering, or water leaving, the plant. The changes in turgor pressure that open and close the pore result mainly from the reversible uptake and loss of potassium ions (and thus water) by the guard cells.

Stomatal pore open

K+ enters the guard cells from the epidermal cells (active transport coupled to a proton pump).

H_2O
K^+

Water follows K+ by osmosis.

H_2O
K^+

Guard cell swells and becomes turgid.

Thickened ventral wall

Pore opens

K^+
H_2O
K^+
H_2O

Nucleus of guard cell

Stomatal pore closed

K+ leaves the guard cell and enters the epidermal cells.

Water follows K+ by osmosis.
H_2O

H_2O
K^+

The guard cells become flaccid.

Pore closes

H_2O
K^+
H_2O
K^+

Ψguard cell < Ψepidermal cell: water enters the guard cells
Stomata open when the guard cells actively take up K+ from the neighboring epidermal cells. The ion uptake causes the water potential (Ψ) to become more negative in the guard cells. As a consequence, water is taken up by the cells and they swell and become turgid. The walls of the guard cells are thickened more on the inside surface (the ventral wall) than the outside wall, so that when the cells swell they buckle outward, opening the pore.

ψepidermal cell < ψguard cell: water leaves the guard cells
Stomata close when K+ leaves the guard cells. The loss causes the water potential (Ψ) to become less negative in the guard cells, and more negative in the epidermal cells. As a consequence, water is lost by osmosis and the cells sag together and close the pore. The K+ movements in and out of the guard cells are thought to be triggered by blue-light receptors in the plasma membrane, which activate the active transport mechanisms involved.

1. Describe two adaptive features of leaves:

   (a) _____

   (b) _____

2. With respect to a mesophytic, terrestrial flowering plant:

   (a) Describe the **net** gas exchanges between the air and the cells of the mesophyll in the dark (no photosynthesis):

   _____

   (b) Explain how this situation changes when a plant is photosynthesizing: _____

   _____

3. Describe two ways in which the continuous air spaces through the plant facilitate gas exchange:

   (a) _____

   (b) _____

4. Outline the role of stomata in gas exchange in an angiosperm: _____

   _____

5. Summarize the mechanism by which the guard cells bring about:

   (a) Stomatal opening: _____

   _____

   _____

   _____

   (b) Stomatal closure: _____

   _____

   _____

   _____

61 Leaf and Stem Adaptations

Key Idea: The shape and cellular structure of leaves are adaptations for collecting the optimum amount of light and carbon dioxide.

In order to photosynthesize, plants must obtain a regular supply of carbon dioxide gas (CO_2); the raw material for the production of carbohydrate. In green plants, the systems for gas exchange and photosynthesis are linked. Without a regular supply of CO_2, photosynthesis ceases. The leaf, as the primary photosynthetic organ, is adapted to maximize light capture and facilitate the entry of CO_2, while minimizing water loss. There are various ways in which plant leaves are adapted to do this. The ultimate structure of the leaf reflects the environment of the leaf (sun or shade, terrestrial or aquatic), its resistance to water loss, and the importance of the leaf relative to other parts of the plant that may be photosynthetic, such as the stem.

Sun plant

A **sun leaf**, when exposed to high light intensities, can absorb much of the light available to the cells.

Intense light Thick leaves

Palisade mesophyll layer often 2 or 3 cells thick

Chloroplasts are mostly restricted to palisade mesophyll cells (few in spongy mesophyll).

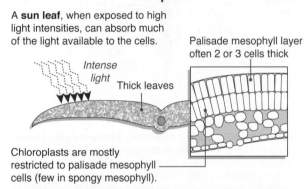

Sun leaves

Sun plants are adapted for growth in full sunlight. They have higher levels of respiration but can produce sugars at rates high enough to compensate for this. Sun plants include many weed species found on open exposed grassland. They expend more energy on the construction and maintenance of thicker leaves than do shade plants. The benefit of this investment is that they can absorb the higher light intensities available and grow rapidly.

Shade plant

A **shade leaf** can absorb the light available at lower light intensities. If exposed to high light, most would pass through.

Low light intensity Thin leaves

Palisade mesophyll layer only 1 cell thick

Chloroplasts occur throughout the mesophyll (as many in the spongy as palisade mesophyll).

Shade leaves

Shade plants typically grow in forested areas, partly shaded by the canopy of larger trees. They have lower rates of respiration than sun plants, mainly because they build thinner leaves. The fewer number of cells need less energy for their production and maintenance. In competition with sun plants, they are disadvantaged by lower rates of sugar production, but in low light environments this is offset by their lower respiration rates.

1. (a) Identify the structures in leaves that facilitate gas exchange: _____

   (b) Explain their critical role in plant nutrition: _____

   _____

   _____

2. (a) State which type of plant (sun or shade adapted) has the highest level of respiration: _____

   (b) Explain how the plant compensates for the higher level of respiration: _____

   _____

3. Discuss the adaptations of leaves in sun and shade plants: _____

   _____

   _____

   _____

   _____

   _____

©2017 **BIOZONE** International
ISBN: 978-1-927309-65-0
Photocopying Prohibited

Adaptations for photosynthesis and gas exchange in plants

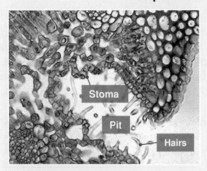

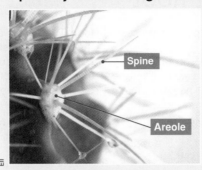

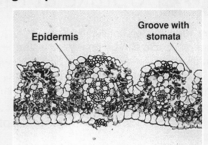

Cross section through a grass leaf

Oleander (above) is an arid-adapted plant (**xerophyte**) and displays many water conserving features. The stomata are found at the bottom of pits on the underside of the leaf. The pits restrict water loss to a greater extent than they reduce CO_2 uptake.

In most cacti, the enlarged stem is the only part of the plant that is photosynthetic and the leaves have been reduced to spines, which are produced from areoles, a kind of highly reduced branch. Spines lack stomata and greatly reduce water loss.

Grass leaves show properties of xerophytes, with several water conserving features. This cross section shows the stomata housed in grooves. When the leaf loses water, it may fold up, closing the grooves and reducing water loss.

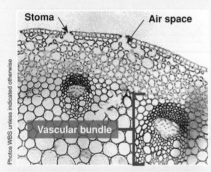

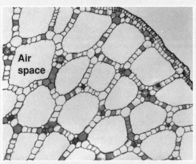

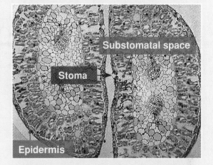

Some plants (e.g. buttercup above) have photosynthetic stems, and CO_2 enters freely into the stem tissue through stomata in the epidermis. The air spaces in the cortex are more typical of leaf mesophyll than stem cortex.

Aquatic plants (hydrophytes) such as *Potamogeton* (above) have stems with massive air spaces. The presence of air in the stem means that they remain floating in the zone of light availability and photosynthesis is not compromised.

This transverse view of the twin leaves of a two-needle pine shows the sunken stomata and substomatal spaces. This adaptation for arid conditions reduces water loss by creating a region of high humidity around the stoma.

4. Describe two adaptations in plants for reducing water loss while maintaining entry of gas into the leaf:

(a) _____

_____

(b) _____

_____

5. Describe two adaptations of photosynthetic stems that are not present in non-photosynthetic stems, and explain the reasons for these:

(a) _____

_____

(b) _____

_____

6. The example of a photosynthetic stem above is from a buttercup, a plant in which the leaves are still the primary organs of photosynthesis.

(a) Identify an example of a plant where the stem is the **only** photosynthetic organ: _____

(b) Describe the structure of the leaves in your example and suggest a reason for their particular structure:

_____

_____

7. Describe one role of the air spaces in the stems of *Potamogeton* related to maintaining photosynthesis: _____

_____

©2017 **BIOZONE** International
ISBN: 978-1-927309-65-0
Photocopying Prohibited

62 Uptake at the Root

Key Idea: Water uptake by the root is a passive process. Mineral uptake can be passive or active.

Plants need to take up water and minerals constantly. They must compensate for the continuous loss of water from the leaves and provide the materials the plant needs to make food. The uptake of water and minerals is mostly restricted to the younger, most recently formed cells of the roots and the root hairs. Water uptake occurs by osmosis, whereas mineral ions enter the root by diffusion and active transport. Pathways for water movements through the plant are outlined below.

Water and mineral uptake by roots

Root hairs have a thin cuticle, so water enters the root easily

Cortex cells of root

Epidermal cell

Xylem

Stele (vascular cylinder). The outer layer of the stele, the pericycle, is next to the endodermis.

Root hair

Water moves by osmosis

Schematic cross-section through a dicot root

The endodermis is the central, innermost layer of the cortex. It is a single layer of cells with a waterproof band of suberin, called the **Casparian strip**, which encircles each cell.

Root hairs are extensions of the root epidermal cells and provide a large surface area for absorbing water and nutrients.

Paths for water movement through the plant

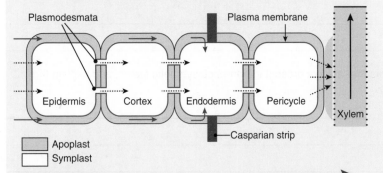

Plasmodesmata

Plasma membrane

Epidermis Cortex Endodermis Pericycle

Xylem

Casparian strip

☐ Apoplast
☐ Symplast

Higher water potential
May be due to fully turgid cells, higher wall pressure, or lower concentration of dissolved substances

Lower water potential
May be due to less turgid cells, lower wall pressure, or higher concentration of dissolved substances

The uptake of water through the roots occurs by osmosis, i.e. the diffusion of water from a higher (less negative) to a lower (more negative) water potential. Most water travels through the **apoplast**, i.e. the spaces within the cellulose cell walls, the water-filled spaces of dead cells, and the hollow tubes of xylem vessels. A smaller amount moves through the **symplast** (the cytoplasm of cells). A very small amount travels through the plant vacuoles.

Some dissolved mineral ions enter the root passively with water. Minerals that are in very low concentration in the soil are taken up by active transport. At the waterproof Casparian strip, water and dissolved minerals must pass into the symplast, so the flow of materials into the stele can be regulated.

1. (a) What two mechanisms do plants use to absorb nutrients?

_____

_____

(b) Describe the two main pathways by which water moves through a plant:

_____

_____

_____

_____

2. Plants take up water constantly to compensate for losses due to transpiration. Describe a benefit of a large water uptake:

_____

_____

_____

_____

_____

3. (a) How does the Casparian strip affect the route water takes into the stele?

_____

_____

_____

_____

(b) Why might this feature be an advantage in terms of selective mineral uptake?

_____

_____

_____

_____

_____

_____

_____

_____

_____

_____

©2017 **BIOZONE** International
ISBN: 978-1-927309-65-0
Photocopying Prohibited

63 Transpiration

Key Idea: Water moves through the xylem primarily as a result of evaporation from the leaves and the cohesive and adhesive properties of water molecules.

Plants lose water all the time through their stomata as a consequence of gas exchange. Approximately 99% of the water a plant absorbs from the soil is lost by evaporation from the leaves and stem. This loss is called **transpiration** and the flow of water through the plant is called the **transpiration**

stream. Plants rely on a gradient in water potential (ψ) from the roots to the air to move water through their cells. Water flows passively from soil to air along a gradient of decreasing water potential. The gradient is the driving force for the movement of water up a plant. Transpiration has benefits to the plant because evaporative water loss is cooling and the transpiration stream helps the plant to take up minerals. Factors contributing to water movement are described below.

Air
Evaporative loss of water from the leaves as water vapor
ψ = −30,000 kPa

Leaves
Highest solute concentration
Lowest free water concentration
ψ = −1200 kPa

Water flows passively from a high water potential to areas where there is a lower (more negative) water potential. This gradient is the driving force in the transport of water up a plant.

The continuous flow of water is called the **transpiration stream**. It is primarily responsible for water moving up the plant.

Soil
Highest free water concentration
Lowest solute concentration
ψ = −10 kPa

Water

Water
Solute particle

Xylem

Water

The role of stomata

Water loss occurs mainly through stomata (pores in the leaf). The rate of water loss can be regulated by specialized guard cells (G) each side of the stoma (S), which open or close the pore.

► Stomata open: gas exchange and transpiration rate increase.

► Stomata closed: gas exchange and transpiration rates decrease.

G

G

S

EII

1. (a) What is transpiration? _____

_____

(b) Describe one benefit of the transpiration stream for a plant: _____

_____

_____

_____

2. Why is transpiration an inevitable consequence of gas exchange? _____

_____

©2017 **BIOZONE** International
ISBN: 978-1-927309-65-0
Photocopying Prohibited

PRACTICES   PRACTICES   CONNECT   WEB

67
AP1

63

KNOW

Processes involved in moving water through the xylem

1 Transpiration pull

Water is lost from the air spaces by evaporation through stomata and is replaced by water from the mesophyll cells. The constant loss of water to the air (and production of sugars) creates a lower (more negative) water potential in the leaves than in the cells further from the evaporation site. Water is pulled through the plant because of a **decreasing gradient in water potential**.

2 Cohesion-tension

The transpiration pull is assisted by the special **cohesive** properties of water. Water molecules cling together as they are pulled through the plant. They also **adhere** to the walls of the xylem (**adhesion**). This creates one **unbroken column of water** through the plant. The upward pull on the cohesive sap creates a tension (a negative pressure). This helps water uptake and movement up the plant.

3 Root pressure

Water entering the stele from the soil creates a **root pressure**; a weak 'push' effect for the water's upward movement through the plant. Root pressure can force water droplets from some small plants under certain conditions (**guttation**), but generally it plays a minor part in the ascent of water.

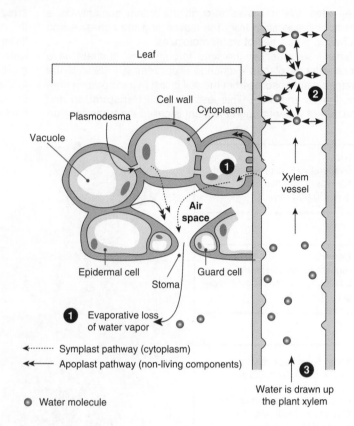

Leaf

Cell wall
Cytoplasm
Plasmodesma
Vacuole
1
Xylem vessel
Air space
Epidermal cell
Guard cell
Stoma

1 Evaporative loss of water vapor

←······· Symplast pathway (cytoplasm)
◄◄—— Apoplast pathway (non-living components)

◉ Water molecule

2
3
Water is drawn up the plant xylem

3. How does the plant regulate the amount of water lost from the leaves? _____

_____

4. (a) What would happen if too much water was lost from the leaves? _____

_____

(b) When might this happen? _____

5. Describe the three processes that assist the transport of water from the roots of the plant upward:

(a) _____

_____

_____

(b) _____

_____

_____

(c) _____

_____

_____

6. The maximum height water can move up the xylem by cohesion-tension alone is about 10 m. How then does water move up the height of a 40 m tall tree?

_____

_____

_____

©2017 **BIOZONE** International
ISBN: 978-1-927309-65-0
Photocopying Prohibited

64 Investigating Plant Transpiration

Key Idea: Factors affecting the rate of transpiration from leaves, including physical factors and features of the leaves themselves, can be investigated using a potometer. Different kinds of plants have different shape and sizes of leaves. Comparing their rate of transpiration requires a comparison of the leaf area and the number of stomata per square centimeter. The volume of water transpired can be measured using a potometer.

The potometer

A potometer is a simple instrument for investigating transpiration rate (water loss per unit time). The equipment is simple to use and easy to obtain. A basic potometer, such as the one shown right, can easily be moved around so that transpiration rate can be measured under different environmental (physical) conditions.

Physical factors that can affect transpiration rate include:

▶ Humidity or vapor pressure (high or low)

▶ Temperature (high or low)

▶ Air movement (still or windy)

▶ Light level (high or low)

▶ Water supply

It is also possible to compare transpiration rates in plants with different adaptations, e.g. comparing transpiration rates in plants with rolled leaves vs rates in plants with broad leaves. If possible, experiments like these should be conducted simultaneously using replicate equipment. If conducted sequentially, care should be taken to keep the environmental conditions the same for all plants used.

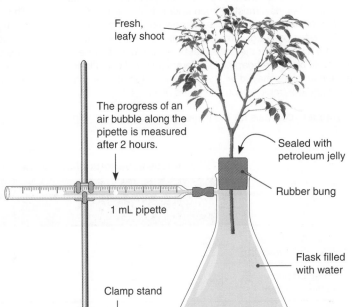

Fresh, leafy shoot

The progress of an air bubble along the pipette is measured after 2 hours.

Sealed with petroleum jelly

Rubber bung

1 mL pipette

Flask filled with water

Clamp stand

Measuring leaf area

Leaf area can be measured by tracing the leaves onto graph paper and counting the squares, or by tracing or photocopying the leaves onto a paper of a known mass per area, then cutting out the shapes and weighing them. For both methods, multiply by 2 for both leaf surfaces.

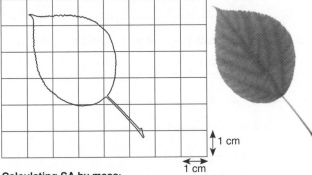

1 cm

1 cm

Calculating SA by mass:
Photocopying leaves onto paper with a known gsm (grams per square meter) allows you to calculate the surface area from the mass of paper they cover.

Calculating SA by leaf trace method:
Count entire squares covered by the leaf. Estimate the area of the partial squares by counting those that are at least half covered by the leaf and disregarding those that are less than half covered.

Determining the number of stomata per mm²

The number of stomata per mm² on the surface of a leaf can be determined by counting the stomata visible under a microscope. Painting clear nail polish over the surface of a leaf and leaving it to dry creates a layer with impressions of the leaf surface. This can be peeled off and viewed under the microscope to count stomata (below).

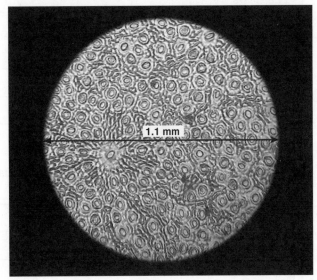

1.1 mm

1. (a) Determine the area of the leaf traced onto the blue grid above: _____

   (b) Twenty leaves from plant A were taped to paper and photocopied on to 80 gsm paper. The shapes were cut out and weighed on a digital balance. The total weight of shapes was 3.21 grams. Calculate the surface area of the leaves.

   _____

2. Calculate the number of stomata per square millimeter in the microscope view of the leaf above: _____

   _____

Investigation 1

The aim

To evaluate the relationship between transpiration rate and stomatal density by examining a variety of plant species.

Background

Plants lose water all the time by evaporation from the leaves and stem. This loss, mostly through pores in the leaf surfaces, is called **transpiration**. Despite the adaptations of plants to reduce water loss (e.g. waxy leaf cuticle), 99% of the water a plant absorbs from the soil is lost by evaporation. Different species of plant are adapted to different physical conditions. These conditions may affect the number of stomata per mm^2 of leaf and the transpiration rate of the plant.

The method

Six plant species from a range of habitats were chosen for use. The stems of several specimens of each species were cut while submerged and set up in a potometer similar to that described on the previous page but with a larger capacity. The temperature was measured at 21°C. The plants were left to transpire in still air at 70% relative humidity for 2 hours and the volume of water transpired was recorded. The surface area of the leaves was also determined as was the number of stomata per mm^2.

3. Write a hypothesis for the investigation: _____

_____

Table 1: Water loss in various plant species over 2 hours

	Total leaf area (cm^2)	Total water lost (μL)	Transpiration rate ($\mu L\ cm^{-2}\ h^{-1}$)	Number of stomata per mm^2 upper surface	Number of stomata per mm^2 lower surface	Total number of stomata per mm^2
Sunflower: *Helianthus annus*	2000	6081		71	172	
Busy Lizzie: *Impatiens sultani*	620	3017		29	143	
Geranium: *Pelargonium zonale*	3800	3721		19	52	
Garden bean: *Phaseolus vulgaris*	1340	4147		40	250	
Caster oil plant: *Ricinus communis*	860	3609		52	121	
Corn: *Zea mays*	4100	6402		60	101	

4. Complete the table by calculating the transpiration rate and total number of stomata per mm^2 for each plant in table 1:

5. (a) Which plant has the highest transpiration rate? _____

   (b) Which plant has the lowest transpiration rate? _____

6. (a) Which plant has the highest stomatal density? _____

   (b) Which plant has the lowest stomatal density? _____

7. (a) Is there a relationship between the number of stomata per mm^2 and the transpiration rate?

   _____

   _____

   (b) Explain your answer: _____

   _____

   _____

   _____

8. (a) Where are the majority of stomata located In a typical dicot leaf? _____

   (b) Suggest why this might be the case: _____

   _____

   _____

©2017 **BIOZONE** International
ISBN: 978-1-927309-65-0
Photocopying Prohibited

Investigation 2

A second investigation focused on the effect of the environment on transpiration rate in one species of plant. Conditions investigated were ambient conditions (still, 20°C, indirect lighting), wind, direct bright light, and high humidity. This time a potometer fitted with a very thin graduated pipette (0.01-0.5 mL) was used to measure the water loss. The results are shown in the table below.

Table 2. Potometer readings in μL^3 water loss

Time (min) Treatment	0	3	6	9	12	15	18	21	24	27	30
Ambient	0	2	5	8	12	17	22	28	32	36	42
Wind	0	2.5	54	88	112	142	175	208	246	283	325
High humidity	0	2	4	6	8	11	14	18	19	21	24
Bright light	0	2.1	42	70	91	112	141	158	183	218	239

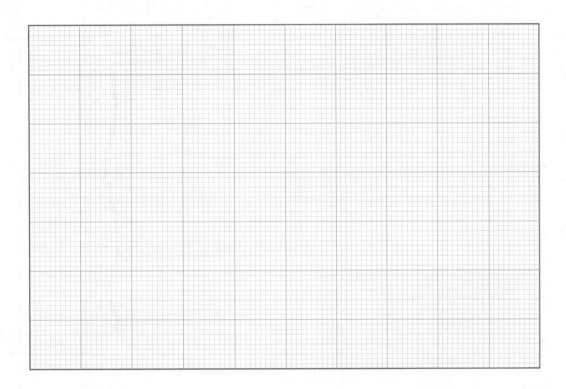

9. (a) Plot the potometer data from Table 2 on the grid provided:

    (b) Identify the independent variable: _____

10. (a) Identify the control: _____

    (b) Explain the purpose of including a control in an experiment: _____

    _____

    _____

    (c) Which factors increased water loss? _____

    (d) How does each environmental factor influence water loss? _____

    _____

    _____

    (e) Explain why the plant lost less water in humid conditions: _____

    _____

    _____

©2017 **BIOZONE** International
ISBN: 978-1-927309-65-0
Photocopying Prohibited

65 Translocation

Key Idea: Phloem transports the organic products of photosynthesis (sugars) through the plant in an active, energy-requiring process called translocation.

In angiosperms, the sugar moves through the sieve-tube members, which are arranged end-to-end and perforated with sieve plates. Apart from water, phloem sap comprises mainly sucrose. It may also contain minerals, hormones, and amino acids, in transit around the plant. Movement of sap in the phloem is from a **source** (a plant organ where sugar is made or mobilized) to a **sink** (a plant organ where sugar is stored or used). Loading sucrose into the phloem at a source involves energy expenditure; it is slowed or stopped by high temperatures or respiratory inhibitors. In some plants, unloading the sucrose at the sinks also requires energy, although in others, diffusion alone is sufficient to move sucrose from the phloem into the cells of the sink organ.

Phloem transport

Phloem sap moves from source to sink at rates as great as 100 m h^{-1}, which is too fast to be accounted for by cytoplasmic streaming. The most acceptable model for phloem movement is the **mass flow hypothesis** (also know as the pressure flow hypothesis). Phloem sap moves by bulk flow, which creates a pressure (hence the term "pressure-flow"). The key elements in this model are outlined below and right. For simplicity, the cells that lie between the source (and sink) cells and the phloem sieve-tube have been omitted.

1 Loading sugar into the phloem from a source (e.g. leaf cell) increases the solute concentration (decreases the water potential, ψ) inside the sieve-tube cells. This causes the sieve-tubes to take up water from the surrounding tissues by osmosis.

2 The water uptake creates a hydrostatic pressure that forces the sap to move along the tube, just as pressure pushes water through a hose.

3 The pressure gradient in the sieve tube is reinforced by the active unloading of sugar and consequent loss of water by osmosis at the sink (e.g. root cell).

4 Xylem recycles the water from sink to source.

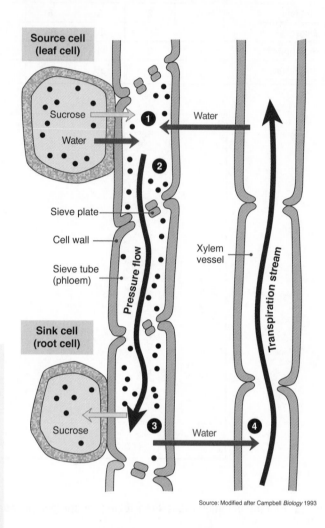

Source: Modified after Campbell *Biology* 1993

Measuring phloem flow
Aphids can act as natural phloem probes to measure phloem flow. The sucking mouthparts (stylet) of the insect penetrates the phloem sieve-tube cell. While the aphid feeds, it can be severed from its stylet, which remains in place and continues to exude sap. Using different aphids, the rate of flow of this sap can be measured at different locations on the plant.

1. (a) From what you know about osmosis, explain why water follows the sugar as it moves through the phloem:

_____

_____

_____

_____

(b) What is meant by '**source to sink**' flow in phloem transport?_____

_____

_____

2. Why does a plant need to move food around, particularly from the leaves to other regions? _____

_____

_____

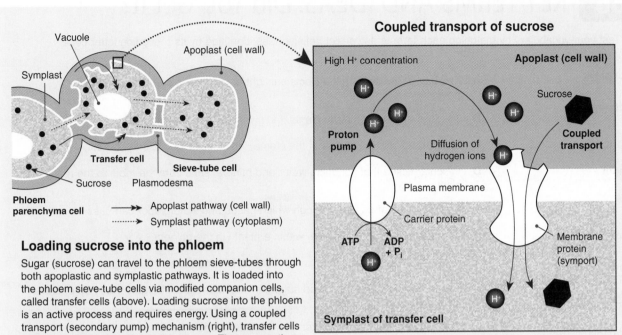

Loading sucrose into the phloem

Sugar (sucrose) can travel to the phloem sieve-tubes through both apoplastic and symplastic pathways. It is loaded into the phloem sieve-tube cells via modified companion cells, called transfer cells (above). Loading sucrose into the phloem is an active process and requires energy. Using a coupled transport (secondary pump) mechanism (right), transfer cells expend energy to accumulate the sucrose. The sucrose then passes into the sieve tube through plasmodesmata. The transfer cells have wall ingrowths that increase surface area for the transport of solutes. Using this mechanism, some plants can accumulate sucrose in the phloem to 2-3 times the concentration in the mesophyll.

Above: Proton pumps generate a hydrogen ion gradient across the membrane of the transfer cell. This process requires expenditure of energy. The gradient is then used to drive the transport of sucrose, by coupling the sucrose transport to the diffusion of hydrogen ions back into the cell.

3. In your own words, describe what is meant by the following:

(a) Translocation: _____

_____

(b) Pressure-flow movement of phloem: _____

_____

(c) Coupled transport of sucrose: _____

_____

4. Briefly explain how sucrose is transported into the phloem: _____

_____

_____

5. Explain the role of the companion (transfer) cell in the loading of sucrose into the phloem: _____

_____

_____

_____

6. (a) What does the flow of phloem sap from a severed aphid stylet indicate?_____

_____

(b) Where would you expect the flow rate to be greatest and why? _____

_____

(c) Why do you think aphid stylets are particularly useful for studying the rate of flow in phloem? _____

_____

_____

66 KEY TERMS AND IDEAS: Did You Get It?

1. Test your vocabulary by matching each term to its correct definition, as identified by its preceding letter code.

guard cells ....

hydrophyte ....

leaf ....

phloem ....

root ....

stem ....

stomata ....

translocation ....

transpiration ....

xerophyte ....

xylem ....

A The part of the plant that collects sunlight and contains cells that carry out photosynthesis.

B The loss of water vapor by plants, mainly from leaves via the stomata.

C Specialized cells each side of the stoma, which open or close the pore.

D Vascular tissue that conducts water and mineral salts from the roots to the rest of the plant.

E Pores in the leaf surface through which gases and water vapor can pass.

F The active transport of sugars within a plant via the phloem.

G A plant that has adaptations to survive in arid conditions.

H Tissue that conducts dissolved sugars in vascular plants. Largely made up of sieve tubes and companion cells.

I One of the main structural axes of a vascular plant.

J Part of the plant, usually found underground that both anchors the plant in place and absorbs water and nutrients from the soil.

K A plant that has adaptations to living either partially or fully submerged in water.

2. (a) What is the name given to the loss of water vapor from plant leaves and stems? _____

(b) What plant tissue is involved in this process? _____

(c) Is this tissue alive or dead? _____

(d) Does this process require energy? _____

3. An experiment was performed to investigate transpiration from a hydrangea shoot in a mass potometer. The experiment was set up and the plant left to stabilize (environmental conditions: still air, light shade, 20°C). The plant was then placed in different environmental conditions and the water loss was measured each hour. Finally, the plant was returned to original conditions, allowed to stabilize and transpiration rate measured again. The data are presented below:

Experimental conditions	Temperature (°C)	Humidity (%)	Transpiration rate (g h^{-1})
(a) Still air, light shade, room temperature	20	70	1.20
(b) Moving air, light shade	20	70	1.60
(c) Still air, bright sunlight	23	70	3.75
(d) Still air and dark, moist chamber	19.5	100	0.05

(a) What conditions acted as the control in this experiment? _____

(b) Which factors increased transpiration rate and why? _____

_____

_____

_____

(c) Why did the plant have such a low transpiration rate in humid, dark conditions? _____

_____

_____

©2017 **BIOZONE** International
ISBN: 978-1-927309-65-0

TEST

Enduring Understanding 2.D 4.A, 4.B

Comparing Animal Systems

Key terms

ammonia

circulatory system

common ancestry

contractile vacuole

countercurrent system

digestive system

excretion

excretory system

gills

homeostasis

kidney

lungs

nephridia

nephron

osmoregulation

protonephridia

respiratory system

thermoregulation

urea

uric acid

2.D.2 Homeostatic mechanisms reflect common ancestry and divergence due to adaptation in different environments

Essential knowledge

Activity number

(a) Continuity of homeostatic mechanisms reflects common ancestry, while changes may occur in response to different environments

☐ 1 Use examples to show that homeostatic mechanisms in organisms reflect common ancestry but there is adaptation to environment, while individuals show physiological changes in response to changing environments. *[also 2.D.2.c]*   68 70 75
76 77 83
84 89 90

(b) Organisms have various mechanisms for obtaining nutrients and eliminating wastes

☐ 1 Use examples to show that animals have various mechanisms for obtaining nutrients and eliminating wastes in different environments. Examples include:
- Gas exchange systems in aquatic and terrestrial animals.   68-76
- Digestive systems in different animal phyla.   80-85
- Production and excretion of nitrogenous wastes in aquatic and terrestrial animals.   86-92

(c) Homeostatic control systems across phyla support common ancestry

☐ 1 Describe examples to illustrate the common ancestry of homeostatic control systems in animals. Examples include:
- Excretory systems of flatworms, earthworms, insects, and vertebrates.   88 91 92
- Osmoregulation in bacteria, fish, and protists. *[also 2.D.3.a]*   86
- Circulatory systems in fish, amphibians, and mammals.   78 79
- Thermoregulation in aquatic and terrestrial animals *[also 2.A.1.d.1]*.   77

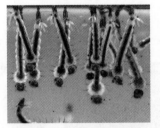  

4.A.4 Organisms exhibit complex properties due to interactions between their component parts

Essential knowledge

Activity number

(a) Interactions and coordination between organs provide essential biological activities

☐ 1 Describe how interactions between organs in animals provide essential activities such as digestion of food and excretion of wastes.   81-85
86-88

(b) Interactions and coordination between systems provide essential biological activities

☐ 1 Describe how interactions between the organ systems provide essential activities such as exchange and transport of gases (respiratory and circulatory systems) and movement and locomotion (nervous and muscular systems).   72-74
also 118

4.B.2 Cooperative interactions within organisms promote efficiency in the use of energy and matter

Essential knowledge

Activity number

(a) Organisms have areas or compartments that perform a subset of functions related to energy and matter and these parts contribute to the whole

☐ 1 Recall how compartmentalization within cells contributes to overall function.   68 86 87

☐ 2 Using examples, explain how specialization of organs within animals contributes to overall functioning. Examples include specialization for exchange of gases, digestion of food, circulation of fluids, and excretion of wastes.   69-76
78-92

67 The Need for Gas Exchange

Key Idea: Cellular respiration (the breaking down of glucose to release energy) requires the exchange of gases between the respiring cells and the environment.
To meet the demands of aerobic metabolism, organisms must exchange gases with the environment. Some organisms can exchange gases directly across their body surface, but most organisms have specialized gas exchange systems that are adapted to operate in their specific environment.

Cellular respiration creates a demand for oxygen

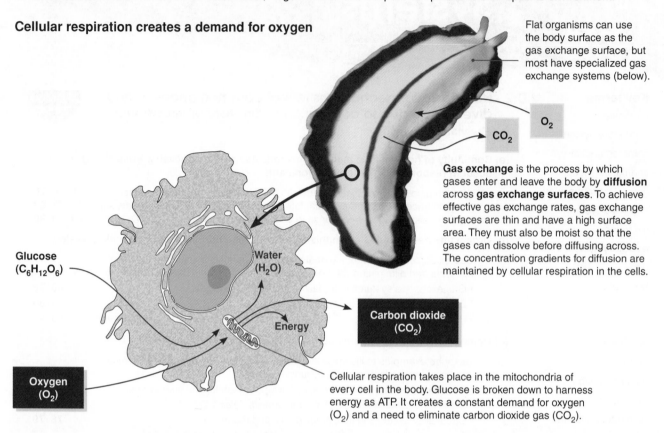

Flat organisms can use the body surface as the gas exchange surface, but most have specialized gas exchange systems (below).

O_2

CO_2

Gas exchange is the process by which gases enter and leave the body by **diffusion** across **gas exchange surfaces**. To achieve effective gas exchange rates, gas exchange surfaces are thin and have a high surface area. They must also be moist so that the gases can dissolve before diffusing across. The concentration gradients for diffusion are maintained by cellular respiration in the cells.

Glucose
($C_6H_{12}O_6$)

Water
(H_2O)

Energy

Carbon dioxide
(CO_2)

Oxygen
(O_2)

Cellular respiration takes place in the mitochondria of every cell in the body. Glucose is broken down to harness energy as ATP. It creates a constant demand for oxygen (O_2) and a need to eliminate carbon dioxide gas (CO_2).

Gas exchange systems and environment

The environment presents different gas exchange challenges to animals. In air, gas exchange surfaces will dry out. In water, the oxygen content is much lower than in air.

Spiracles

Insects exchange gases with the air through a system of tubes called **tracheae**, which penetrate deep into their tissues. Air enters through spiracles (openings) in the body wall.

The **lungs** of **air breathers**, such as mammals, are internalized to prevent drying out. They are linked to the air outside by airways that carry the air to and from the gas exchange surface.

In **water**, the gas exchange membranes are in direct contact with the environment. Fish **gills** are very efficient at extracting oxygen from water, which compensates for water's relatively low oxygen content.

1. What is the purpose of gas exchange? _____

_____

2. Name the respiratory gases: _____

3. How are gases exchanged with the environment? _____

4. Contrast air and water in terms of the challenges they present for gas exchange: _____

_____

_____

_____

©2017 **BIOZONE** International
ISBN: 978-1-927309-65-0
Photocopying Prohibited

68 Gas Exchange in Animals

Key Idea: Animal gas exchange systems are suited to the animal's environment, body form, and metabolic needs.

The way an animal exchanges gases with its environment is influenced by the animal's body form and by the environment in which the animal lives. Small or flat organisms in moist or aquatic environments, such as sponges and flatworms, require no specialized structures for gas exchange. Larger or more complex animals have specialized systems to supply the oxygen to support their metabolic activities. The type and complexity of the exchange system reflects the demands of metabolism for gas exchange (oxygen delivery and carbon dioxide removal) and the environment (aquatic or terrestrial).

1. Describe two reasons for the development of gas exchange structures and systems in animals:

(a) _____

_____

_____

(b) _____

_____

_____

2. Describe two ways in which air breathers manage to keep their gas exchange surfaces moist:

(a) _____

_____

(b) _____

_____

3. Explain why gills would not work in a terrestrial environment:

_____

_____

_____

_____

_____

_____

4. Explain why mammals must ventilate their lungs (breathe in and out):

_____

_____

_____

_____

_____

_____

Representative gas exchange systems

Oxygen
Carbon dioxide

Simple organisms
The high surface area to volume ratio of very flat or very small organisms, such as this nematode, enables them to use the body surface as the gas exchange surface.

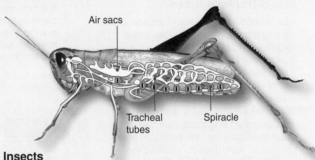

Air sacs

Tracheal tubes   Spiracle

Insects
Insects transport gases via a system of branching tubes called **tracheae** (or tracheal tubes). The tracheae deliver oxygen directly to the tissues. Larger insects can increase the air moving in and out of these tubes by contracting and expanding the abdomen.

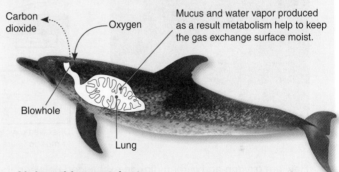

Carbon dioxide   Oxygen   Mucus and water vapor produced as a result metabolism help to keep the gas exchange surface moist.

Blowhole

Lung

Air breathing vertebrates
The gas exchange surface in mammals and other air breathing vertebrates is located in internal **lungs**. Their internal location within the body keeps the exchange surfaces moist and prevents them from drying out. The many alveoli of the lungs provide a large surface area for maximizing gas exchange. For example, human lungs have 600 million alveoli with a total surface area of 100 m^2.

Oxygen

Carbon dioxide

Gills under gill cover (operculum).

Bony fish, sharks, and rays
Fish extract oxygen dissolved in water using **gills**. Gills achieve high extraction rates of oxygen from the water which is important because there is less oxygen in water than air. Bony fish ventilate the gill surfaces by movements of the gill cover. The water supports the gills, and the gill lamellae (the gas exchange surface) can be exposed directly to the environment without drying out.

©2017 **BIOZONE** International
ISBN: 978-1-927309-65-0
Photocopying Prohibited

CONNECT **62** AP1   CONNECT **40** AP2   CONNECT **29** AP2   WEB **68**

KNOW

69 Gas Exchange in Insects

Key Idea: Insects transport air throughout their bodies via a system of tracheal tubes. Spiracles allow air to enter and leave the body.

Terrestrial air breathers lose water from their gas exchange surfaces to the environment. Most terrestrial insects have a large surface area to volume ratio and so are at risk of drying out. They minimize water losses with a waxy outer layer to their exoskeleton and a system of **tracheal tubes** for gas exchange that loses very little water to the environment.

Tracheal systems, which open to the air via paired openings (**spiracles**) in the body wall, are the most common gas exchange organs of insects. Filtering devices stop the system clogging and valves control the degree to which the spiracles are open. In small insects, diffusion is the only mechanism needed to exchange gases, because it occurs so rapidly through the air-filled tubules. Larger, active insects, such as locusts, have air sacs, which can be compressed and expanded to assist in moving air through the tubules.

Insect tracheal tubes

Insects (and some spiders) transport gases via a system of branching tubes called tracheae or tracheal tubes. Respiratory gases move by diffusion across the moist lining directly to and from the tissues. The end of each tube contains a small amount of fluid in which the respiratory gases are dissolved. The fluid is drawn into the muscle tissues during their contraction, and is released back into the tracheole when the muscle rests. Insects ventilate their tracheal system by making rhythmic body movements to help move the air in and out of the tracheae.

Spiracle openings on the abdomen

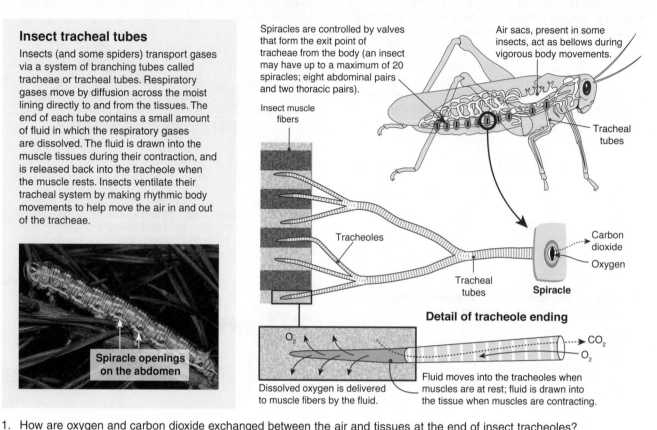

Spiracles are controlled by valves that form the exit point of tracheae from the body (an insect may have up to a maximum of 20 spiracles; eight abdominal pairs and two thoracic pairs).

Insect muscle fibers

Air sacs, present in some insects, act as bellows during vigorous body movements.

Tracheal tubes

Tracheoles

Carbon dioxide

Oxygen

Tracheal tubes

Spiracle

Detail of tracheole ending

O_2

CO_2

O_2

Dissolved oxygen is delivered to muscle fibers by the fluid.

Fluid moves into the tracheoles when muscles are at rest; fluid is drawn into the tissue when muscles are contracting.

1. How are oxygen and carbon dioxide exchanged between the air and tissues at the end of insect tracheoles?

_____

_____

_____

_____

2. Valves in the spiracles can regulate the amount of air entering the tracheal system. Suggest a reason for this adaptation:

_____

_____

3. How is ventilation achieved in a terrestrial insect? _____

_____

_____

4. Even though most insects are small, they have evolved an efficient and intricate gas exchange system that is independent of diffusion across the body surface. What are the adaptive advantages of this evolutionary innovation?

_____

_____

_____

©2017 **BIOZONE** International
ISBN: 978-1-927309-65-0
Photocopying Prohibited

70 Adaptations of Aquatic Insects

Key Idea: Insects living in water exchange gases with the environment via tracheae. Air can be carried with the insect or it can diffuse into the tracheal system from the water.

Aquatic insects exchange gases via a system of air-filled **tracheae**. Oxygen enters this system in various ways. Aquatic insect larvae rely on diffusion across the body surface, with or without gills. Adult insects carry air with them when submerged, either as a bubble or trapped by hairs. A thin film of air trapped by hairs is called a **plastron**. It acts as a diffusion gill, into which oxygen can diffuse from the water.

Gas exchange in aquatic invertebrates

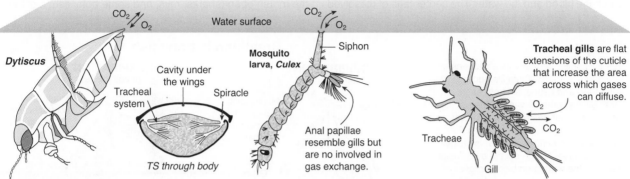

Surface air breathers

The **diving beetle**, *Dytiscus*, traps air from the surface beneath its wings where it forms a compressible gill. The spiracles open into the air space and lead to the tracheal tubes. As the submerged insect respires, the oxygen is used up and the bubble decreases in size. A **mosquito larva** penetrates the water surface with a siphon extending from a spiracle at the tip of the abdomen. The larva hangs at the surface and gas exchange occurs by diffusion from the air, so is independent of the water's O_2 content of the water.

Tracheal gills

In the larvae of many aquatic insects, gas exchange occurs by diffusion across the body surface. This is enhanced by the presence of **tracheal gills** (abdominal or anal) which may account for 20-70% of O_2 uptake depending on their surface area.

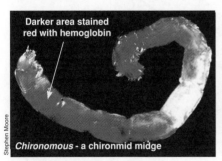

Anisops - the backswimmer

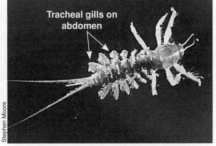

Dytiscus, great diving beetle

Water scavenger beetle

Anisops carries only a small air mass when diving but can exploit oxygen-poor waters because it has large hemoglobin-filled cells in its abdomen.

Dytiscus (above) carries its air supply from the surface, but in other aquatic beetles, such as the water scavenger (above right), the hydrophobic body hairs form a nonwettable surface called a plastron. Water cannot penetrate the hairs, which trap air permanently. The plastron thus forms a non-compressible gill into which oxygen diffuses from the water.

Chironomous - a chironmid midge

Tracheal gills on abdomen

Tracheal gills at the tip of the abdomen

The blood of a few insect larvae, e.g. *Chironomus,* contains the oxygen-carrying pigment **hemoglobin**, which allows them to survive when oxygen levels fall.

The tracheal gills of this spiny gilled mayfly (a very active species), are located on the abdomen. Like other insects with gills, they are intolerant of low oxygen.

The anal tracheal gills of this damselfly larva are located at the tip of the abdomen. Mayfly larvae generally are indicators of high water quality as they do not tolerate low oxygen.

1. Briefly describe one structural adaptation of freshwater insects for gas exchange: _____

_____

2. Describe one physiological adaptation of aquatic insects for gas exchange: _____

_____

3. What advantage would the water scavenger have over *Dytiscus* in terms of gas exchange: _____

_____

©2017 **BIOZONE** International
ISBN: 978-1-927309-65-0
Photocopying Prohibited

CONNECT **62** AP1   WEB **70**

KNOW

71 Gas Exchange in Fish

Key Idea: Fish gills are thin, vascular structures just behind the head. Countercurrent flow enables efficient exchange of gases between the water and the blood in the gill capillaries. Fish obtain the oxygen they need from the water using gills, which are membranous structures supported by cartilaginous or bony struts. Gill surfaces are very large and as water flows over the gill surface, respiratory gases are exchanged between the blood and the water. The percentage of dissolved oxygen in a volume of water is much less than in the same volume of air. Air is 21% oxygen, whereas in water, dissolved oxygen is about 1% by volume. Active organisms with gills must therefore be able to extract oxygen efficiently from the water. In fish, high oxygen extraction rates are achieved using countercurrent exchange and by pumping water across the gill surface (most bony fish) or swimming continuously with the mouth open (sharks, rays, and some bony fish, e.g. tuna).

Fish gills

The gills of fish are very thin, filamentous structures, with individual filaments supported and kept apart from each other by the water. This gives them a high surface area for gas exchange. The outer surface of the gill is in contact with the water, and blood flows in vessels inside the gill. Gas exchange occurs by diffusion between the water and blood across the gill membrane and capillaries. The operculum (gill cover) permits exit of water and acts as a pump, drawing water past the gill filaments. The gills of fish are very efficient and achieve an 80% extraction rate of oxygen from water; over three times the rate of human lungs from air.

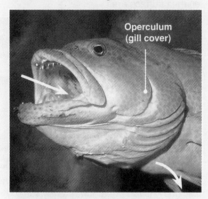

Operculum
(gill cover)

Ventilation of the gills

Most bony fish ventilate the gills by opening and closing the mouth in concert with opening and closing the operculum. The mouth opens, increasing the volume of the buccal (mouth) cavity, causing water to enter. The operculum bulges slightly, moving water into the opercular cavity. The mouth closes and the operculum opens and water flows out over the gills. These pumping movements keep oxygenated water flowing over the gills, maintaining the concentration gradient for diffusion. Other fish (e.g. sharks and tuna) must swim continuously to achieve the same gill ventilation.

Breathing in bony fish

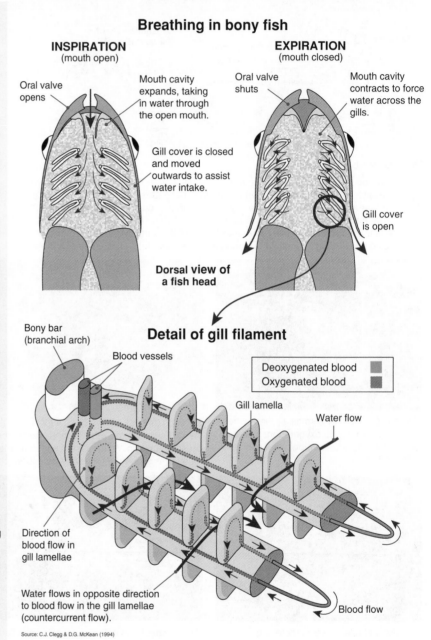

INSPIRATION
(mouth open)

Oral valve opens

Mouth cavity expands, taking in water through the open mouth.

Gill cover is closed and moved outwards to assist water intake.

EXPIRATION
(mouth closed)

Oral valve shuts

Mouth cavity contracts to force water across the gills.

Gill cover is open

Dorsal view of a fish head

Detail of gill filament

Bony bar (branchial arch)

Blood vessels

Deoxygenated blood
Oxygenated blood

Gill lamella

Water flow

Direction of blood flow in gill lamellae

Water flows in opposite direction to blood flow in the gill lamellae (countercurrent flow).

Blood flow

Source: C.J. Clegg & D.G. McKean (1994)

---

1. Describe three features of a fish gas exchange system (gills and related structures) that facilitate gas exchange:

   (a) _____

   (b) _____

   (c) _____

2. Why do fish need to ventilate their gills? _____

   _____

   _____

WEB
71

CONNECT
77
AP2

CONNECT
62
AP1

KNOW

©2017 **BIOZONE** International
ISBN: 978-1-927309-65-0
Photocopying Prohibited

Countercurrent flow

▶ The structure of fish gills and their physical arrangement in relation to the blood flow maximizes gas exchange rates. A constant stream of oxygen-rich water flows over the gill filaments in the opposite direction to the blood flowing through the gill filaments.

▶ This is called countercurrent flow (below left) and it is an adaptation for maximizing the amount of O_2 removed from the water. Blood flowing through the gill capillaries encounters water of increasing oxygen content.

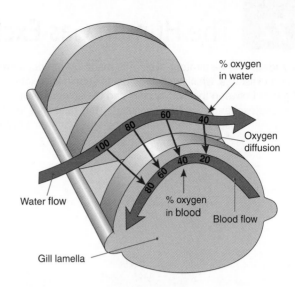

% oxygen in water

Oxygen diffusion

Water flow

% oxygen in blood

Blood flow

Gill lamella

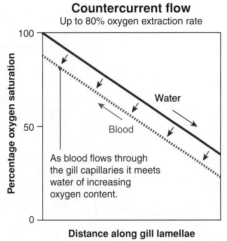

Countercurrent flow
Up to 80% oxygen extraction rate

Water

Blood

As blood flows through the gill capillaries it meets water of increasing oxygen content.

Parallel current flow
Up to 50% oxygen extraction rate

Water

Blood

At this point, blood and water have the same O_2 concentration and no more O_2 exchange takes place.

Percentage oxygen saturation

Distance along gill lamellae

The concentration gradient (for oxygen uptake) across the gill is maintained across the entire distance of the gill lamella and oxygen continues to diffuse into the blood (CO_2 diffuses out at the same time).

A parallel current flow could not achieve the same oxygen extraction rates because the concentrations across the gill would quickly equalize (far right).

3. Describe how fish achieve adequate ventilation of the gills through:

(a) Pumping (mouth and operculum): _____

_____

(b) Continuous swimming (mouth open): _____

_____

4. Describe countercurrent flow: _____

_____

_____

5. (a) How does the countercurrent system in a fish gill increase the efficiency of oxygen extraction from the water?

_____

_____

_____

(b) Explain why parallel flow would not achieve the same rates of oxygen extraction: _____

_____

_____

6. In terms of the amount of oxygen available in the water, explain why fish are very sensitive to increases in water temperature or suspended organic material in the water:

_____

_____

©2017 **BIOZONE** International
ISBN: 978-1-927309-65-0
Photocopying Prohibited

72 The Human Gas Exchange System

Key Idea: The respiratory system is made up of specialized cells and tissues, which work together to enable the exchange of gases between the body's cells and the environment.
The gas exchange system consists of the passages of the mouth and nose, the trachea, and the tubes and air sacs of the lungs. Cooperation with the muscles of the diaphragm and ribcage contribute to its function. Each region is specialized to perform a particular role in the organ system's overall function, which is to exchange respiratory gases (O_2 and CO_2) between the body's cells and the environment.

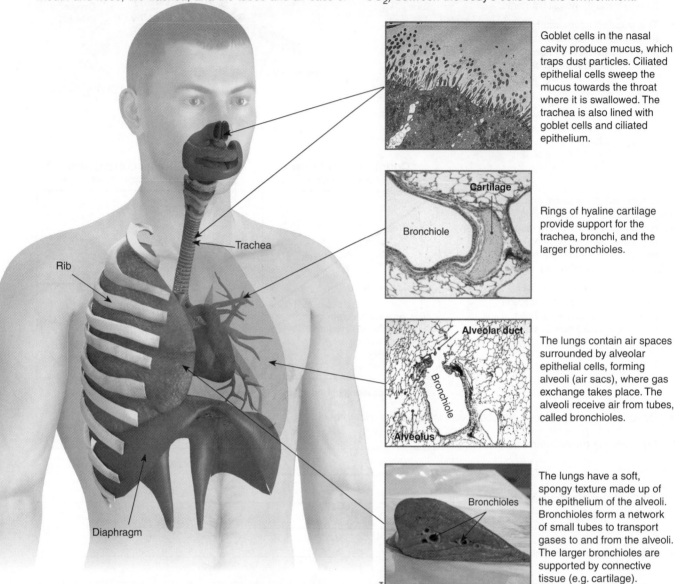

Goblet cells in the nasal cavity produce mucus, which traps dust particles. Ciliated epithelial cells sweep the mucus towards the throat where it is swallowed. The trachea is also lined with goblet cells and ciliated epithelium.

Rings of hyaline cartilage provide support for the trachea, bronchi, and the larger bronchioles.

The lungs contain air spaces surrounded by alveolar epithelial cells, forming alveoli (air sacs), where gas exchange takes place. The alveoli receive air from tubes, called bronchioles.

The lungs have a soft, spongy texture made up of the epithelium of the alveoli. Bronchioles form a network of small tubes to transport gases to and from the alveoli. The larger bronchioles are supported by connective tissue (e.g. cartilage).

1. Name three types of cells in the respiratory system and their function:

   (a) _____

   _____

   (b) _____

   _____

   (c) _____

   _____

2. Which cells form the alveoli? _____

3. What is the purpose of the hyaline cartilage in the respiratory system? _____

   _____

4. Where does gas exchange take place in the lungs? _____

©2017 **BIOZONE** International
ISBN: 978-1-927309-65-0
Photocopying Prohibited

73 The Lungs

Key Idea: Lungs are internal sac-like organs connected to the outside by a system of airways. The smallest airways end in thin-walled alveoli, where gas exchange occurs.
The respiratory system includes all the structures associated with exchanging respiratory gases with the environment. In mammals, the gas exchange organs are paired lungs connected to the outside air by way of a system of tubular passageways: the trachea, bronchi, and bronchioles.

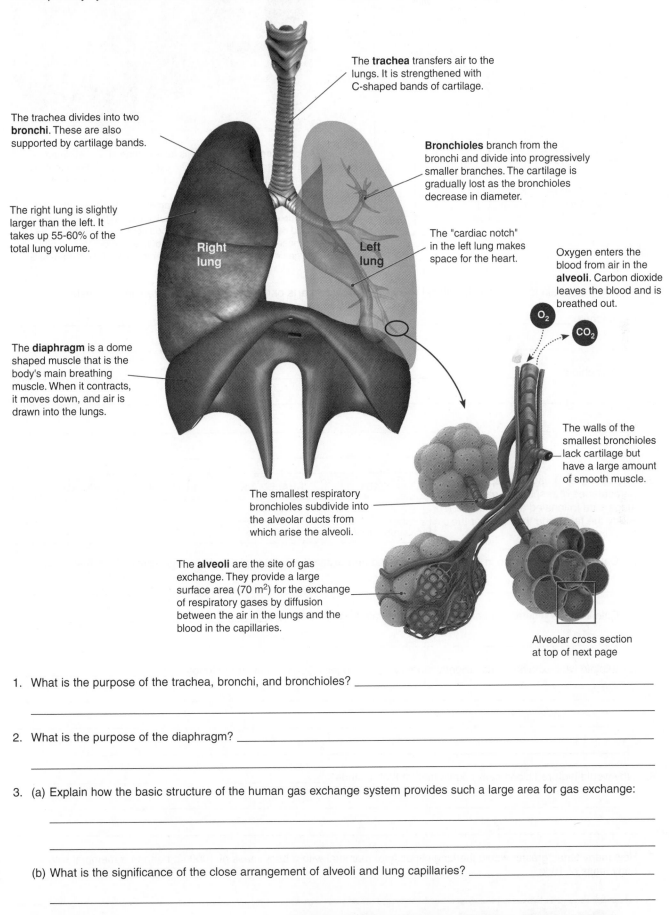

The **trachea** transfers air to the lungs. It is strengthened with C-shaped bands of cartilage.

The trachea divides into two **bronchi**. These are also supported by cartilage bands.

Bronchioles branch from the bronchi and divide into progressively smaller branches. The cartilage is gradually lost as the bronchioles decrease in diameter.

The right lung is slightly larger than the left. It takes up 55-60% of the total lung volume.

Right lung

Left lung

The "cardiac notch" in the left lung makes space for the heart.

Oxygen enters the blood from air in the **alveoli**. Carbon dioxide leaves the blood and is breathed out.

O_2

CO_2

The **diaphragm** is a dome shaped muscle that is the body's main breathing muscle. When it contracts, it moves down, and air is drawn into the lungs.

The walls of the smallest bronchioles lack cartilage but have a large amount of smooth muscle.

The smallest respiratory bronchioles subdivide into the alveolar ducts from which arise the alveoli.

The **alveoli** are the site of gas exchange. They provide a large surface area (70 m^2) for the exchange of respiratory gases by diffusion between the air in the lungs and the blood in the capillaries.

Alveolar cross section at top of next page

1. What is the purpose of the trachea, bronchi, and bronchioles? _____

_____

2. What is the purpose of the diaphragm? _____

_____

3. (a) Explain how the basic structure of the human gas exchange system provides such a large area for gas exchange:

_____

_____

(b) What is the significance of the close arrangement of alveoli and lung capillaries? _____

_____

CONNECT	CONNECT	CONNECT	CONNECT	CONNECT	WEB
62	**98**	**96**	**94**	**40**	**73**
AP1	AP2	AP2	AP2	AP2	

KNOW

In mammals, lung size is a regular function of body size. Small mammals with high metabolic rates obtain sufficient oxygen with lungs of the same relative size as in larger mammals.

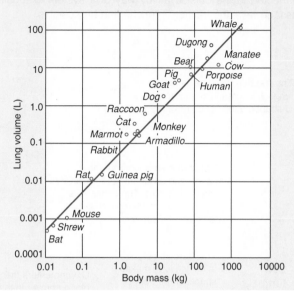

Air pressure at high altitude is far less than at sea level and less oxygen is available for gas exchange. Above 2000 meters, breathing in humans becomes labored. Above 5000 meters, oxygen equipment is needed.
Llamas, vicunas, and Bactrian camels are well adapted to high altitude life. Vicunas and llamas, which live in the Andes, have high blood cell counts and their red blood cells live almost twice as long as those in humans. Their hemoglobin also picks up and off-loads oxygen more efficiently than the hemoglobin of most mammals.

4. Describe the main features and functional role of the following parts of the gas exchange system in mammals:

   (a) Trachea: _____

   _____

   (b) Bronchioles: _____

   _____

   (c) Alveoli: _____

   _____

5. The table (right) gives the approximate percentages of respiratory gases in the lungs, and in inhaled and exhaled air. Study the table and then answer the following questions.

GAS	INHALED AIR	AIR IN LUNGS	EXHALED AIR
Oxygen	21%	15%	16%
Carbon dioxide	0.04%	5.5%	3.6%

   (a) Calculate the difference in CO_2 between inhaled and exhaled air and explain where this 'extra' CO_2 comes from:

   _____

   (b) Calculate the difference in oxygen between inhaled and exhaled air: _____

   _____

   (c) Explain why exhaled air has slightly more oxygen and less CO_2 than air in the lungs: _____

   _____

   _____

   _____

6. Why would high red blood cells counts help at high altitude? _____

   _____

   _____

7. How many times greater would the lung capacity of mammal with a body mass of 1000 kg be than a mammal with a body mass of 10 kg?

   _____

©2017 **BIOZONE** International
ISBN: 978-1-927309-65-0
Photocopying Prohibited

74 Ventilation in Mammalian Lungs

Key Idea: Breathing provides a continual supply of air to the lungs to maintain the concentration gradients for gas exchange. Different muscles are used in inspiration and expiration to move air in and out of the lungs.

Breathing (ventilation) provides a continual supply of oxygen-rich air to the lungs and expels air high in carbon dioxide. Together with the cardiovascular system, which transports respiratory gases between the alveolar and the cells of the body, breathing maintains concentration gradients for gas exchange. Breathing is achieved by the action of muscles.

1. Explain the purpose of breathing: _____

_____

_____

2. In general terms, how is breathing achieved?

_____

_____

_____

_____

3. (a) Describe the sequence of events involved in quiet breathing:

_____

_____

_____

_____

_____

(b) What is the essential difference between this and the situation during forced breathing:

_____

_____

_____

4. During inspiration, which muscles are:

(a) Contracting: _____

_____

(b) Relaxed: _____

_____

5. During forced expiration, which muscles are:

(a) Contracting: _____

_____

(b) Relaxed: _____

_____

6. Explain the role of antagonistic muscles in breathing:

_____

_____

_____

Breathing and muscle action

Muscles can only do work by contracting, so they can only perform movement in one direction. To achieve motion in two directions, muscles work as antagonistic pairs. Antagonistic pairs of muscles have opposing actions and create movement when one contracts and the other relaxes. Breathing in humans involves two sets of antagonistic muscles. The external and internal intercostal muscles of the ribcage, and the diaphragm and abdominal muscles.

Inspiration (inhalation or breathing in)

During quiet breathing, inspiration is achieved by increasing the thoracic volume (therefore decreasing the pressure inside the lungs). Air then flows into the lungs in response to the decreased pressure inside the lung. Inspiration is always an active process involving muscle contraction.

① External intercostal muscles contract causing the ribcage to expand and move up. Diaphragm contracts and moves down.

② Thoracic volume increases, lungs expand, and the pressure inside the lungs decreases.

③ Air flows into the lungs in response to the pressure gradient.

Intercostal muscles

Diaphragm contracts and moves down

Expiration (exhalation or breathing out)

In quiet breathing, expiration is a passive process, achieved when the external intercostals and diaphragm relax and thoracic volume decreases. Air flows passively out of the lungs to equalize with the air pressure. In active breathing, muscle contraction is involved in bringing about both inspiration and expiration.

① In **quiet breathing**, external intercostals and diaphragm relax. The elasticity of the lung tissue causes recoil.

In **forced breathing**, the internal intercostals and abdominal muscles contract to compress the thoracic cavity and increase the force of the expiration.

② Thoracic volume decreases and the pressure inside the lungs increases.

③ Air flows passively out of the lungs in response to the pressure gradient.

Diaphragm relaxes and moves up

©2017 **BIOZONE** International
ISBN: 978-1-927309-65-0
Photocopying Prohibited

CONNECT
62
AP1

WEB
74

KNOW

75 Gas Exchange in Birds

Key Idea: In contrast to the tidal system of mammals, air flows in one direction through the avian gas exchange system.

In birds, the finest branches of the bronchi end in open-ended tubelike structures, called **parabronchi**, which allow movement of air in only one direction. The one-way movement of air across the gas exchange surface maximizes the efficiency of oxygen uptake, especially at altitude (e.g. during flight). The lungs of birds are ventilated by a series of air sacs, which act as bellows to pump air through the lungs. Unlike mammals, birds do not have a diaphragm and their breathing movements rely on rocking motion of the sternum to create local areas of reduced pressure to supply the air sacs.

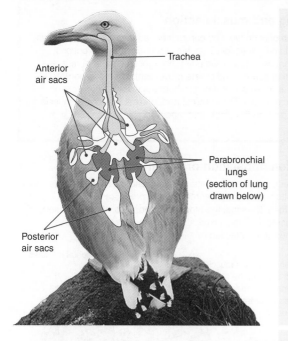

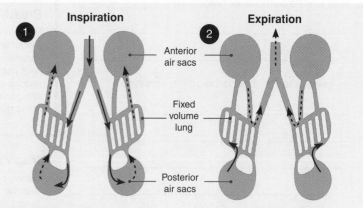

During inspiration, the sternum moves forward and down, expanding the air sacs and lowering the pressure so that air moves into them. Air from the trachea and bronchi moves into the posterior air sacs at the same time as air moves from the lungs into the anterior air sacs.

During expiration, the sternum moves back and up, reducing the volume of the air sacs and causing air to move out. Air from the posterior sacs moves into the lungs at the same time as air from the anterior sacs moves into the trachea and out of the body.

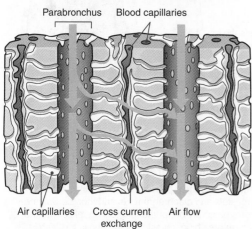

The air in a bird's lungs moves in one direction through **parabronchi**. The parabronchi are connected by tiny tubules called **air capillaries**. Blood capillaries flow past these air capillaries in a **cross current exchange**.

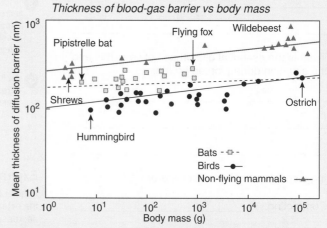

Thickness of blood-gas barrier vs body mass

To meet the high oxygen demands of flight muscles, birds not only have efficient parabronchial lungs but also a thinner respiratory (blood-air) membrane. A thinner diffusion barrier ensures a faster rate of gas exchange relative to mammals of comparable size and metabolic rate (above).

1. Describe how the flow of air in bird lungs differs from that in mammalian lungs: _____
   _____

2. State how many full respiratory cycles are required to move air through its complete path in a bird lung: _____

3. (a) Explain how birds achieve high gas exchange efficiencies: _____
   _____
   _____

   (b) Explain the adaptive value of this to birds in terms of their lifestyles: _____
   _____
   _____

©2017 **BIOZONE** International
ISBN: 978-1-927309-65-0
Photocopying Prohibited

76 Adaptations of Diving Mammals

Key Idea: Being able to dive efficiently requires the ability to store oxygen in muscle (not the lungs) and prevent nitrogen from entering the blood and tissues.

All air breathing animals that dive must cope with the problem of oxygen supply to the tissues. This is particularly a problem for mammals because of their high metabolic rate and high oxygen demand. In addition, resurfacing from deep dives

of 20 meters or more carries the risk of decompression sickness (commonly called **the bends**). Primates (including humans) are one of the few orders of mammals without diving representatives. The problems that humans encounter when diving while breathing compressed air (e.g. the bends) are the result of continuing to breathe during the dive. Animals adapted for diving do not do this.

Diving mammals (e.g. dolphins, whales, seals) are among the most well adapted divers. They store oxygen in muscle myoglobin and, before diving, exhale any air left in the lungs. At depth, the lungs are compressed and only the trachea contains air. This stops nitrogen entering the blood (nitrogen causes problems by coming out of solution in the tissues when surfacing from dives). During dives, heart rate slows and blood flow is redistributed to supply critical organs. Diving mammals have high levels of muscle myoglobin and their muscles also function well anaerobically. Sperm whales are the deepest divers (3000 m). Weddell seals dive to 1000 m for 40 minutes or more. During these dives, heart rate drops to 4% of the rate at the surface.

Humans are poorly adapted for diving. They lack adequate body fat for long periods underwater and they inhale before diving. Divers using compressed air (SCUBA) may stay submerged for much longer but must take care when ascending in order to equalize the pressure of their blood gases with those on the surface.

1. (a) Describe an advantage gained from breathing out before diving: _____

   _____

   _____

   (b) Explain how this behavior is different from a human diving (unaided by equipment): _____

   _____

   (c) Describe the adaptive advantage of reducing heart rate during a dive: _____

   _____

2. Remaining submerged for long periods of time requires an ability to maintain oxygen supply to the tissues. This depends on oxygen stores. This table compares the oxygen in different regions of the body during a dive in a small seal and a human (not on scuba). In the spaces, calculate the amount of oxygen (in mL) per kilogram of body weight for both.

Location of oxygen in the body	Seal (30 kg)		Human (70 kg)	
	Amount of oxygen (mL)	Oxygen (mL kg^{-1})	Amount of oxygen (mL)	Oxygen (mL kg^{-1})
Alveolar air	55		720	
Blood	1125		1000	
Muscle	270		240	
Tissue water	100		200	
Total	1550	51.67	2160	30.86

3. (a) Describe the most striking difference between a seal and a human in terms of the oxygen stores during a dive:

   _____

   (b) With respect to diving adaptations, suggest why this is the case: _____

   _____

   _____

©2017 **BIOZONE** International
ISBN: 978-1-927309-65-0
Photocopying Prohibited

PRACTICES CONNECT CONNECT WEB

207 AP1 **35** AP2 **76**

KNOW

77 Countercurrent Exchanges in Thermoregulation

Key Idea: Temperature can be regulated and maintained efficiently by countercurrent exchange systems.

Countercurrent systems are found in many homeostatic systems including in the gills of fish and the kidneys and peripheral circulation of mammals. The operation of these systems depends on the animal's environment. Mammals in cold environments use countercurrent exchange to reduce heat losses to the environment. Those in hot environments may use countercurrent exchange to cool arterial blood supply to the brain during intense activity.

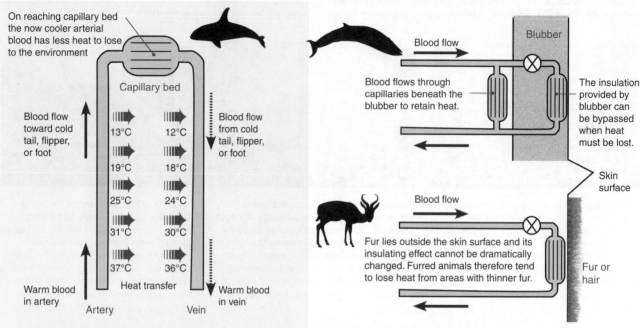

Countercurrent heat exchange systems occur in both aquatic and terrestrial animals as an adaptation to maintaining a stable core temperature. In the flippers and fins of whales and dolphins, and the legs of aquatic birds, they minimize heat loss. In some terrestrial animals adapted to hot climates, the heat exchangers works in the opposite way to prevent the brain from overheating: venous blood cools the arterial blood before it supplies the brain.

Control of blood flow: The blubber in marine mammals provides good insulation against heat loss but presents a problem in warmer waters or during exertion when a lot of metabolic heat is generated. In these situations, blood flows through the blubber to the skin surface where excess heat is dissipated. Cold adapted land mammals have insulation outside the skin and have thinly covered areas on the face and feet, where heat can be lost during exertion.

1. (a) Explain how countercurrent heat exchangers help retain body heat in marine mammals: _____

    _____

    _____

    _____

    _____

    (b) Explain the thermoregulatory changes a marine mammal makes when moving from colder to warmer waters:

    _____

    _____

2. Explain why cold adapted terrestrial mammals have regions of the body with thinner fur: _____

    _____

    _____

3. How would a countercurrent heat exchanger be used in a desert mammal to prevent the brain overheating?

    _____

    _____

    _____

4. In groups, discuss how you might create a physical model of countercurrent heat exchange using flexible PVC piping and thermochromatic pigment. Draw an annotated diagram of your model and staple it to this page. How well does your model represent the biological system? What are its limitations?

©2017 **BIOZONE** International
ISBN: 978-1-927309-65-0
Photocopying Prohibited

78 Single and Double Circulatory Systems

Key Idea: Closed circulatory systems occur as single or double circuit systems. The single circuit system of fish operates at lower pressure than the double circuit system of mammals.

All vertebrates have closed circulatory systems in which the body's blood flows entirely within blood vessels. Exchanges between the blood and tissues occur by diffusion across thin capillary walls. In fish, blood moves in a single circuit, leaving the gills at low pressure to flow around the body before returning to the heart. In all other vertebrates, there is a double circuit system and blood passes from the heart to the lungs (the pulmonary circuit), returning to the heart before being pumped to the body's tissues (the systemic circuit). This double pump system produces much higher pressure in the systemic circuit than in the pulmonary circuit, preventing fluid accumulation in the lungs and providing the pressure to supply the brain and maintain kidney filtration rates.

Closed, single circulatory systems

▶ In the **single circulatory system** of fish, the blood is pumped from the heart to the gills, then directly to the body.

▶ The blood loses pressure at the gills and flows at low pressure around the body. The low pressure reduces blood flow through the body and thus reduces the rate of oxygen delivery to the body's cells. However most fish have relatively low metabolic rates and this system adequately meets their needs.

▶ Water has much lower oxygen content than air (about 12 parts per million maximum compared with 210,000 ppm in air). This limits the amount of oxygen fish can extract even with efficient gills. The low oxygen content in the water would not support a higher metabolic rate, but because most fish do not use metabolism to maintain body temperature (a large energy cost), a lower metabolic rate still allows a relatively active lifestyle.

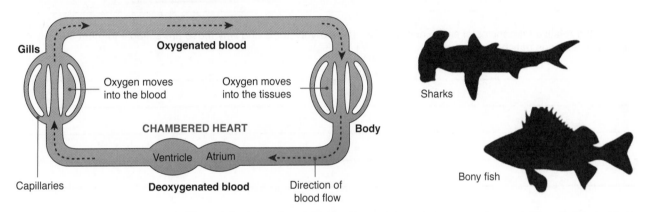

Gills · Oxygenated blood · Oxygen moves into the blood · Oxygen moves into the tissues · Body · CHAMBERED HEART · Ventricle · Atrium · Capillaries · Deoxygenated blood · Direction of blood flow · Sharks · Bony fish

Closed, double circulatory systems

▶ Because oxygen is relatively abundant in the air, metabolic rates in air breathing animals can be relatively high (although they are not necessarily so). **Double circulatory systems** develop higher pressure than single circuit systems, delivering oxygenated blood to the body at a rate sufficient to meet higher metabolic demands.

▶ Double circulatory systems occur in all vertebrates other than fish. They are most efficient in mammals and birds where the heart is fully divided into two halves and the two circuits are completely divided. These animals rely on metabolism to maintain body temperature, so their metabolic demands are necessarily high.

▶ Double circulatory systems have two distinct circuits, the **pulmonary circuit**, which circulates blood between the lungs and the heart, and the **systemic circuit**, which pumps oxygenated blood to the rest of the body. The return of oxygenated blood from the lungs to the heart means that the blood can be pumped to the rest of the body at the higher pressures needed to supply organs and maintain kidney filtration rates, while the blood in the lungs (the pulmonary circuit) remains at a low pressure, suitable for facilitating gas exchange.

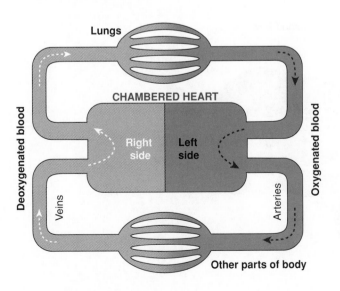

Lungs · CHAMBERED HEART · Right side · Left side · Deoxygenated blood · Oxygenated blood · Veins · Arteries · Other parts of body

Mammals · Birds · Amphibians · Reptiles

Double systems are also found in birds, amphibians, and reptiles. Birds, like mammals, use metabolism to maintain body temperature (a high energy cost) and maintain high metabolic rates.

CONNECT **29** AP2 · WEB **78**

KNOW

1. Compare and contrast the basic structure of the heart and circulatory system in a fish and a mammal: _____

_____

_____

_____

2. Describe where the blood flows to immediately after it has passed through the gills in a fish: _____

_____

3. Describe where the blood flows immediately after it has passed through the lungs in a mammal: _____

_____

4. Explain why the blood in the body of a fish is at a lower pressure than the blood in the body of a mammal: _____

_____

_____

_____

5. Discuss the relative efficiencies of single and double circulatory systems and why these are suitable to meet the metabolic demands of fish and mammals:

_____

_____

_____

_____

_____

_____

_____

_____

_____

_____

_____

_____

_____

6. (a) Label the diagram of the fish circulatory system below with the following labels: *heart, capillaries in gills, capillaries in body, low pressure blood, high pressure blood.*

   (b) Draw arrows on the diagram to show the general direction of blood flow:

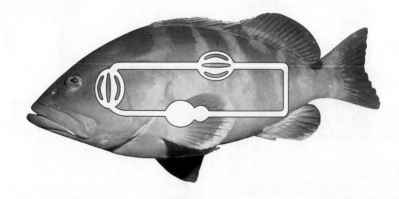

©2017 **BIOZONE** International
ISBN: 978-1-927309-65-0
Photocopying Prohibited

79 The Comparative Anatomy of the Heart

Key Idea: In vertebrates, the heart shows a sequential increase in complexity from fish through to mammals.

In fish, the heart is linear and contains two major chambers in series. On the venous side there is an enlarged chamber or **sinus** on the vein (the sinus venosus), which collects blood before it enters the atrium. In mammals, the heart comprises four chambers (two pumps side by side) and there are large pressure differences between the pulmonary (lung) and systemic (body) circulations. The three chambered heart of amphibians reflects, in part, their incomplete shift to terrestrial life. Although the ventricle is undivided, a baffle-like spiral valve at the exit point of the ventricle helps to separate the arterial and venous flows and there is limited mixing of oxygenated and deoxygenated blood. The pulmonary circuit in amphibians also sends branches to the skin, reflecting its importance in oxygen uptake.

Fish heart

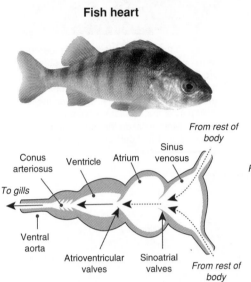

The fish heart is linear, with the chambers in series. There are two main chambers (atrium and ventricle) as well as an entry (the sinus venosus) and sometimes a smaller exit chamber (the conus). Blood from the body first enters the heart through the sinus venosus, then passes through the chambers in series. One-way valves between the chambers prevents reverse blood flow. Blood leaving the heart travels to the gills.

Amphibian heart

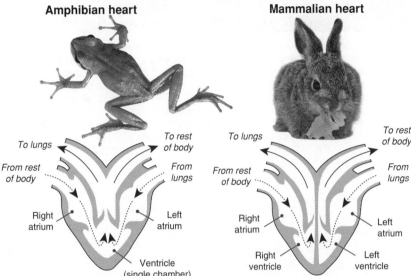

Amphibian hearts are three chambered. The atrium is divided into left and right chambers, but the ventricle lacks an internal dividing wall. Although this allows mixing of oxygenated and deoxygenated blood, the spongy nature of the ventricle reduces mixing. Amphibians are able to tolerate this because much of their oxygen uptake occurs across their moist skin, and not their lungs.

Mammalian heart

In birds and mammals, the heart is fully partitioned into two halves, resulting in four chambers. Blood circulates through two circuits, with no mixing of the two. Oxygenated blood from the lungs is kept separated from the deoxygenated blood returning from the rest of the body.

Heart size and rate in mammals

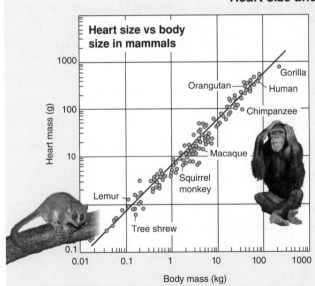

Heart size in mammals (above) increases with body size, but relative to body size, small and large mammals have the same heart size. Irrespective of body size, the size of the heart in mammals is 0.59% of body size. Humans and other primates fall within the range of other mammals.

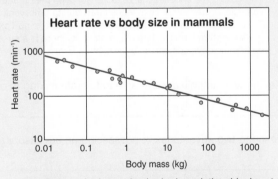

In contrast to the heart size-body size relationship, heart rate (above) is inversely related to body size. An elephant (3000 kg) has a resting pulse of around 25 beats per minute, compared with a 3 g shrew (the smallest mammal), which has a resting heart rate of over 600 beats per minute. The relationship is identical to that between body mass and oxygen consumption per unit body weight. The information from these two figures tell us that, in mammals:

1. The size of the heart (the pump) remains a constant percentage of body size, and…
2. The increase in heart rate in smaller mammals is in exact proportion to the need for oxygen.

PRACTICES   CONNECT   WEB
    102 AP2   79

KNOW

Evolution of the heart and circulatory systems in vertebrates

The evolution of vertebrate heart structure is a response to an enlarged body and an active lifestyle. Modifications to it reflect the major changes in lifestyle of the vertebrates, most importantly from living in water to living on land.

Swim bladder of a bony fish (rudd)

Lungfish

Uwe Gille CC 3.0

Amphibian (frog)

The evolution of accessory air breathing organs was a major step in the evolution of the heart. Outgrowths from the stomachs of fishes provided a way to adjust buoyancy and increase oxygen supply in poorly oxygenated waters. In fish, these outgrowths evolved into the **swim bladder**. In ancestral amphibians, they evolved into lungs. **Lungfish** show a link between double and single circulatory systems. Blood from the lungs follows a partially separated circuit. Modifications to the heart in **amphibians** produced three chambers in which blood from both the lungs and the body enter the one ventricle. In early reptiles, further modifications produced incomplete division within the ventricle.

Birds and mammals have four chambered hearts with a completely divided double circulatory system. The arrangement of their circulatory systems reflects their separate ancestries. Birds have lost what was the left hand part of the circulatory system in early reptiles. Mammals, on the other hand, have lost the right hand part of the system (below).

Basic heart and circulatory patterns in vertebrates

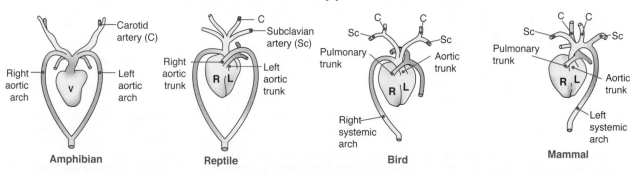

Amphibian — Carotid artery (C), Right aortic arch, Left aortic arch, V

Reptile — C, Subclavian artery (Sc), Right aortic trunk, Left aortic trunk, R, L

Bird — Sc, C, C, Sc, Pulmonary trunk, Aortic trunk, R, L, Right systemic arch

Mammal — Sc, C, C, Sc, Pulmonary trunk, R, L, Aortic trunk, Left systemic arch

▶ For an ancestral vertebrate, having a supplementary organ to provide air assisted survival in poorly oxygenated environments (such as swamps). During the evolution of lungs, the circulatory system began to take more blood through the lungs and less through the gills, which were eventually lost. The evolution of a double circulatory system, formed by separating the ventricle, enabled much more efficient extraction and delivery of oxygen. This evolutionary step was a response to selection pressure on birds and mammals to meet the oxygen demands of highly active lifestyles.

▶ The Tbx5 gene encodes the Tbx5 transcription factor, and has a role in the development of the **septum**, the wall that divides the ventricle. A gradient of Tbx5 in the heart during embryonic development is needed for the septum to form. In embryonic lizards, Tbx5 is expressed evenly across the heart, there is no Tbx5 gradient, so no septum forms. Turtles have a weak gradient so a partial septum forms. In bird and mammals, there is a strong gradient between the left and right sides of the ventricle during embryonic development, so a full septum forms.

1. Explain why amphibians are able to tolerate mixing of oxygenated and deoxygenated blood in the heart:

_____

2. Describe two advantages of possessing an accessory air breathing organ for a vertebrate living in an ancient swamp:

_____

_____

3. How is the heart mass: body mass relationship different to the heart rate: body mass relationship:

_____

_____

_____

4. What is the significance of the Tbx5 gene in heart development? _____

_____

_____

80 Obtaining Food

Key Idea: Animals are heterotrophs, displaying a wide range of feeding modes and adaptations to exploit different diets. All animals are heterotrophs, meaning they feed on other organisms (dead or alive). They display a wide range of feeding modes and show structural and behavioral adaptations (inherited specializations) related to these feeding modes. Animals may feed on solid or fluid food and may suck, bite, lap, or swallow it whole.

Bulk feeding

Bulk feeding is the most common mode of feeding, especially in large animals.
- It involves feeding on large food masses which are ingested whole or in pieces after chewing.
- Bulk feeders occur across all phyla and may be carnivores, herbivores, or omnivores.
- Adaptations for bulk feeding include mouthparts for cutting, tearing, or chewing. These include jaws and teeth in mammals, beaks in birds, and mandibles in insects.
- Teeth in mammals are variously specialized for holding and shearing (as in carnivores) or nipping and grinding (as in herbivores).

Filter feeding

Basking shark · Mysid shrimp

Filter feeders remove food that is suspended in the water using specialized filtering structures. Filter feeding can be an extremely effective way of feeding.
- Krill and many other crustaceans are filter feeders. Krill are vital in marine food chains, providing biomass that can be utilized by larger animals, including the world's largest filter feeders, the baleen whales and filter feeding sharks such as the whale shark and basking shark.
- Adaptations for filter feeding include the flagellated chambers of sponges, the filter baskets of crustaceans (above right), the gills of bivalve molluscs, the comb-like plates (baleen) suspended from the upper jaw of baleen whales, and the modified gill rakers of filter feeding fish.

Fluid feeding

Fluid feeding is an uncommon form of feeding. However, a number of insect species (e.g. flies, moths, butterflies, aphids), spiders (right), and some fish and mammals are fluid feeders.

- Fluid feeding involves obtaining nutrients by sucking or lapping the fluids of another organism. These fluids include blood or body fluids, plant sap, or nectar.
- Adaptations for fluid feeding include piercing mouthparts (e.g. mosquitoes) or small sharp teeth to remove skin (e.g. vampire bats).
- Spiders are specialized fluid feeders, and have chelicerae with fangs, but lack jaws. They inject their prey with venom and enzymes, and feed on the liquefied tissues.

Deposit feeding

Deposit feeders sift the substrate (e.g. mud, silt) and remove the food particles it contains. Deposit feeding works best in areas of fertile sediment such as rich soils or the mud of estuaries or river beds.

- Many invertebrates, but relatively few insects, are deposit feeders. Typical deposit feeders include millipedes, springtails, woodlice, slugs, many terrestrial worms, sea stars, sea cucumbers, and some crabs.
- Deposit feeders such as earthworms have an important role in turning over and conditioning the soil.
- Fish such as mullet, catfish, and carp suck up mud and ingest the plant and animal material in it, expelling the inorganic material back out their mouths.

Vampire bats

Striped catfish feeding

CONNECT **40** AP2   CONNECT **29** AP2   WEB **80**

KNOW

1. Describe one **structural** adaptation for obtaining food in each of the following:

   (a) A blood sucking mosquito: _____

   _____

   (b) A filter feeding whale: _____

   _____

   (c) A mammalian herbivore: _____

   _____

   (d) A leaf chewing caterpillar: _____

   _____

   (e) A spider: _____

   _____

2. Compare and contrast the types of adaptations for feeding in a bulk feeder, a filter feeder, and a fluid feeder, including examples to illustrate your answer from insects, fish, and mammals:

   _____

   _____

   _____

   _____

   _____

   _____

   _____

   _____

   _____

   _____

   _____

   _____

   _____

   _____

   _____

   _____

   _____

   _____

3. Carp are a deposit feeding fish. In many places where they have been introduced they are responsible for the deterioration of water quality in some lakes and rivers. Based on their method of feeding, suggest why carp have such a negative effect on the environment:

   _____

   _____

   _____

   _____

©2017 **BIOZONE** International
ISBN: 978-1-927309-65-0
Photocopying Prohibited

81 Food Vacuoles and Simple Guts

Key Idea: Intracellular digestion occurs within food vacuoles. Extracellular digestion occurs outside the cells by enzymes. The simplest form of digestion occurs inside cells (intracellularly) within food vacuoles. This process is slow and digestion is exclusively intracellular only in protozoa and sponges. In animals with simple, sac-like guts, digestion begins extracellularly (with secretion of enzymes) and is completed intracellularly.

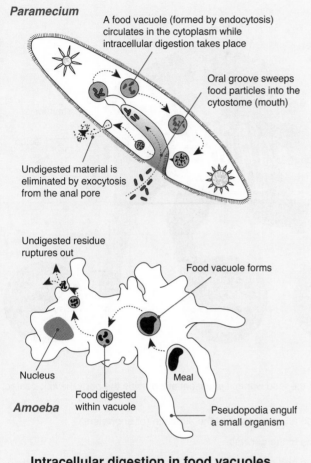

Paramecium

A food vacuole (formed by endocytosis) circulates in the cytoplasm while intracellular digestion takes place

Oral groove sweeps food particles into the cytostome (mouth)

Undigested material is eliminated by exocytosis from the anal pore

Undigested residue ruptures out

Food vacuole forms

Nucleus

Meal

Amoeba

Food digested within vacuole

Pseudopodia engulf a small organism

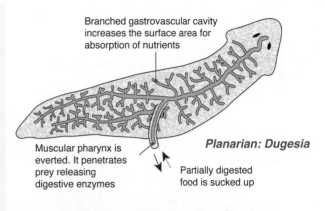

Branched gastrovascular cavity increases the surface area for absorption of nutrients

Planarian: Dugesia

Muscular pharynx is everted. It penetrates prey releasing digestive enzymes

Partially digested food is sucked up

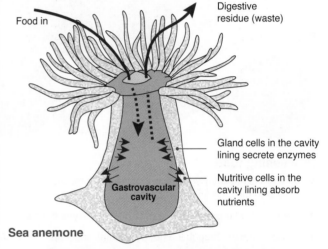

Digestive residue (waste)

Food in

Gland cells in the cavity lining secrete enzymes

Nutritive cells in the cavity lining absorb nutrients

Gastrovascular cavity

Sea anemone

Intracellular digestion in food vacuoles

EXAMPLES: *Protozoans (above), sponges*

The simplest digestive compartments are food vacuoles: organelles where a single cell can digest its food without the digestive enzymes mixing with the cell's own cytoplasm. Sponges and protozoans (e.g. *Paramecium* and *Amoeba*) digest food in this way. *Paramecium* sweeps food into a food groove, from where vacuoles form. *Amoeba* engulfs food using cytoplasmic extensions called pseudopodia. Digestion is intracellular, occurring within the cell itself.

Digestion in a gastrovascular cavity

EXAMPLES: *Cnidarians (above), flatworms (top)*

Some of the simplest animals have a digestive sac or gastrovascular cavity with a single opening through which food enters and digested waste passes out. In organisms with this system, digestion is both extra- and intracellular. Digestion begins (using secreted enzymes) either in the cavity (in cnidarians) or outside it (flatworms). In both these groups, the digestion process is completed intracellularly within the vacuoles in cells.

1. Describe two ways in which simple sac-like gastrovascular cavities differ from tubelike guts:

(a) _____

(b) _____

2. (a) Distinguish between intracellular and extracellular digestion: _____

_____

(b) Why is intracellular digestion unsuitable as the only means of digestion for most animals? _____

_____

_____

3. What is the main difference between extracellular digestion in sea anemones and *Dugesia*? _____

_____

©2017 **BIOZONE** International
ISBN: 978-1-927309-65-0
Photocopying Prohibited

WEB

81

KNOW

82 Diversity in Tube Guts

Key Idea: The tube guts of animals have a similar basic plan, but have specific adaptations related to taxon as well as diet. Most animals have digestive tubes running between two openings, a mouth and an anus. One-way movement of food allows the gut to become regionally specialized for processing food. Tube guts are quite uniform in general structure, with regions for storing, digesting, absorbing, and eliminating the food. Specific adaptations are related to diet and mode of feeding, as well as the taxonomic group to which the animal belongs. The processes in a tube gut are described below.

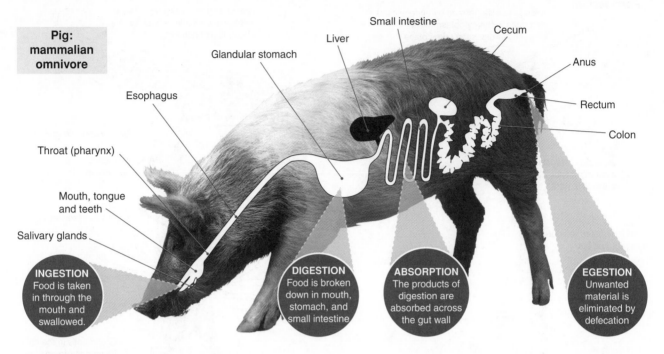

Pig: mammalian omnivore

INGESTION
Food is taken in through the mouth and swallowed.

DIGESTION
Food is broken down in mouth, stomach, and small intestine.

ABSORPTION
The products of digestion are absorbed across the gut wall

EGESTION
Unwanted material is eliminated by defecation

Processing food

▶ The gut is essentially a hollow, open-ended, muscular tube, and the food within it is essentially outside the body, having contact only with the cells lining the tract.

▶ Usually, food **ingested** at the mouth and pharynx passes through an esophagus to the stomach (or equivalent).

▶ Food is moved through the gut by rhythmic muscular contractions of the gut wall.

▶ During **digestion**, food is first physically broken down by chewing and mixed by the muscular activity of the stomach. It is then broken down in stages by enzymes contained within digestive secretions. This is called chemical digestion.

▶ The products of chemical digestion are then **absorbed** across the gut wall. Absorbed molecules can then be assimilated (taken up by the cells). The liver is a large organ associated with the gut but not part of it. Like the pancreas, it contributes digestive secretions but has other non-digestive functions as well.

▶ Undigested wastes are **egested** through an anus (or cloaca in most non-mammalian vertebrates except fish).

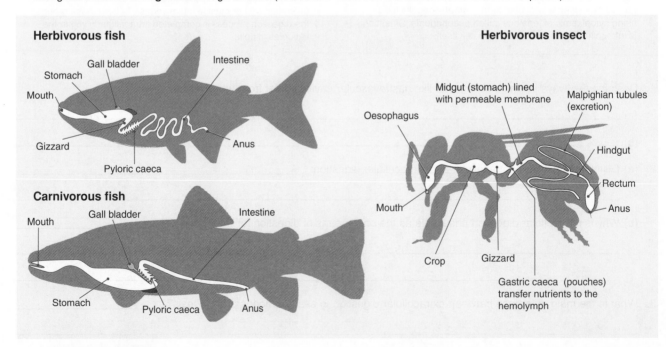

Herbivorous fish

Carnivorous fish

Herbivorous insect

©2017 **BIOZONE** International
ISBN: 978-1-927309-65-0
Photocopying Prohibited

Recognizing organs in a dissection of a rat

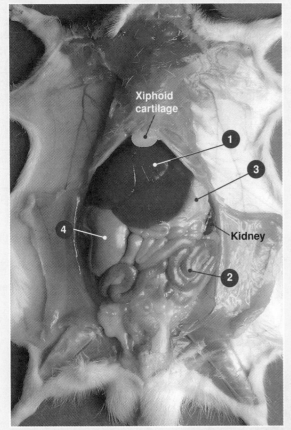 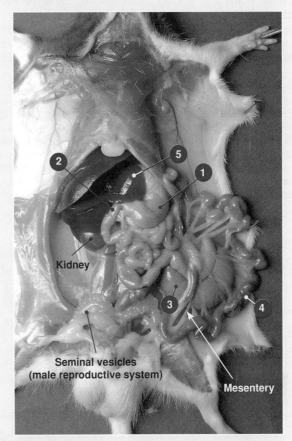

A: Undisturbed organs in the abdominal cavity   **B: Abdominal organs partially dissected**

Dissections are often required when studying animal systems. The images above show two stages of a rat dissection. Use the information provided in this activity to help you identify the structures indicated. Most, but not all, are digestive organs.

1. In the dissection of the rat (above), label each of the structures indicated in the spaces provided (photo **A: 1-4**, photo **B: 1-5**). Some structures are the same, but the same numbers may not indicate the same structure. *List of structures: liver, large intestine (colon), jejunum (second part of small intestine), duodenum (first part of small intestine), stomach.*

A. 1 _____   B. 1 _____

2 _____   2 _____

3 _____   3 _____

4 _____   4 _____

5 _____

2. Some structures have a similar function in different animals. What is the general function of the following gut structures?

(a) Gizzard: _____

(b) Stomach or crop: _____

(c) Intestine (midgut in insects): _____

3. What is the most obvious difference between the gut of a herbivorous fish and the gut of a carnivorous fish?

_____

_____

4. Describe an advantage of having a one way digestive system: _____

_____

83 Cellulose Digestion in a Ruminant

Key Idea: Ruminants rely on bacteria in the rumen of the stomach to break down cellulose.

Mammals lack the enzymes needed to break the bonds joining the glucose molecules in cellulose. Herbivorous mammals are able to digest cellulose by making use of mutualistic microorganisms in their gut. Ruminants, such as cattle, goats, and sheep, rely on microbes in the rumen to break down the cellulose, regurgitating, rechewing, and reswallowing the partially digested material so that digestion can continue.

Grass is a lower energy diet than meat, because mammals lack the enzymes to break the bonds joining the glucose molecules in cellulose. However, many herbivores are able to digest cellulose by making use of **mutualistic** microorganisms in their gut. Ruminants, such as cattle, goats, and sheep, rely on microbes in the **rumen** (a large chamber of the stomach) to break down the cellulose. The food is partially digested by the microbes, and is then regurgitated, rechewed, and reswallowed so that the microbes can complete their digestion of the material. The digested slurry then passes to the rest of the gut.

Region of stomach	Functional role	% of total volume
Rumen	Microbial fermentation of cellulose to produce volatile fatty acids (VFAs)	~84%
Reticulum		
Omasum	Removal of water	12%
Abomasum	Gastric digestion	4%

The rumen and reticulum act as a single chamber and occupy more than 80% of the volume of the stomach.

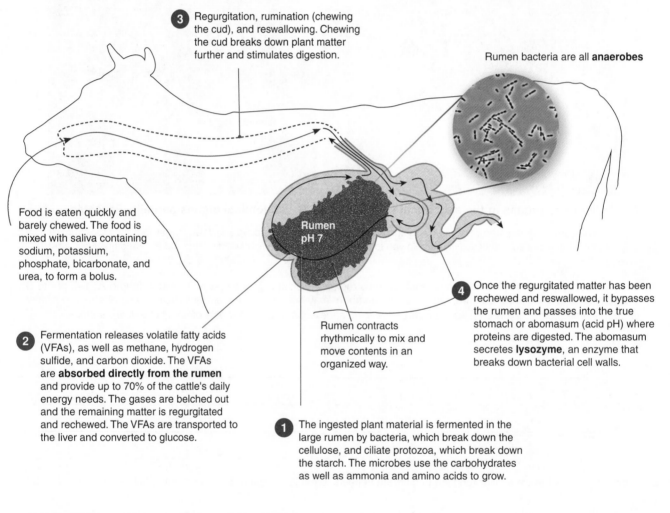

③ Regurgitation, rumination (chewing the cud), and reswallowing. Chewing the cud breaks down plant matter further and stimulates digestion.

Rumen bacteria are all **anaerobes**

Food is eaten quickly and barely chewed. The food is mixed with saliva containing sodium, potassium, phosphate, bicarbonate, and urea, to form a bolus.

Rumen pH 7

② Fermentation releases volatile fatty acids (VFAs), as well as methane, hydrogen sulfide, and carbon dioxide. The VFAs **are absorbed directly from the rumen** and provide up to 70% of the cattle's daily energy needs. The gases are belched out and the remaining matter is regurgitated and rechewed. The VFAs are transported to the liver and converted to glucose.

Rumen contracts rhythmically to mix and move contents in an organized way.

④ Once the regurgitated matter has been rechewed and reswallowed, it bypasses the rumen and passes into the true stomach or abomasum (acid pH) where proteins are digested. The abomasum secretes **lysozyme**, an enzyme that breaks down bacterial cell walls.

① The ingested plant material is fermented in the large rumen by bacteria, which break down the cellulose, and ciliate protozoa, which break down the starch. The microbes use the carbohydrates as well as ammonia and amino acids to grow.

1. What relationship do ruminants form with their bacteria? _____

2. How does this relationship enable the ruminant to meet its energy needs? _____

_____

_____

3. What is the purpose of chewing the cud? _____

_____

4. Suggest why the rumen pH is 7, whereas the abomasum (true stomach) pH is 2? _____

_____

©2017 **BIOZONE** International
ISBN: 978-1-927309-65-0
Photocopying Prohibited

84 Cellulose Digestion in a Hindgut Fermenter

Key Idea: In hindgut fermenters, microbial breakdown of cellulose occurs in the large intestine.
Hindgut fermenters, such as horses and rodents, house their cellulose-digesting microorganisms in the hindgut, rather than the foregut like ruminants. Rabbits have a further strategy, called **cecotrophy**, which enhances their nutritional gain from the food ingested. In cecotrophy, some feces are reingested to be processed for a second time.

▶ Unlike ruminants, **hindgut fermenters**, such as horses and rabbits, house their cellulose-digesting microorganisms in the hindgut, rather than in a foregut rumen. As in ruminants, the microbes in this mutualistic relationship benefit by obtaining food (carbohydrate) and are provided with a suitable environment for growth and reproduction.

▶ Rabbits are small active mammals. Like all hindgut fermenters, they require a high throughput of material, so they eat often. Rabbits increase the nutritional gain from their browse by eating soft, nutrient-rich fecal pellets. This behavior is called **cecotrophy** and it enables the rabbit to meet its requirements for vitamins and protein. Rabbits also produce hard pellets which are not eaten.

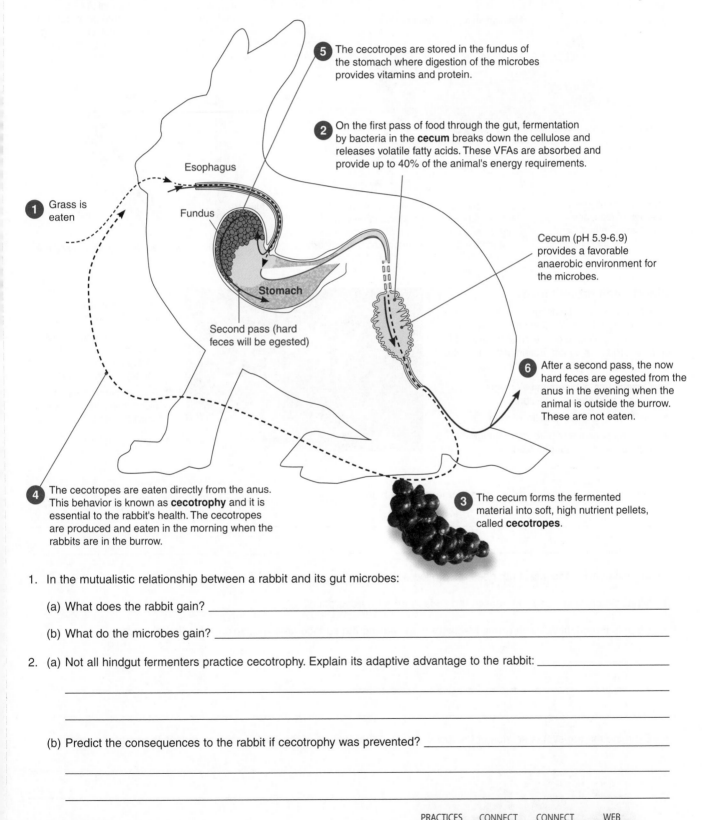

5 The cecotropes are stored in the fundus of the stomach where digestion of the microbes provides vitamins and protein.

2 On the first pass of food through the gut, fermentation by bacteria in the **cecum** breaks down the cellulose and releases volatile fatty acids. These VFAs are absorbed and provide up to 40% of the animal's energy requirements.

Esophagus

1 Grass is eaten

Fundus

Cecum (pH 5.9-6.9) provides a favorable anaerobic environment for the microbes.

Stomach

Second pass (hard feces will be egested)

6 After a second pass, the now hard feces are egested from the anus in the evening when the animal is outside the burrow. These are not eaten.

4 The cecotropes are eaten directly from the anus. This behavior is known as **cecotrophy** and it is essential to the rabbit's health. The cecotropes are produced and eaten in the morning when the rabbits are in the burrow.

3 The cecum forms the fermented material into soft, high nutrient pellets, called **cecotropes**.

1. In the mutualistic relationship between a rabbit and its gut microbes:

   (a) What does the rabbit gain? _____

   (b) What do the microbes gain? _____

2. (a) Not all hindgut fermenters practice cecotrophy. Explain its adaptive advantage to the rabbit: _____

   _____

   _____

   (b) Predict the consequences to the rabbit if cecotrophy was prevented? _____

   _____

   _____

©2017 **BIOZONE** International
ISBN: 978-1-927309-65-0
Photocopying Prohibited

PRACTICES

CONNECT
218
AP2

CONNECT
161
AP2

WEB
84

KNOW

85 Digestion in Insects

Key Idea: In insect guts, nutrient absorption can only occur in the midgut. Gastric ceca increase the midgut's surface area. The small molecules that are produced by the digestion of food must be absorbed across the gut wall before they can be assimilated (taken up by the body's cells). The rate of nutrient transport across any type of gut wall depends partly on its surface area. In insects, pouches from the midgut, called gastric ceca, increase the surface area available for absorption of nutrients, which are taken up by diffusion or active transport. Absorption is only possible in the midgut region because the hindgut and foregut regions are lined with a chitinous cuticle, which is shed each time the animal molts. Within the midgut, the epithelium is protected by a semipermeable peritrophic membrane, which allows the products of digestion to pass through but prevents undigested material clogging the epithelium.

Generalized insect digestive tract

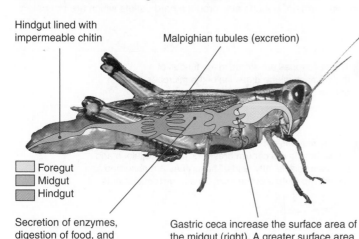

Hindgut lined with impermeable chitin

Malpighian tubules (excretion)

Foregut
Midgut
Hindgut

Secretion of enzymes, digestion of food, and absorption of nutrients occurs in the midgut.

Gastric ceca increase the surface area of the midgut (right). A greater surface area increases the rate of nutrient absorption so that more food can be processed more quickly.

Insect gastric ceca

In grasshoppers, the gastric ceca are midgut pouches just behind the gizzard. The ceca improve absorption by transferring nutrients to the hemolymph ('blood'). The **thin peritrophic** membrane lining the midgut is continually lost, surrounding the fecal pellets when they are passed out. The membrane is replaced by the underlying epithelial cells (as in human skin).

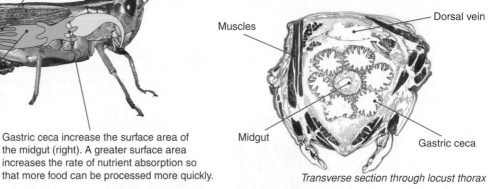

Muscles

Dorsal vein

Midgut

Gastric ceca

Transverse section through locust thorax

Blood (high protein fluid)

Insects that feed on fluids must be able to store large volumes during feeding and remove the water from the fluid to concentrate it before it is digested. Blood is a high protein, low bulk fluid. The problems with processing it include:

▶ Coagulation of blood and blockage of mouthparts during ingestion.

▶ Storage of a large quantity of fluid.

▶ Slowing passage of bulk food through the gut so that it can be digested.

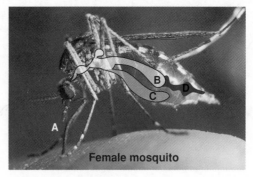

Female mosquito

A Piercing mouthparts inject powerful anticoagulants to keep the blood flowing.

B Three sequential regions in the midgut (1) absorb water (2) secrete protease enzymes, and (3) absorb nutrients.

C Enlarged crop stores the blood, releasing it slowly in small amounts to the midgut.

D Hindgut (egestion)

1. Describe the mosquito's adaptations for feeding on blood: _____

_____

_____

2. In which part of the insect gut does absorption occur? _____

3. What structure in insects increases the surface area for absorption? _____

4. What is the adaptive value (the advantage) of increasing the surface area available for nutrient absorption?

_____

_____

_____

5. Examine the transverse section of the locust thorax. What feature of the gastric ceca increases their surface area?

_____

_____

©2017 **BIOZONE** International
ISBN: 978-1-927309-65-0
Photocopying Prohibited

86 Osmoregulation

Key Idea: Osmoregulators are able to tightly regulate salt and water balance. Osmoconformers match the osmolarity of their environment. All osmoconformers are marine.

Osmoregulation is the process of managing fluid and ion balance to maintain the homeostasis of the body's water content (osmolarity). Osmoconformers are marine organisms that match the osmolarity of their bodily fluids with that of the external environment, although the ionic composition of their body fluids may be different. In contrast, osmoregulators maintain constant water and solute concentrations even when the environmental conditions vary. Osmoregulation thus represents a considerable energy cost to the organism.

Osmoregulators vs osmoconformers

Nhobgood cc 3.0

Anemones are osmoconformers

Osmoconformers

The osmolarity of the body fluids of osmoconformers fluctuates with the osmolarity of the environment, although the composition of their body fluids may be different. Most marine invertebrates are osmoconformers and many rely on a relatively stable external osmotic environment for survival. Others, e.g. some intertidal species, can tolerate frequent dilutions of normal seawater.

Sharks tolerate high tissue urea

Marine sharks are osmoconformers, but ion regulators, and generate osmotic concentrations in their body fluids similar to seawater by tolerating high urea and absorbing seawater directly. Excess salt from the diet is excreted via a salt gland in the rectum. Osmoconformers don't need to expend as much energy as osmoregulators in order to regulate ion gradients but a small amount of energy is still expended on ion transport in order to maintain the correct ion balances for normal function.

Osmoregulators

Animals that regulate their salt and water fluxes independently of the environment, such as fish and marine mammals, are osmoregulators. Marine bony fish lose water osmotically and counter the loss by drinking salt water and excreting the excess salt across the gill surfaces. Marine mammals produce a urine that is high in both salt and urea.

Marine bony fish osmoregulate

Freshwater animals, such as the freshwater crayfish left, have body fluids that are osmotically more concentrated than the water they live in and all are osmoregulators. Water tends to enter their tissues by osmosis and must be expelled to avoid flooding the body. Freshwater crayfish osmoregulate by excreting the excess water.

Osmoregulation in freshwater protists

Paramecium and other freshwater protists live in environments with low solute levels. Water is constantly diffusing into the cell and must be expelled if the cell is not to burst. This is achieved using contractile vacuoles (CV) and active transport. The CV expands as it absorbs water from the cytoplasm. When full, it contracts to force water out of the cell through a pore in the plasma membrane.

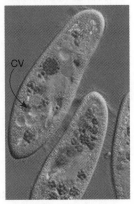

Ion and water fluxes in bacteria

Bacteria have a similar problem to protists in that water tends to enter the cell by osmosis. Unlike freshwater protists, bacteria regulate water fluxes by retaining salts or pumping them out of the cell. They respond to osmotic stress by rapidly accumulating ions or organic solutes via membrane transporters which are stimulated by increases in osmolarity in the environment.

When the solute level is lower outside the cell than inside it, a bacterium will pump salts out of the cell so that water will tend to leave by osmosis (rather than enter).

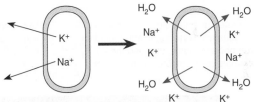

If the solute level is higher outside the cell, a bacterium will retain salts, so that water will tend to enter (rather than leave) the cell.

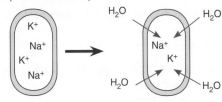

1. (a) What is the main osmoregulatory problem of animals living in freshwater: _____

   (b) Why are all freshwater organisms osmoregulators? _____

   _____

2. (a) How does a freshwater protist get rid of excess water?_____

   (b) Would you expect a marine protist to do this? Explain: _____

3. Explain how a bacterium reduces the amount of water entering the cell: _____

   _____

   _____

CONNECT	CONNECT	CONNECT	CONNECT	WEB
62 AP1	**115** AP2	**40** AP2	**29** AP2	**86**

KNOW

87 Nitrogenous Wastes in Animals

Key Idea: Nitrogenous wastes are produced from the breakdown of nitrogen containing compounds. They must be excreted before they accumulate to toxic levels.

The process of removing the waste products of cellular metabolism is called excretion. These waste products include toxic nitrogenous wastes from the metabolism of amino acids and nucleic acids, as well as water, CO_2, and excess ions. The simplest breakdown product of nitrogen-containing compounds is ammonia, a highly toxic molecule that cannot be retained in the body for long. Most aquatic animals excrete ammonia immediately into the water where it is washed away. Other animals convert the ammonia to a less toxic form (urea or uric acid) that can remain in the body for a short time before being excreted. The form of the excretory product in terrestrial animals depends on the type of organism and its life history. Terrestrial animals that lay eggs produce uric acid rather than urea, because it is non-toxic and very insoluble. It remains as an inert solid mass in the egg until hatching.

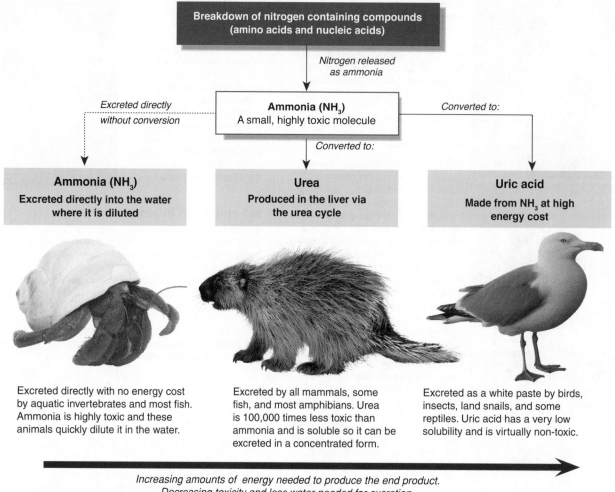

Breakdown of nitrogen containing compounds (amino acids and nucleic acids)

Nitrogen released as ammonia

Ammonia (NH₃)
A small, highly toxic molecule

Excreted directly without conversion

Converted to:

Converted to:

Ammonia (NH₃)
Excreted directly into the water where it is diluted

Urea
Produced in the liver via the urea cycle

Uric acid
Made from NH₃ at high energy cost

Excreted directly with no energy cost by aquatic invertebrates and most fish. Ammonia is highly toxic and these animals quickly dilute it in the water.

Excreted by all mammals, some fish, and most amphibians. Urea is 100,000 times less toxic than ammonia and is soluble so it can be excreted in a concentrated form.

Excreted as a white paste by birds, insects, land snails, and some reptiles. Uric acid has a very low solubility and is virtually non-toxic.

Increasing amounts of energy needed to produce the end product.
Decreasing toxicity and less water needed for excretion.

1. What is the main source of nitrogen-containing wastes in animals? _____

_____

2. (a) Describe one advantage of uric acid as an excretory product (relative to urea and ammonia): _____

_____

    (b) Describe one disadvantage of ammonia as an excretory product: _____

_____

3. Discuss the relationship between the type of excretory product and an animal's environment: _____

_____

_____

_____

_____

_____

WEB
87

CONNECT
89
AP2

CONNECT
111
AP2

PRACTICES

KNOW

©2017 **BIOZONE** International
ISBN: 978-1-927309-65-0
Photocopying Prohibited

88 Osmoregulation and Excretion in Fish

Key Idea: Fish must cope with excess water loss or gain from the environment. In fish, most excretion occurs via the gills. Fish face different osmoregulatory problems depending on their environment. Fish kidneys cannot produce a urine that is more concentrated than the body fluids and nearly all their nitrogenous waste is excreted via diffusion across the gills, which also have an important role in salt balance. In freshwater fish, excess water is lost in copious amounts of dilute urine. In marine fish, the kidneys produce a scanty urine with the same osmolarity as the blood.

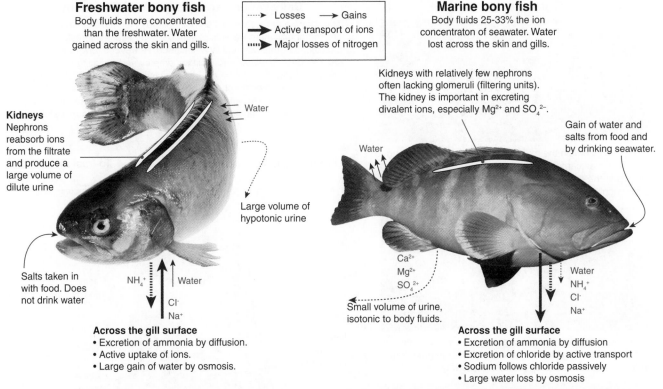

Freshwater bony fish
Body fluids more concentrated than the freshwater. Water gained across the skin and gills.

- - - → Losses ⟶ Gains
⟹ Active transport of ions
⟫⟫⟫ Major losses of nitrogen

Marine bony fish
Body fluids 25-33% the ion concentraton of seawater. Water lost across the skin and gills.

Kidneys
Nephrons reabsorb ions from the filtrate and produce a large volume of dilute urine

Water

Large volume of hypotonic urine

Salts taken in with food. Does not drink water

NH_4^+ Water

Cl^-
Na^+

Across the gill surface
- Excretion of ammonia by diffusion.
- Active uptake of ions.
- Large gain of water by osmosis.

Kidneys with relatively few nephrons often lacking glomeruli (filtering units). The kidney is important in excreting divalent ions, especially Mg^{2+} and SO_4^{2-}.

Gain of water and salts from food and by drinking seawater.

Water

Ca^{2+}
Mg^{2+}
SO_4^{2+}

Small volume of urine, isotonic to body fluids.

Water
NH_4^+
Cl^-
Na^+

Across the gill surface
- Excretion of ammonia by diffusion
- Excretion of chloride by active transport
- Sodium follows chloride passively
- Large water loss by osmosis

Fish in freshwater must excrete **excess water** gained through osmosis and they must excrete **nitrogenous waste**. Their kidneys excrete large amounts of dilute urine; valuable ions are lost because of the large urine volumes produced. The kidneys reabsorb salts from the filtrate through active transport mechanisms, and the gills take up ions from the water.

Marine fish must excrete **excess salt** gained through diet as well as **nitrogenous waste**. The urine is isotonic and excess salts are actively excreted across the gill surface into the water (against a concentration gradient). Note that, unlike bony fish, sharks and rays tolerate high urea levels in their tissues and excrete excess salts via a salt gland in the rectum.

1. Describe the contrasting problems of excretion and osmoregulation for bony fish in fresh and salt water environments:

_____

_____

_____

_____

2. What is the primary organ for nitrogen excretion in fish? _____

3. With reference to their ion and water regulation, explain why:

   (a) Marine bony fish must drink vast quantities of salt water: _____

   _____

   _____

   _____

   (b) Freshwater bony fish do not drink water at all: _____

   _____

   _____

   _____

©2017 **BIOZONE** International
ISBN: 978-1-927309-65-0
Photocopying Prohibited

PRACTICES   CONNECT **71** AP1   CONNECT **62** AP1   CONNECT **40** AP2   WEB **88**

KNOW

89 Managing Fluid Balance on Land

Key Idea: Terrestrial animals have adaptations to obtain enough water to maintain their fluid and ion balance, either through drinking or the metabolism of foodstuffs.

All organisms, whether terrestrial or aquatic, must maintain their water and solute concentrations at levels that support their life processes. For animals on land, the main challenges to fluid and ion regulation arise from a dependence on water, which is often in short supply. They show specific adaptations for obtaining and conserving water in an environment where water loss is a constant problem.

Obtaining water

Most animals obtain the majority of their water by **drinking.** Some, such as camels, can retain relatively large volumes of water in the gut, but most will need to regularly visit a water supply.

Obtaining water from **food** is important in dry environments where free standing water is limited. Many large predators obtain a large amount of their water in this way.

Some desert animals, such as the kangaroo rat, do not need to drink. 90% of their water comes from metabolism (oxidation of glucose to ATP, CO_2, and water). The rest comes from the small amount of water present in the food.

Bcexp CC 4.0

Amphibians can take up water directly through the skin, which is water permeable. When they need water, they can acquire it by osmosis while submerged or resting on a damp surface. Desert adapted frogs burrow underground and spread wax over their skin from epidermal wax glands.

Daily water transfers in an adult human

The daily water losses must be balanced by the intake of water. In environments where water is plentiful, e.g. tropical rainforests these balances are relatively easily maintained. However in environments such as deserts, water is difficult to come by and reducing the amount of water lost to the environment is as important as obtaining it. The illustration below shows the typical daily water transfers in a human and the daily water losses and gains. Various secretions in the digestive tract require water, but much of this water is reabsorbed in the gut. The lumen is the space within the digestive tract.

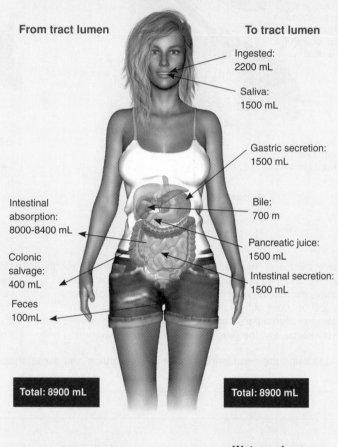

From tract lumen

Intestinal absorption: 8000-8400 mL

Colonic salvage: 400 mL

Feces 100mL

Total: 8900 mL

To tract lumen

Ingested: 2200 mL

Saliva: 1500 mL

Gastric secretion: 1500 mL

Bile: 700 m

Pancreatic juice: 1500 mL

Intestinal secretion: 1500 mL

Total: 8900 mL

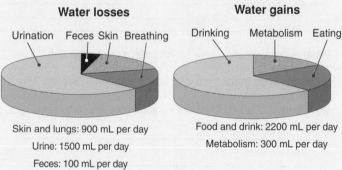

Water losses

Urination   Feces   Skin   Breathing

Skin and lungs: 900 mL per day
Urine: 1500 mL per day
Feces: 100 mL per day

Water gains

Drinking   Metabolism   Eating

Food and drink: 2200 mL per day
Metabolism: 300 mL per day

1. Name four ways in which water can be obtained: _____

   _____

2. How does metabolism provide water for the body's activities? _____

   _____

WEB **89**   CONNECT **40** AP2   CONNECT **87** AP2   PRACTICES

KNOW

©2017 **BIOZONE** International
ISBN: 978-1-927309-65-0
Photocopying Prohibited

General adaptations associated with major routes of water loss in animals (vertebrates and arthropods)	
An insulated body covering	Body coverings reduce but do not totally eliminate water loss. They form a moderate barrier against water loss and may be thickened or covered with insulation or wax to limit water loss.
Using metabolic water	All animals produce water through metabolism. Metabolizing fat yields more water than metabolizing carbohydrate and this can be utilized to maintain water balance when no liquid water is available.
Changing behavior	Behavioral strategies to reduce water losses are common and often associated with physiological adaptations to take up water from humid environment.
Minimizing losses from the excretory system	Water balance and excretion are tightly linked because excretion of nitrogenous wastes represents a major route for water loss. Being able to produce a concentrated urine is a feature of terrestrial life.

Arthropods

Reptiles

Birds

The permeability of the waxy cuticle of desert adapted arthropods is very low, greatly reducing water loss across the body surface. Insects excrete nitrogenous wastes as uric acid with low water loss, and many are able to take up water from the air when humidity increases. Desert darkling beetles (above) can collect water on their hardened forewings, funneling it off their backs to the mouth.

The shared evolutionary history of birds and reptiles is evident in many of their strategies for water conservation. Scales in reptiles and feathers (which are modified scales) in birds reduce water losses across the body surface. Neither animal sweats and although water is lost via the respiratory tract, salt glands help to excrete excess salt with virtually no loss of water. In both, seeking shade through the hottest parts of the day reduces water losses from the respiratory tract and both excrete uric acid with minimal loss of water. Roadrunners have a longer gut than expected for an omnivorous bird and reabsorption of water from food is a major source of water. Desert reptiles too, such as the chuckwalla, obtain most of their water from their food and some are also able to absorb water across the nasal epithelium.

All mammals excrete nitrogenous wastes as urea and lose water by sweating or panting. Arid-adapted mammals show similar adaptations within the constraints of their evolutionary history. All have long loops of Henle in the kidney and produce a very concentrated urine.

Mammals

Camels have further adaptations for life in desert conditions, most directly related to water balance. When they have access to water, a dehydrated camel can drink up to 200 L in 3 minutes. In most animals, this would cause osmotic shock, but the camel's red blood cells can withstand huge fluctuations in body water content. Instead of losing a lot of water through sweat, they tolerate a rise in body temperature (to 41°) during the day and a fall at night (to 34°C). Water is obtained from metabolism of stored fat in the hump and reabsorption from the gut. Even the nostrils, which are convoluted, trap water vapor and return it to body fluids.

3. Identify three ways in which animals lose water to the environment: _____

_____

4. Identify two ways in which each of the following animals conserves water:

 (a) Arthropod: _____

 (b) Amphibian: _____

 (c) Reptile: _____

 (d) Bird: _____

 (e) Mammal: _____

5. What features of fluid and ion homeostasis in reptiles and birds reflects their shared ancestry? _____

_____

_____

6. Compare and contrast the ways in which humans obtain and conserve water with those of a desert adapted mammal:

_____

_____

_____

©2017 **BIOZONE** International
ISBN: 978-1-927309-65-0
Photocopying Prohibited

90 | Adaptations for Conserving Water

Key Idea: Desert mammals have adaptations to conserve water or reduce water intake requirements.

Water loss is a major problem for most mammals. The adaptations of mammals in arid regions enables them to minimize water losses and reduce the amount of water they need to drink. Arid-adapted species typically produce very concentrated urine, so lose very little water this way. In addition, the metabolic breakdown of food contributes a large proportion of daily water needs. Some, like kangaroo rats, do not drink at all, and obtain most of their water metabolically.

Adaptations of small desert mammals

Most desert-dwelling mammals are able to tolerate a low water intake. Arid adapted rodents, such as the kangaroo rats of North America, conserve water by reducing losses to the environment and obtain the balance of their water needs from the oxidation of dry foods (respiratory metabolism). The table below shows the water balance in a kangaroo rat after eating 100 g of dry pearl barley. Note the high urine to plasma concentration ratio (17) which is more than four times that of a human (4).

Water balance in a kangaroo rat
(*Dipodomys spectablis*)

Water gains (cm³)		Water losses (cm³)	
Absorbed from food	6.0	Breathing	43.9
From metabolism	54.0	Urination	13.5
		Defecation	2.6

Urine/plasma concentration ratio = 17

Highly concentrated urine and dry feces

Respiratory moisture is reabsorbed from the nasal passages

Kangaroo rats, and other arid-adapted rodents, tolerate long periods without drinking, meeting their water requirements from the metabolism of dry foods. They dispose of nitrogenous wastes with very little output of water. They do not sweat or pant to keep cool, but are nocturnal and remain underground during the day. Most of their water losses occur through the respiratory tract but some of this condenses in the nasal passages and is reabsorbed when they inhale. The distantly related jerboas of Africa and Asia have similar adaptations, an example of parallel evolution.

Large desert mammals

Florian Prischl cc 3.0

As described in the previous activity, camels are well adapted to desert life. They can tolerate large losses of water (>25% of body weight) without circulatory disturbance because their blood cells flow even when they are dehydrated. In non-desert mammals, losses of ~15% are fatal. The urine is concentrated and the feces are dry enough to be used as fuel. The location of fat in the hump minimizes its heat-trapping effect and metabolism of the fat produces water. The closely related camelids of South America show no similar adaptations.

A suite of adaptations enable kangaroos to exploit the vast arid environment of Australia's semi-deserts. Water is taken in with food but is also derived from metabolism. An elongated large intestine maximizes the removal of water from undigested material, producing very dry feces. The urine is also concentrated, reducing water loss. The hopping mode of locomotion is also very energy efficient, and enables them to cover large distances quickly at relatively low cost in search of food and water.

1. Describe two physiological adaptations shared by desert-dwelling mammals:

    (a) _____

    _____

    (b) _____

    _____

2. Describe the adaptive advantage of being able to meet a large proportion of water needs from metabolism: _____

    _____

3. Suggest why many small desert dwelling animals are nocturnal: _____

    _____

©2017 **BIOZONE** International
ISBN: 978-1-927309-65-0
Photocopying Prohibited

91 Invertebrate Excretory Systems

Key Idea: Invertebrate excretory systems include nephridia and protonephridia, which produce dilute urine by filtration. The simplest excretory organs, found in many primitive invertebrates, are simple ciliated tubes (protonephridia and nephridia) opening to the outside through a pore. Beating cilia create a filtrate by drawing body fluids into tubules. These systems produce a dilute urine so these animals are restricted to environments where water is freely available.

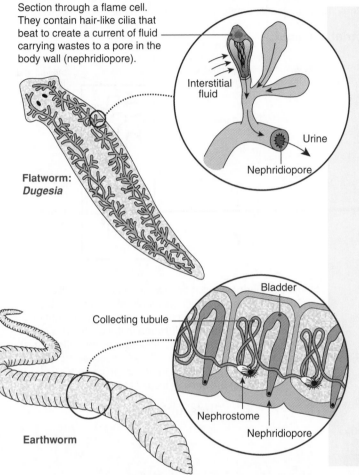

Section through a flame cell. They contain hair-like cilia that beat to create a current of fluid carrying wastes to a pore in the body wall (nephridiopore).

Interstitial fluid

Urine

Nephridiopore

Flatworm: Dugesia

Bladder

Collecting tubule

Nephrostome

Nephridiopore

Earthworm

Platyhelminthes (flatworms)
Excretory system: **protonephridia**

▶ Protonephridia are very simple excretory structures. Each protonephridium consists of a network of dead-end tubules ending in a number of **flame cells**.

▶ Flatworms do not have a circulatory system or fluid-filled inner body spaces. They use their branching network of flame cells to regulate the composition of the fluid bathing the cells (interstitial fluid).

▶ Interstitial fluid enters the flame cell and is propelled along the tubule by beating cilia, which drive wastes through small perforations in the cell and into the tubules. The tubules merge into ducts that expel the urine (as **ammonia**) through **nephridiopores** to the outside. The excretory fluid is very dilute, helping to balance osmotic influx from the moist environment.

Annelids (segmented worms)
Excretory system: **nephridia**

▶ In earthworms, each segment has a pair of excretory organs called **nephridia**, which drain the next segment in front.

▶ Fluid enters the nephrostome and passes through the collecting tubule. These tubules are surrounded by a capillary network of blood vessels (not shown here) which recover valuable salts from the urine as it is forming (this does not happen in protonephridia).

▶ The collecting tubule empties into a storage bladder which expels the dilute urine (a mix of **ammonia** and **urea**) to the outside through the nephridiopore.

1. For each of the following, name the organs for excreting nitrogenous waste and state the form of the waste product:

   (a) Flatworm: _____ Waste: _____

   (b) Earthworm: _____ Waste: _____

2. Compare and contrast the excretory systems of flatworms and earthworms: _____

   _____

   _____

   _____

   _____

   _____

   _____

   _____

3. The fluid that enters the nephridia in earthworms is isotonic with the body fluids. Would you expect the urine to be hypotonic, isotonic, or hypertonic? Explain why:

   _____

   _____

   _____

92 Vertebrate Excretory Systems

Key Idea: Kidneys are the main excretory organs of terrestrial vertebrates.

In vertebrates, the excretory units (**nephrons**) are collected into organs called **kidneys**. The kidneys of most vertebrates are similar in that they produce an excretory fluid called **urine** by filtering the body fluids and then modifying the filtrate by reabsorption and secretion of ions. The kidneys of all vertebrates produce urine, but only birds and mammals can produce a urine that is more concentrated than the body fluids. Mammals and birds have very efficient kidneys that can produce a concentrated urine, excreting nitrogenous wastes whilst conserving water and ions.

Terrestrial mammal

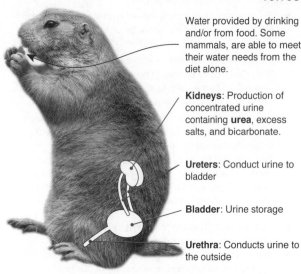

Water provided by drinking and/or from food. Some mammals, are able to meet their water needs from the diet alone.

Kidneys: Production of concentrated urine containing **urea**, excess salts, and bicarbonate.

Ureters: Conduct urine to bladder

Bladder: Urine storage

Urethra: Conducts urine to the outside

Water loss is a major problem for most mammals. The degree to which urine can be concentrated (and water conserved) depends on the number of nephrons present and the length of the loop of Henle. The highest urine concentrations are found in desert-adapted mammals.

Mammalian kidney showing orientation of the nephrons.

Cortex

Medulla

Ureter

Urine flow

Capsule and glomerulus

Tubule

Tubule

Loop of Henle

Collecting duct

Urine flow

Nephron

Mammalian **kidneys** each contain more than one million **nephrons**: the selective filter elements which regulate the composition of the blood and excrete wastes. In the nephron, the initial urine is formed by **filtration** in the glomerulus and Bowman's capsule. The filtrate is modified by **secretion** and **reabsorption** of ions and water. These processes create a salt gradient in the fluid around the nephron, which allows water to be withdrawn from the urine in the collecting duct.

Marine mammals

Marine mammals have a high incidental salt intake with the food they eat. Seawater contains about 35 gL^{-1} of salt (mostly NaCl). For a human to excrete the salt in 1 L of seawater it would require about 1.5 L of urine. Thus humans use more water than they gain from drinking seawater and become more thirsty. In contrast, whales are able to produce urine with a salt concentration of about 44 gL^{-1}, much higher than seawater. Thus they use less water than they gain from drinking seawater. Marine mammals have very large multi-lobed kidneys with a large cortex area.

Birds and reptiles

Birds and reptiles (especially marine birds and reptiles) excrete excess salts via salt glands in the head, rather than the kidneys (which excrete nitrogenous wastes). In birds, the salt glands excrete salt via the nares (nostrils). In turtles, the salt is excreted by glands near the eyes, so that turtles often appear to be weeping. Salt glands are highly efficient. One experiment found that a gull given 10% of its body weight in seawater was able to excrete the salt in around 3 hours. Concentrations of up to 64 gL^{-1} of salt have been recorded in the salt gland secretions of petrels (nearly twice the salt concentration of seawater). Marine iguanas can produce nearly 70 gL^{-1}

1. (a) What is the functional excretory unit of the mammalian kidney? _____

   (b) How are mammals able to produce a concentrated urine? _____

   _____

   _____

   _____

©2017 **BIOZONE** International
ISBN: 978-1-927309-65-0
Photocopying Prohibited

2. What two factors determine the degree to which mammalian urine can be concentrated? Explain your answer:

_____

_____

_____

_____

_____

_____

3. The graph below shows the volume of urine collected from a subject after drinking 1000 cm^3 of distilled water. The subject's urine was collected at 25 minute intervals over a number of hours.

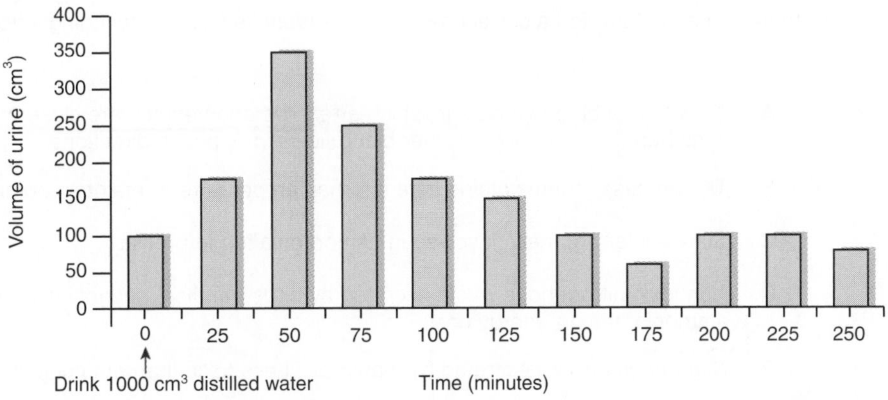

(a) Describe the changes in urine output during the experiment: _____

_____

_____

_____

(b) Explain the difference in the volume of urine collected at 25 minutes and 50 minutes: _____

_____

_____

_____

4. Explain why humans can not drink seawater to quench their thirst: _____

_____

_____

5. What is the role of the salt gland in marine birds and reptiles? _____

_____

_____

_____

6. Both desert and marine mammals produce highly concentrated urine. Contrast the different reasons for this.

_____

_____

_____

_____

93 KEY TERMS AND IDEAS: Did You Get It?

1. Complete the table below. Entering the organ for gas exchange and the method the organ is ventilation

Animal	Organ for gas exchange	Method of ventilation
Flatworm		
Insect		
Mammal		
Fish		

2. Test your vocabulary by matching each term to its correct definition, as identified by its preceding letter code.

ammonia

circulatory system

contractile vacuole

countercurrent heat exchanger

excretion

gills

lungs

nephridia

nephron

osmoregulation

thermoregulation

urea

uric acid

A A system of blood vessels in which arterioles and venules are close enough together to pass heat from one to the other blood flows in opposite directions.

B The process that maintains core internal temperature to maintain homeostasis.

C Sub-cellular organelle involved in osmoregulation in protists.

D Non toxic nitrogenous waste excreted by birds, reptiles, and insects. Produced from ammonia at high energy cost.

E Nitrogenous waste excreted by mammals. Less toxic than ammonia and very soluble.

F Organ system that permits blood and nutrients to circulate about the body in a system of vessels.

G Functional unit of the vertebrate kidney. A selective filtering element that regulates the composition of the blood and extracts wastes.

H Excretory organs of annelids comprising ciliated tubules that pump water carrying surplus ions and wastes out of the organism through openings.

I Highly toxic, very soluble nitrogenous waste excreted by aquatic invertebrates and fish.

J The active regulation of osmotic pressure (through water and ion regulation).

K Internal gas exchange structures found in terrestrial vertebrates.

L The gas exchange organs of most aquatic animals (although not aquatic mammals).

M Elimination (by an organism) of waste products of metabolism.

3. Draw labeled schematics of an amphibian heart and a mammalian heart to compare the blood flow through them:

Enduring Understanding 2.D 4.A, 4.B

Interactions in Physiological Systems

Key terms

bladder

blood buffer

Bohr effect

circulatory system

common ancestry

digestive system

excretion

excretory system

glomerulus

hemoglobin

homeostasis

kidney

large intestine

musculoskeletal system

nephron

nervous system

respiratory system

small intestine

stomach

urea

2.D.2 Homeostatic mechanisms reflect common ancestry and divergence due to adaptation in different environments

Essential knowledge

Activity number

(a) Continuity of homeostatic mechanisms reflects common ancestry, while changes may occur in response to different environments

☐ 1 Use examples to show that homeostatic mechanisms in organisms reflect common ancestry but there is adaptation to environment, while individuals show physiological changes in response to changing environments. *[also 2.D.2.c]*

2.D.3 Biological systems are affected by disruptions to homeostasis

Essential knowledge

Activity number

(a) Disruptions at the molecular and cellular levels affect the health of the organism

☐ 1 Explain how homeostasis can be disrupted by toxic substances.

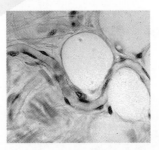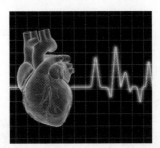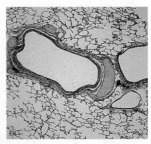

4.A.4 Organisms exhibit complex properties due to interactions between their component parts

Essential knowledge

Activity number

(a) Interactions and coordination between organs provide essential biological activities

☐ 1 Using examples, describe how interactions between organs in animals provide essential activities such as digestion of food (stomach and small intestines) and excretion of wastes (kidney and bladder).

(b) Interactions and coordination between systems provide essential biological activities

☐ 1 Describe how interactions between two or more organ systems provide essential activities, including:
- Exchange and transport of gases (respiratory and circulatory systems)
- Movement and locomotion (nervous and musculoskeletal systems)
- Acid-base homeostasis (respiratory, circulatory, and excretory systems).

4.B.2 Cooperative interactions within organisms promote efficiency in the use of energy and matter

Essential knowledge

Activity number

(a) Organisms have areas or compartments that perform a subset of functions related to energy and matter and these parts contribute to the whole

☐ 1 Recall how compartmentalization within cells contributes to overall function.

☐ 2 Using examples, explain how specialization of organs within animals contributes to overall functioning. Examples include specialization for exchange of gases, digestion of food, circulation of fluids, and excretion of wastes.

94 Transport and Exchange in Animals

Key Idea: Internal transport systems in animals move materials between exchange surfaces by mass transport. Living cells require a constant supply of nutrients and oxygen, and continuous removal of wastes. Simple, small organisms achieve this through simple diffusion across the body surface. As their surface area to volume ratio decreases, larger, more complex organisms require specialized organs and organ systems to facilitate exchanges. Interactions and coordination between these systems are essential for supporting biological activities. **Mass transport** (also called mass flow or bulk flow) describes the movement of materials at equal rates or as a single mass. Mass transport accounts for the long distance transport of fluids in living organisms. It includes the movement of blood in the circulatory systems of animals.

Exchange across a body surface

In unicellular organisms, such as this amoeba, and small flattened organisms, such as planarians, no part of the organism is very far from the environment. Simple diffusion of molecules from the fluid (or moist) environment into and out of the organism provides for all the organism's needs.

O₂

CO₂

Wastes

Nutrients

Amoeba

In some small multicellular organisms, where body depth is not great, diffusion is sufficient to allow adequate exchanges with the environment.

Flow of water

Central cavity where digestion takes place, and nutrients and wastes are exchanged.

Diffusion of nutrients and wastes

Cnidarian (Hydra)

Systems for exchange and transport

The blood circulates within a network of blood vessels, which transport the blood to all the regions of the body.

The blood transports nutrients, wastes, hormones, and respiratory gases. It moves by **mass transport** between the exchange surfaces at the tissues and the fills (or lungs).

Specialized exchange surfaces at the gills or lungs, enable the gases to be exchanged with the environment by diffusion.

The heart is a pumping device to circulate blood through a network of blood vessels. The heart may be a simple tube or have several chambers.

Gray reef shark

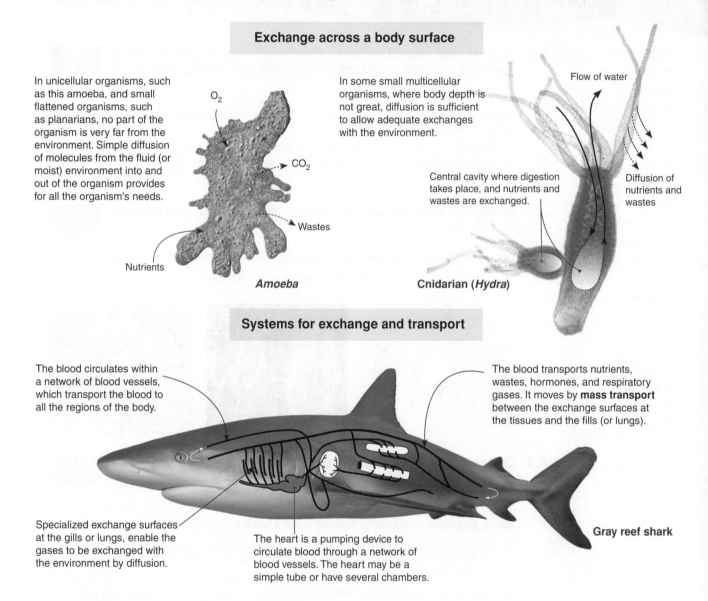

1. Why do animals above a certain size or level of complexity require specialized systems for transport and exchange?

_____

_____

_____

2. (a) How do materials move within the circulatory system of a vertebrate? _____

_____

(b) Contrast this with how materials are transported in a flatworm or single celled eukaryote: _____

_____

_____

(c) Identify two exchange sites in a vertebrate: _____

©2017 **BIOZONE** International
ISBN: 978-1-927309-65-0
Photocopying Prohibited

95 Circulatory Fluids

Key Idea: Circulatory fluid transports nutrients, wastes, hormones, and often respiratory gases around the body.
The internal transport system of most animals includes a circulating fluid. In animals with closed systems, the fluid in the blood vessels is distinct from the tissue fluid outside the vessels and is called **blood**. Blood can have many different appearances, depending on the animal group, but it usually

consists of cells and cell fragments suspended in a watery fluid. It serves many functions, including transporting dissolved gases, nutrients, wastes, and hormones. In animals with open systems, there is no difference between the fluid in the vessels and that in the sinuses (hemocoel) so the circulating fluid is called **hemolymph**. In insects, the hemolymph carries nutrients but not respiratory gases.

Insect hemolymph

Molted exoskeleton

Emerged cricket

Fluid pressure is used to facilitate molting in insects. Overwintering insects even have antifreezes, such as glycerol, in the plasma so they don't freeze during winter.

Hemolymph may make up between 11% and 40% of the total body mass of an insect

Hemolymph is a blood-like substance found in all invertebrates with open circulatory systems. The hemolymph fills the hemocoel and surrounds all cells.

About 90% of insect hemolymph is a watery fluid plasma, which is usually clear. Compared to vertebrate blood, it contains relatively high concentrations of amino acids, proteins, sugars, and inorganic ions.

The remaining 10% of hemolymph volume is made up of various cell types (hemocytes). These are involved in clotting and internal defense. Unlike vertebrate blood, insect hemolymph lacks red blood cells and (with a few exceptions) lacks respiratory pigment, because oxygen is delivered directly to tissues by the tracheal system.

Mammalian blood

Non-cellular components

The non-cellular part of the blood is the plasma. It is 92% water and also contains proteins, electrolytes, hormones, and dissolved gases.

Plasma makes up 50-60% of blood volume. The water in the plasma transports blood cells, distributes heat and helps maintain blood volume.

Cellular components

The formed elements of blood float in the plasma. They include white blood cells, platelets, and red blood cells.

White blood cells (WBCs) and platelets
2-3% of the total blood volume. WBCs are involved in internal defense. They include lymphocytes and granulocytes. Platelets are small, membrane-bound cell fragments with a role in blood clotting.

Red blood cells (RBCs)
38-48% of total blood volume. RBCs transport oxygen (carried bound to hemoglobin) and a small amount of carbon dioxide. Red blood cells in mammals are different to all other vertebrates in that they are not nucleated.

Platelets   Red blood cells   Granulocyte   Lymphocyte

1. (a) What is the main function of circulatory fluids? _____

(b) Describe two other common functions of circulatory fluids: _____

2. Describe one function of mammalian blood not performed by insect hemolymph: _____

3. Describe one function of insect hemolymph not performed by mammalian blood: _____

4. Compare the composition between blood and hemolymph: _____

CONNECT **122** AP2   WEB **95**   **KNOW**

96 Exchanges at the Lungs

Key Idea: The exchange of respiratory gases between the lungs and the blood occurs in the alveoli. This involves interaction between the respiratory and circulatory systems. The alveoli are grape-like sacs at the end of the terminal airways in lungs. They are the site of gas exchange, gases diffuse across a gas exchange membrane formed by the close arrangement of the capillary wall and the epithelium of the alveolus. The gas exchange membrane is very thin and provides a large surface area so that the respiratory gases (O_2 and CO_2) are exchanged efficiently between the lungs and the blood in the capillaries. Carbon dioxide in the blood diffuses across the membrane and enters the lungs where it is breathed out. Inhaled oxygen diffuses from the lungs into the blood for distribution around the body by red blood cells.

Cross section through lung tissue

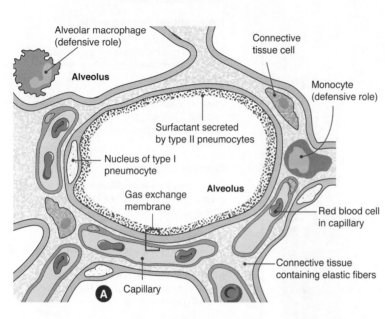

The gas exchange membrane

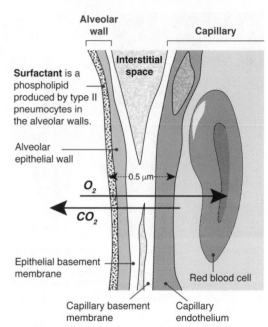

This diagram shows the physical relationship of the alveoli to the capillaries through which the blood moves. The alveolus is lined with alveolar epithelial cells called pneumocytes. Phagocytes (monocytes and macrophages) are also present to protect the lung tissue. Elastic connective tissue gives the alveoli their ability to expand and recoil. A surfactant helps to prevent alveolar collapse after each exhalation by decreasing surface tension.

The **gas exchange membrane** is the layered junction between the alveolar epithelial cells, the endothelial cells of the capillary, and their associated basement membranes (thin connective tissue layers under the epithelia). Gases move freely across this membrane.

1. With respect to gas exchange across the alveoli, discuss how the respiratory and circulatory systems interact to perform an essential life function:

_____

_____

_____

_____

_____

2. (a) Identify the structure labeled **A** in the cross section through the lung: _____

   (b) Explain why it is important that the gas exchange membrane is very thin and very extensive: _____

_____

_____

_____

   (c) What process maintains the diffusion gradient for respiratory gases: _____

_____

3. What would happen to the surface area of the lungs if no surfactant was secreted by the type II pneumocytes? _____

_____

©2017 **BIOZONE** International
ISBN: 978-1-927309-65-0
Photocopying Prohibited

97 Gas Transport in Humans

Key Idea: Hemoglobin is a respiratory pigment in red blood cells, which binds oxygen and increases the efficiency of its transport and delivery to tissues throughout the body. The transport of respiratory gases around the body is the role of the blood and its respiratory pigment. Most of the carbon dioxide in the blood is carried as bicarbonate in the plasma. Oxygen does not dissolve in blood easily, so in vertebrates, e.g humans, it is transported throughout the body chemically bound to the respiratory pigment **hemoglobin** (Hb) inside the red blood cells.

In the muscles, oxygen from hemoglobin is transferred to and retained by **myoglobin**, a molecule that is chemically similar to hemoglobin except that it consists of only one heme-globin unit. Myoglobin has a greater affinity for oxygen than hemoglobin and acts as an oxygen store within muscles, releasing the oxygen during periods of prolonged or extreme muscular activity.

Air movement

Bronchiole

Capillary

Alveoli

Each **alveolus** is a cup-shaped pouch surrounded by lung capillaries.

Area of contact with lung capillary enlarged below

CO_2 O_2

Gas exchange membrane: Formed by the epithelial cells of the alveolus and capillary. It is only 0.5 μm thick so gases diffuse rapidly across.

Most CO_2 in the blood (85%) is carried as bicarbonate (HCO_3^-) formed in the red blood cells from CO_2 in a reversible, enzyme-catalyzed reaction. HCO_3^- diffuses out of the red blood cells and into the plasma where it contributes to the buffer capacity of the blood.

HCO_3^-

When oxygen levels are high (lungs and surrounding blood vessels) hemoglobin binds with a lot of oxygen (the Hb is saturated).

HbO_2

Most oxygen in the blood (97%) is carried in the red blood cells by the protein hemoglobin (Hb). Hb increases the amount of oxygen the blood can carry by binding oxygen in a reversible reaction.

HCO_3^-

When CO_2 levels rise too quickly, H^+ can accumulate in the blood, reducing pH. This provides a strong stimulus to increase breathing rate.

Body tissue capillary: The capillaries in the tissues are very close to the body's cells, allowing for rapid diffusion back and forth.

HbO_2 HbO_2

When carbon dioxide levels are high the hemoglobin releases its oxygen to the tissues.

CO_2 CO_2 O_2 O_2

Carbon dioxide diffuses from the body's cells into the capillary.

Oxygen diffuses into the body's cells from the capillary.

Body cells

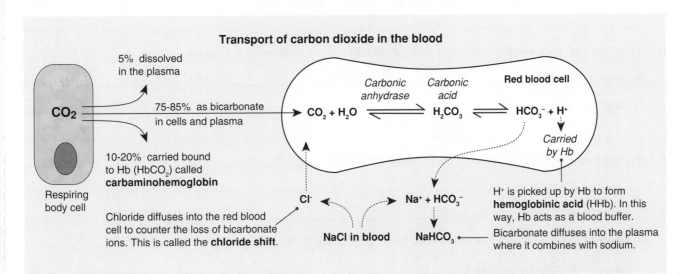

Transport of carbon dioxide in the blood

CO_2

5% dissolved in the plasma

75-85% as bicarbonate in cells and plasma

10-20% carried bound to Hb ($HbCO_2$) called **carbaminohemoglobin**

Respiring body cell

Chloride diffuses into the red blood cell to counter the loss of bicarbonate ions. This is called the **chloride shift**.

Carbonic anhydrase   Carbonic acid   **Red blood cell**

$CO_2 + H_2O \rightleftharpoons H_2CO_3 \rightleftharpoons HCO_3^- + H^+$

Carried by Hb

Cl^- $Na^+ + HCO_3^-$

NaCl in blood $NaHCO_3$

H^+ is picked up by Hb to form **hemoglobinic acid** (HHb). In this way, Hb acts as a blood buffer.

Bicarbonate diffuses into the plasma where it combines with sodium.

©2017 **BIOZONE** International
ISBN: 978-1-927309-65-0
Photocopying Prohibited

PRACTICES   CONNECT   CONNECT   WEB

62 AP1   117 AP2   97

KNOW

Respiratory pigments and the transport of oxygen

Fig.1: Dissociation curves for hemoglobin and myoglobin at normal body temperature for fetal and adult human blood.

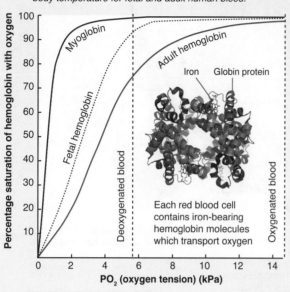

- ▶ The most important factor determining how much oxygen is carried by hemoglobin (Hb) is the level of oxygen in the blood. The greater the oxygen tension, the more oxygen will combine with Hb.

- ▶ This relationship can be illustrated in an **oxyhemoglobin dissociation curve** (left). In the lung capillaries (high O_2), a lot of oxygen is picked up and bound by Hb. In the tissues (low O_2), oxygen is released.

- ▶ Myoglobin in skeletal muscle has a very high affinity for oxygen and will take up oxygen from Hb in the blood. It can therefore act as an oxygen store.

- ▶ Fetal Hb has a high affinity for oxygen and carries 20-30% more than maternal Hb.

- ▶ The release of oxygen to the tissues is enhanced by the effect of pH. As pH increases (lower CO_2), more oxygen combines with Hb. As the blood pH decreases (higher CO_2), Hb binds less oxygen and releases more to the tissues. This is called the Bohr effect.

1. (a) Identify two regions in the body where oxygen levels are relatively high: _____

   _____

   (b) Identify two regions where carbon dioxide levels are relatively high: _____

   _____

2. Explain the significance of the reversible binding of oxygen by hemoglobin (Hb): _____

   _____

3. (a) How is hemoglobin saturation affected by the oxygen level in the blood? _____

   _____

   _____

   (b) What is the significance of this relationship to oxygen delivery to the tissues? _____

   _____

   _____

4. (a) How is the behavior of fetal Hb different to adult Hb? _____

   (b) Explain the significance of this difference: _____

   _____

5. At low blood pH, less oxygen is bound by hemoglobin and more is released to the tissues:

   (a) Name this effect: _____

   (b) What is its significance? _____

   _____

6. Why is the very high affinity of myoglobin for oxygen important? _____

   _____

7. (a) How is most CO_2 carried in the blood? _____

   (b) Identify the two main contributors to the buffer capacity of the blood: _____

   _____

©2017 **BIOZONE** International
ISBN: 978-1-927309-65-0
Photocopying Prohibited

98 Physiological Responses to High Altitude

Key Idea: Short and long term physiological changes help the body adjust to the low oxygen conditions at high altitude. The circulatory and respiratory systems respond to environmental changes (e.g. increased altitude) to ensure sufficient oxygen is delivered to the tissues. Air pressure decreases with altitude so the pressure (therefore amount) of oxygen in the air also decreases. Many of the physiological effects of high altitude arise from the low oxygen pressure, not the low air pressure in itself. Humans can make short and long term physiological adjustments (acclimation) to altitude.

Physiological adjustments to altitude

The body makes several physiological adjustments to compensate for the low oxygen pressure at altitude, a process called acclimation (or acclimatization). However, sudden exposure to an altitude of 2000 m causes breathlessness, and ascending above 4500 m too rapidly results in mountain sickness. The symptoms include breathlessness and nausea. Continuing to ascend with mountain sickness can result in fatal accumulation of fluid on the lungs and brain.

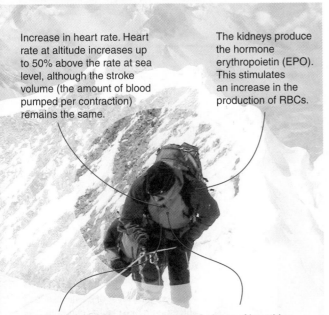

Increase in heart rate. Heart rate at altitude increases up to 50% above the rate at sea level, although the stroke volume (the amount of blood pumped per contraction) remains the same.

The kidneys produce the hormone erythropoietin (EPO). This stimulates an increase in the production of RBCs.

Acid-base readjustment and an increase in red blood cells (RBCs) are longer term changes. Hyperventilation increases O_2 in the blood, but it also reduces CO_2. This makes body fluids more alkaline. The kidneys respond to this by removing bicarbonate from the blood.

Increased rate of breathing (hyperventilation). Normally, the rate of breathing is regulated by a sensitivity to blood pH (CO_2 level). However, low oxygen pressures (pO_2) in the blood induce a hypoxic response, stimulating oxygen-sensitive receptors in the aorta and inducing hyperventilation.

Time scales of physiological changes at altitude

Some physiological adjustments to altitude take place almost immediately (e.g. increased breathing and heart rates). Other adjustments, such as increasing the number of red blood cells (RBCs) and associated hemoglobin level, may take weeks. These responses all increase the rate at which oxygen is supplied to the tissues. When more permanent adjustments to physiology are made (increased blood cells and capillaries), heart and breathing rates return to normal.

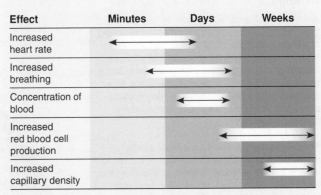

Effect	Minutes	Days	Weeks
Increased heart rate			
Increased breathing			
Concentration of blood			
Increased red blood cell production			
Increased capillary density			

Effects of altitude on hemoglobin levels

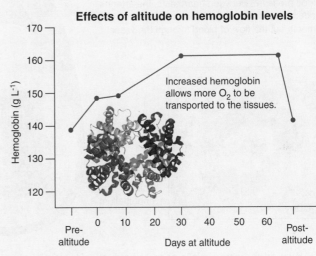

Increased hemoglobin allows more O_2 to be transported to the tissues.

Hemoglobin (g L^{-1}) vs Days at altitude (Pre-altitude ... Post-altitude)

1. The body has a range of physiological responses to the stresses of high altitude.

   (a) Identify two immediate responses and explain how they help to maintain homeostasis: _____

   _____

   _____

   (b) Identify two responses that occur over several weeks and explain how they help to maintain homeostasis:

   _____

   _____

2. (a) What happens to hemoglobin levels after several weeks at altitude?_____

   (b) What is the biological significance of this? _____

   _____

©2017 **BIOZONE** International
ISBN: 978-1-927309-65-0
Photocopying Prohibited

CONNECT WEB

32 AP1 **98**

KNOW

99 Blood Vessels

Key Idea: The blood vessels of the circulatory system connect the body's cells to the organs that exchange gases, absorb nutrients, and dispose of wastes.

In vertebrates, **arteries** are the blood vessels that carry blood away from the heart to the capillaries within the tissues. The large arteries that leave the heart divide into medium-sized (distributing) arteries. Within the tissues and organs, these distributing arteries branch to form **arterioles**, which deliver blood to capillaries. Blood flow to the tissues is altered by contraction (**vasoconstriction**) or relaxation (**vasodilation**) of the blood vessel walls. Vasoconstriction increases blood pressure whereas vasodilation has the opposite effect. **Veins** are the blood vessels that return blood to the heart from the tissues. The smallest veins (**venules**) return blood from the capillaries to the veins. Veins and their branches contain about 59% of the blood in the body. The structural differences between veins and arteries are mainly associated with differences in the relative thickness of the vessel layers and the diameter of the lumen (space within the vessel). These, in turn, are related to the vessel's functional role.

Arteries

Arteries, regardless of size, can be recognized by their well-defined rounded lumen (internal space) and the muscularity of the vessel wall. Arteries have an elastic, stretchy structure that gives them the ability to withstand the high pressure of blood being pumped from the heart. At the same time, they help to maintain pressure by having some contractile ability themselves (a feature of the central muscle layer).

Arteries nearer the heart have more elastic tissue to resist the higher pressures of the blood leaving the left ventricle. Arteries further from the heart have more muscle to help them maintain blood pressure. Between heartbeats, the arteries undergo elastic recoil and contract. This tends to smooth out the flow of blood through the vessel.

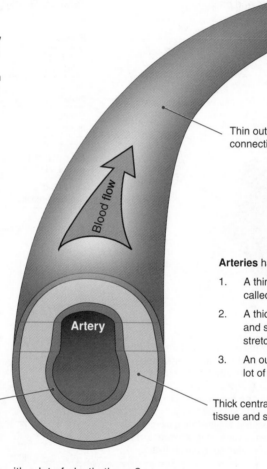

Blood flow

Thin outer layer connective tissue.

Arteries have three regions (left):

1. A thin inner layer of epithelial cells called the endothelium lines the artery.

2. A thick central layer of elastic tissue and smooth muscle that can both stretch and contract.

3. An outer connective tissue layer has a lot of elastic tissue.

Artery

Thin endothelium

Thick central layer of elastic tissue and smooth muscle.

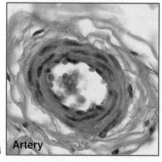

Artery

1. Why do the artery walls need to be thick with a lot of elastic tissue? _____

_____

2. What is the purpose of smooth muscle in the artery walls? _____

_____

3 (a) What is the effect of vasodilation on the diameter of an artery? _____

(b) What would the effect of this be on blood pressure? _____

4. Describe the structure of a capillary and explain the purpose of this structure: _____

_____

_____

_____

_____

©2017 **BIOZONE** International
ISBN: 978-1-927309-65-0
Photocopying Prohibited

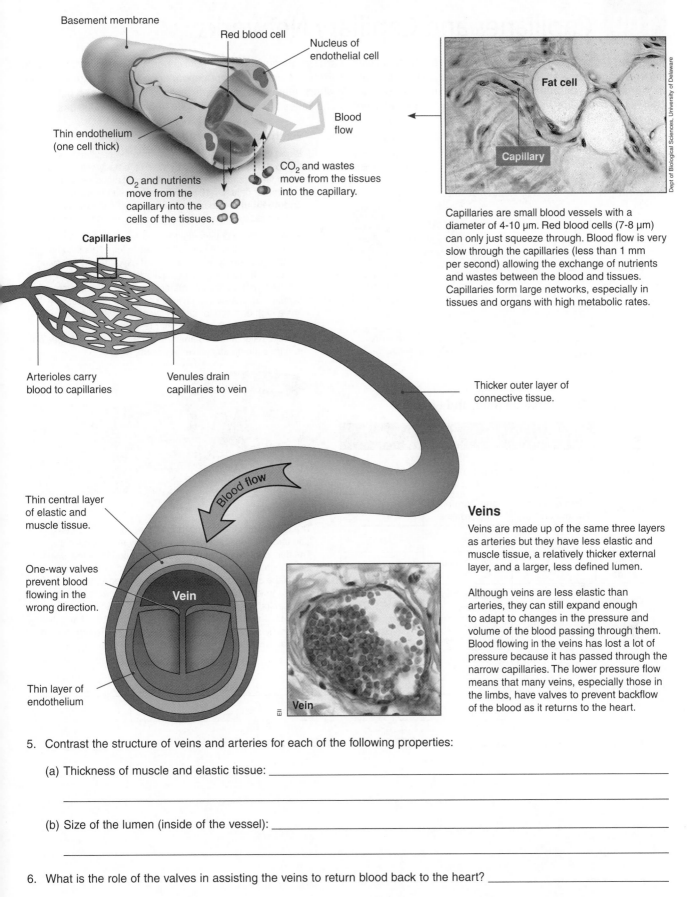

Basement membrane

Red blood cell

Nucleus of endothelial cell

Blood flow

Thin endothelium (one cell thick)

O_2 and nutrients move from the capillary into the cells of the tissues.

CO_2 and wastes move from the tissues into the capillary.

Capillaries

Fat cell

Capillary

Dept of Biological Sciences, University of Delaware

Capillaries are small blood vessels with a diameter of 4-10 μm. Red blood cells (7-8 μm) can only just squeeze through. Blood flow is very slow through the capillaries (less than 1 mm per second) allowing the exchange of nutrients and wastes between the blood and tissues. Capillaries form large networks, especially in tissues and organs with high metabolic rates.

Arterioles carry blood to capillaries

Venules drain capillaries to vein

Thicker outer layer of connective tissue.

Blood flow

Thin central layer of elastic and muscle tissue.

One-way valves prevent blood flowing in the wrong direction.

Vein

Thin layer of endothelium

Vein

Eli

Veins

Veins are made up of the same three layers as arteries but they have less elastic and muscle tissue, a relatively thicker external layer, and a larger, less defined lumen.

Although veins are less elastic than arteries, they can still expand enough to adapt to changes in the pressure and volume of the blood passing through them. Blood flowing in the veins has lost a lot of pressure because it has passed through the narrow capillaries. The lower pressure flow means that many veins, especially those in the limbs, have valves to prevent backflow of the blood as it returns to the heart.

5. Contrast the structure of veins and arteries for each of the following properties:

(a) Thickness of muscle and elastic tissue: _____

_____

(b) Size of the lumen (inside of the vessel): _____

_____

6. What is the role of the valves in assisting the veins to return blood back to the heart? _____

_____

_____

7. Why does blood ooze from a venous wound, rather than spurting as it does from an arterial wound?_____

_____

_____

100 Capillaries and Capillary Networks

Key Idea: Capillaries are small, thin-walled vessels that allow the exchange of material between the blood and the tissues. In vertebrates, **capillaries** are very small vessels that connect arterial and venous circulation and allow efficient exchange of nutrients and wastes between the blood and tissues. Capillaries form networks or beds and are abundant where metabolic rates are high. Fluid that leaks out of the capillaries has an essential role in bathing the tissues.

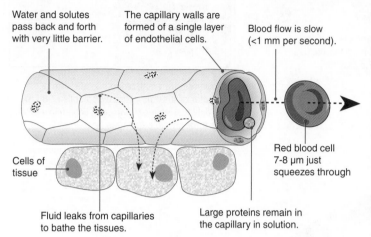

Water and solutes pass back and forth with very little barrier.

The capillary walls are formed of a single layer of endothelial cells.

Blood flow is slow (<1 mm per second).

Cells of tissue

Red blood cell 7-8 μm just squeezes through

Fluid leaks from capillaries to bathe the tissues.

Large proteins remain in the capillary in solution.

Exchanges in capillaries

Blood passes from the arterioles into the capillaries where the exchange of materials between the body cells and the blood takes place. Capillaries are small blood vessels with a diameter of just 4-10 μm. The only tissue present is an **endothelium** of squamous epithelial cells. Capillaries are so numerous that no cell is more than 25 μm from any capillary.

Blood pressure causes fluid to leak from capillaries through small gaps where the endothelial cells join. This fluid bathes the tissues, supplying nutrients and oxygen, and removing wastes (left). The density of capillaries in a tissue is an indication of that tissue's metabolic activity. For example, cardiac muscle relies heavily on oxidative metabolism. It has a high demand for blood flow and is well supplied with capillaries. Smooth muscle is far less active than cardiac muscle, relies more on anaerobic metabolism, and does not require such an extensive blood supply.

Blood, tissue fluid, and lymph

	Blood	Tissue fluid	Lymph
Cells	Erythrocytes, leukocytes, platelets	Some leukocytes	Lymphocytes
Proteins	Hormones and plasma proteins	Some hormones and proteins	None
Glucose	High	None	Low
Amino acids	High	Used by body cells	Low
Oxygen	High	Used by body cells	Low
Carbon dioxide	Low	Produced by body cells	High

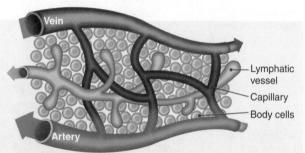

Vein

Lymphatic vessel

Capillary

Body cells

Artery

The pressure at the arterial end of the capillaries forces fluid through gaps between the capillary endothelial cells. The fluid contains nutrients and oxygen and is called tissue fluid. Some of this fluid returns to the blood at the venous end of the capillary bed, but some is drained by lymph vessels to form lymph. Blood transports nutrients, wastes, and respiratory gases to and from the tissues. Tissue fluid facilitates the transport of these between the blood and the tissues. Lymph drains excess tissue fluid and returns it to the general circulation, and it has a role in the immune system.

1. What is the role of capillaries? _____

_____

2. Describe the structure of a capillary, contrasting it with the structure of a vein and an artery: _____

_____

_____

_____

_____

_____

3. Distinguish between blood, tissue fluid, and lymph: _____

_____

_____

_____

_____

©2017 **BIOZONE** International
ISBN: 978-1-927309-65-0
Photocopying Prohibited

The flow of blood through a capillary bed is called **microcirculation**. In most parts of the body, there are two types of vessels in a capillary bed: the true capillaries, where exchanges take place, and a vessel called a vascular shunt, which connects the arteriole and venule at either end of the bed. The shunt diverts blood past the true capillaries when the metabolic demands of the tissue are low. When tissue activity increases, the entire network fills with blood.

4. Describe the structure of a capillary network:

_____

_____

_____

_____

_____

_____

A

When the sphincters contract (close), blood is diverted via the vascular shunt to the postcapillary venule, bypassing the exchange capillaries.

5. Explain the role of the smooth muscle sphincters and the vascular shunt in a capillary network:

_____

_____

_____

_____

_____

B

When the sphincters are relaxed (open), blood flows through the entire capillary bed allowing exchanges with the cells of the surrounding tissue.

6. (a) Describe a situation where the capillary bed would be in the condition labeled **A**:

_____

_____

_____

_____

_____

Connecting capillary beds
The role of portal venous systems

(b) Describe a situation where the capillary bed would be in the condition labeled **B**:

_____

_____

_____

_____

7. How does a portal venous system differ from other capillary systems?

_____

_____

_____

_____

_____

_____

A portal venous system occurs when a capillary bed drains into another capillary bed through veins, without first going through the heart. Portal systems are relatively uncommon. Most capillary beds drain into veins which then drain into the heart, not into another capillary bed. The diagram depicts the **hepatic portal system**, which includes both capillary beds and the blood vessels connecting them.

101 Adaptations of Vertebrate Blood

Key Idea: The variations seen in vertebrate blood reflect the divergence of organisms into specific environments.

Blood varies in composition and function throughout the vertebrate orders. This variation reflects how animals with a shared common ancestry have adapted to the selection pressures of a specific environment. For example, the blood of some Antarctic animals contains anti-freeze components that allow them to function at sub-zero temperatures.

Oxygen capacity and hemoglobin

Hemoglobin is a respiratory pigment (a pigmented protein) capable of combining reversibly with oxygen, hence increasing the amount of oxygen that can be carried by the blood. All vertebrates use hemoglobin to transport blood through the body.

Taxon	Oxygen capacity ($cm^3 O_2$ per 100 cm^3 blood)	Pigment
Fishes	2 - 4	Hemoglobin
Reptiles	7 - 12	Hemoglobin
Birds	20 - 25	Hemoglobin
Mammals	15 - 30	Hemoglobin

The oxygen carrying capacity of hemoglobin is not the same for all vertebrates. Variations in its structure change both its capacity to carry oxygen and its ability to take up and release oxygen. In general, the oxygen carrying capacity of blood in vertebrates increases from fish, to reptiles, to birds and mammals and is correlated with metabolic rate. Fish blood frequently possesses several hemoglobins. This allows them to move between different environments while maintaining the oxygen carrying capacity of their blood.

Photo: Professor Dr. habil. Uwe Kils CC3.0

Crocodile icefish larva

The Antarctic icefish (Channichthyidae) are the only known vertebrates without hemoglobin. They survive because the extremely cold water (as low as -2°C) contains high concentrations of O_2. This diffuses directly across their scaleless skin into the blood plasma and is transported throughout the body. Many icefish contain antifreeze glycoproteins in their blood allowing the blood to remain fluid at freezing temperatures.

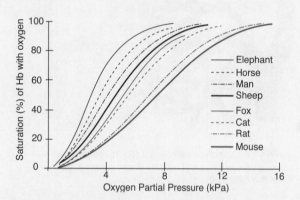

The hemoglobin of small mammals releases oxygen more readily than larger mammals because it has a lower affinity for oxygen. This is related to the higher per mass metabolic rate of small mammals. They consume oxygen at a higher rate than larger mammals and so require a faster delivery of oxygen.

The graph shows Saturation (%) of Hb with oxygen versus Oxygen Partial Pressure (kPa) for: Elephant, Horse, Man, Sheep, Fox, Cat, Rat, Mouse.

Kenneth Catania, Vanderbilt University NSF

Adaptations of hemoglobin have enabled some mammals to live in places with low O_2 content. Llamas and their relatives live at high altitudes with low O_2 pressure. Moles live underground where CO_2 levels can be high and O_2 levels low. The hemoglobin in some moles has lost the ability to bind the molecule DPG (which inhibits O_2 binding in deoxygenated tissues). This increases the amount of O_2 and CO_2 that can be transported in the blood.

1. What evidence does blood provide that there is a shared common ancestry between the vertebrate taxa? _____

_____

2. Describe the relationship in mammals between hemoglobin's affinity for oxygen and body size and give a likely reason for this relationship:

_____

_____

_____

3. Using examples, describe how vertebrate blood reflects evolution in different environments: _____

_____

_____

_____

_____

©2017 **BIOZONE** International
ISBN: 978-1-927309-65-0
Photocopying Prohibited

102 The Human Heart

Key Idea: Humans have a four chambered heart divided into left and right halves. It acts as a double pump.

The heart is the center of the human cardiovascular system. It is a hollow, muscular organ made up of four chambers (two **atria** and two **ventricles**) that alternately fill and empty of blood, acting as a double pump. The left side (systemic circuit) pumps blood to the body tissues and the right side (pulmonary circuit) pumps blood to the lungs. The heart lies between the lungs, to the left of the midline, and is surrounded by a double layered pericardium of connective tissue, which prevents over distension of the heart and anchors it within the central compartment of the thoracic cavity.

Human heart structure

(sectioned, anterior view)

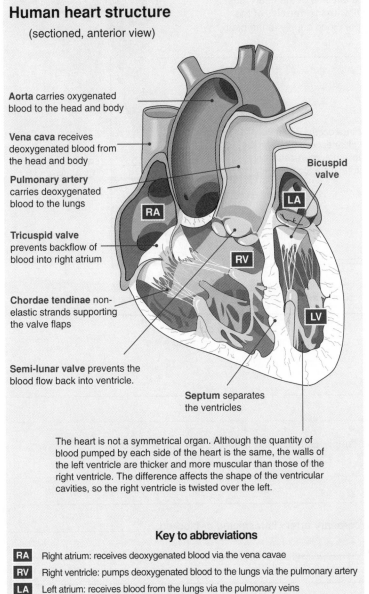

Aorta carries oxygenated blood to the head and body

Vena cava receives deoxygenated blood from the head and body

Pulmonary artery carries deoxygenated blood to the lungs

Tricuspid valve prevents backflow of blood into right atrium

Chordae tendinae non-elastic strands supporting the valve flaps

Semi-lunar valve prevents the blood flow back into ventricle.

Bicuspid valve

RA RV LA LV

Septum separates the ventricles

The heart is not a symmetrical organ. Although the quantity of blood pumped by each side of the heart is the same, the walls of the left ventricle are thicker and more muscular than those of the right ventricle. The difference affects the shape of the ventricular cavities, so the right ventricle is twisted over the left.

Key to abbreviations

RA	Right atrium: receives deoxygenated blood via the vena cavae
RV	Right ventricle: pumps deoxygenated blood to the lungs via the pulmonary artery
LA	Left atrium: receives blood from the lungs via the pulmonary veins
LV	Left ventricle: pumps oxygenated blood to the head and body via the aorta

Top view of a heart in section to show valves

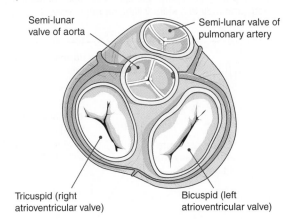

Semi-lunar valve of aorta

Semi-lunar valve of pulmonary artery

Tricuspid (right atrioventricular valve)

Bicuspid (left atrioventricular valve)

Anterior view of heart to show coronary arteries

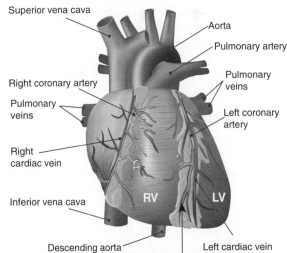

Superior vena cava

Aorta

Pulmonary artery

Right coronary artery

Pulmonary veins

Pulmonary veins

Left coronary artery

Right cardiac vein

Inferior vena cava

RV LV

Descending aorta

Left cardiac vein

The high oxygen demands of the heart muscle are met by a dense capillary network branching from the coronary arteries. The coronary arteries (left and right) arise from the aorta and spread over the surface of the heart supplying the cardiac muscle with oxygenated blood. The left carries 70% of the coronary blood supply and the right the remaining 30%. Deoxygenated blood is collected by the cardiac veins and returned to the right atrium via a large coronary sinus.

1. In the schematic diagram of the heart, below, label the four chambers and the main vessels entering and leaving them. The arrows indicate the direction of blood flow. Use large colored circles to mark the position of each of the four valves.

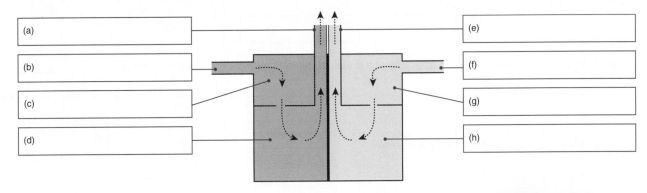

(a) (b) (c) (d) (e) (f) (g) (h)

CONNECT
79
AP2

WEB
102

KNOW

Pressure changes and the asymmetry of the heart

Aorta, 100 mg Hg

The heart is not a symmetrical organ. The left ventricle and its associated arteries are thicker and more muscular than the corresponding structures on the right side. This asymmetry is related to the necessary pressure differences between the pulmonary (lung) and systemic (body) circulations (not to the distance over which the blood is pumped *per se*). The graph below shows changes in blood pressure in each of the major blood vessel types in the systemic and pulmonary circuits (the horizontal distance not to scale). The pulmonary circuit must operate at a much lower pressure than the systemic circuit to prevent fluid from accumulating in the alveoli of the lungs. The left side of the heart must develop enough "spare" pressure to enable increased blood flow to the muscles of the body and maintain kidney filtration rates without decreasing the blood supply to the brain.

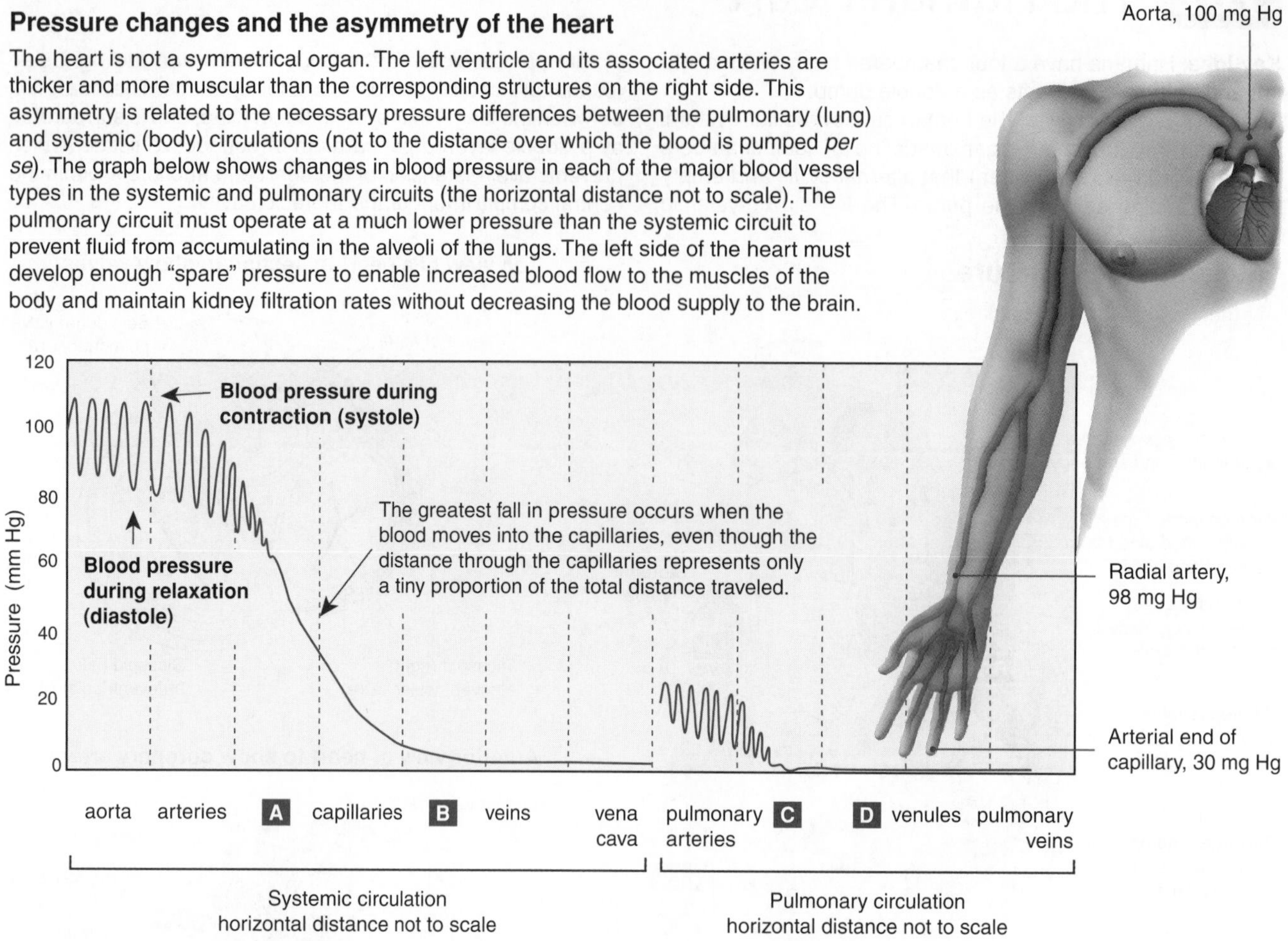

Blood pressure during contraction (systole)

Blood pressure during relaxation (diastole)

The greatest fall in pressure occurs when the blood moves into the capillaries, even though the distance through the capillaries represents only a tiny proportion of the total distance traveled.

Radial artery, 98 mg Hg

Arterial end of capillary, 30 mg Hg

aorta   arteries   **A**   capillaries   **B**   veins   vena cava   pulmonary arteries   **C**   **D**   venules   pulmonary veins

Systemic circulation
horizontal distance not to scale

Pulmonary circulation
horizontal distance not to scale

2. What is the purpose of the valves in the heart? _____

_____

3. The heart is full of blood, yet it requires its own blood supply. Suggest two reasons why this is the case:

(a) _____

(b) _____

4. Predict the effect on the heart if blood flow through a coronary artery is restricted or blocked: _____

_____

_____

5. Identify the vessels corresponding to the letters **A-D** on the graph above:

A: _____   B: _____   C: _____   D: _____

6. (a) Why must the pulmonary circuit operate at a lower pressure than the systemic system?_____

_____

_____

(b) Relate this to differences in the thickness of the wall of the left and right ventricles of the heart: _____

_____

_____

7. What are you recording when you take a pulse? _____

_____

©2017 **BIOZONE** International
ISBN: 978-1-927309-65-0
Photocopying Prohibited

Key Idea: Heartbeat is initiated by the sinoatrial node which acts as a pacemaker by setting the basic heart rhythm.

The heartbeat is **myogenic**, meaning it originates within the cardiac muscle itself. The heartbeat is regulated by a conduction system consisting of the pacemaker (**sinoatrial**

node) and a specialized conduction system of Purkyne tissue. The pacemaker sets the basic heart rhythm, but this rate can be influenced by hormones and by the cardiovascular control center. Changing the rate and force of heart contraction is the main mechanism for controlling cardiac output.

Generation of the heartbeat

The basic rhythmic heartbeat is myogenic. The nodal cells (SAN and atrioventricular node) spontaneously generate rhythmic action potentials without neural stimulation. The normal resting rate of self-excitation of the SAN is about 50 beats per minute. The amount of blood ejected from the left ventricle per minute is called the **cardiac output**. It is determined by the **stroke volume** (the volume of blood ejected with each contraction) and the **heart rate** (number of heart beats per minute).

Cardiac muscle responds to stretching by contracting more strongly. The greater the blood volume entering the ventricle, the greater the force of contraction. This relationship is important in regulating stroke volume in response to demand. The hormone **epinephrine** also influences cardiac output, increasing heart rate in preparation for vigorous activity. Changing the rate and force of heart contraction is the main mechanism for controlling cardiac output in order to meet changing demands.

TEM of cardiac muscle showing striations in a fiber (muscle cell). The Z lines that delineate the contractile units of the rod-like units of the. The fibers are joined by specialized electrical junctions called intercalated discs, which allow impulses to spread rapidly through the heart muscle.

Sinoatrial node (SAN) is the heart's **pacemaker**. It is a small mass of specialized muscle cells on the wall of the right atrium, near the entry point of the superior vena cava. It starts the cardiac cycle, spontaneously generating **action potentials** that cause the atria to contract. The SAN sets the basic heart rate, but this rate is influenced by hormones and impulses from the autonomic nervous system.

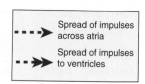

Spread of impulses across atria

Spread of impulses to ventricles

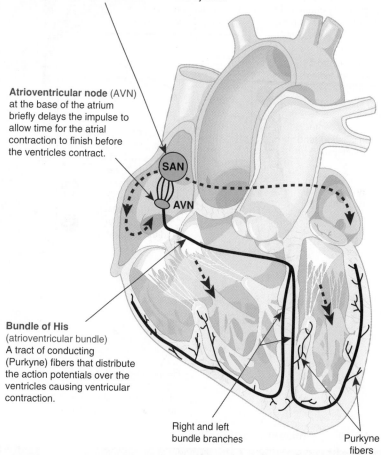

Atrioventricular node (AVN) at the base of the atrium briefly delays the impulse to allow time for the atrial contraction to finish before the ventricles contract.

Bundle of His (atrioventricular bundle) A tract of conducting (Purkyne) fibers that distribute the action potentials over the ventricles causing ventricular contraction.

Right and left bundle branches

Purkyne fibers

---

1. Describe the role of each of the following in heart activity:

   (a) The sinoatrial node: _____

   (b) The atrioventricular node: _____

   (c) The bundle of His: _____

   (d) Intercalated discs: _____

2. What is the significance of delaying the impulse at the AVN? _____

_____

3. What is the advantage of the physiological response of cardiac muscle to stretching? _____

_____

4. The heart-beat is intrinsic. Why is it important to be able to influence the basic rhythm via the central nervous system?

_____

CONNECT **80** AP1   CONNECT **181** AP2   WEB **103**

KNOW

04 Responding to Changes in Oxygen Demand

Key Idea: Control centers in the brainstem alter breathing and heart rates in response to demand.

Increased heart and breathing rates during exercise are brought about by control centers in the brainstem and occur in response to changes in blood pH caused by changes in the blood's concentration of CO_2. A drop in blood pH indicates a rise in CO_2 and an increased demand for oxygen. The control centers are the respiratory and cardiovascular centers and their output influences the muscles responsible for breathing and heart contraction respectively. They monitor sensory information from cells in the lungs and the vessels of the heart and adjust the breathing and heart rates as required.

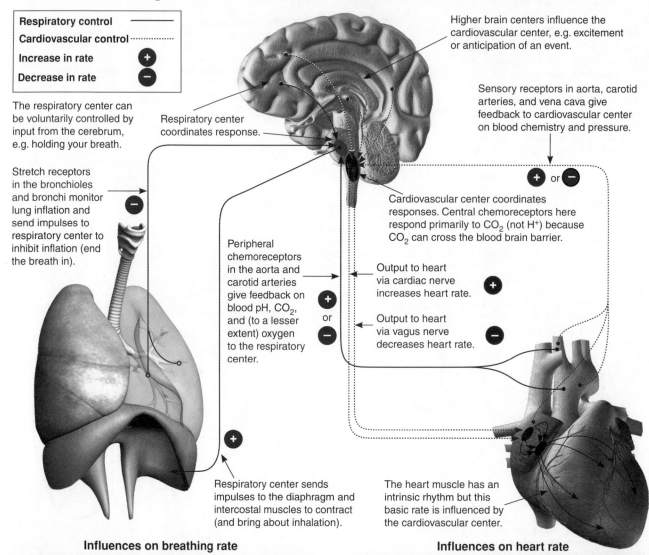

Respiratory control ——
Cardiovascular control ········
Increase in rate ⊕
Decrease in rate ⊖

Higher brain centers influence the cardiovascular center, e.g. excitement or anticipation of an event.

The respiratory center can be voluntarily controlled by input from the cerebrum, e.g. holding your breath.

Respiratory center coordinates response.

Sensory receptors in aorta, carotid arteries, and vena cava give feedback to cardiovascular center on blood chemistry and pressure.

Stretch receptors in the bronchioles and bronchi monitor lung inflation and send impulses to respiratory center to inhibit inflation (end the breath in).

Peripheral chemoreceptors in the aorta and carotid arteries give feedback on blood pH, CO_2, and (to a lesser extent) oxygen to the respiratory center.

Cardiovascular center coordinates responses. Central chemoreceptors here respond primarily to CO_2 (not H^+) because CO_2 can cross the blood brain barrier.

Output to heart via cardiac nerve increases heart rate.

Output to heart via vagus nerve decreases heart rate.

Respiratory center sends impulses to the diaphragm and intercostal muscles to contract (and bring about inhalation).

The heart muscle has an intrinsic rhythm but this basic rate is influenced by the cardiovascular center.

Influences on breathing rate

Increase	Decrease
Voluntary control	Voluntary control
Increase in H^+ or CO_2 concentrations in blood	Decrease in H^+ or CO_2 concentrations in blood
Increased body temperature	Decrease in body temperature
Decrease in blood pressure	Increase in blood pressure
Increased physical activity	Decreased physical activity

Influences on heart rate

Increase	Decrease
Increased physical activity	Decreased physical activity
Increase in H^+ or CO_2 concentrations in blood	Decrease in H^+ or CO_2 concentrations in blood
Secretion of epinephrine	Re-uptake and metabolism of epinephrine
Decrease in blood pressure	Increase in blood pressure

1. (a) What is the primary stimulus for an increase in heart rate and breathing rate? _____

   (b) What is the connection between this stimulus and physiological state? _____

   _____

   (c) Why is it adaptive to have voluntary control over breathing rate? _____

   _____

2. Both heart rate and breathing rate show the same responses to H^+ and blood pressure. Suggest why: _____

   _____

©2017 **BIOZONE** International
ISBN: 978-1-927309-65-0
Photocopying Prohibited

105 Exercise and Blood Flow

Key Idea: Breathing rate and heart rate both increase during exercise to meet the body's increased metabolic demands. During exercise, the body's metabolic rate increases and the demand for oxygen increases. Oxygen is required for cellular respiration and ATP production. Increasing the rate of breathing delivers more oxygen to working tissues and enables them to make the ATP they need to keep working. An increased breathing rate also increases the rate at which carbon dioxide is expelled from the body. Heart rate increases so blood can be moved around the body more quickly. This allows for faster delivery of oxygen and removal of carbon dioxide.

1. The table gives data for the rate of blood flow to parts of the body at rest and during strenuous exercise.
   (a) Calculate the percentage of the total blood flow that each organ or tissue receives under each regime.
   (b) In the final column, calculate the percentage change using the formula: $(E - R) \div R \times 100$

Organ or tissue	At rest		Strenuous exercise		% change
	cm³ min⁻¹	% of total (R)	cm³ min⁻¹	% of total (E)	
Brain	700	14	750	4.2	
Heart	200		750		
Lung tissue	100		200		
Kidneys	1100		600		
Liver	1350		600		
Skeletal muscles	750		12,500		
Bone	250		250		
Skin	300		1900		
Thyroid gland	50		50		
Adrenal glands	25		25		
Other tissue	175		175		
TOTAL	5000	100	17,800	100	

2. (a) Calculate approximately how many times the total rate of blood flow increases between rest and exercise: _____

   _____

   (b) How does the body increase the rate of blood flow during exercise? _____

   _____

   (c) What is the purpose of this increase? _____

   _____

   (d) What else has to happen to achieve this? _____

   _____

3. (a) What tissue shows the greatest percentage increase in blood flow between rest and exercise? _____

   (b) Explain why this occurs:_____

   _____

   _____

4. Explain why there is a large increase in blood flow to the skin between rest and exercise: _____

   _____

   _____

5. Why do you think some organs or tissues might show a fall in the percentage of blood flow they receive?

   _____

   _____

©2017 **BIOZONE** International
ISBN: 978-1-927309-65-0
Photocopying Prohibited

PRACTICES PRACTICES PRACTICES CONNECT

44
AP2

KNOW

106 The Physiological Effects of Smoking

Key Idea: The toxins and carcinogens in cigarettes disrupt homeostasis and are harmful to human health.
Cigarettes contain 600 ingredients, and when burned produce thousands of chemicals, many of which cause cancer or are toxic and directly or indirectly cause physiological changes. Cigarette smoking has harmful effects at the tissue, cellular, and molecular level. It alters body system responses, disrupts homeostatic mechanisms, and harms long term health.

Changes at the molecular level

Cigarette smoke contains a number of carcinogens (cancer causing substances), which can cause mutations. The presence of other chemicals in cigarettes (e.g. chromium and arsenic) can help the carcinogens bind strongly to DNA or inhibit the repair process. These increase the chances of cancerous cells forming.

Carcinogens, such as benzo(a)pyrene (BaP), cause changes in the DNA (right). If errors are not repaired, the damaged DNA is replicated and may give rise to cancerous cells, especially if errors accumulate. For example, the mutation caused by BaP binding to guanine is called a transversion. Scientists have found G-T transversions at several key places in the p53 tumor-suppressor gene of smokers.

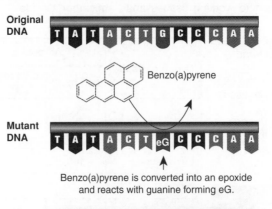

Original DNA

Benzo(a)pyrene

Mutant DNA

Benzo(a)pyrene is converted into an epoxide and reacts with guanine forming eG.

mRNA

The altered guanine (eG) is misread by RNA polymerase as thymine. The corresponding base in the mRNA is therefore adenine.

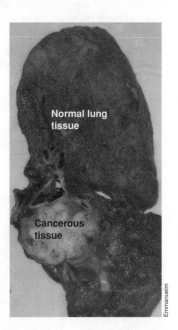

Normal lung tissue

Cancerous tissue

Smoking can disrupt homeostasis

The body's ability to transport oxygen is reduced in smokers. Inhaled smoke contains high levels of carbon monoxide (CO). Both CO and oxygen cross the gas exchange membrane to enter the blood within the capillaries. The CO is preferentially picked up by hemoglobin in the red blood cells.

CO has a much higher affinity for Hb than oxygen, so when CO is inhaled, it displaces oxygen from Hb. As a result, less oxygen is supplied to the tissues. High levels of CO are fatal.

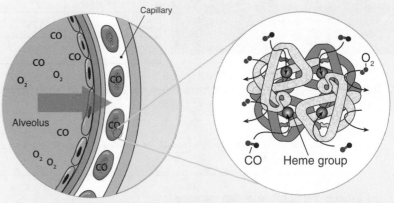

The blood protein hemoglobin (Hb) binds and carries oxygen within the red blood cells to supply the cells and tissues of the body. CO displaces oxygen from Hb (above),

1. How can smoking cause disruption to an organism at the DNA level? _____

_____

_____

2. (a) Explain why carbon monoxide reduces the amount of oxygen transported around the body: _____

_____

(b) How would this affect a person? _____

_____

_____

3. Smoking causes the lung tissue to lose its elasticity and tar from the tobacco smoke clogs the airways and damages the alveoli. What effect would this have on gas exchange capability?

_____

_____

©2017 **BIOZONE** International
ISBN: 978-1-927309-65-0
Photocopying Prohibited

107 The Human Digestive Tract

Key Idea: Interactions between the tissues and organs in the different regions of the digestive tract maximize the efficiency of digestion and absorption.

The human digestive system (gut) is a tubular tract, which is regionally specialized into a complex series of organs and glands that function in sequence to process food. Specialization of organs, so that each carries out a specific role, maximizes the efficiency with which food is processed.

1. What is the function of the digestive tract? _____
   _____
   _____
   _____
   _____

2. (a) Describe the basic structure of the digestive tract:
   _____
   _____
   _____
   _____

   (b) Describe how compartmentalization of the digestive tract is achieved:
   _____
   _____
   _____
   _____

   (c) How does this contribute to metabolic efficiency?
   _____
   _____
   _____
   _____
   _____

3. The digestive systems of some simple invertebrates have a single opening serving as both a mouth and an anus.

   (a) Describe a limitation of this type of system:
   _____
   _____
   _____

   (b) Describe the opportunities arising as a result of the evolution of a tube gut:
   _____
   _____
   _____
   _____

Image labels: Salivary glands, Esophagus, Liver, Gall bladder, Pancreas, Stomach, Large intestine, Small intestine

▶ Collectively, the organs of the digestive tract (above) carry out the physical and chemical breakdown (digestion) of food, absorption of nutrients, and elimination of undigested material.

▶ The gut is a hollow, open-ended, muscular tube. The food within the gut is essentially outside the body, it is only in contact with the cells lining the tract. Food movement through the system is one way only allowing food to be sequentially and efficiently processed.

▶ The regional specialization of the digestive tract provides a degree of compartmentalization, with each region carrying out a highly specific role. The physical and chemical environments in each region can vary widely. As a result, the enzymes located in each region and the metabolic reactions they carry out are specific to the region and its environment.

▶ External to the digestive tract are several accessory organs and glands, which add enzymes to the food to aid digestion. Examples include the pancreas and liver.

PRACTICES   CONNECT   WEB

29 AP2   107

KNOW

108 Moving Food Through the Gut

Key Idea: Solid food is chewed into a small mass called a bolus and swallowed. Further digestion produces chyme. Food is moved through the gut by waves of muscular contraction called peristalsis.

Ingested food is chewed and mixed with saliva to form a small mass called a bolus. Wave-like muscular contractions called **peristalsis** moves the food, first as a bolus and then as semi-fluid chyme, through the digestive tract as described below.

Peristalsis

The process of moving food through the esophagus by waves of muscular contractions is called peristalsis.

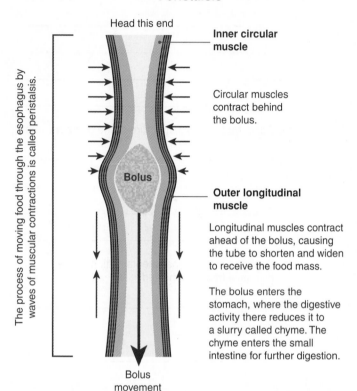

Head this end

Inner circular muscle

Circular muscles contract behind the bolus.

Bolus

Outer longitudinal muscle

Longitudinal muscles contract ahead of the bolus, causing the tube to shorten and widen to receive the food mass.

The bolus enters the stomach, where the digestive activity there reduces it to a slurry called chyme. The chyme enters the small intestine for further digestion.

Bolus movement

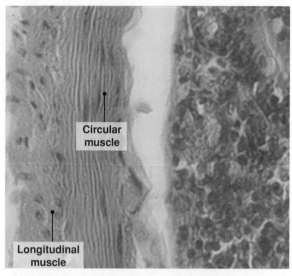

Circular muscle

Longitudinal muscle

Cross section through the small intestine

A cross section through the small intestine shows the outer longitudinal and inner circular muscles involved in peristalsis. In a cross sectional view, the longitudinal muscles appear circular because they are viewed end on, whereas the circular muscle appears in longitudinal section.

Peristaltic movement in the colon

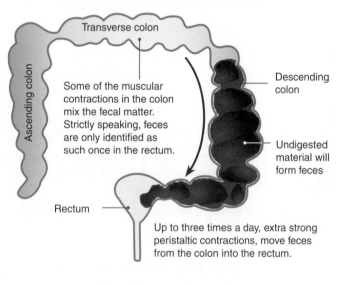

Transverse colon

Ascending colon

Some of the muscular contractions in the colon mix the fecal matter. Strictly speaking, feces are only identified as such once in the rectum.

Descending colon

Undigested material will form feces

Rectum

Up to three times a day, extra strong peristaltic contractions, move feces from the colon into the rectum.

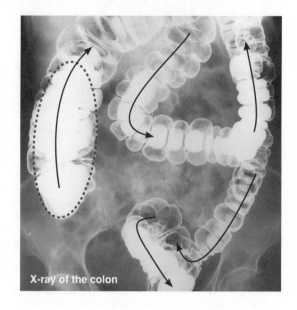

X-ray of the colon

1. Describe how peristalsis moves food through the gut: _____

_____

_____

2. What are the two main functions of peristalsis? _____

_____

3. In the X-ray of the colon, what is represented by:

   (a) The dashed oval? _____ (b) The arrows? _____

©2017 **BIOZONE** International
ISBN: 978-1-927309-65-0
Photocopying Prohibited

109 The Stomach and Small Intestine

Key Idea: The stomach produces acid and a protein-digesting enzyme, which break food down into a slurry, called chyme. The **stomach** is a hollow, muscular organ between the esophagus and small intestine. In the stomach, food is mixed in an acidic environment to produce a semi-fluid mixture called chyme. The low pH of the stomach destroys microbes, denatures proteins, and activates a protein-digesting enzyme precursor. There is very little absorption in the stomach, although small molecules (glucose, alcohol) are absorbed across the stomach wall into the surrounding blood vessels.

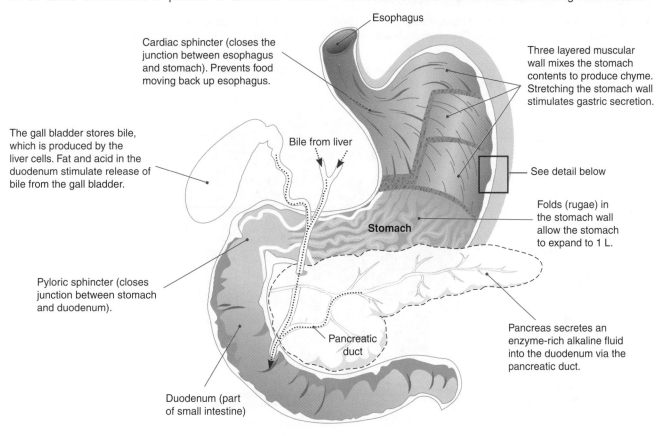

Esophagus

Cardiac sphincter (closes the junction between esophagus and stomach). Prevents food moving back up esophagus.

Three layered muscular wall mixes the stomach contents to produce chyme. Stretching the stomach wall stimulates gastric secretion.

The gall bladder stores bile, which is produced by the liver cells. Fat and acid in the duodenum stimulate release of bile from the gall bladder.

Bile from liver

See detail below

Stomach

Folds (rugae) in the stomach wall allow the stomach to expand to 1 L.

Pyloric sphincter (closes junction between stomach and duodenum).

Pancreatic duct

Pancreas secretes an enzyme-rich alkaline fluid into the duodenum via the pancreatic duct.

Duodenum (part of small intestine)

Detail of a gastric gland (stomach wall)

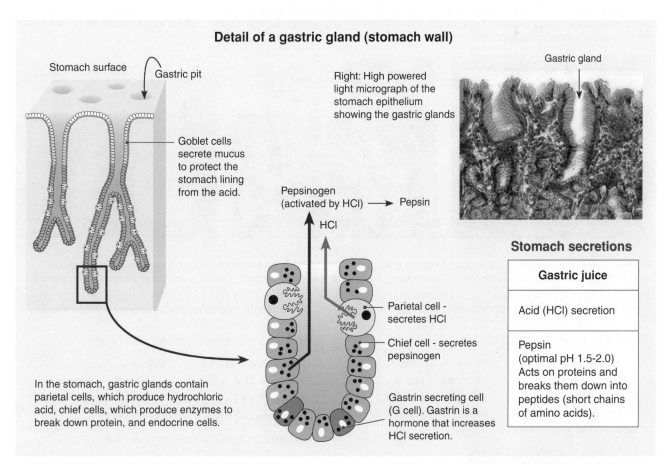

Stomach surface

Gastric pit

Gastric gland

Right: High powered light micrograph of the stomach epithelium showing the gastric glands

Goblet cells secrete mucus to protect the stomach lining from the acid.

Pepsinogen (activated by HCl) → Pepsin

HCl

Parietal cell - secretes HCl

Chief cell - secretes pepsinogen

Gastrin secreting cell (G cell). Gastrin is a hormone that increases HCl secretion.

In the stomach, gastric glands contain parietal cells, which produce hydrochloric acid, chief cells, which produce enzymes to break down protein, and endocrine cells.

Stomach secretions

Gastric juice
Acid (HCl) secretion
Pepsin (optimal pH 1.5-2.0) Acts on proteins and breaks them down into peptides (short chains of amino acids).

CONNECT
29
AP2

CONNECT
23
AP2

WEB
109

KNOW

154

The small intestine

▶ The small intestine receives the chyme directly from the stomach. It is divided into three regions, which are distinguished by the cell types present: the **duodenum**, where most chemical digestion occurs, and then the **jejunum** and the **ileum**. Most absorption occurs in the jejunum and ileum.

▶ The intestinal lining is folded into many **intestinal villi**, which project into the gut lumen (the space enclosed by the gut). The villi increase the surface area for nutrient absorption. The **epithelial cells** that make up the lining of each villus in turn have a **brush-border** of many **microvilli**, which are primarily responsible for nutrient absorption. The membrane of the microvilli is packed with enzymes that break down food molecules for absorption.

▶ Enzymes bound to the microvilli of the epithelial cells, and in the pancreatic and intestinal juices, break down fats, peptides, and carbohydrates (see tables below). The small molecules produced by this digestion are then absorbed into the underlying blood and lymph vessels.

▶ Tubular exocrine glands and goblet cells secrete alkaline fluid and mucus into the lumen, neutralizing the acidity of the chyme entering the small intestine from the stomach and protecting the lining of the intestine from damage.

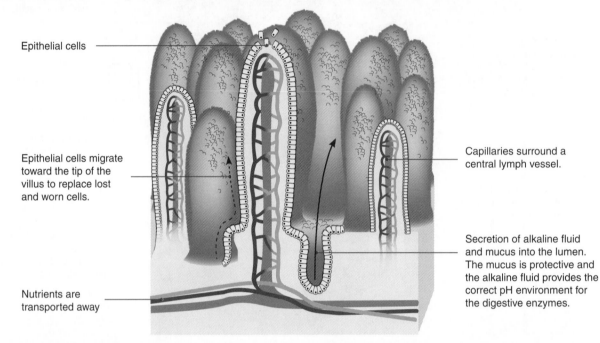

Epithelial cells

Epithelial cells migrate toward the tip of the villus to replace lost and worn cells.

Nutrients are transported away

Capillaries surround a central lymph vessel.

Secretion of alkaline fluid and mucus into the lumen. The mucus is protective and the alkaline fluid provides the correct pH environment for the digestive enzymes.

Photographs below: The intestinal villi are shown projecting into the gut lumen in a scanning electron micrograph (left image) and in a light microscope image (center image). The microvilli forming the brush border of a single intestinal epithelial cell are shown in the transmission electron micrograph (right image).

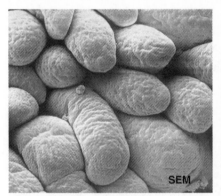
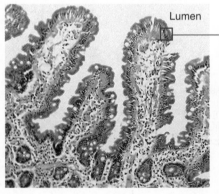
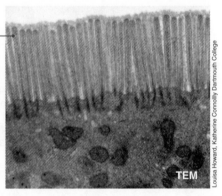

Lumen

Louisa Howard, Katherine Connolly Dartmouth College

Enzymes in the small intestine break down food into small molecules that can be absorbed through the gut wall. Enzymes are present in the pancreatic juice added to the duodenum, in intestinal juice, and bound to the surfaces of the intestinal epithelial cells.

Enzymes in pancreatic juice		Enzymes in intestinal juice (IJ) and epithelium (E)	
Enzymes in duodenum (optimal pH)		Enzymes in small intestine (location, optimal pH)	
1. Pancreatic amylase (6.7-7.0)	1. Starch → maltose	1. Maltase (E, 6.0-6.5)	1. Maltose → glucose
2. Trypsin* (7.8-8.7)	2. Protein → peptides	2. Peptidases (IJ, E, ~ 8.0)	2. Polypeptides → amino acids
3. Chymotrypsin* (7.8)	3. Protein → peptides	3. Sucrase (E, ~6.0)	3. Sucrose → fructose & glucose
4. Pancreatic lipase (8.0)	4. Fats → fatty acids & glycerol	4. Enteropeptidase (IJ 8.0)	4. Activates trypsin*
* secreted in an inactive form		*Once activated, trypsin activates chymotrypsin	

©2017 **BIOZONE** International
ISBN: 978-1-927309-65-0
Photocopying Prohibited

1. Summarize the structure and role of each of the following regions of the human digestive tract:

   (a) Stomach: _____

   _____

   _____

   _____

   (b) Small intestine: _____

   _____

   _____

   _____

2. (a) What is the purpose of the hydrochloric acid produced by the parietal cells of the stomach? _____

   _____

   _____

   (b) Explain why protein-digesting enzymes (e.g. pepsin) are secreted in an inactive form and then activated after release:

   _____

   _____

3. Identify an endocrine cell in the stomach epithelium and state its purpose: _____

   _____

4. How does the stomach achieve the mixing of acid and enzymes with food? _____

   _____

5. (a) What is the purpose of the intestinal villi? _____

   _____

   (b) What is the purpose of the microvilli (brush border) on intestinal epithelial cells? _____

   _____

   _____

6. Identify two sites for secretion of enzymes active in the small intestine. Identify an enzyme produced there and its role:

   (a) Site: _____ Enzyme: _____

   Enzyme's role: _____

   _____

   (b) Site: _____ Enzyme: _____

   Enzyme's role: _____

   _____

   (c) In general, do the enzymes act in acidic or alkaline conditions? _____

   _____

   (d) How is this pH environment generated? _____

7. Suggest why the small intestine is so long: _____

   _____

©2017 **BIOZONE** International
ISBN: 978-1-927309-65-0
Photocopying Prohibited

110 The Large Intestine

Key Idea: The large intestine absorbs water and solidifies the indigestible material before eliminating the undigested waste as feces from the anus (a process called egestion).

After most of the nutrients have been absorbed in the small intestine, the remaining semi-fluid contents pass into the large intestine (appendix, cecum, colon, and rectum). The large intestine's main role is to reabsorb water and electrolytes, consolidating the waste material into semi-solid masses called feces, which are collected in the rectum before being expelled from the anus (a process called egestion).

▶ After most of the nutrients have been absorbed in the small intestine, the remaining semi-fluid contents pass into the large intestine. This mixture includes undigested or indigestible food, (such as **cellulose**), bacteria, dead cells, mucus, bile, ions, and water. In humans and other omnivores, the large intestine's main role is to reabsorb water and electrolytes, and form the consolidated material into feces.

▶ The appendix is a blind ending sac off the cecum. It may have a minor immune function although it is not needed for normal gut function.

▶ The rectum is the final part of the large intestine and stores the fecal material before it is discharged out the anus. Fullness in the rectum produces the urge to defecate. If too little water is absorbed, the feces will be watery as in diarrhea. If too much water is absorbed the feces will become compacted and difficult to pass.

▶ Defecation is controlled by the anal sphincters, whose usual state is to be contracted (closing the orifice). Defecation is under nervous control.

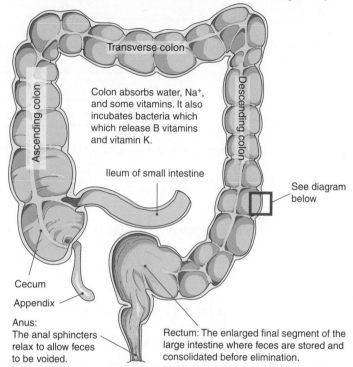

Transverse colon

Ascending colon

Descending colon

Colon absorbs water, Na⁺, and some vitamins. It also incubates bacteria which which release B vitamins and vitamin K.

Ileum of small intestine

See diagram below

Cecum

Appendix

Anus:
The anal sphincters relax to allow feces to be voided.

Rectum: The enlarged final segment of the large intestine where feces are stored and consolidated before elimination.

Lining of the large intestine

The lining of the large intestine has a simple epithelium containing tubular glands (crypts) with many mucus-secreting cells. The mucus lubricates the colon wall and helps to form and move the feces. In the photograph, some of the crypts are in cross section (XS) and some are in longitudinal section (LS).

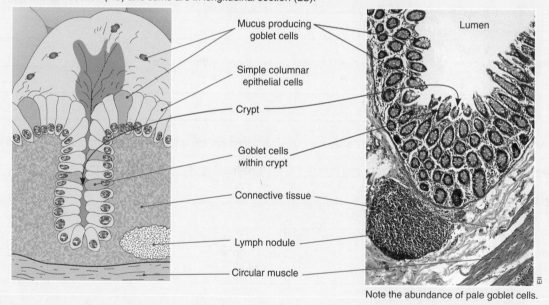

Mucus producing goblet cells

Simple columnar epithelial cells

Crypt

Goblet cells within crypt

Connective tissue

Lymph nodule

Circular muscle

Lumen

Note the abundance of pale goblet cells.

1. What is the main purpose of the large intestine? _____

_____

_____

2. What are the effects of absorbing too little and too much water in the large intestine? _____

_____

WEB   CONNECT   CONNECT
110   29   66
            AP2   AP1

KNOW

©2017 **BIOZONE** International
ISBN: 978-1-927309-65-0
Photocopying Prohibited

111 The Liver's Homeostatic Role

Key Idea: The liver is the body's largest homeostatic organ and interacts with other systems to maintain homeostasis.
The liver is the body's largest homeostatic organ, performing a vast number of functions that affect all other body systems. The liver has a unique double blood supply and up to 20% of the total blood volume flows through it at any one time. 25% of the blood flowing through the liver comes from the hepatic artery and 75% comes from the hepatic portal vein. This rich vascularization makes it the central organ for regulating activities associated with the blood and circulatory system.

Homeostatic functions of the liver

The liver is one of the largest and most complex organs in the body, with a large number of homeostatic functions. The liver:

1. Secretes bile, important in emulsifying fats in digestion.
2. Metabolizes amino acids, fats, and carbohydrates (below).
3. Synthesizes glucose from non-carbohydrate sources when glycogen stores are exhausted (gluconeogenesis).
4. Stores iron, copper, and some vitamins (A, D, E, K, B_{12}).
5. Converts unwanted amino acids to urea (urea cycle, right).
6. Manufactures heparin and plasma proteins (e.g. albumin) essential in maintaining oncotic pressure (this is the opposing force to hydrostatic pressure in the movement of fluids across membranes, e.g into and out of capillaries).
7. Detoxifies poisons or turns them into less harmful forms.
8. Some liver cells phagocytose worn-out blood cells.
9. Synthesizes cholesterol from acetyl coenzyme A.

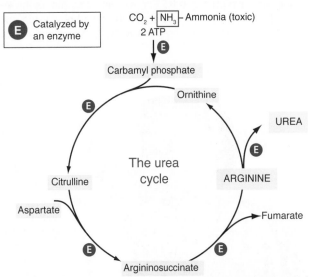

Nutrient processing in the liver

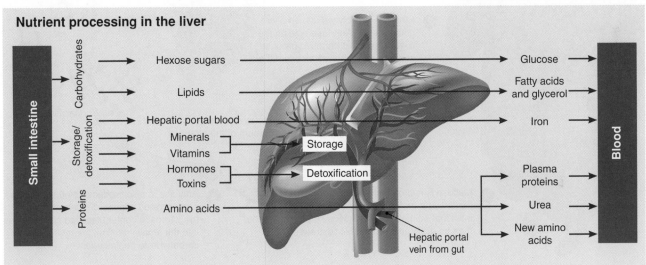

1. Explain how the liver's interactions with other body systems make it central to homeostatic regulation: _____

_____

_____

_____

_____

2. (a) Explain how the liver's rich blood supply enables it to regulate activities associated with the circulatory system:

_____

_____

(b) What is the significance of the liver receiving blood from the gut via a portal system? _____

_____

_____

3. Predict what would happen if the enzyme converting ammonia to carbamyl phosphate was inhibited: _____

_____

©2017 **BIOZONE** International
ISBN: 978-1-927309-65-0
Photocopying Prohibited

CONNECT
100
AP2

CONNECT
45
AP2

WEB
111

KNOW

112 Digestion, Absorption, and Transport

Key Idea: Food must be digested into components small enough to be absorbed by the body's cells and assimilated. Nutrient absorption involves both active and passive transport. Digestion breaks down food molecules into small molecules that can pass through the intestinal lining into the underlying blood and lymph vessels. For example, starch is broken down first into maltose and short chain carbohydrates such as dextrose, before being hydrolyzed to the simple sugar glucose (below). Breakdown products of other foodstuffs include amino acids (from proteins), and fatty acids, glycerol, and acylglycerols (from fats). The passage of these molecules from the gut into the blood or lymph is called absorption. Nutrients are then transported directly or indirectly to the liver for storage or processing. After they have been **absorbed** nutrients can be **assimilated**, i.e incorporated into the substance of the body itself.

Digestion of starch

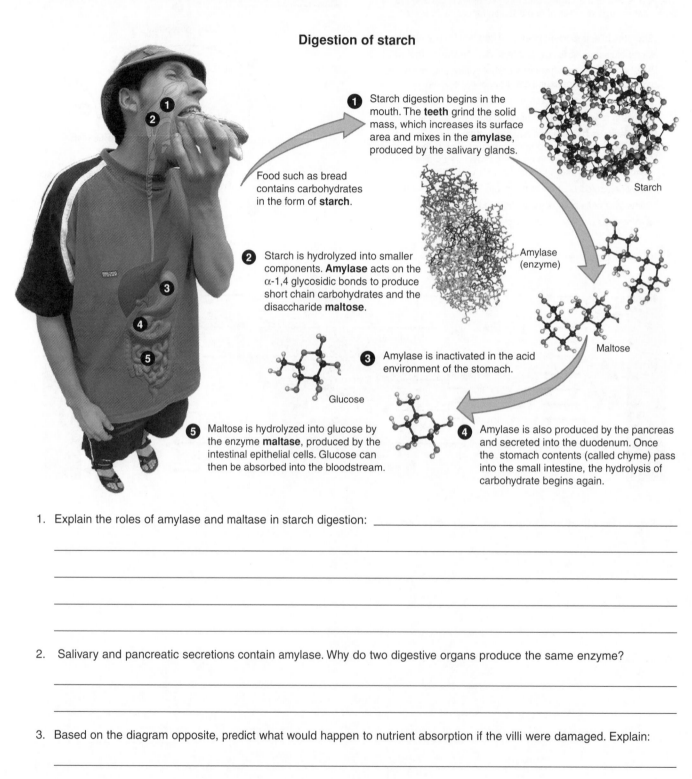

1 Starch digestion begins in the mouth. The **teeth** grind the solid mass, which increases its surface area and mixes in the **amylase**, produced by the salivary glands.

Food such as bread contains carbohydrates in the form of **starch**.

2 Starch is hydrolyzed into smaller components. **Amylase** acts on the α-1,4 glycosidic bonds to produce short chain carbohydrates and the disaccharide **maltose**.

3 Amylase is inactivated in the acid environment of the stomach.

4 Amylase is also produced by the pancreas and secreted into the duodenum. Once the stomach contents (called chyme) pass into the small intestine, the hydrolysis of carbohydrate begins again.

5 Maltose is hydrolyzed into glucose by the enzyme **maltase**, produced by the intestinal epithelial cells. Glucose can then be absorbed into the bloodstream.

Starch

Amylase (enzyme)

Maltose

Glucose

1. Explain the roles of amylase and maltase in starch digestion: _____

_____

_____

_____

_____

2. Salivary and pancreatic secretions contain amylase. Why do two digestive organs produce the same enzyme?

_____

_____

3. Based on the diagram opposite, predict what would happen to nutrient absorption if the villi were damaged. Explain:

_____

_____

_____

_____

©2017 **BIOZONE** International
ISBN: 978-1-927309-65-0
Photocopying Prohibited

Nutrient absorption by intestinal villi

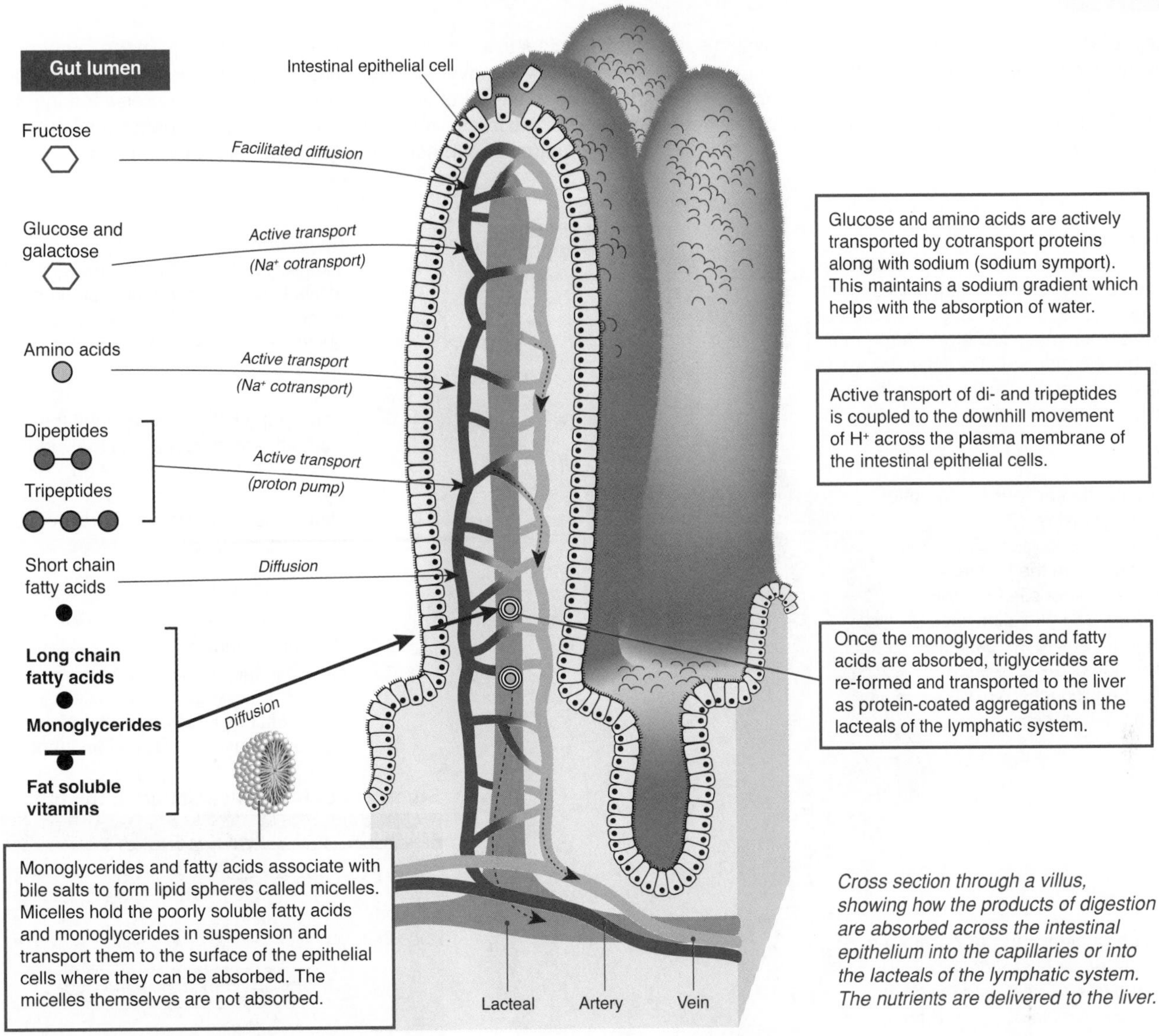

Gut lumen

Intestinal epithelial cell

Fructose — *Facilitated diffusion*

Glucose and galactose — *Active transport (Na⁺ cotransport)*

Amino acids — *Active transport (Na⁺ cotransport)*

Dipeptides

Tripeptides — *Active transport (proton pump)*

Short chain fatty acids — *Diffusion*

Long chain fatty acids

Monoglycerides

Fat soluble vitamins

Diffusion

Glucose and amino acids are actively transported by cotransport proteins along with sodium (sodium symport). This maintains a sodium gradient which helps with the absorption of water.

Active transport of di- and tripeptides is coupled to the downhill movement of H⁺ across the plasma membrane of the intestinal epithelial cells.

Once the monoglycerides and fatty acids are absorbed, triglycerides are re-formed and transported to the liver as protein-coated aggregations in the lacteals of the lymphatic system.

Monoglycerides and fatty acids associate with bile salts to form lipid spheres called micelles. Micelles hold the poorly soluble fatty acids and monoglycerides in suspension and transport them to the surface of the epithelial cells where they can be absorbed. The micelles themselves are not absorbed.

Lacteal   Artery   Vein

Cross section through a villus, showing how the products of digestion are absorbed across the intestinal epithelium into the capillaries or into the lacteals of the lymphatic system. The nutrients are delivered to the liver.

4. Describe how each of the following nutrients are absorbed by the intestinal villi:

(a) Glucose: _____

(b) Fructose: _____

(c) Amino acids: _____

(d) Di- and tripeptides: _____

5. Describe the two purposes of the sodium symport in the intestinal epithelium:_____

_____

_____

6. What is the role of micelles in the absorption of lipids? _____

_____

_____

7. How are concentration gradients maintained for the absorption of nutrients by diffusion? _____

_____

_____

113 Control of Digestion

Key Idea: The endocrine and nervous systems are both involved in the regulation of digestion.

The majority of digestive juices are secreted only when there is food in the gut and both nervous and hormonal mechanisms are involved in coordinating and regulating this activity. The digestive system is innervated by branches of the autonomic nervous system. Hormonal regulation is achieved through the activity of several hormones, which are released into the bloodstream in response to nervous or chemical stimuli and influence the activity of gut and associated organs.

Feeding center:
The feeding center in the hypothalamus continuously monitors metabolites in the blood and stimulates hunger when these metabolites reach low levels. After a meal, the neighboring satiety center suppresses the activity of the feeding center for a period of time.

Pancreatic secretions and bile:
Cholecystokinin (CCK) stimulates secretion of enzyme-rich fluid from the pancreas and release of bile from the gall bladder. Secretin stimulates the pancreas to increase its secretion of alkaline fluid and the production of bile from the liver cells.

Vagus nerve

Gastrin

CCK and secretin

Intestinal secretion of hormones:
The entry of chyme (especially fat and gastric acid) into the small intestine stimulates the intestinal mucosa to secrete the hormones cholecystokinin (CCK) and secretin.

Salivation:
Entirely under nervous control. Some saliva is secreted continuously. Food in the mouth stimulates the salivary glands to increase their secretions.

Parasympathetic stimulation of the stomach and pancreas via the vagus nerve increases their secretion. Sympathetic stimulation has the opposite effect. These are reflexes in response to the sight, smell, or taste of food.

Gastric secretion:
Physical distension and the presence of food in the stomach causes release of the hormone gastrin from cells in the gastric mucosa. Gastrin in the blood increases gastric secretion and motility.

Summary of hormones acting in the gut

Hormone	Organ	Effect
Secretin	Pancreas	Increases secretion of alkaline fluid
Secretin	Liver	Increases bile production
CCK	Pancreas	Increases enzyme secretion
CCK	Liver	Stimulates release of bile
Gastrin	Stomach	Increases stomach motility and secretion

1. Describe the role of each of the following stimuli in the control of digestion, identifying both the response and its effect:

   (a) Presence of food in the mouth: _____

   _____

   (b) Presence of fat and acid in the small intestine: _____

   _____

   (c) Stretching of the stomach by the presence of food: _____

   _____

2. Outline the role of the vagus nerve in regulating digestive activity: _____

   _____

3. Describe the role of nerves and hormones in controlling digestion: _____

   _____

   _____

   _____

WEB
113

CONNECT
181
AP2

CONNECT
80
AP1

KNOW

©2017 **BIOZONE** International
ISBN: 978-1-927309-65-0
Photocopying Prohibited

114 The Urinary System

Key Idea: The urinary system filters the blood and removes wastes, producing urine. The functional unit of the kidney is a selective filter element called the nephron.

The urinary system consists of the kidneys and bladder, and their associated blood vessels and ducts. The **kidneys** have a plentiful blood supply from the renal artery. The blood plasma is filtered by the **kidney nephrons** to form urine. Urine is produced continuously, passing along the **ureters** to the **bladder**, a hollow muscular organ lined with smooth muscle and stretchable epithelium. Each day the kidneys filter about 180 dm³ of plasma. Most of this is reabsorbed, leaving a daily urine output of about 1 dm³. By adjusting the composition of the fluid excreted, the kidneys help to maintain the body's internal chemical balance. Human kidneys are very efficient, producing a urine that is concentrated to varying degrees depending on requirements.

Urinary system

Vena cava
Dorsal aorta
Kidney
Ureter
Bladder
Urethra

Kidneys *in-situ* (rat)

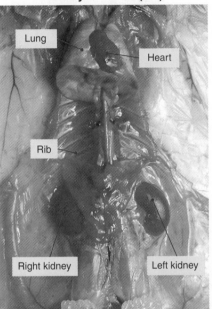

Lung
Heart
Rib
Right kidney
Left kidney

Eli

Sagittal section of kidney (pig)

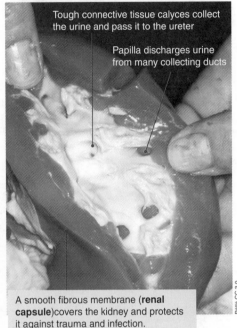

Tough connective tissue calyces collect the urine and pass it to the ureter

Papilla discharges urine from many collecting ducts

Pöllö CC 3.0

A smooth fibrous membrane (**renal capsule**) covers the kidney and protects it against trauma and infection.

The kidneys of humans (above), rats (dissection, above center), and many other mammals (e.g. pig above right) are distinctive, bean shaped organs that lie at the back of the abdominal cavity to either side of the spine. The kidneys lie outside the peritoneum of the abdominal cavity (**retroperitoneal**) and are partly protected by the lower ribs (see kidneys *in-situ* above center).

Human kidneys are ~100-120 mm long and 25 mm thick. A cut through in a sagittal plane (see photo above right), reveals numerous tough connective tissue calyces. These collect the urine from the papillae where it is discharged and drain it into the ureter.

The kidneys and their blood supply

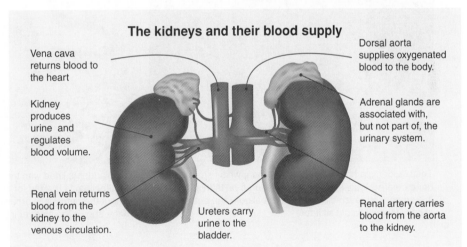

Vena cava returns blood to the heart

Kidney produces urine and regulates blood volume.

Renal vein returns blood from the kidney to the venous circulation.

Ureters carry urine to the bladder.

Dorsal aorta supplies oxygenated blood to the body.

Adrenal glands are associated with, but not part of, the urinary system.

Renal artery carries blood from the aorta to the kidney.

1. State the function of each of the following components of the urinary system:

    (a) Kidney: _____

    (b) Ureters: _____

    (c) Bladder: _____

    (d) Urethra: _____

    (e) Renal artery: _____

    (f) Renal vein: _____

    (g) Renal capsule: _____

PRACTICES

CONNECT
87
AP2

CONNECT
86
AP2

WEB
114

KNOW

Internal structure of the human kidney

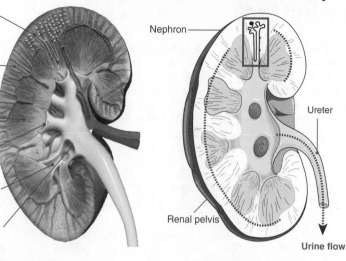

Nephrons are arranged with all the collecting ducts pointing towards the renal pelvis.

Outer cortex contains the renal corpuscles and convoluted tubules.

Inner medulla is organized into pyramids.

Each pyramid ends in a papilla or opening.

Urine enters the **calyces** Urine collects in a space near the ureter called the renal pelvis, before leaving the kidney via the ureter.

Nephron

Ureter

Renal pelvis

Urine flow

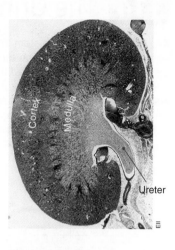

Ureter

The functional units of the kidney are selective filter elements called **nephrons**. Each kidney contains more than 1 million nephrons and they are precisely aligned so that urine is concentrated as it flows towards the ureter (model and diagram above). The alignment gives the kidney tissue a striated (striped) appearance and makes it possible to accommodate all the filtering units needed.

The outer cortex and inner medulla can be seen in a low power LM of the kidney. The ureter is seen extending into the fat and connective tissue surrounding and protecting the kidney.

The bladder

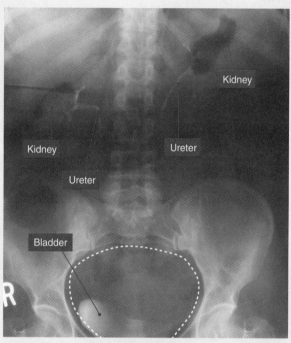

Kidney

Kidney

Ureter

Ureter

Bladder

R

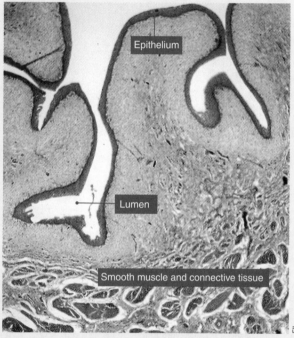

Epithelium

Lumen

Smooth muscle and connective tissue

The bladder is a hollow stretchable organ, which stores the urine before it leaves the body via the urethra. In this X-ray, it is empty and resembles a deflated balloon. The dotted line shows where it would sit if full.

The bladder is lined with transitional epithelium. This type of epithelium is layered, or stratified, so it can be stretched without the outer cells breaking apart from each other. This image shows the bladder in a deflated state.

2. Calculate the percentage of the plasma reabsorbed by the kidneys: _____

3. (a) What is a nephron? _____

   (b) What is its role in excretion? _____

   _____

4. (a) Where would you find transitional epithelium in the urinary system: _____

   (b) Why do you find this type of epithelium here? _____

   _____

5. In adults, the opening of the urethra is regulated by a voluntary sphincter muscle. What is the purpose of this sphincter?

   _____

©2017 **BIOZONE** International
ISBN: 978-1-927309-65-0
Photocopying Prohibited

115 The Physiology of the Kidney

Key Idea: Each nephron comprises a renal corpuscle and its associated tubules and ducts. It produces the urine by ultrafiltration, selective reabsorption, and secretion.

Ultrafiltration, i.e. forcing fluid and dissolved substances through a membrane by pressure, occurs in the first part of the nephron, across the membranes of the capillaries and the glomerular capsule. The passage of water and solutes into the nephron and the formation of the glomerular filtrate depends on the pressure of the blood entering the afferent arteriole (below). If it increases, filtration rate increases. When it falls, glomerular filtration rate also falls. This process is so precisely regulated that, in spite of fluctuations in arteriolar pressure, glomerular filtration rate per day stays constant. After formation of the initial filtrate, the **urine** is modified through secretion and tubular reabsorption according to physiological needs at the time.

Nephron structure and function

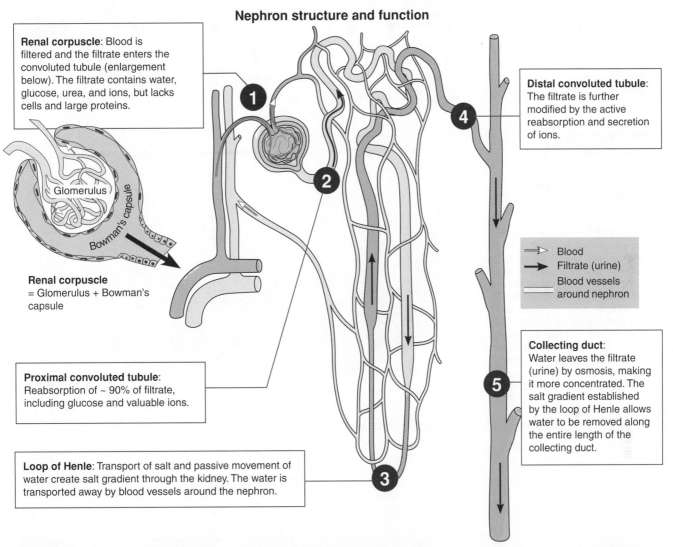

Renal corpuscle: Blood is filtered and the filtrate enters the convoluted tubule (enlargement below). The filtrate contains water, glucose, urea, and ions, but lacks cells and large proteins.

Glomerulus

Bowman's capsule

Renal corpuscle
= Glomerulus + Bowman's capsule

Proximal convoluted tubule: Reabsorption of ~ 90% of filtrate, including glucose and valuable ions.

Loop of Henle: Transport of salt and passive movement of water create salt gradient through the kidney. The water is transported away by blood vessels around the nephron.

Distal convoluted tubule: The filtrate is further modified by the active reabsorption and secretion of ions.

Blood
Filtrate (urine)
Blood vessels around nephron

Collecting duct: Water leaves the filtrate (urine) by osmosis, making it more concentrated. The salt gradient established by the loop of Henle allows water to be removed along the entire length of the collecting duct.

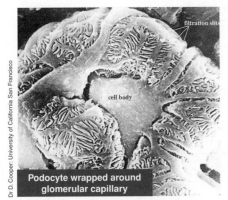

filtration slits

cell body

Podocyte wrapped around glomerular capillary

Dr D. Cooper: University of California San Francisco

The epithelium of Bowman's capsule is made up of specialized cells called **podocytes**. The finger-like cellular processes of the podocytes wrap around the capillaries of the glomerulus, and the plasma filtrate passes through the filtration slits between them.

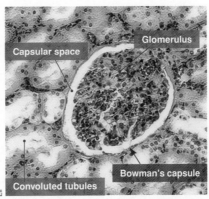

Glomerulus

Capsular space

Convoluted tubules

Bowman's capsule

Bowman's capsule is a double walled cup, lying in the cortex of the kidney. It encloses a dense capillary network called the **glomerulus**. The capsule and its enclosed glomerulus form a **renal corpuscle**. In this section, the convoluted tubules can be seen surrounding the renal corpuscle.

Normal, fresh urine is clear and pale to dark yellow or amber in color. A urine dipstick test is a fast and convenient way to make a qualitative analysis of urine to diagnose a medical problem. The presence of specific molecules in the urine (e.g. glucose) are indicated by a color change on the dipstick.

PRACTICES

CONNECT **66** AP1

CONNECT **100** AP2

CONNECT **87** AP2

CONNECT **86** AP2

CONNECT **77** AP2

WEB **115**

KNOW

Summary of activities in the kidney nephron

Urine formation begins by **ultrafiltration** of the blood, as fluid is forced through the capillaries of the glomerulus, forming a filtrate similar to blood but lacking cells and proteins. The filtrate is then modified by **secretion** and **reabsorption** to add or remove substances (e.g. ions). The processes involved in urine formation are summarized below. The loop of Henle acts as a **countercurrent multiplier**, establishing and increasing the salt gradient through the medullary region. This is possible because the descending loop is freely permeable to water but the ascending loop is not.

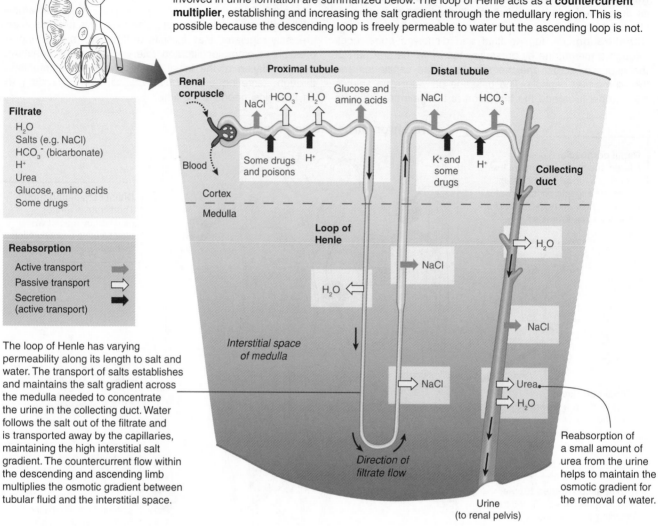

Filtrate

H_2O
Salts (e.g. NaCl)
HCO_3^- (bicarbonate)
H^+
Urea
Glucose, amino acids
Some drugs

Reabsorption

Active transport
Passive transport
Secretion (active transport)

The loop of Henle has varying permeability along its length to salt and water. The transport of salts establishes and maintains the salt gradient across the medulla needed to concentrate the urine in the collecting duct. Water follows the salt out of the filtrate and is transported away by the capillaries, maintaining the high interstitial salt gradient. The countercurrent flow within the descending and ascending limb multiplies the osmotic gradient between tubular fluid and the interstitial space.

Reabsorption of a small amount of urea from the urine helps to maintain the osmotic gradient for the removal of water.

1. Why does the kidney receive blood at a higher pressure than other organs? _____

_____

2. Explain the importance of the following in the production of urine in the kidney nephron:

    (a) Filtration of the blood at the glomerulus: _____

    _____

    (b) Active secretion: _____

    _____

    (c) Reabsorption: _____

    _____

    (d) Osmosis: _____

    _____

3. (a) What is the purpose of the salt gradient in the kidney? _____

    _____

    (b) How is this salt gradient produced? _____

    _____

    _____

©2017 **BIOZONE** International
ISBN: 978-1-927309-65-0
Photocopying Prohibited

116 Control of Urine Output

Key Idea: The body's balance of fluid and electrolytes is regulated by varying the composition and volume of urine. This is achieved through the action of the hormones antidiuretic hormone (ADH) and aldosterone.

The body regulates the composition and volume of the blood to compensate for variations in salt and water intake, and environmental conditions. This is achieved by varying the volume and composition of the urine and is under hormonal control. Antidiuretic hormone (ADH), from the posterior pituitary, regulates water reabsorption from the kidney collecting duct. Aldosterone, from the adrenal cortex, regulates sodium absorption from the kidney tubules.

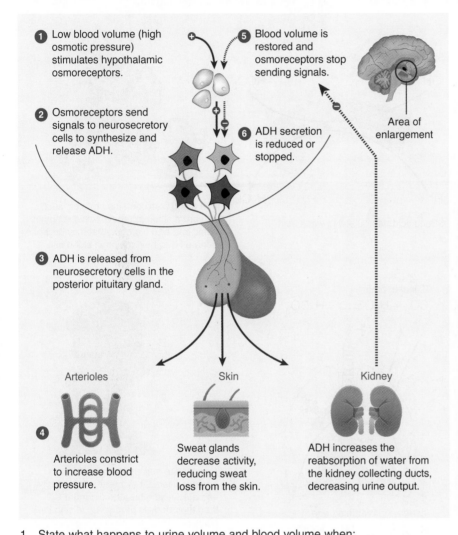

1 Low blood volume (high osmotic pressure) stimulates hypothalamic osmoreceptors.

2 Osmoreceptors send signals to neurosecretory cells to synthesize and release ADH.

3 ADH is released from neurosecretory cells in the posterior pituitary gland.

5 Blood volume is restored and osmoreceptors stop sending signals.

6 ADH secretion is reduced or stopped.

Area of enlargement

Arterioles

4 Arterioles constrict to increase blood pressure.

Skin

Sweat glands decrease activity, reducing sweat loss from the skin.

Kidney

ADH increases the reabsorption of water from the kidney collecting ducts, decreasing urine output.

Osmoreceptors in the **hypothalamus** of the brain respond to changes in blood volume. A fall in blood volume stimulates the synthesis and secretion of the hormone ADH (antidiuretic hormone), which is released from the posterior pituitary into the blood. ADH increases the permeability of the kidney collecting duct to water so that more water is reabsorbed and urine volume decreases. A second hormone, aldosterone, helps by increasing sodium reabsorption.

Factors causing ADH release

▶ Low blood volume
  = More negative water potential
  = High blood sodium levels
  = Low fluid intake
▶ Nicotine and morphine

Factors inhibiting ADH release

▶ High blood volume
  = Less negative water potential
  = Low blood sodium levels
▶ High fluid intake
▶ Alcohol consumption

Factors causing the release of aldosterone

Low blood volumes also stimulate secretion of aldosterone from the adrenal cortex. This is mediated through a complex pathway involving osmoreceptors near the kidney glomeruli and the hormone renin from the kidney.

1. State what happens to urine volume and blood volume when:

   (a) ADH secretion increases: _____

   (b) ADH secretion decreases: _____

2. Diabetes insipidus is caused by a lack of ADH. From what you know about ADH, describe the symptoms of this disease:

   _____

   _____

3. Explain why alcohol consumption (especially to excess) causes dehydration and thirst: _____

   _____

4. (a) State the effect of aldosterone on the kidney nephron: _____

   (b) What would be the net result of this effect: _____

5. Explain the role of negative feedback in the regulation of blood volume and urine output: _____

   _____

   _____

©2017 **BIOZONE** International
ISBN: 978-1-927309-65-0
Photocopying Prohibited

CONNECT
80
AP1

WEB
116

KNOW

117 Cooperating Systems: Acid–Base Balance

Key Idea: The body's acid-base balance is maintained by interactions between three body systems.

Normal functioning of the body requires that the pH of the body's fluids are maintained between pH 7.35 and 7.45. The products of metabolic activity are generally acidic and could alter pH considerably without a buffer system to counteract pH changes. The carbonic acid-bicarbonate buffer works throughout the body to maintain the pH of blood plasma close to 7.40. The body maintains the buffer by eliminating either the acid (carbonic acid) or the base (bicarbonate ions). The blood buffers, the lungs, and the kidneys interact to maintain pH homeostasis. Changes in breathing rate bring about rapid changes in pH. The renal system acts more slowly, controlling pH by either excreting or retaining ions.

The blood buffer system

A buffer is able to resist changes to the pH of a fluid when either an acid or base is added to it. The bicarbonate ion (HCO_3^-) and its acid, carbonic acid (H_2CO_3), work in the following way:

$$H^+ + HCO_3^- \rightleftharpoons H_2CO_3$$

$$H_2CO_3 \rightleftharpoons H^+ + HCO_3^-$$

If a strong acid (such as HCl) is added to the system a weak acid is formed and thus the pH falls only slightly.

Strong base neutralized to weak base

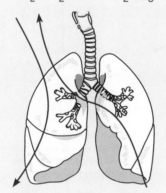

$$OH^- \qquad HCO_3^-$$

$$H^+ \qquad H_2CO_3$$

Strong acid neutralized to weak acid

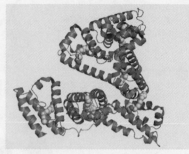

The blood also contains proteins, such as serum albumin (above) which contain basic and acidic groups that may accept or donate H^+ to help maintain blood pH.

The respiratory system

Carbon dioxide (CO_2) in the blood, an end-product of cellular respiration, forms carbonic acid (H_2CO_3) which dissociates to form H^+ and bicarbonate (HCO_3^-).

As CO_2 rises in the blood so too does the H^+ concentration. Chemoreceptors in the brain detect the rise in H^+ ions and increase the rate of breathing to expel the CO_2.

Low levels of CO_2 have the effect of depressing the respiratory system so that H^+ builds up and the pH is once again restored.

Signal to brain

$$CO_2 + H_2O \rightleftharpoons H_2CO_3$$

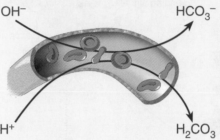

Increase in breathing rate

$$H_2CO_3 \rightleftharpoons H^+ + HCO_3^-$$

Anxiety can make some people hyperventilate. They breathe too deeply and quickly and breathe out more CO_2 than their body is producing, raising their blood pH. They can reverse the effects by breathing into a paper bag.

The renal system

A net loss of HCO_3^- effectively results in the gain of H^+.

Bicarbonate is reabsorbed by the kidney tubules all the time so pH is regulated mainly through retaining or secreting H^+. When blood pH rises, H^+ is retained by the tubule cells. When blood pH falls, H^+ is actively secreted into the kidney tubules. The kidneys can also produce HCO_3^- which enters the body fluids.

Urine pH normally varies from 4.5 to 8.0, reflecting the ability of the renal tubules to lose or retain ions to maintain blood pH homeostasis.

Rise in pH stimulates:

Retain H^+

Fall in pH stimulates:

Removal H^+

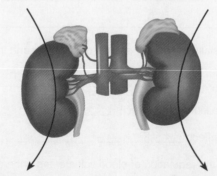

Equates to removal of HCO_3^-

Equates to gain of HCO_3^-

Urine has a slightly acid pH 6 but can range from pH 4.5-8.0. Diet, and certain disease processes and medications can alter urine pH showing that the kidneys are working to retain or excrete H^+ to regulate blood pH.

WEB   CONNECT   CONNECT   PRACTICES

117   **95** AP2   **97** AP2

KNOW

©2017 **BIOZONE** International
ISBN: 978-1-927309-65-0
Photocopying Prohibited

1. Why must the blood must be kept at a pH between 7.35 and 7.45? _____

_____

_____

_____

2. A drop in the blood pH to below 7.35 is called metabolic acidosis. If prolonged, it can be life threatening:

(a) From information on the previous page, explain how metabolic acidosis might arise: _____

_____

_____

(b) What would you expect the levels of bicarbonate ions to be in the blood of someone with metabolic acidosis?

_____

3 (a) How does the blood buffer system maintain blood pH? _____

_____

_____

_____

_____

(b) What happens when a base (e.g. ingestion of alkaline substances) is added to the system? _____

_____

_____

4. (a) Describe the respiratory response to excess H^+ in the blood: _____

_____

(b) Where do these H^+ ions come from? _____

_____

5. An abnormal increase in blood CO_2 is called respiratory acidosis.

(a) Explain the consequences to blood pH of increased CO_2: _____

(b) Explain how respiratory acidosis might arise: _____

_____

_____

6. (a) What would happen to the blood pH of someone who was hyperventilating during an anxiety attack? _____

_____

(b) Why does breathing into a paper bag help someone who is hyperventilating?

_____

7. Explain the role of the renal system in maintaining the pH of the blood: _____

_____

_____

_____

_____

118 Cooperating Systems: Generating Movement

Key Idea: The nervous and muscular systems interact to bring about movement through muscle contraction.

Movement is an essential activity requiring coordination between nervous and muscular systems to bring about muscle contraction. Nerve impulses originating at the central nervous system are carried by neurons, which terminate in a specialized junction (synapse) at the muscle fiber called the neuromuscular junction. The arrival of a nerve impulse at the neuromuscular junction results in release of the neurotransmitter acetylcholine and contraction of the fiber. A motor neuron and all the muscle fibers it innervates is called a motor unit. The response of a single muscle fiber is all-or-none, meaning it contracts maximally or not at all. However, the strength of the contraction in the entire muscle can be varied (below). This way, muscles can produce contractions of varying force, suitable for different tasks.

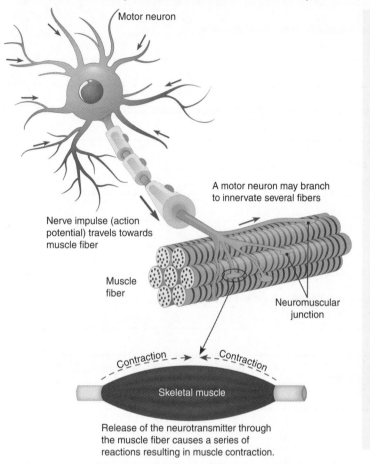

Motor neuron

Nerve impulse (action potential) travels towards muscle fiber

A motor neuron may branch to innervate several fibers

Muscle fiber

Neuromuscular junction

Contraction →   ← Contraction

Skeletal muscle

Release of the neurotransmitter through the muscle fiber causes a series of reactions resulting in muscle contraction.

Muscles have graded responses

Muscle fibers respond to an action potential by contracting maximally or not all. This response is called the **all or none law** of muscle contraction. However, skeletal muscles as a whole can produce contractions of varying force. This is achieved by changing the frequency of stimulation (more rapid arrival of action potentials) and by changing the number of fibers active at any one time (recruiting more motor units). A stronger muscle contraction is produced when a large number of muscle fibers are recruited (e.g. lifting weights), whereas less strenuous movements, such as picking up a pen, require fewer active fibers.

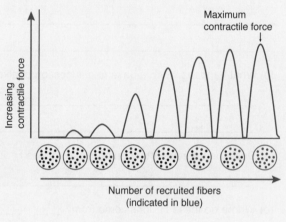

Increasing contractile force

Maximum contractile force

Number of recruited fibers (indicated in blue)

1. How do the nervous system and muscular system work together to bring about movement? _____

_____

_____

_____

_____

2. The release of acetylcholine to cause muscle contraction is an example of what type of signaling (circle correct answer):

Autocrine signaling / cell-to-cell communication / local regulation / endocrine signaling

3. (a) What is meant by an all-or-none response of a muscle fiber? _____

_____

_____

(b) How can a muscle produce contractions of varying force (e.g. effort required to pick up a pen versus the effort required to lift a heavy object?

_____

_____

_____

WEB 118   CONNECT 182 AP2   CONNECT 80 AP1   PRACTICES

KNOW

©2017 **BIOZONE** International
ISBN: 978-1-927309-65-0
Photocopying Prohibited

119 KEY TERMS AND IDEAS: Did You Get It?

1. (a) What process moves food through the gut? _____

   (b) In what region of the digestive system does most nutrient absorption occur? _____

   (c) What structures from the mammalian gastrointestinal tract are shown in the photograph (right)?

   _____

   (d) What is their function? _____

   (e) What organ secretes amylase into the small intestine?

   _____

2. (a) What type of blood vessel transports blood away from the heart? _____

   (b) What type of blood vessel transports blood to the heart? _____

   (c) What type of blood vessel enables exchanges between the blood and tissues? _____

3. (a) Add the following labels to the cross section of the heart right: *right ventricle, wall of left ventricle, left ventricle, wall of right ventricle.*

   (b) Explain why the heart has one ventricle with a thick wall and one with a thinner wall:

   _____

   _____

   _____

   _____

   _____

4. (a) What component of vertebrate blood is involved in carrying oxygen?

   _____

   (b) What cell types in vertebrate blood are involved in defense against pathogens? _____

   _____

5. (a) Name the excretory organ of mammals: _____

   (b) Name its selective filtering element: _____

6. Test your vocabulary by matching each term to its correct definition, as identified by its preceding letter code.

circulatory system	A	This organ system is responsible for the exchange of gases between the environment and the circulatory fluid.
digestive system	B	This system breaks down food into its components and processes it so that they can be absorbed by the body. Waste products are egested.
excretory system	C	Comprising the muscular and skeletal systems, this system interacts with the nervous system to produce locomotion.
musculoskeletal system	D	This system circulates blood through the body, delivering essential nutrients and gases to the tissues and removing waste. It consists of the heart, blood vessels, and circulatory fluid.
nervous system	E	This system coordinates and responds to changes in the environment by sending nerve impulses to effector organs.
respiratory system	F	The organ of this system filters materials from the blood, retaining useful components and eliminating waste products in urine. It has an important homeostatic role.

©2017 **BIOZONE** International
ISBN: 978-1-927309-65-0
Photocopying Prohibited

TEST

Defense Mechanisms

Key terms

active immunity

allergen

allergic response

antigen

antibody

B cell

cell-mediated
response

complement

cytotoxic T cell

humoral response

hypersensitive
reaction (plant)

hypersensitivity

immunity

immune response

inflammation

innate immune
system

memory cell

MHC antigens

passive immunity

pathogen

phagocytosis

plasma cell

resistance

specific immune
system

T cell

vaccination

2.D.3 Biological systems are affected by disruptions to homeostasis
Essential knowledge

Activity
number

(a) Disruptions at the molecular and cellular levels affect the health of the organism

☐ 1 Describe how homeostasis can be disrupted by allergens and pathogens.   **120 133**
Understand that immunological responses to foreign antigens can be appropriate   **134**
(as in internal defense) or inappropriate (hypersensitivity reactions to allergens).

2.D.4 Plants and animals have a variety of chemical defenses against infections that affect dynamic homeostasis
Essential knowledge

Activity
number

(a) Plants and animals have multiple, nonspecific defenses

☐ 1 Understand what is meant by an antigen and distinguish between self and non-   **120 121**
self antigens. Describe some of the many nonspecific defenses that plants and
animals have against physical damage and pathogens. Examples include:
- Invertebrate immune systems (nonspecific and pathogen specific responses).   **122**
- Plant defenses (molecular recognition systems with systemic responses and   **122**
  hypersensitive responses that result in localization of damage). *[also 3.D.2.b]*
- Nonspecific defense mechanisms in vertebrates (innate immune system)   **124 123**
  including anatomical barriers, inflammation, complement, and phagocytosis.   **125**

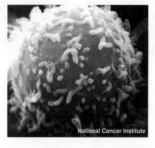
National Cancer Institute

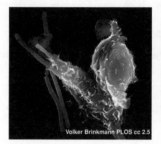
Volker Brinkmann PLOS cc 2.5

(b) Mammals use specific immune responses triggered by natural or artificial agents that disrupt dynamic homeostasis

☐ 1 Explain what is meant by specific immunity. Distinguish between the cell   **123 127**
mediated and humoral responses of the mammalian specific immune system.   **128 130**

☐ 2 Describe the cell-mediated response including the role and action of cytotoxic T   **126 127**
cells. Include reference to the role of antigenic presentation in this response.   **130**

☐ 3 Describe the humoral response, including the role of plasma (B) cells, memory   **127 128**
cells, and antibodies.   **130-132**

☐ 4 Explain how antigens are recognized by antibodies.   **132**

☐ 5 Describe clonal selection and explain how, through clonal selection, B cells   **131**
produce antibodies against specific antigens.

☐ 6 Explain why a second exposure to an antigen results in a faster and greater   **129**
immune response. Explain how this property of the immune system underlies the
use of vaccination programs to protect against infectious diseases.

4.C.1 Molecular variation provides cells with a range of functions
Essential knowledge

Activity
number

(a) Variations within molecular classes provide a wider range of functions

☐ 1 Using examples from the immune system, explain how molecular variations   **120 132**
within a class of molecules provide a wider range of functions. Examples include
MHC antigens and the diversity of antibodies in response to antigens.

120 The Nature of Antigens

Key Idea: Antigens are substances capable of producing an immune response. It is important that the body can distinguish its own tissues from foreign material so that it does not attack itself.

An **antigen** is any substance that evokes an immune response in an organism. Most antigens are **non-self antigens**, i.e. they are foreign and originate from outside the organism. Sometimes an organism will react to its own cells

and tissues. Antigens that originate from within the body are called **self-antigens**. Normally, because of the development of self-tolerance, the body recognizes and does not attack its own tissues. However, in some instances, the immune system may mistakenly destroy its own tissues. Such a response is called an autoimmune disorder. **Allergens** are a specific type of antigen, they produce a vigorous hypersensitive allergic response.

Distinguishing self from non-self

▶ Every type of cell has unique protein markers (antigens) on its surface. The type of antigen varies greatly between cells and between species. The immune system uses these markers to identify its own cells (self) from foreign cells (non-self). If the immune system recognizes the antigen markers, it will not attack the cell. If the antigen markers are unknown, the cell is attacked and destroyed.

▶ In humans, the system responsible for this property is the major histocompatibility complex (**MHC**). The MHC is a cluster of tightly linked genes on chromosome 6. These genes code for protein molecules (MHC antigens) that are attached to the surface of body cells. The main role of MHC antigens is to bind to antigenic fragments and display them on the cell surface so that they can be recognized by the cells of the immune system.

▶ Class I MHC antigens are found on the surfaces of almost all human cells. Class II MHC antigens occur only on macrophages and B-cells of the immune system.

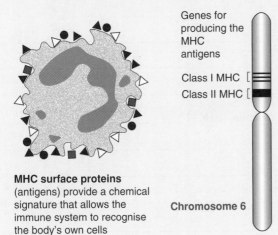

MHC surface proteins (antigens) provide a chemical signature that allows the immune system to recognise the body's own cells

Genes for producing the MHC antigens

Class I MHC
Class II MHC

Chromosome 6

Tolerance towards foreign bodies

▶ The human body has a very large population of resident microbes. Under normal conditions, *E.coli* in the gut form a protective layer preventing the colonization of pathogenic bacteria. The microbial cells have foreign antigens but they are not attacked by the immune system because **tolerance** (the prevention of an immune response) has developed.

E.coli

▶ During pregnancy, specific features of the self recognition system are suppressed to allow the mother to tolerate a nine month relationship with a foreign body (the fetus).

Intolerance to tissue transplants

The MHC is responsible for the rejection of tissue grafts and organ transplants. Foreign MHC molecules on the transplanted tissue are viewed as antigenic, causing the immune system to respond and the tissue to be rejected. To minimize rejection, attempts are made to match the MHC of the organ donor to that of the recipient as closely as possible. Immunosuppressant drugs are also used to minimize the immune response.

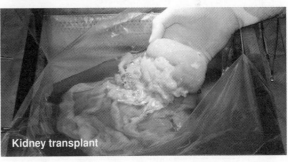

Kidney transplant

1. (a) What is an antigen? _____

_____

(b) Distinguish between non-self antigens, self antigens, and allergens: _____

_____

_____

_____

(c) Why is it important that the body detects foreign antigens? _____

_____

_____

_____

Types of antigens

Non-self antigens

Any foreign material provoking an immune response is termed a non-self antigen. Disease-causing organisms (pathogens) such as bacteria, viruses, and fungi are non-self antigens. The body recognizes them as foreign and will attack and destroy them before they cause harm.

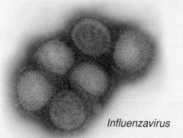

CDC

Influenzavirus

Pathogens have ways of avoiding detection. Mutations result in new surface antigens, delaying the immune response and allowing the pathogen to reproduce in its host undetected for a time (e.g. the flu virus, above). Some pathogens, e.g. the malaria-causing *Plasmodium*, switches off its surface antigens in order to enter cells undetected.

Self antigens

The body is usually tolerant of its own antigens. However, sometimes the self-tolerance system fails and the body attacks its own cells and tissues as though they were foreign. This can result in an autoimmune disorder in which tissue is destroyed, grows abnormally, or changes in function.

Autoimmune disorders, such as multiple sclerosis and rheumatoid arthritis, may be triggered by infection by a pathogen. Similarity in pathogen and self antigens is thought to be behind this failure of self recognition.

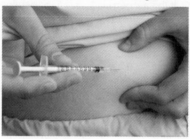

Type 1 diabetes is the result of autoimmune destruction of the insulin-producing pancreatic cells. Patients must inject insulin to maintain normal blood glucose levels.

Allergens

Antigens that cause allergic reactions are called allergens. An allergic reaction is a very specific type of immune response in which the immune system overreacts to a normally harmless substance. An allergic response can produce minor symptoms (itching, sneezing, rashes, swelling) or life-threatening anaphylaxis (respiratory and cardiovascular distress).

Common allergens include dust, chemicals, mold, pet hair, food proteins, or pollen grains.

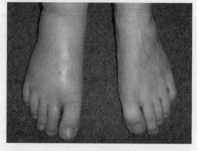

Kent Pryor

The swelling on the foot in the left of the photograph is a result of an allergic reaction to a bee sting.

2. How can pathogens avoid detection by the immune system? _____

_____

_____

3. (a) What is the nature and purpose of the major histocompatibility complex (MHC)? _____

_____

_____

_____

(b) Why is a self-recognition system important? _____

_____

4. (a) What is immune tolerance? _____

_____

(b) When might tolerance to foreign antigens be beneficial or necessary? _____

_____

_____

5. Using examples, describe what happens when the body develops an inappropriate response to:

(a) Self-antigens: _____

_____

(b) Normally non-antigenic substances: _____

_____

©2017 **BIOZONE** International
ISBN: 978-1-927309-65-0
Photocopying Prohibited

121 Blood Group Antigens

Key Idea: Blood groups classify blood according to the different marker molecules (antigens) on the surface of red blood cells (RBCs). These antigens determine the ability of RBCs to provoke an immune response.

Blood group typing is essential for safe **blood transfusion**. Transfusion with incompatible blood types will cause clumping of the red blood cells and cell lysis (hemolysis). Although human RBCs have more than 500 known antigens, fewer than 30 (in 9 blood groups) are regularly tested for when blood is donated for transfusion. The **ABO** and **rhesus** (Rh) blood group antigens are the best known. Where the father of a baby is Rh-positive and the mother is Rh-negative, a second baby, if Rh-positive, will suffer from **hemolytic disease of the newborn**. This severe immune reaction is caused by the mother's acquired antibodies attacking the fetal blood cells.

	Blood type A	**Blood type B**	**Blood type AB**	**Blood type O**
Antigens present on the red blood cells	antigen **A**	antigen **B**	antigens **A** and **B**	Neither antigen **A** nor **B**
Antibodies present in the plasma	Contains **anti-B** antibodies; but no antibodies that would attack its own antigen **A**	Contains **anti-A** antibodies; but no antibodies that would attack its own antigen **B**	Contains neither **anti-A** nor **anti-B** antibodies	Contains both **anti-A** and **anti-B** antibodies

Rh blood groups and HDN

Like the ABO grouping, the Rh system is based on antigens on the surfaces of red blood cells (RBCs). People whose RBCs have the Rh antigens on their surface are Rh⁺. Those who lack the Rh antigen are Rh⁻. About 84% of people in the US are Rh⁺. Normally, human plasma does not contain antibodies against the Rh antigen. However, if an Rh⁻ mother encounters Rh⁺ blood from the baby during delivery, her body will make antibodies against the antigen. If, in a second pregnancy, the fetus is Rh⁺, these antibodies will pass across the placenta and react with and destroy the baby's blood cells. This condition is called **hemolytic disease of the newborn** (HDN).

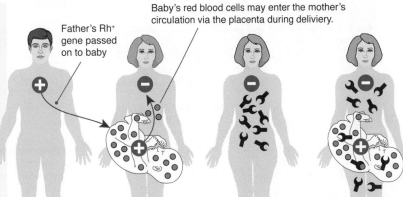

Baby's red blood cells may enter the mother's circulation via the placenta during delivery.

Father's Rh⁺ gene passed on to baby

Father is Rh⁺ (positive)

First pregnancy Rh⁻ mother is pregnant with Rh⁺ fetus. Rh⁺ antigens pass to the mother during labor.

After exposure to the Rh⁺ antigens, the mother makes anti-Rh antibodies.

Second pregnancy Mother's anti-Rh antibodies cross the placenta to the fetal blood. An Rh⁺ baby will develop HDN.

1. Complete the table below to show the antibodies and antigens present in each blood group, and donor blood types:

Blood Type	Freq. in US		Antigen	Antibody	Can donate blood to:	Can receive blood from:
	Rh⁺	**Rh⁻**				
A	34%	6%	A	anti-B	A, AB	A, O
B	9%	2%				
AB	3%	1%				
O	38%	7%				

2. What causes hemolytic disease of the newborn? _____

_____

_____

_____

©2017 **BIOZONE** International
ISBN: 978-1-927309-65-0
Photocopying Prohibited

CONNECT **148** AP1 WEB **121**

KNOW

122 Chemical Defenses In Plants and Animals

Key Idea: Plants and animals have chemical defenses against pathogens. Some mechanisms are always present while others are stimulated by the presence of a pathogen. Living organisms are under constant attack from pathogens. As a result plants and animals have evolved a wide range of chemical defenses to protect themselves from pathogens and limit the damage they can do. Some defense mechanisms are always present, while others (including the adaptive immune responses of animals) are stimulated only by an attack. The chemical defenses of plants not only protect them from attack by pathogens, but they also act to prevent browser damage and inhibit the growth of competitor plants.

Chemical defenses in animals

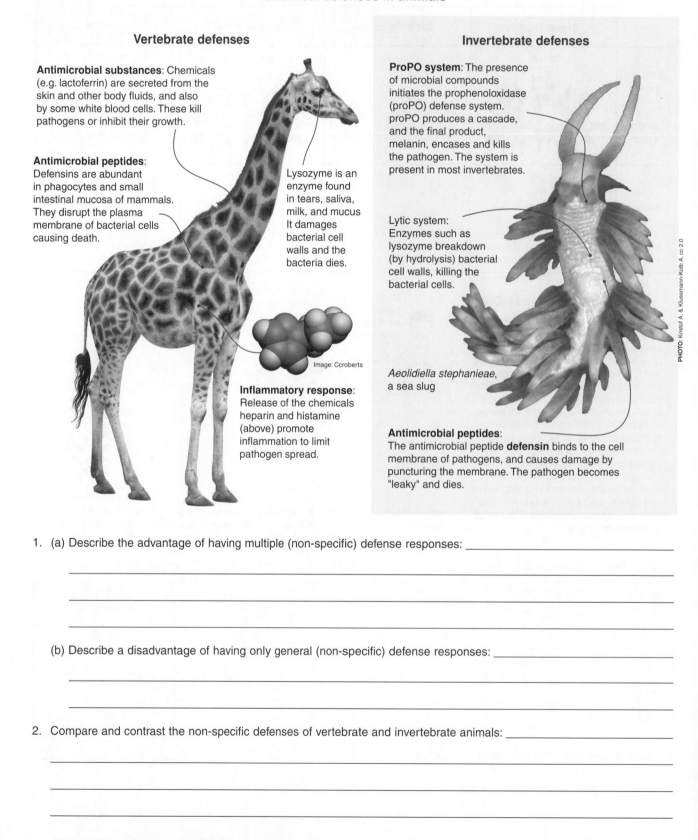

Vertebrate defenses

Antimicrobial substances: Chemicals (e.g. lactoferrin) are secreted from the skin and other body fluids, and also by some white blood cells. These kill pathogens or inhibit their growth.

Antimicrobial peptides: Defensins are abundant in phagocytes and small intestinal mucosa of mammals. They disrupt the plasma membrane of bacterial cells causing death.

Lysozyme is an enzyme found in tears, saliva, milk, and mucus It damages bacterial cell walls and the bacteria dies.

Image: Ccroberts

Inflammatory response: Release of the chemicals heparin and histamine (above) promote inflammation to limit pathogen spread.

Invertebrate defenses

ProPO system: The presence of microbial compounds initiates the prophenoloxidase (proPO) defense system. proPO produces a cascade, and the final product, melanin, encases and kills the pathogen. The system is present in most invertebrates.

Lytic system: Enzymes such as lysozyme breakdown (by hydrolysis) bacterial cell walls, killing the bacterial cells.

Aeolidiella stephanieae, a sea slug

PHOTO: Kristof A. & Klussmann-Kolb A. cc 2.0

Antimicrobial peptides: The antimicrobial peptide **defensin** binds to the cell membrane of pathogens, and causes damage by puncturing the membrane. The pathogen becomes "leaky" and dies.

1. (a) Describe the advantage of having multiple (non-specific) defense responses: _____

_____

_____

_____

(b) Describe a disadvantage of having only general (non-specific) defense responses: _____

_____

_____

2. Compare and contrast the non-specific defenses of vertebrate and invertebrate animals: _____

_____

_____

_____

_____

WEB
122

CONNECT
40
AP2

CONNECT
80
AP1

PRACTICES

KNOW

©2017 **BIOZONE** International
ISBN: 978-1-927309-65-0
Photocopying Prohibited

Chemical defenses in plants

Passive defenses

Passive defenses are always present and are not the result of contact with a pathogen or grazer. Plants have both physical and chemical defenses to deter pathogens. For example, the thick waxy surface of many leaves (right) acts as a physical barrier to limit pathogen entry. However, if the physical defense is breached, the chemical defenses protect the plant against further damage.

The powdery mildew infecting this plant is a fungus

Many plants produce a range of antimicrobial and antifungal chemicals and enzymes to kill or inhibit the growth of pathogens. Some of these compounds cover the surface of the plant, killing pathogens before they enter the plant. Other compounds act internally.

Many herbs have antimicrobial properties. These compounds are sometimes extracted for human use.

Active defenses

Once infected, a plant responds actively to prevent any further damage. **Active defenses** are invoked only after a pathogen has been recognized, or after wounding or attack by a herbivore. This makes biological sense because active defenses are costly to produce and maintain. Active defenses work through a variety of mechanisms including slowing pathogen growth, puncturing the cell wall, disrupting metabolism, or killing cells by release of reactive oxygen species such as hydrogen peroxide (H_2O_2).

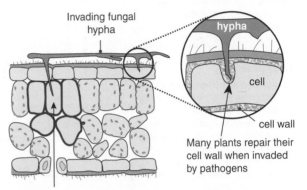

Invading fungal hypha

hypha

cell

cell wall

Many plants repair their cell wall when invaded by pathogens

Many plants produce an enzyme-activated **hypersensitive response** when invaded by pathogens. This leads to the production of reactive nitric oxide and cell death. Cell death in the infected region limits the spread of the pathogen.

Sealing off infected areas gives rise to abnormal swellings called **galls** (oak gall, left and bulls-eye galls on a maple leaf, right). These galls limit the spread of the parasite or the infection in the plant.

3. (a) Distinguish between passive and active defense mechanisms in plants: _____

_____

_____

_____

(b) Why are most plant defensive chemicals produced only after a pathogen is detected? _____

_____

4. How are galls effective in reducing the spread of infection in some plants? _____

_____

_____

5. What similarities are there between the active defense mechanisms of plants and the immune responses of animals?

_____

_____

_____

©2017 **BIOZONE** International
ISBN: 978-1-927309-65-0
Photocopying Prohibited

123 Our Body's Defenses

Key Idea: The human body has a tiered system of defenses that provides resistance against disease.

The body has a suite of physical, chemical, and biological defenses against pathogens, collectively called **resistance**. The first line of defense consists of external barriers to prevent pathogen entry. If this fails, a second line of defense targets any foreign bodies that enter. Lastly, the specific immune response provides targeted defense against the pathogen. The defense responses of the body fall into two broad categories, the innate and the adaptive immune responses. The **innate** (non-specific) response (the first and second lines of defense) protects against a broad range of non-specific pathogens. It involves blood proteins (e.g. complement) and phagocytic white blood cells. The **adaptive** (or specific) immune response (the third line of defense) is specific to identified pathogens. It involves defense by specific T-cells (**cellular immunity**) as well as antibodies, which neutralize foreign antigens (**humoral immunity**).

Most microorganisms find it difficult to get inside the body. If they succeed, they face a range of other defenses.

The natural populations of harmless microbes living on the skin and mucous membranes inhibit the growth of most pathogenic microbes

Microorganisms are trapped in sticky mucus and expelled by cilia (tiny hairs that move in a wavelike fashion).

1st line of defense

The skin provides a physical barrier to the entry of pathogens. Healthy skin is rarely penetrated by microorganisms. Its low pH is unfavorable to the growth of many bacteria and its chemical secretions (e.g. sebum, antimicrobial peptides) inhibit growth of bacteria and fungi. Tears, mucus, and saliva also help to wash bacteria away.

2nd line of defense

A range of defense mechanisms operate inside the body to inhibit or destroy pathogens. These responses react to the presence of any pathogen, regardless of which species it is. White blood cells are involved in most of these responses. It includes the **complement system** whereby plasma proteins work together to bind pathogens and induce inflammation to help fight infection.

3rd line of defense

Once the pathogen has been identified by the immune system, **lymphocytes** launch a range of specific responses to the pathogen, including the production of defensive proteins called **antibodies**. Each type of antibody is produced by a B-cell clone and is specific against a particular antigen.

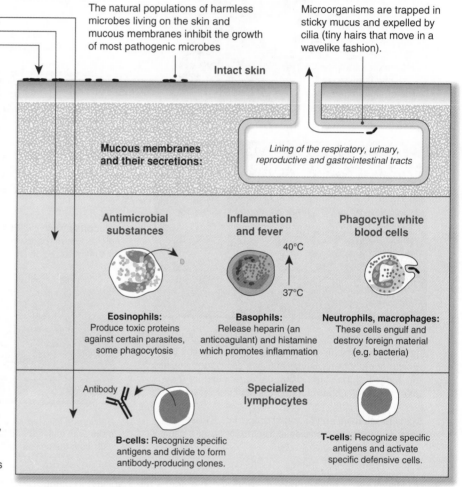

Intact skin

Mucous membranes and their secretions:

Lining of the respiratory, urinary, reproductive and gastrointestinal tracts

Antimicrobial substances

Eosinophils: Produce toxic proteins against certain parasites, some phagocytosis

Inflammation and fever

40°C
↑
37°C

Basophils: Release heparin (an anticoagulant) and histamine which promotes inflammation

Phagocytic white blood cells

Neutrophils, macrophages: These cells engulf and destroy foreign material (e.g. bacteria)

Antibody

Specialized lymphocytes

B-cells: Recognize specific antigens and divide to form antibody-producing clones.

T-cells: Recognize specific antigens and activate specific defensive cells.

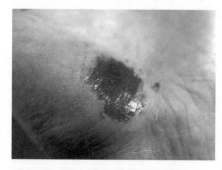

Intact skin provides a physical barrier to stop pathogens entering the body. Cuts or abrasions (above) allow pathogens to enter.

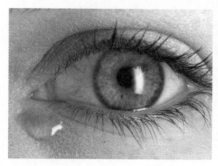

Antimicrobial chemicals are present in many bodily secretions, including tears. Tears also wash away contaminants from the eyes.

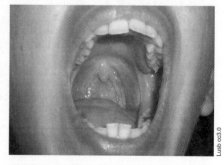

Harmless microbes colonize certain areas of the body (e.g. skin, gut, mouth, nose and throat) and prevent pathogens establishing.

1. Distinguish between the adaptive and innate immune responses: _____

_____

_____

_____

©2017 **BIOZONE** International
ISBN: 978-1-927309-65-0
Photocopying Prohibited

The importance of the first line of defense

The skin is the largest organ of the body. It forms an important physical barrier against the entry of pathogens into the body. A natural population of harmless microbes live on the skin, but most other microbes find the skin inhospitable. The continual shedding of old skin cells (arrow, right) physically removes bacteria from the surface of the skin. Sebaceous glands in the skin (top right) produce sebum, which has antimicrobial properties, and the slightly acidic secretions of sweat inhibit microbial growth.

Cilia line the epithelium of the nasal passage (below right). Their wave-like movement sweeps foreign material out and keeps the passage free of microorganisms, preventing them from colonizing the body.

Antimicrobial chemicals are present in many bodily secretions. Tears, saliva, nasal secretions, and human breast milk all contain **lysozymes** and **phospholipases**. Lysozymes kill bacterial cells by catalyzing the hydrolysis of cell wall linkages, whereas phospholipases hydrolyse the phospholipids in bacterial cell membranes, causing bacterial death. Low pH gastric secretions also inhibit microbial growth, and reduce the number of pathogens establishing colonies in the gastrointestinal tract.

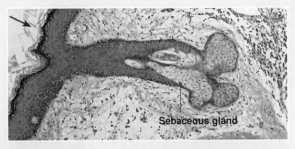

Sebaceous gland

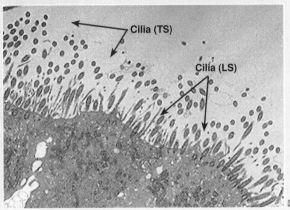

Cilia (TS)

Cilia (LS)

2. How does the skin act as a barrier to prevent pathogens entering the body? _____

_____

_____

3. Describe the role of each of the following in non-specific defense:

(a) Phospholipases: _____

_____

(b) Cilia: _____

_____

(c) Sebum: _____

_____

4. Describe the functional role of each of the following defense mechanisms:

(a) Phagocytosis by white blood cells: _____

_____

(b) Antimicrobial substances: _____

_____

(c) Antibody production: _____

_____

5. Explain the value of a three tiered system of defense against microbial invasion: _____

_____

_____

_____

124 The Innate Immune Response

Key Idea: The innate immune response provides a rapid response to contain and destroy pathogens. Inflammation is an important part of the response.

The innate immune system provides protection against a pathogen, even if it has never encountered it before. The innate response is very fast and provides general protection (it is not antigen specific), but does not provide long lasting immunity. Many different cells and processes are involved. The primary outcome is to destroy and remove the cause of infection. This is achieved through containing the infection through inflammation and then recruitment of immune cells to destroy the pathogen. During this process a series of biochemical reactions (the complement system) are activated to destroy the pathogen and recruit immune cells to the site.

Phagocytic cells of the innate immune system

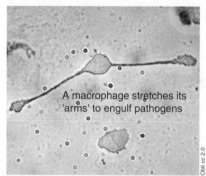
A macrophage stretches its 'arms' to engulf pathogens
Obli cc 2.0

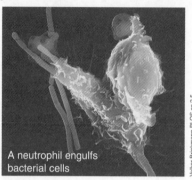
A neutrophil engulfs bacterial cells
Volker Brinkmann PLOS cc 2.5

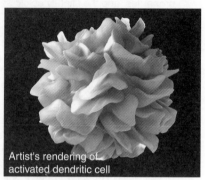
Artist's rendering of activated dendritic cell

Macrophage

Macrophages are very large and are highly efficient phagocytes. They are found throughout the body and move using an amoeboid movement (above) to hunt down and destroy pathogens. Macrophages also have a role in recruiting other immune cells to an infection site.

Neutrophil

Neutrophils are the most abundant type of phagocyte and are usually the first cells to arrive at the site of an infection. They contain toxic substances that kill or inhibit the growth of bacteria and fungal pathogens. Neutrophils release cytokines which amplify the immune response and recruit other cells to the infection site.

Dendritic cell

Dendritic cells are present in tissue that are in contact with the external environment (e.g. skin, and linings of the nose, lungs, and digestive tract). They act as messengers between the innate and adaptive immune system by presenting antigen materials to the T cells of the immune system.

Other cells and processes of the innate immune response

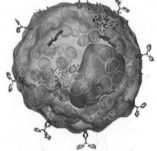

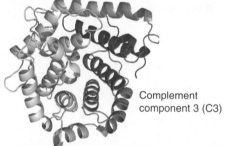
Complement component 3 (C3)

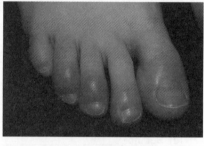

Mast cells

Mast cells contain a lot of histamine, a chemical involved in both inflammation and allergic responses. When activated, histamine is released from the mast cell causing the blood vessels to dilate and become leaky. The increased permeability allows phagocytes to reach the site of infection.

Complement proteins

The complement system comprises a number of different proteins. The proteins circulate as inactive precursors until they are activated. Complement proteins have three main roles: phagocytosis, attracting macrophages and neutrophils to the infection site, and rupturing the membranes of foreign cells.

The process of inflammation

The inflammatory process is a protective response to pathogen invasion. It has several functions: (1) to destroy the cause of the infection and remove it and its products from the body; (2) if this fails, to limit the effects on the body by confining the infection to a small area; (3) replacing or repairing tissue damaged by the infection.

1. Outline the role of the following phagocytes in the innate immune response:

   (a) Macrophages: _____

   _____

   (b) Neutrophils: _____

   _____

   (c) Dendritic cells: _____

   _____

©2017 **BIOZONE** International
ISBN: 978-1-927309-65-0
Photocopying Prohibited

The inflammatory response

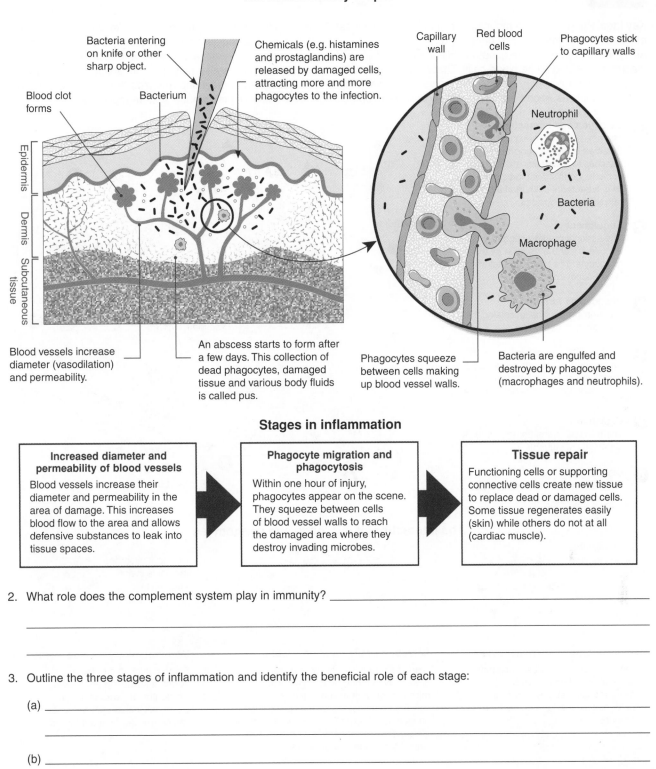

Bacteria entering on knife or other sharp object.

Chemicals (e.g. histamines and prostaglandins) are released by damaged cells, attracting more and more phagocytes to the infection.

Blood clot forms

Bacterium

Capillary wall

Red blood cells

Phagocytes stick to capillary walls

Neutrophil

Bacteria

Macrophage

Epidermis

Dermis

Subcutaneous tissue

Blood vessels increase diameter (vasodilation) and permeability.

An abscess starts to form after a few days. This collection of dead phagocytes, damaged tissue and various body fluids is called pus.

Phagocytes squeeze between cells making up blood vessel walls.

Bacteria are engulfed and destroyed by phagocytes (macrophages and neutrophils).

Stages in inflammation

Increased diameter and permeability of blood vessels

Blood vessels increase their diameter and permeability in the area of damage. This increases blood flow to the area and allows defensive substances to leak into tissue spaces.

Phagocyte migration and phagocytosis

Within one hour of injury, phagocytes appear on the scene. They squeeze between cells of blood vessel walls to reach the damaged area where they destroy invading microbes.

Tissue repair

Functioning cells or supporting connective cells create new tissue to replace dead or damaged cells. Some tissue regenerates easily (skin) while others do not at all (cardiac muscle).

2. What role does the complement system play in immunity? _____

_____

_____

3. Outline the three stages of inflammation and identify the beneficial role of each stage:

(a) _____

_____

(b) _____

_____

(c) _____

_____

4. What role do mast cells play in inflammation? _____

_____

_____

5. Why does pus form at the site of infection? _____

_____

©2017 **BIOZONE** International
ISBN: 978-1-927309-65-0
Photocopying Prohibited

125 The Action of Phagocytes

Key Idea: Phagocytes are mobile white blood cells that ingest microbes and digest them by phagocytosis.

All types of phagocytes (e.g. neutrophils, dendritic cells, and macrophages) are white blood cells. These specialized cells have receptors on their surfaces that can detect antigenic material, such as microbes. They then ingest the microbes and digest them by phagocytosis. As well as destroying microbes, phagocytes also release cytokines that help to coordinate the overall response to an infection. Macrophages and dendritic cells also play a role in antigen presentation in processing and presenting antigens from ingested microbes to other cells of the immune system (opposite).

1 Detection and interaction
Microbe coated in opsonins is detected by the phagocyte and attaches to it. Opsonins are molecules in the blood that coat foreign material (e.g. a bacterial cell), marking it as a target for phagocytosis.

2 Engulfment
The opsonin markers trigger engulfment of the microbe by the phagocyte. The microbe is taken in by endocytosis.

3 Phagosome forms
A phagosome forms, enclosing the microbe in a membrane.

4 Fusion with lysosome
Phagosome fuses with a lysosome containing powerful antimicrobial proteins. The fusion forms a phagolysosome.

5 Digestion
The microbe is broken down into its chemical constituents.

6 Discharge
Indigestible material is discharged from the phagocyte.

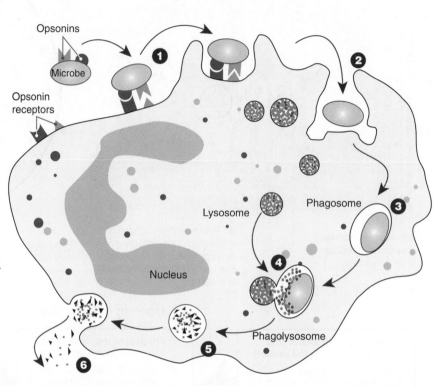

The interaction of microbes and phagocytes

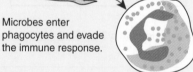

Some microbes kill phagocytes
Some microbes produce toxins that can actually kill phagocytes, e.g. toxin-producing staphylococci and the dental plaque-forming bacteria *Actinobacillus*.

Microbes evade immune system
Some microbes can evade the immune system by entering phagocytes. The microbes prevent fusion of the lysosome with the phagosome and multiply inside the phagocyte, almost filling it. Examples include *Chlamydia*, *Mycobacterium tuberculosis*, *Shigella*, and malarial parasites.

Dormant microbes hide inside
Some microbes can remain dormant inside the phagocyte for months or years at a time. Examples include the microbes that cause brucellosis and tularemia.

1. Identify the white blood cells capable of phagocytosis: _____

2. Explain the role of opsonins and phagocyte receptors in enhancing phagocytosis: _____

_____

_____

3. Explain how some microbes can overcome phagocytic cells and use them to their advantage: _____

_____

_____

_____

WEB   CONNECT   CONNECT   PRACTICES
125   **40**   **74**
          AP2   AP1

KNOW

©2017 **BIOZONE** International
ISBN: 978-1-927309-65-0
Photocopying Prohibited

126 Processing Antigens

Key Idea: Antigen processing prepares and displays antigens for presentation to the T-cells of the immune system.
Antigen presenting cells (APCs) process and present antigens for recognition by T-cells. During antigen processing, the APC digests the foreign antigen into smaller peptide fragments. These fragments are then displayed on the surface of the APC by MHC receptors. The immune response evoked by the T-cells depends on which MHC receptor (MHCI or MHCII) is activated. Antigen presentation is necessary for T-cells to recognize infection or abnormal growth and activate other cells of the immune system. Dendritic cells, macrophages, and B-cells are APCs.

The role of MHC receptors

Recall there are two types of MHC receptors, class I and class II (right). Both have similar functions in that they display antigenic peptides on cell surfaces so antigens can be recognized and processed by the T-cells of the immune system. T-cells can only recognize antigenic peptides if they are displayed by the MHC receptors. MHC receptors presenting no foreign antigenic peptides are ignored by T-cells, because they are signalling that the cell is healthy. Only MHC receptors with foreign antigenic peptides bound to them will attract T-cells and evoke an immune response.

The source of the antigenic peptides bound to each class of MHC receptor differs. Class I MHC receptors display antigenic peptides of intracellular parasites such as viruses. Class II MHC receptors display antigenic peptides originating from outside of the cell (such as those from ingested microbes).

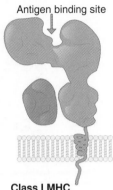

Class I MHC
Intracellular antigens, e.g. viral proteins

Class II MHC
Extracellular antigens, e.g. proteins from phagocytosed microbes

An overview of antigen processing

The diagram on the right represents antigen processing of an extracellular peptide antigen via a class II MHC receptor.

1. An APC encounters an antigen.

2. The antigen is engulfed via phagocytosis and digested into short peptide fragments.

3. Class II MHC receptors bind the fragments and form a MHC-antigen complex.

4. The MHC-antigen complex is displayed on the surface of the APC.

5. A receptor on the T helper cell recognizes the peptide as foreign. It binds and a series of events stimulate the adaptive immune response.

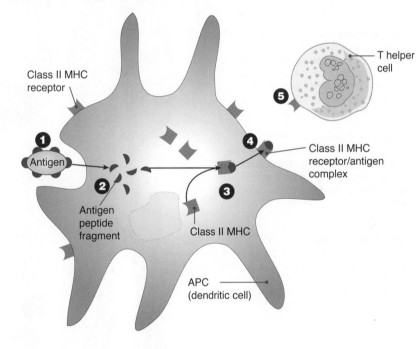

1. What is the purpose of antigen processing? _____

_____

_____

2. Why do MHC receptors with no antigenic peptide bound not cause an immune response?_____

_____

_____

3. Describe the differences between class I and class II MHC receptors: _____

_____

_____

_____

PRACTICES   PRACTICES   CONNECT **81** AP1 CONNECT **80** AP1 WEB **126**

KNOW

127 The Lymphatic System

Key Idea: Defensive white blood cells are transported in lymph through the lymphatic system and are concentrated in the lymph nodes.

The lymphatic system is a network of tissues and organs that collects the tissue fluid leaked from the blood vessels and returns it to the heart. The lymphatic system has an important role in immunity because the fluid transported around the body by the lymphatic system (lymph) is rich in infection-fighting white blood cells. The thymus is a primary lymphoid organ and the site of T-cell maturation. Secondary lymphoid tissues (spleen and lymph nodes) are important as the site of lymphocyte (T and B cell) activation.

Tonsils
A collection of secondary lymphoid tissues in the throat. They provide defense against ingested or inhaled pathogens and produce activated B and T-cells.

Thymus
A primary lymphoid organ located above the heart. It is large in infants and shrinks after puberty to a fraction of its original size. Important for maturation of **T-cells**.

Spleen
The largest mass of lymphatic tissue in the body. It stores and releases blood in case of demand (e.g. in severe bleeding), produces mature B-cells and antibodies and removes antibody-coated antigenic material.

Lymph nodes
Ovoid masses of lymph tissue where lymphocytes are concentrated. Each node receives lymph through several narrow afferent (entry) vessels and exits via wider efferent (exit) vessels.

Red bone marrow
A primary lymphoid tissue where all the different kinds of blood cells (including white blood cells) are produced by cellular differentiation from stem cells. B-cells also mature here.

Lymphatic vessels
When the fluid leaking from capillaries is picked up by lymph capillaries, it is called **lymph**. The lymph, carrying leukocytes, flows in lymphatic vessels through the secondary lymphoid tissues.

The lymphatic system and immunity

The fluid circulating through the lymphatic system passes through the secondary lymphoid tissues, including the **lymph nodes**. These are ovoid organs, which are present throughout the lymphatic system. Lymph nodes receive lymph via incoming (afferent) vessels and are the site of lymphocyte activation.

Lymphocytes in circulation are constantly moving between sites where antigens may be encountered. These antigens are presented to T-cells in the secondary lymphoid tissues. Recognition of the antigen leads to activation and proliferation of both T and B cells, vastly increasing the number of lymphocytes. After several days, antigen-activated lymphocytes begin leaving the lymphoid tissue.

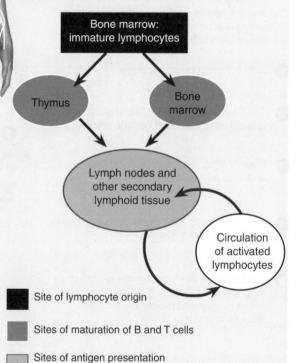

Site of lymphocyte origin

Sites of maturation of B and T cells

Sites of antigen presentation and activation of B and T cells

1. What is the general role of the lymphatic system in immunity? _____

_____

_____

2. (a) What is the role of the secondary lymphoid tissue, e.g. lymph nodes, in the immune response? _____

_____

_____

(b) Why do you think lymph nodes become swollen when someone has an infection? _____

_____

©2017 **BIOZONE** International
ISBN: 978-1-927309-65-0
Photocopying Prohibited

128 Acquired Immunity

Key Idea: Acquired immunity is a resistance to specific pathogens acquired over the life-time of an organism.

We are born with natural or **innate resistance** which provides non-specific immunity to certain illnesses. In contrast, **acquired immunity** is protection developed over time to specific antigens. **Active immunity** develops after the immune system responds to being exposed to microbes or foreign substances. **Passive immunity** is acquired when antibodies are transferred from one person to another. Immunity may also be naturally acquired, through natural exposure to microbes, or artificially acquired as a result of medical treatment (below).

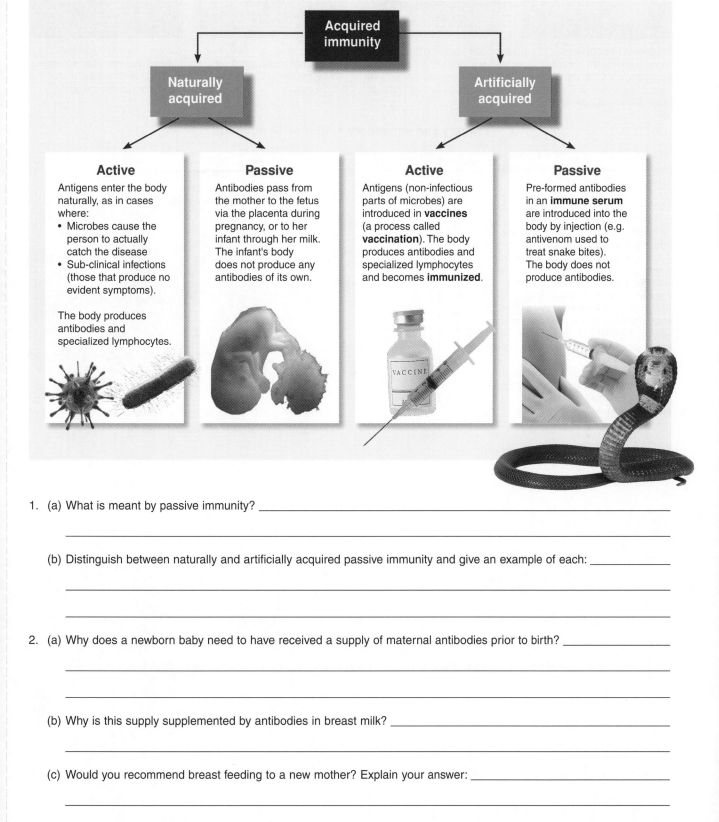

Acquired immunity

Naturally acquired

Active

Antigens enter the body naturally, as in cases where:
- Microbes cause the person to actually catch the disease
- Sub-clinical infections (those that produce no evident symptoms).

The body produces antibodies and specialized lymphocytes.

Passive

Antibodies pass from the mother to the fetus via the placenta during pregnancy, or to her infant through her milk. The infant's body does not produce any antibodies of its own.

Artificially acquired

Active

Antigens (non-infectious parts of microbes) are introduced in **vaccines** (a process called **vaccination**). The body produces antibodies and specialized lymphocytes and becomes **immunized**.

Passive

Pre-formed antibodies in an **immune serum** are introduced into the body by injection (e.g. antivenom used to treat snake bites). The body does not produce antibodies.

VACCINE

---

1. (a) What is meant by passive immunity? _____

_____

(b) Distinguish between naturally and artificially acquired passive immunity and give an example of each: _____

_____

_____

2. (a) Why does a newborn baby need to have received a supply of maternal antibodies prior to birth? _____

_____

_____

(b) Why is this supply supplemented by antibodies in breast milk? _____

_____

(c) Would you recommend breast feeding to a new mother? Explain your answer: _____

_____

_____

_____

CONNECT
129
AP2

WEB
128

KNOW

If a person has not been immunized against a disease, exposure to the pathogen causes them to become ill and their body forms antibodies against it.

Antibodies passing from the mother's milk to her newborn baby provide protection until the baby develops its own antibodies.

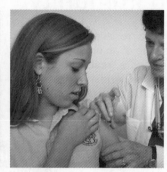

Vaccines provide immunity to specific pathogens and greatly reduce the risk of contracting the disease.

Without treatment with the appropriate preformed antibodies in antivenom, the bites of many snake species can be fatal.

Primary and secondary response to antigens

When the B-cells encounter antigens and produce antibodies, the body develops active immunity against that antigen.

The initial response to antigenic stimulation, caused by the sudden increase in B-cell clones, is called the primary response. Antibody levels as a result of the primary response peak a few weeks after the response begins and then decline. However, because the immune system develops an immunological memory of that antigen, it responds much more quickly and strongly when presented with the same antigen subsequently (the secondary response).

This forms the basis of immunization programmes where one or more booster shots are provided following the initial vaccination.

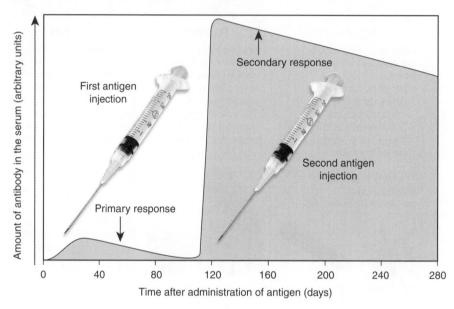

3. (a) What is active immunity? _____

_____

_____

   (b) Distinguish between naturally and artificially acquired active immunity and give an example of each: _____

_____

_____

_____

4. (a) Describe two differences between the primary and secondary responses to presentation of an antigen: _____

_____

_____

_____

   (b) Why is the secondary response so different from the primary response? _____

_____

_____

_____

©2017 **BIOZONE** International
ISBN: 978-1-927309-65-0
Photocopying Prohibited

129 Vaccines and Vaccination

Key Idea: A vaccine is a suspension of antigens that is deliberately introduced into the body to protect against disease. If enough of the population are vaccinated, herd immunity provides protection to unvaccinated individuals.

A **vaccine** is a preparation of a harmless foreign antigen that is deliberately introduced into the body to protect against a specific disease. The antigen in the vaccine is usually some part of the pathogen and it triggers the immune system to produce antibodies against the antigen, but it does not cause the disease. The immune system remembers its response and will produce the same antibodies if it encounters the antigen again. If enough of the population are vaccinated, herd immunity (indirect protection) provides unvaccinated individuals in the population with a measure of protection against the disease. There are two basic types of vaccine, subunit vaccines and whole-agent vaccines (below).

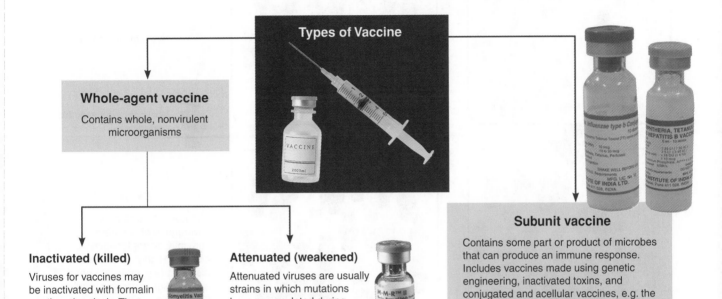

Types of Vaccine

Whole-agent vaccine

Contains whole, nonvirulent microorganisms

Subunit vaccine

Contains some part or product of microbes that can produce an immune response. Includes vaccines made using genetic engineering, inactivated toxins, and conjugated and acellular vaccines, e.g. the diphtheria-tetanus-pertussis vaccine and the vaccine against bacterial meningitis.

Inactivated (killed)

Viruses for vaccines may be inactivated with formalin or other chemicals. They present no risk of infection, e.g. most influenza vaccines, Salk polio vaccine.

Attenuated (weakened)

Attenuated viruses are usually strains in which mutations have accumulated during culture. These live viruses can back-mutate to a virulent form, e.g. MMR vaccine.

Why are vaccinations given?

Vaccines against common diseases are given at various stages during childhood according to an immunization schedule. Vaccination has been behind the decline of some once-common childhood diseases, such as mumps and measles.

Most vaccinations are given in childhood, but adults may be vaccinated against a disease (e.g. TB, tetanus) if they are in a high risk group (e.g. the elderly or farmers) or to provide protection against seasonal diseases such as influenza.

Tourists may need specific vaccines if the country they are visiting has a high incidence of a certain disease. For example, travellers to South America should be immunized against yellow fever, a disease that does not occur in the United States.

1. (a) What is a vaccine? _____

_____

(b) Provide some examples of when vaccinations are needed: _____

_____

_____

_____

Vaccination can provide herd immunity

Herd immunity occurs when the vaccination of a significant portion of a population provides some protection for individuals who have not developed immunity (e.g. have not been vaccinated and are not immunized). In order to be effective for any particular disease, a high percentage of the population needs to be vaccinated against that disease. High vaccination rates make it difficult for the disease to spread because there are very few susceptible people in the population. Herd immunity is important for people who cannot be vaccinated (e.g. the very young, people with immune system disorders, or people who are very sick, such as cancer patients).

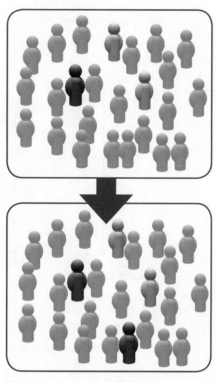

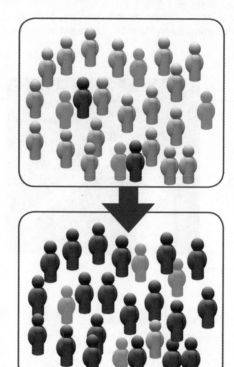

Immunized and healthy

Not immunized and healthy

Not immunized, sick and contagious

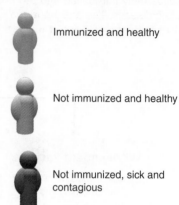

DID YOU KNOW?

The level of vaccination coverage to obtain herd immunity differs for each disease. Highly contagious diseases (e.g. measles) need a much higher vaccine uptake (95%) than a less contagious disease such as polio (80-85%).

High herd immunity: Most of the population is immunized. The spread of the disease is limited. Only a few people are susceptible and become infected.

Low herd immunity: Only a small proportion of the population is immunized. The disease spreads more readily through the population infecting many more people.

2. Attenuated viruses provide long term immunity to their recipients and generally do not require booster shots. Why do you think attenuated viruses provide such effective long-term immunity when inactivated viruses do not?

_____

_____

_____

_____

3. (a) What is herd immunity? _____

_____

_____

(b) Why are health authorities concerned when the vaccination rates for an infectious disease fall? _____

_____

_____

4. Some members of the population are unable to be vaccinated. Give an example and explain why herd immunity is very important to them?

_____

_____

_____

©2017 **BIOZONE** International
ISBN: 978-1-927309-65-0
Photocopying Prohibited

130 The Adaptive Immune System

Key Idea: Antigens, such as the cell walls of microbial cells, when processed by antigen-presenting cells, activate the B and T cells of the immune system against specific pathogens. There are two main components of the adaptive immune system: the humoral and the cell-mediated responses. They work separately and together to protect against disease. The **humoral immune response** is associated with the serum (the non-cellular part of the blood) and involves the action of antibodies secreted by B-cells (B lymphocytes). Antibodies

are found in extracellular fluids including lymph, plasma, and mucus secretions and protect against viruses, and bacteria and their toxins. The **cell-mediated immune response** is associated with the production of specialized lymphocytes called **T-cells**. Antigens are recognized by T-cells only after antigen processing. The antigen is first engulfed by an antigen-presenting cell, which processes the antigen and presents it on its surface. T-helper cells can then recognize the antigen and activate other cells of the immune system.

Lymphocytes and their functions

Bone marrow
B-cells mature in the bone marrow in the shaft of the long bones (e.g. the femur). They migrate from here to the lymphatic organs.

Stem cell
Stem cells in the bone marrow and fetal liver give rise to T-cells and B-cells.

Thymus gland
The thymus gland is located above the heart. It is large in infants but regresses with age. Immature T-cells move to the thymus to mature before migrating to other lymphatic organs.

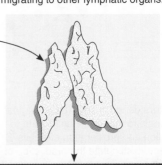

B cell

B-cells recognize and bind antigens. Each B cell recognizes one specific antigen. Helper T cells recognize specific antigens on B cell surfaces and induce their maturation and proliferation. A mature B-cell may carry as many as 100,000 antigenic receptors embedded in its surface membrane. B-cells defend against bacteria and viruses outside the cell and toxins produced by bacteria (free antigens).

Free antigen

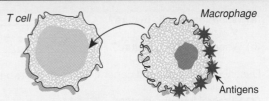

T cell

T cell   *Macrophage*

Antigens

T-cells respond only to antigen fragments that have been processed and presented by infected cells or macrophages (phagocytic cells) (see opposite). They defend against:
- Intracellular bacteria and viruses
- Protozoa, fungi, flatworms, and roundworms
- Cancerous cells and transplanted foreign tissue

Differentiate into two kinds of cells

Antibody

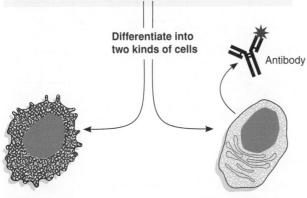

Differentiate into various kinds of cells:

T_H

T_C

Memory cells
Some B-cells differentiate into long-lived memory cells (see Clonal Selection). When these cells encounter the same antigen again (even years or decades later), they rapidly differentiate into antibody-producing plasma cells.

Plasma cells
When stimulated by an antigen (see Clonal Selection), some B-cells differentiate into plasma cells, which secrete **antibodies** into the bloodstream. The antibodies then inactivate the circulating antigens.

T helper cell activates T cytotoxic cells and other helper T-cells. They are needed for B cell activation.

T cytotoxic cell destroys target cells on contact. Recognizes tumour or virus-infected cells by their surface markers. Also called T killer cells.

There are also other types of T-cells:
T memory cells have encountered specific antigens before and can respond quickly and strongly when the same antigen is encountered again.
T regulator cells control immune response by turning it off when no more antigen is present. They are important in the development of self tolerance.

CONNECT
81
AP1

CONNECT
80
AP1

WEB
130

KNOW

Dendritic cells stimulate the activation and proliferation of lymphocytes

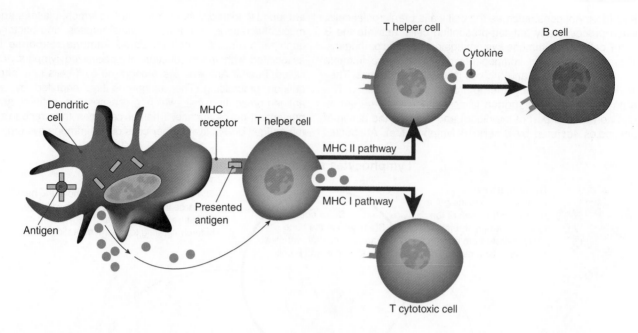

Immature dendritic cells (DCs) originate in the bone marrow and migrate throughout the body. Once they have processed an antigen they begin to mature. They migrate to lymph nodes and, through antigen presentation and secretion of cytokines, stimulate the activation and proliferation of T-cells. DCs exhibiting MHC I receptors stimulate the production of T cytotoxic cells. DCs exhibiting MHC II receptors stimulate the production of T helper cells. These in turn go on to stimulate the production of antibody-producing B-cells.

1. Where do B-cells and T-cells originate (before maturing)? _____

2. (a) Where do B-cells mature? _____

   (b) Where do T-cells mature? _____

3. Describe the nature and general action of the two major divisions in the immune system:

   (a) Humoral immune system: _____

   _____

   _____

   (b) Cell-mediated immune system: _____

   _____

   _____

4. Explain how an antigen causes the activation and proliferation of T-cells and B-cells, including the role of dendritic cells:

   _____

   _____

   _____

   _____

5. In what way do dendritic cells act as messengers between the innate and the adaptive immune systems?

   _____

   _____

6. Describe the function of each of the following cells in the immune system response:

   (a) T helper cells: _____

   (b) T cytotoxic cells: _____

©2017 **BIOZONE** International
ISBN: 978-1-927309-65-0
Photocopying Prohibited

131 Clonal Selection

Key Idea: Clonal selection theory explains how lymphocytes can respond to a large and unpredictable range of antigens. The **clonal selection theory** explains how the immune system can respond to the large and unpredictable range of potential antigens in the environment. The diagram below describes clonal selection after antigen exposure for B-cells. In the same way, a T-cell stimulated by a specific antigen will multiply and develop into different types of T-cells. Clonal selection and differentiation of lymphocytes provide the basis for **immunological memory.**

Five (a-e) of the many B-cells generated during development. Each one can recognize only one specific antigen.

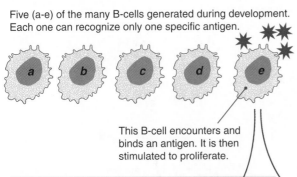

This B-cell encounters and binds an antigen. It is then stimulated to proliferate.

Clonal selection theory

Millions of B-cells form during development. Antigen recognition is randomly generated, so collectively they can recognize many antigens, including those that have never been encountered. Each B-cell has receptors on its surface for specific antigens and produces antibodies that correspond to these receptors. When a B-cell encounters its antigen, it responds by proliferating and producing many clones that produce the same kind of antibody. This is called clonal selection because the antigen selects the B cells that will proliferate.

Memory cells

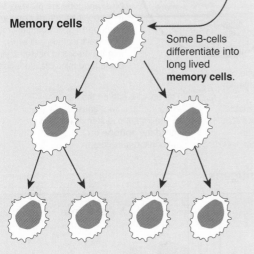

Some B-cells differentiate into long lived **memory cells.**

Plasma cells

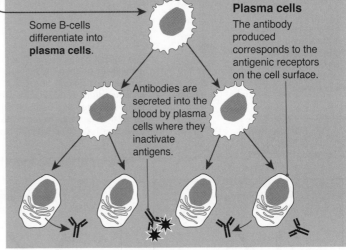

Some B-cells differentiate into **plasma cells.**

The antibody produced corresponds to the antigenic receptors on the cell surface.

Antibodies are secreted into the blood by plasma cells where they inactivate antigens.

Some B-cells differentiate into long lived **memory cells.** These are retained in the lymph nodes to provide future immunity (**immunological memory**). In the event of a second infection, memory B-cells react more quickly and vigorously than the initial B-cell reaction to the first infection.

Plasma cells secrete antibodies specific to the antigen that stimulated their development. Each plasma cell lives for only a few days, but can produce about 2000 antibody molecules per second. Note that during development, any B-cells that react to the body's own antigens are selectively destroyed in a process that leads to **self tolerance** (acceptance of the body's own tissues).

1. Describe how clonal selection results in the proliferation of one particular B-cell clone: _____

_____

_____

_____

2. (a) What is the function of the plasma cells in the immune system response? _____

_____

(b) What is the significance of B-cells producing antibodies that correspond to (match) their antigenic receptors?

_____

_____

3. (a) Explain the basis of immunological memory: _____

_____

(b) Why are B memory cells able to respond so rapidly to an encounter with an antigen long after an initial infection?

_____

©2017 **BIOZONE** International
ISBN: 978-1-927309-65-0
Photocopying Prohibited

PRACTICES

CONNECT
130
AP1

WEB
131

KNOW

132 Antibodies

Key Idea: Antibodies are large, Y-shaped proteins made by plasma cells, which destroy specific antigens.

Antibodies and antigens play key roles in the response of the immune system. **Antigens** are foreign molecules which promote a specific immune response. Antigens include pathogenic microbes and their toxins, as well as substances such as pollen grains, blood cell surface molecules, and the surface proteins on transplanted tissues. **Antibodies** (or immunoglobulins) are proteins made in response to antigens. They are secreted from B-cells into the plasma where they can recognize, bind to, and help destroy antigens. There are five classes of antibodies, each plays a different role in the immune response. Each type of antibody is specific to only one particular antigen.

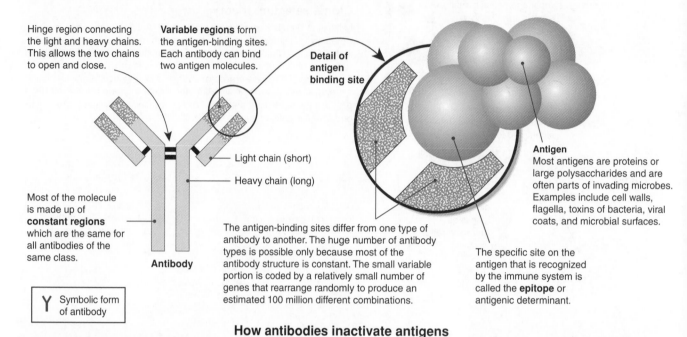

Hinge region connecting the light and heavy chains. This allows the two chains to open and close.

Variable regions form the antigen-binding sites. Each antibody can bind two antigen molecules.

Detail of antigen binding site

Light chain (short)

Heavy chain (long)

Most of the molecule is made up of **constant regions** which are the same for all antibodies of the same class.

Antibody

Antigen
Most antigens are proteins or large polysaccharides and are often parts of invading microbes. Examples include cell walls, flagella, toxins of bacteria, viral coats, and microbial surfaces.

The antigen-binding sites differ from one type of antibody to another. The huge number of antibody types is possible only because most of the antibody structure is constant. The small variable portion is coded by a relatively small number of genes that rearrange randomly to produce an estimated 100 million different combinations.

The specific site on the antigen that is recognized by the immune system is called the **epitope** or antigenic determinant.

Y Symbolic form of antibody

How antibodies inactivate antigens

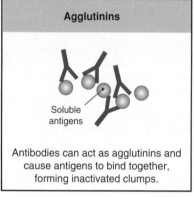

Agglutinins

Soluble antigens

Antibodies can act as agglutinins and cause antigens to bind together, forming inactivated clumps.

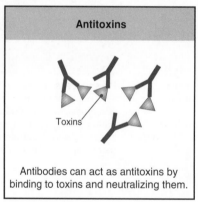

Antitoxins

Toxins

Antibodies can act as antitoxins by binding to toxins and neutralizing them.

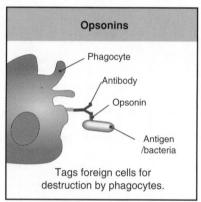

Opsonins

Phagocyte

Antibody

Opsonin

Antigen /bacteria

Tags foreign cells for destruction by phagocytes.

1. Describe the structure of an antibody, identifying the specific features of its structure that contribute to its function:

_____

_____

_____

2. Explain how the following actions by antibodies enhance the immune systems ability to stop infections:

(a) Acting as agglutinins: _____

_____

(b) Acting as antitoxins: _____

_____

(c) Working with opsonins: _____

_____

©2017 **BIOZONE** International
ISBN: 978-1-927309-65-0
Photocopying Prohibited

133 Failures of Defense: Allergies

Key Idea: Hypersensitivity occurs when the immune system overreacts to an antigen or reacts to the wrong substance. Histamine plays a significant role in this response.

Sometimes the immune system may overreact, or react to the wrong substances instead of responding appropriately. This response is called an allergy, and it consists of two phases. The first is **sensitivity**, in which the body becomes primed to respond to an allergen but there are no symptoms. This is followed by a **reaction phase** in which the person is re-exposed to the same allergen and experiences symptoms. In some cases, the response causes only minor discomfort (e.g. hayfever) but in extreme cases it can cause death.

Sensitization phase

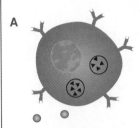

Dendritic cell

Pollen allergen

MHC II receptor

Activated T helper cell

Inactivated T cell

Antibody

B cell

Histamine

Mast cell

When a new allergen is encountered, it is phagocytosed by a dendritic cell and digested into smaller antigenic fragments. The fragments are displayed by the dendritic cell on MHC class II receptors. When an inactivated (or naive) T-cell encounters the novel allergen it becomes an activated T helper cell. Cytokines called interleukins produced by the T helper cell activate B-cells to produce IgE antibodies. IgE antibodies bind to the surface of mast cells where they remain. No allergic symptoms occur.

Reaction phase

A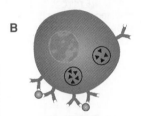

The same allergen to which a person is already sensitized encounters a mast cell. This second encounter can be weeks or months after the initial encounter.

B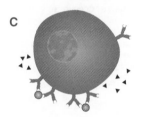

The allergen binds to specific IgE antibodies on the mast cell. Once bound, the allergen becomes cross-linked to neighboring antibodies.

C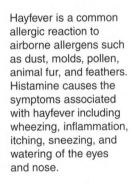

Several chemicals, including histamine are released from the mast cell. Several reactions occur in response to histamine including dilation of blood vessels, inflammation, and mucus secretion.

Hayfever is a common allergic reaction to airborne allergens such as dust, molds, pollen, animal fur, and feathers. Histamine causes the symptoms associated with hayfever including wheezing, inflammation, itching, sneezing, and watering of the eyes and nose.

1. What happens when a person becomes sensitized to an allergen? _____

_____

_____

2. What is the role of histamine in hypersensitivity responses? _____

_____

_____

3. In what way is the hypersensitivity reaction a malfunction of the immune system? _____

_____

_____

CONNECT **81** AP1   CONNECT **80** AP1   WEB **133**

KNOW

134 Failures of Defense: HIV/AIDS

Key Idea: The human immunodeficiency virus infects lymphocyte cells, eventually causing AIDS, a fatal disease, which acts by impairing the immune system.

HIV (human immunodeficiency virus) is a retrovirus, a single-stranded RNA virus which infects lymphocytes called helper T-cells. Over time, a disease called **AIDS** (acquired immunodeficiency syndrome) develops and the immune system loses its ability to fight off infections as more T helper cells are destroyed. There is no cure or vaccine for HIV, but some drugs can slow the progress of the disease. Antibiotics are ineffective against viruses because they only target specific aspects of bacterial metabolism.

HIV infects lymphocytes

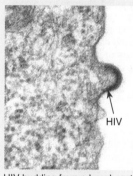

CDC

HIV budding from a lymphocyte

HIV infects T helper cell lymphocytes. It uses the cells to replicate itself in great numbers, then the newly formed viral particles exit the cell to infect more T helper cells. Many T helper cells are destroyed in the process of HIV replication.

T helper cells are part of the body's immune system, so when their levels become too low, the immune system can no longer fight off infections.

The graph below shows the relationship between the level of HIV infection and the number of T helper cells in an individual.

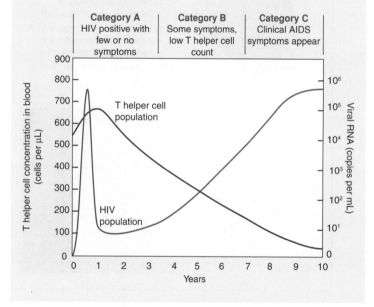

AIDS: The end stage of an HIV infection

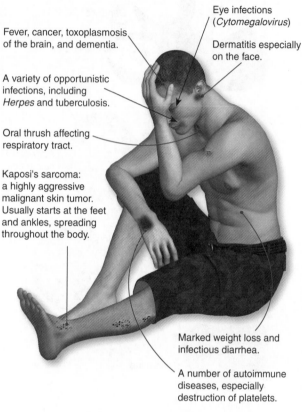

Eye infections (*Cytomegalovirus*)

Dermatitis especially on the face.

Fever, cancer, toxoplasmosis of the brain, and dementia.

A variety of opportunistic infections, including *Herpes* and tuberculosis.

Oral thrush affecting respiratory tract.

Kaposi's sarcoma: a highly aggressive malignant skin tumor. Usually starts at the feet and ankles, spreading throughout the body.

Marked weight loss and infectious diarrhea.

A number of autoimmune diseases, especially destruction of platelets.

The range of symptoms resulting from HIV infection is huge, but are not the result of the HIV infection directly. The symptoms arise from secondary infections that gain a foothold in the body due to the weakened immune system (due to the reduced number of T helper cells). People with healthy immune systems can be exposed to pathogens and not suffer serious effects because their immune system fights them off. However, people with HIV are susceptible to all pathogens because their immune system is too weak to fight them off. As the immune system become progressively weaker, the infected person becomes sicker.

1. (a) What type of cells does HIV infect? _____

   (b) What effect does HIV have on the body's immune system? _____

   _____

   _____

   _____

2. Study the graph above showing how HIV affects the number of T helper cells. Describe how the viral population changes with the progression of the disease:

   _____

   _____

©2017 **BIOZONE** International
ISBN: 978-1-927309-65-0
Photocopying Prohibited

135 KEY TERMS AND IDEAS: Did You Get It?

1. Test your vocabulary by matching each term to its correct definition, as identified by its preceding letter code.

antigen ....

antibody ....

cell-mediated response ....

humoral response ....

immunity ....

inflammation ....

pathogen ....

vaccination ....

A Immune response against antigens involving the activation of macrophages, specific T cells, and cytokines.

B A disease-causing organism.

C The collective defenses of the body that provide resistance to infection or disease.

D A molecule, usually a protein, that is capable of inducing an immune response.

E The delivery of antigenic material to produce immunity to a disease (produce an immunized individual).

F The protective response of vascular tissues to harmful stimuli, such as irritants, pathogens, or damaged cells.

G Immune response that is mediated by secreted antibodies.

H Gamma globulin protein in the blood or other bodily fluids, which identifies and neutralizes foreign material, such as bacteria and viruses.

2. Contrast the innate and the adaptive immune responses with reference to the basic action and the cells involved:

_____

_____

_____

_____

3. The photograph on the right shows the effect of a pathogen infecting a human.

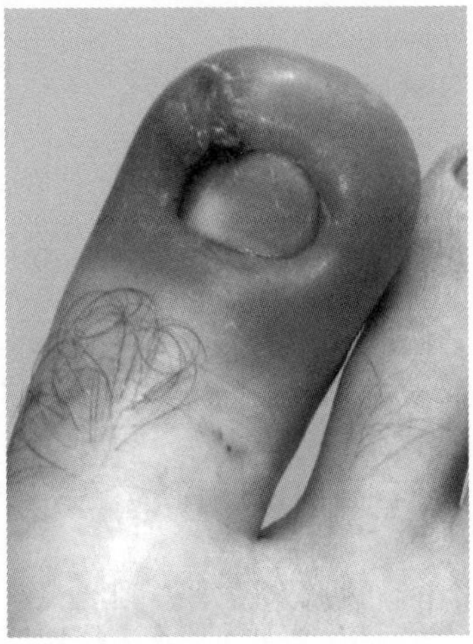

(a) Name the defensive response occurring: _____

(b) What is happening to the blood vessels at this location? _____

_____

(c) Name the substance responsible for the change in the blood vessels:

_____

(d) What type of cell is the substance released from? _____

(e) During this response, the number of white blood cells increases/decreases (delete one).

(f) The process occurring here is an example of innate immunity / adaptive immunity (delete one).

4. The diagram below shows a hypersensitivity sequence. Label the following components on the diagram: *mast cell, antibody, histamine, antigen, B cell, plasma B cell*. In the space below each drawing, briefly state what is happening.

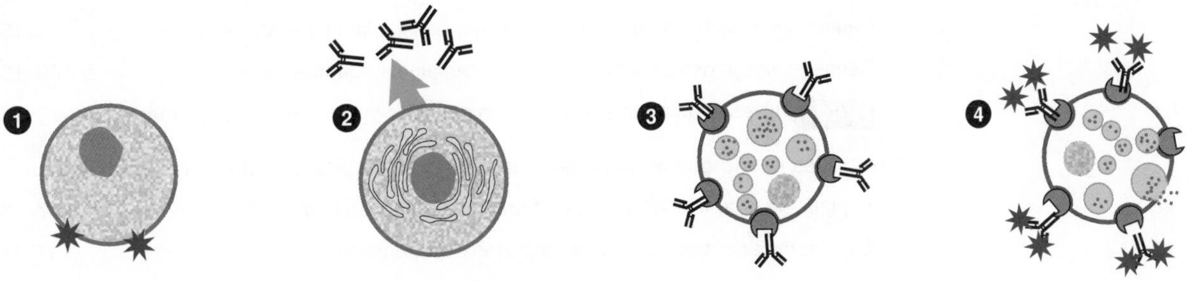

❶   ❷   ❸   ❹

_____   _____   _____   _____

_____   _____   _____   _____

_____   _____   _____   _____

PRACTICES

TEST

Enduring Understanding
2.C
2.E

Timing and Coordination

2.C.2 Organisms respond to changes in their external environment

Essential knowledge

(a) Organisms respond to changes in their environment through behavioral and physiological mechanisms

☐ 1 Describe examples to show that organisms use various behavioral and physiological mechanisms to respond to changes in their environment:
• Plants: photoperiodism and phototropism
• Animals: Hibernation and migration. Taxes and kineses. Circadian rhythms.

137-145
147-152
154 155

2.E.2 Multiple mechanisms regulate the timing and coordination of physiological events

Essential knowledge

(a) Physiological events in plants involve interaction between environmental stimuli and internal molecular signals

☐ 1 Describe phototropism in plants, including its adaptive role and the role of auxin.   137-140

☐ 2 Describe the photoperiodic responses of plants to changes in night length.   137 141

(b) In animals, internal and external signals regulate a variety of physiological responses that synchronize with environmental cycles and cues

☐ 1 Using examples, explain how internal signals (e.g. hormones) and external signals (e.g. daylength, pheromones) regulate a range of physiological responses in animals that synchronize with environmental cycles and cues. Examples include • endogenous circadian rhythms • diurnal and nocturnal activity patterns and sleep/wake cycles • jet lag in humans (causes and consequences) • seasonal responses (migration, and hibernation, torpor, and estivation) • release and reaction to pheromones *[also 3.E.1.b.1]*.

142-148
154-157
also 169

2.E.3 Timing and coordination of behavior are regulated by various mechanisms and are important in natural selection

Essential knowledge

(a) Individuals can act on information and communicate it to others

☐ 1 Define innate behavior, describe examples, and explain its adaptive value.   150

☐ 2 Define learning, describe examples, and explain its adaptive value.   158 159

☐ **PR-12** ▶ Investigate adaptive orientation behavior in invertebrates, e.g. fruit flies.   153

(b) Responses to and communication of information are vital to natural selection

☐ 1 Explain the adaptive value of the phototropic response in plants.   137-139

☐ 2 Explain the adaptive value of photoperiodic responses in plants.   137 141

☐ 3 Describe behaviors in animals that are triggered by environmental cues and are adaptive in that they are vital to survival and reproduction. Examples include:
• hibernation • estivation • migration • courtship.

142-152
154-157
also 170

☐ 4 Using examples, explain how cooperative behavior within or between populations contributes to population survival. Examples include: • fruiting body formation in fungi and bacteria related to resource availability • niche and resource partitioning • mutualistic relationships, including pollination relationships.   160-164

136 Timing and Coordination in Simple Organisms

Key Idea: Single celled organisms are able to coordinate their behavior to enhance survival.

Survival in a changing environment requires the ability to synchronize various physiological functions with the environment. The timing and coordination of physiological functions requires the ability both to receive and act upon information from the environment, and to communicate with other individuals of the same species. Communication may occur through the use of chemical signals secreted into the surrounding environment or by cell to cell contact. Often a physiological event is triggered by the density of a population reaching a certain critical level.

Timing of virulence in *Pseudomonas*

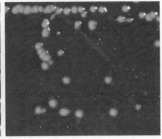

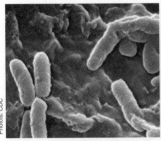

Photos: CDC

Pseudomonas aeruginosa is a common and usually relatively harmless soil bacterium. However, it is also an extremely versatile opportunistic pathogen and can cause fatal infections in people who have had at least part of their immune system compromised.

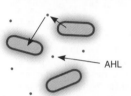

During its initial colonization phase, *P. aeruginosa* does not secrete toxins but devotes resources to increasing population size.

AHL

Individual bacteria produce signal molecules called acyl homoserine lactones (AHLs) that are received by others.

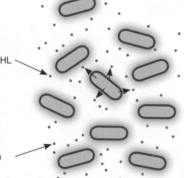

AHL

When the concentration of AHLs is high enough, they trigger the expression of virulence genes in individual bacteria and the production of toxins.

Toxin

The switch to the virulent phase is regulated by quorum sensing. AHLs provide a way to coordinate the regulation of gene expression. When the population reaches a critical density *P. aeruginosa* suddenly switches to secreting an array of toxins and becomes highly virulent.

Chemotaxis and coordination in *Myxococcus*

The soil bacteria *Myxococcus xanthus* normally live in small colonies as a biofilm. Coordination between single cells in a colony in response to environmental cures can help in survival by the production of spores.

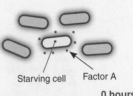

Starving cell   Factor A

0 hours

When resources dwindle, the cells begin to starve and signal to other bacteria using a chemical called factor A.

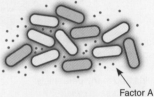

Factor A

4 hours

Fruiting body assembling

When factor A reaches high density (many cells are starving) the cells congregate and form a fruiting body, which can eventually contain about 100,000 cells.

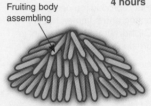

20 hours

Fruiting body containing spores

The fruiting body develops thousands of spores that are able to withstand the lack of resources. The spores may then be transported by wind or animals to new territory where they can germinate and produce a new colony.

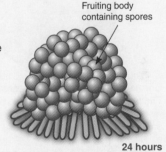

24 hours

1. Explain how the following events are coordinated:

   (a) The timing of virulence in *Pseudomonas aeruginosa*: _____

   _____

   _____

   (b) The production of the fruiting body in *Myxococcus xanthus*: _____

   _____

   _____

2. Explain how coordinating these events gives a survival advantage to *P. aeruginosa* and *M. xanthus*: _____

   _____

   _____

   _____

©2017 **BIOZONE** International
ISBN: 978-1-927309-65-0
Photocopying Prohibited

PRACTICES   CONNECT   WEB

79 AP1   **136**

KNOW

137 Timing and Coordination in Plants

Key Idea: Plant hormones play crucial roles in the timing of activities including fruit ripening and breaking dormancy.

In plants, responses to day length and temperature are mediated by hormones. The ability to detect changes in day length allows plants to respond to annual cycles by making the appropriate structures at the appropriate time, e.g. flowers or frost tolerant buds. Plant hormones are chemicals that act as signal molecules to regulate plant growth and responses. Alone and together, plant hormones target specific parts of a plant and produce a specific effect. Many have roles in coordinating timing responses in plants including promoting and breaking bud dormancy, seed germination, and fruit ripening. In addition these rhythms are often linked to temperature.

Hormones, plant growth, and fruiting

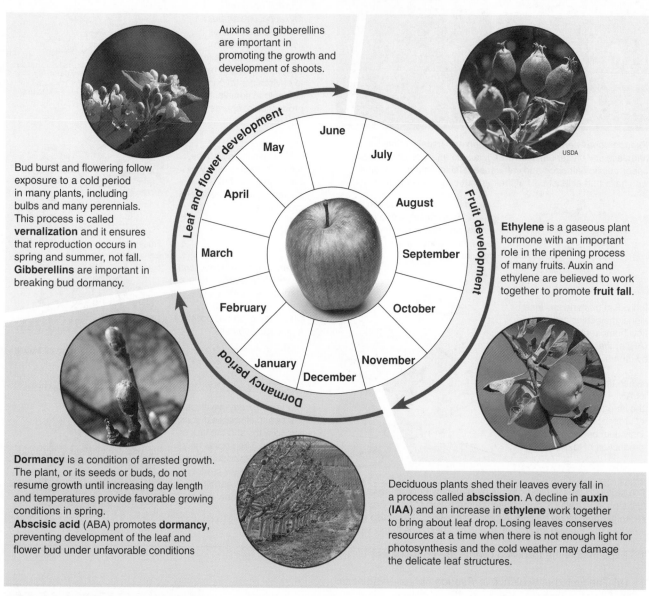

Auxins and gibberellins are important in promoting the growth and development of shoots.

Bud burst and flowering follow exposure to a cold period in many plants, including bulbs and many perennials. This process is called **vernalization** and it ensures that reproduction occurs in spring and summer, not fall. **Gibberellins** are important in breaking bud dormancy.

Ethylene is a gaseous plant hormone with an important role in the ripening process of many fruits. Auxin and ethylene are believed to work together to promote **fruit fall**.

Dormancy is a condition of arrested growth. The plant, or its seeds or buds, do not resume growth until increasing day length and temperatures provide favorable growing conditions in spring. **Abscisic acid (ABA)** promotes **dormancy**, preventing development of the leaf and flower bud under unfavorable conditions

Deciduous plants shed their leaves every fall in a process called **abscission**. A decline in **auxin (IAA)** and an increase in **ethylene** work together to bring about leaf drop. Losing leaves conserves resources at a time when there is not enough light for photosynthesis and the cold weather may damage the delicate leaf structures.

Seed germination and hormones

The seeds of many cold-climate plants will not germinate until they have been exposed to a period of wet, cold (5°C) conditions. This is called **cold stratification**.

ABA accumulates in seeds during fruit production and is important in seed dormancy. A high level of ABA in the seed embryo promotes dormancy.

Gibberellins are required for **seed germination**. They stimulate cell division and cell elongation, allowing the root to penetrate the seed coat.

©2017 **BIOZONE** International
ISBN: 978-1-927309-65-0
Photocopying Prohibited

Daily rhythm in tulips

Many flowers, including tulips, show **sleep movements**. In most species, these are triggered by day length, but in tulips the environmental cue is temperature. This series of photographs shows the sleep movements of a single tulip flower over one 12 hour period during spring. Sleep movements may prevent flower damage, stop the entry of non-pollinating insects, or stop the pollen becoming wet with dew.

| 7.00 am | 9.30 am | 11.00 am | 5.00 pm | 7.00 pm |

All photos: RA

1. (a) Describe the adaptive value of dormancy in plants: _____

    _____

    _____

    (b) What cues are likely to be involved in breaking dormancy? _____

    _____

2. How does vernalization ensure a plant will not flower in fall? _____

    _____

    _____

    _____

3. Describe the adaptive value of leaf abscission: _____

    _____

    _____

    _____

4. Why is it likely that the same hormones are responsible for both leaf fall (abscission) and fruit fall in plants?

    _____

    _____

    _____

5. Describe the advantage of cold stratification in plant seeds: _____

    _____

    _____

    _____

6. (a) Describe the sleep movements of tulips in response to temperature: _____

    _____

    (b) How might these movements be adaptive? _____

    _____

    (c) What is the name of this type of response? _____

138 Tropisms and Growth Responses

Key Idea: Tropisms are directional growth responses to external stimuli. They may be positive (towards a stimulus) or negative (away from a stimulus).

Tropisms are plant growth responses to external stimuli, in which the stimulus direction determines the direction of the growth response. Tropisms are identified according to the stimulus involved and direction (positive or negative). Stimuli are identified as photo- (light), gravi- (gravity), hydro- (water), chemo- (chemicals), and thigmo- (touch). Tropisms act to position the plant in the most favorable growth environment.

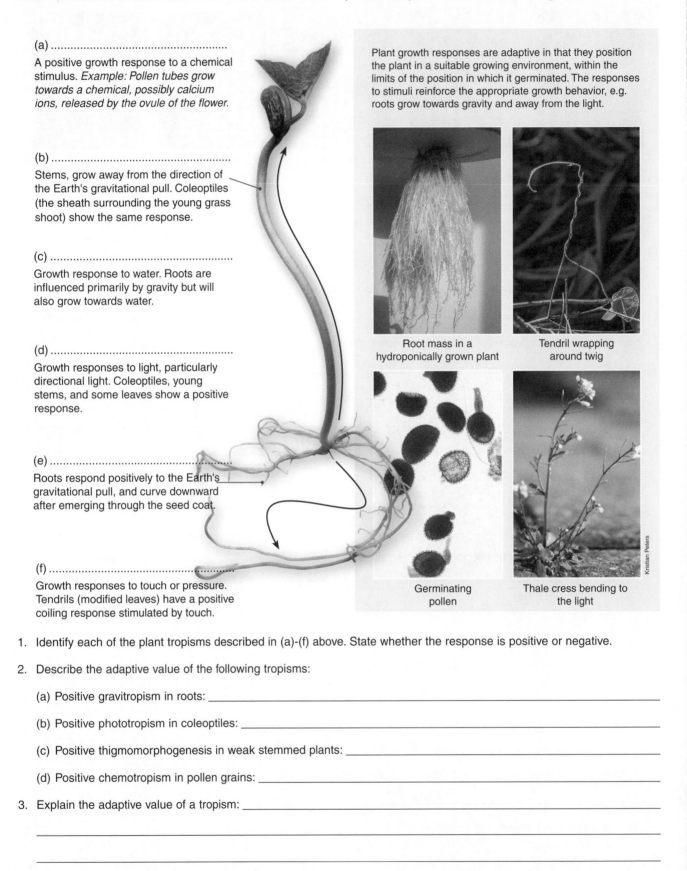

(a) ....

A positive growth response to a chemical stimulus. *Example: Pollen tubes grow towards a chemical, possibly calcium ions, released by the ovule of the flower.*

(b) ....

Stems, grow away from the direction of the Earth's gravitational pull. Coleoptiles (the sheath surrounding the young grass shoot) show the same response.

(c) ....

Growth response to water. Roots are influenced primarily by gravity but will also grow towards water.

(d) ....

Growth responses to light, particularly directional light. Coleoptiles, young stems, and some leaves show a positive response.

(e) ....

Roots respond positively to the Earth's gravitational pull, and curve downward after emerging through the seed coat.

(f) ....

Growth responses to touch or pressure. Tendrils (modified leaves) have a positive coiling response stimulated by touch.

Plant growth responses are adaptive in that they position the plant in a suitable growing environment, within the limits of the position in which it germinated. The responses to stimuli reinforce the appropriate growth behavior, e.g. roots grow towards gravity and away from the light.

Root mass in a hydroponically grown plant

Tendril wrapping around twig

Germinating pollen

Thale cress bending to the light

1. Identify each of the plant tropisms described in (a)-(f) above. State whether the response is positive or negative.

2. Describe the adaptive value of the following tropisms:

   (a) Positive gravitropism in roots: _____

   (b) Positive phototropism in coleoptiles: _____

   (c) Positive thigmomorphogenesis in weak stemmed plants: _____

   (d) Positive chemotropism in pollen grains: _____

3. Explain the adaptive value of a tropism: _____

   _____

   _____

   _____

©2017 **BIOZONE** International
ISBN: 978-1-927309-65-0
Photocopying Prohibited

139 Plant Hormones as Signal Molecules

Key Idea: Auxin is a plant hormone involved in the differential growth responses of plants to environmental stimuli. Auxin promotes apical growth in plants and inhibits the growth of lateral (side) buds.

Auxins are plant hormones with a central role in a range of growth and developmental responses in plants. Auxins are responsible for apical dominance in shoots and are produced in the shoot tip. Indole-acetic acid (IAA) is the most potent native auxin in intact plants. The response of a plant tissue to IAA depends on the tissue itself, the hormone concentration, the timing of its release, and the presence of other hormones. Gradients in auxin concentration during growth prompt differential responses in specific tissues and contribute to directional growth.

Light is an important growth requirement for all plants. Most plants show an adaptive response of growing towards the light. This growth response is called phototropism.

The bending of the plants shown on the right is a phototropism in response to light shining from the left and is caused by the plant hormone **auxin**. Auxin causes the elongation of cells on the shaded side of the stem, causing it to bend (photo right).

Auxin is produced in the shoot tip and is responsible for apical dominance by suppressing growth of the lateral (side) buds.

Auxin movement through the plant is polar. It moves from the shoot tip down the plant.

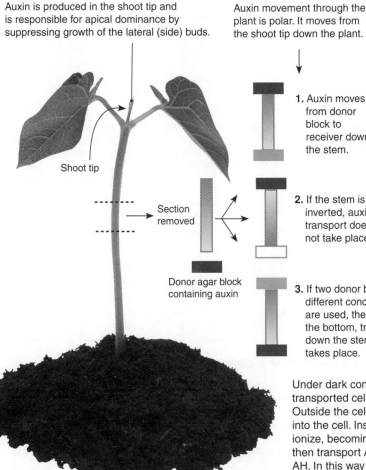

Shoot tip

Section removed

Donor agar block containing auxin

1. Auxin moves from donor block to receiver down the stem.

2. If the stem is inverted, auxin transport does not take place.

3. If two donor blocks of different concentration are used, the higher at the bottom, transport down the stem still takes place.

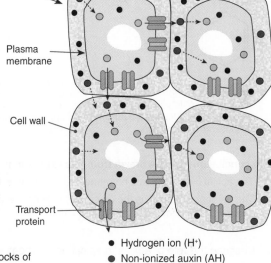

Plasma membrane

Cell wall

Transport protein

- ● Hydrogen ion (H⁺)
- ● Non-ionized auxin (AH)
- ● Ionized auxin (A⁻)
- ····▶ Diffusion
- ──▶ Active transport

Under dark conditions auxin moves evenly down the stem. It is transported cell to cell by diffusion and transport proteins (above right). Outside the cell auxin is a non-ionized molecule (AH) which can diffuse into the cell. Inside the cell the pH of the cytoplasm causes auxin to ionize, becoming A⁻ and H⁺. Transport proteins at the basal end of the cell then transport A⁻ out of the cell where it reacquires an H⁺ ion and reforms AH. In this way auxin is transported in one direction through the plant. When plant cells are illuminated by light from one direction transport proteins in the plasma membrane on the shaded side of the cell are activated and auxin is transported to the shaded side of the plant.

1. What is the term given to the tropism being displayed in the photo (top right)? _____

2. Describe one piece of evidence that demonstrates the transport of auxin is polar: _____

_____

_____

_____

3. What is the effect of auxin on cell growth? _____

_____

©2017 **BIOZONE** International
ISBN: 978-1-927309-65-0
Photocopying Prohibited

PRACTICES   PRACTICES   PRACTICES   CONNECT   CONNECT   WEB

80 AP1   **78** AP1   **139**

KNOW

Auxin and apical dominance

Auxin is produced in the shoot tip and diffuses down to inhibit the development of the lateral (side) buds. The effect of auxin on preventing the development of lateral buds can be demonstrated by removing the source of the auxin and examining the outcome (below).

▶ In many plants the growth of the shoot apex inhibits the growth of side (lateral) buds. As a result, plants tend to grow a single main stem upwards, which dominates over lateral branches.

▶ This response is called **apical dominance**.

▶ The hormone responsible for this response is **auxin**. It acts by stimulating cell elongation.

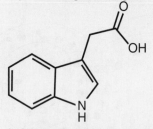

Indole-acetic acid (above) is the only known naturally occurring auxin. It is produced in the apical shoot and young leaves.

No treatment
Apical bud is left intact.

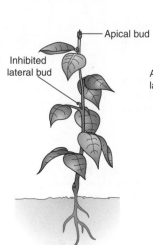

Apical bud

Inhibited lateral bud

In an intact plant, the plant stem elongates and the lateral buds remain inactive. No side growth occurs.

Treatment one
Apical bud is removed; no auxin is applied.

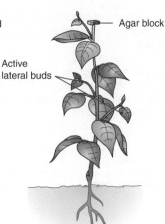

Agar block

Active lateral buds

The apical bud is removed and an agar block without auxin is placed on the cut surface. The seedling begins to develop lateral buds.

Treatment two
Apical bud is removed; auxin is applied.

Agar block

Inhibited lateral bud

The apical bud is removed and an agar block containing auxin is placed on the cut surface. Lateral bud development is inhibited.

Two conclusions can be drawn from this experiment:

(1) The apical bud contains a hormone that inhibits lateral growth because its removal promoted lateral growth.

(2) The presence of auxin in the apical bud inhibits lateral growth because auxin applied to a cut stem tip could inhibit lateral growth and mimic the effect of an intact apical bud.

4. Describe the role of auxins in apical dominance: _____

_____

_____

5. Outline the experimental evidence supporting the role of auxins in apical dominance: _____

_____

_____

_____

_____

6. Study the photo (right) and then answer the following questions:

   (a) Label the apical bud.   (b) Label the lateral bud(s).

   (c) Which buds are the largest? _____

   (d) Why would this be important? _____

   _____

7. If you were a gardener, how would you make your plants bushier?_____

_____

_____

NASA

140 Investigating Phototropism

Key Idea: Experimental evidence supports the hypothesis that auxin is responsible for tropic responses in stems.

Phototropism in plants was linked to a growth promoting substance as early as the 1920s. Early experiments investigating phototropism in severed **coleoptiles** provided evidence for the hypothesis that the plant hormone auxin was responsible for tropic responses in stems. These experiments (outlined below) have been criticized as being too simplistic because coleoptiles are different to stems. However their conclusions have been shown to be valid. **Auxins** (a group of plant hormones) promote cell elongation and are inactivated by light. Thus, when a stem is exposed to directional light, auxin becomes unequally distributed either side of the stem. The stem responds to the unequal auxin concentration by differential growth, i.e. it bends. The mechanisms behind this response are now well understood.

1. **Directional light**: A pot plant is exposed to direct sunlight near a window and as it grows, the shoot tip turns in the direction of the Sun. When the plant was rotated, it adjusted by growing towards the Sun in the new direction.

   (a) What hormone regulates this growth response?

   _____

   (b) What is the name of this growth response?

   _____

   (c) How do the cells behave to bring about this change in shoot direction at:

   Point A? _____

   Point B? _____

   (d) Which side (A or B) would have the highest hormone concentration and why?

   _____

   _____

   (e) In the rectangle on the right, draw a diagram of the cells as they appear across the stem from point A to B.

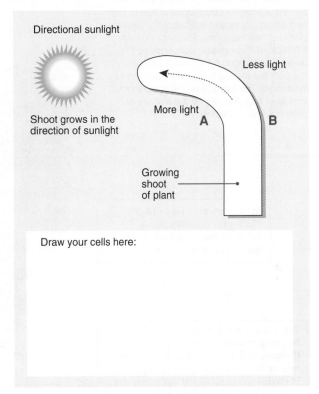

Directional sunlight

Shoot grows in the direction of sunlight

Less light

More light

A B

Growing shoot of plant

Draw your cells here:

2. **Light excluded from shoot tip**: When a tin-foil cap is placed over the top of the shoot tip, light is prevented from reaching the shoot tip. When growing under these conditions, the direction of growth does not change towards the light source, but grows straight up. State what conclusion you can come to about the source and activity of the hormone that controls the growth response:

   _____

   _____

   _____

   _____

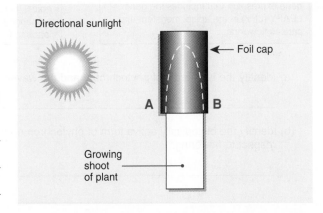

Directional sunlight

Foil cap

A B

Growing shoot of plant

3. **Cutting into the transport system**: Two identical plants were placed side-by-side and subjected to the same directional light source. Razor blades were cut half-way into the stem, thereby interfering with the transport system of the stem. Plant A had the cut on the same side as the light source, while Plant B was cut on the shaded side. Predict the growth responses of:

   Plant **A**: _____

   _____

   Plant **B**: _____

   _____

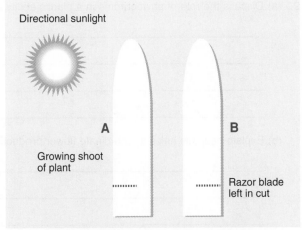

Directional sunlight

A B

Growing shoot of plant

Razor blade left in cut

©2017 **BIOZONE** International
ISBN: 978-1-927309-65-0
Photocopying Prohibited

PRACTICES PRACTICES PRACTICES WEB

140

KNOW

141 Photoperiodism in Plants

Key Idea: Photoperiodism is the response of a plant to the relative lengths of light and dark. It is controlled by the pigment phytochrome, which occurs in two forms P*r* and P*fr*. Flowering is a photoperiodic activity that is dependent on the species' response to light. It is controlled through the action of a pigment called **phytochrome**. Phytochrome acts as a signal for some biological clocks in plants and exists in two forms, P*r* and P*fr*. It is important in initiating flowering in long-day and short-day plants, but is also involved in other light initiated responses, such as germination and shoot growth.

Phytochrome

Phytochrome is a blue-green pigment that acts as a photoreceptor for detection of night and day in plants and is universal in vascular plants. It has two forms: **P*r*** (inactive) and **P*fr*** (active). P*r* is readily converted to P*fr* under natural light. P*fr* converts back to P*r* in the dark but more slowly. P*fr* predominates in daylight. The plant measures daylength (or rather night length) by the amount of phytochrome in each form.

Sunlight

In **daylight** or **red light** (660 nm), P*r* converts rapidly, but reversibly, to P*fr*.

Rapid conversion

Slowly in darkness

P*r*

P*fr*

P*fr* is the physiologically active form of phytochrome. It promotes flowering in long-day plants and inhibits flowering in short-day plants.

Phytochrome interacts with genes collectively called "clock genes" that maintain the plant's biological clock.

In the **dark** or in **far red light** (730 nm) P*fr* reverts slowly, but spontaneously, back to the inactive form of phytochrome P*r*.

Physiologically active

"Clock genes"

Flowering hormone

There is still uncertainty over what the flowering hormone (commonly called **florigen**) is. Recent studies suggested it may be the protein product of the gene FLOWERING LOCUS T (FT) (in long day plants at least) which appears to influence gene expression that includes the gene LEAFY (LFY) in the apical meristem and causes flowering.

The hormone is transported to the apical meristem where it causes a change in gene expression that leads to flowering.

1. (a) Identify the two forms of phytochrome and the wavelengths of light they absorb: _____

   _____

   (b) Identify the biologically active form of phytochrome and how it behaves in long day plants and short day plants with respect to flowering:

   _____

   _____

2. (a) Discuss the role of phytochrome in a plant's ability to measure day length: _____

   _____

   _____

   _____

   _____

   (b) Explain how this helps to coordinate flower production in a plant species? _____

   _____

   _____

   _____

WEB
141

CONNECT
166
AP2

CONNECT
78
AP1

CONNECT
80
AP1

PRACTICES

KNOW

©2017 **BIOZONE** International
ISBN: 978-1-927309-65-0
Photocopying Prohibited

Long day vs short day plants

1. Long-day plants (LDP) flower when the photoperiod is greater than a critical day length. Short-day plants (SDP) flower when the photoperiod is less than a critical day length.

2. Interruption of the long dark period inhibits flowering in SDP but promotes flowering in LDP.

3. Dark must be continuous in SDP but not in LDP.

4. Interruption of the light period inhibits flowering in LDP but not in SDP.

5. Alternating cycles of short light and short dark inhibit flowering in SDP.

6. Plants that do not use day length to initiate flowering are called day-neutral (e.g. cucumber, tomato).

Chrysanthemums

Manipulating flowering in plants

Controlling the light-dark regime has allowed flower growers and horticulturists to produce flowers out of season or to coincide flowering with specific dates. Plants kept in greenhouses can be subjected to artificial lighting or covered to control the amount of light they receive. To be totally effective at controlling flowering, temperature must also be controlled, as this is also an important flowering cue.

For the example of the *Chrysanthemum*, a short-day plant, flowering is can be controlled under the following conditions. The temperature is kept between 16 - 25 °C. The light-dark regime is controlled at 13 hours of light and 11 hours of dark for 4-5 weeks from planting to ensure vegetative growth. Then the regime changes to 10 hours light and 14 hours darkness to induce flowering.

Long-day plants

When subjected to the light regimes on the right, the 'long-day' plants below flowered as indicated:

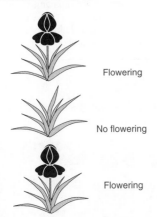

Flowering

No flowering

Flowering

Examples: *lettuce, clover, delphinium, gladiolus, beets, corn, coreopsis*

Photoperiodism in plants

An experiment was carried out to determine the environmental cue that triggers flowering in 'long-day' and 'short-day' plants. The diagram below shows 3 different light regimes to which a variety of long-day and short-day plants were exposed.

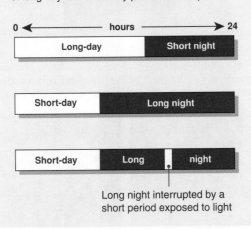

Long night interrupted by a short period exposed to light

Short-day plants

When subjected to the light regimes on the left, the 'short-day' plants below flowered as indicated:

No flowering

Flowering

No flowering

Examples: *potatoes, asters, dahlias, cosmos, chrysanthemums, pointsettias*

---

3. (a) What is the environmental cue that synchronizes flowering in plants? _____

   (b) What is a biological advantage of this synchronization to the plants? _____

   _____

   _____

   _____

4. Study the three light regimes above and the responses of short-day and long-day flowering plants to that light. From this observation, describe the most important factor controlling the onset of flowering in:

   (a) Short-day plants: _____

   (b) Long-day plants: _____

5. What evidence is there for the idea that short-day plants are best described as "long-night plants": _____

   _____

   _____

   _____

   _____

   _____

142 Biological Clocks

Key Idea: A biological clock is the endogenous timing system an organism uses to synchronize its activities with the external environment.

A biological clock is an **endogenous** (internal) timing system that helps to control the physiological responses and activities of an organism. Rhythms established by the biological clock (**endogenous rhythms**) will continue even in the absence of environmental cues, although the period (duration) of the rhythm may be slightly different to the environmental rhythm. Biological clocks have an adaptive function, such as helping anticipate environmental changes and preparing the body for the activities that will predictably follow.

Where is the biological clock located?

The location of the biological clock varies between organisms. In birds, reptiles, and amphibians it is located in the pineal gland (in the brain). In insects each cell has its own biological clock. In mammals the biological clock is located in the hypothalamus.

For most humans, the biological clock runs at about a 25½ hour day. To keep it synchronized with the 24 hour-day cycle it needs to be reset each day, reacting to outside stimuli such as light and dark and meal times. The clock is made up of a collection of cells in the hypothalamus, called the suprachiasmatic nucleus (**SCN**), just behind the eyes. Light from the eyes stimulates the nerve pathways to the SCN and regulates its activity.

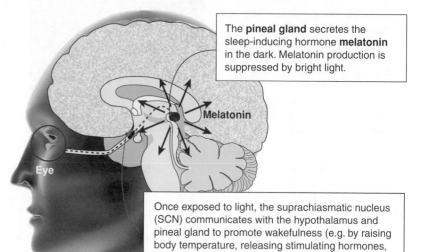

The **pineal gland** secretes the sleep-inducing hormone **melatonin** in the dark. Melatonin production is suppressed by bright light.

Once exposed to light, the suprachiasmatic nucleus (SCN) communicates with the hypothalamus and pineal gland to promote wakefulness (e.g. by raising body temperature, releasing stimulating hormones, and suppressing melatonin production.

Functions of the biological clock

The biological clock helps to control internal rhythms such as heart activity, hormone secretion, blood pressure, oxygen consumption, and metabolic rate. When the rhythms controlled by the biological clock become out-of-sync with the environment various short or long term disorders can occur, e.g. jet lag. Some functions of the biological clock are described below:

Prediction of and preparation for events in the environment (e.g. storing food reserves as fat for periods of torpor or hibernation).

Synchronization of migration, reproduction, or social activities. Animals congregate at breeding grounds at the same time of year.

Synchronizing circadian and annual rhythms, e.g. basking in tuatara and other reptiles, with changes in the environment.

Time compensation in navigation and sun compass orientation using a continuously consulted clock (e.g. honeybee food collection).

1. (a) Where is the biological clock located in mammals? _____

   (b) What is the main stimulus that helps synchronize the biological clock with the environment? _____

   _____

2. Describe two functions of the biological clock and their adaptive value:

   (a) _____

   _____

   _____

   (b) _____

   _____

   _____

©2017 **BIOZONE** International
ISBN: 978-1-927309-65-0
Photocopying Prohibited

143 Biological Clocks and the Environment

Key Idea: External cues synchronize the biological clock with the environment.

Biological clocks stay synchronized with the environment because they are regularly reset by an external environmental cue or **zeitgeber**. The process of resetting the internal clock is known as **entrainment**. Endogenous rhythms that are synchronized to specific environmental cues are adaptive, contributing to fitness by ensuring the success of critical activities such as mating, birth, germination, foraging, and periods of torpor and dormancy.

A simple mechanism for biological clocks

Zeitgeber (e.g sunlight)

Entrainment (environmental cues synchronize the biological clock with the environment).

Biological clock

Output (internal mechanisms produce internal changes that match changes in the environment).

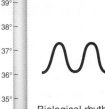

Biological rhythm (e.g. fluctuations in body temperature)

Travelling and biological clocks

Most animals travel slowly enough that their biological clock is never far out of sync with the environment and entrainment by the rising of the Sun each day can reset any variation. Travelling west-east (or east-west) in a plane can result in the biological clock being severely out-of-sync with environmental cues. This phenomenon is called **jet lag**.

Jet lag occurs because the **biological clock** is responsible for regulating the natural sleep-wake cycle, which involves being awake and active during the day and sleeping at night when it is dark. Rapid, long distance air travel can lead to disruption of the normal sleep-wake cycle. When travelling across multiple time zones, the biological clock will not be synchronized with the destination time and must adjust to the new schedule.

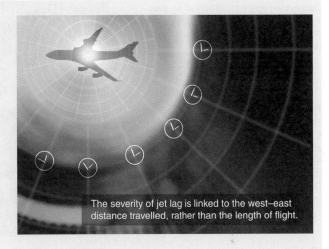

The severity of jet lag is linked to the west–east distance travelled, rather than the length of flight.

1. (a) What is a zeitgeber? _____

_____

(b) Identify a common zeitgeber in animals: _____

2. (a) What is entrainment? _____

_____

(b) Why is entrainment important to an organism? _____

_____

_____

_____

3. (a) Person A travels 5000 km east in 9 hours. Person B travels 500 km east in 9 hours. Which of these people is more likely to experience jet lag?

_____

(b) Explain you answer to (a): _____

_____

_____

_____

CONNECT
166
AP2

WEB
143

KNOW

144 Biological Rhythms

Key Idea: Living organisms show regular cycles of activity that approximate cyclic events in the environment.

The activity patterns of organisms often occur with frequencies that approximate the predictable cyclic events in the environment, such as the light-dark cycle and the changing of the seasons. The length of time it takes to complete the entire cycle is termed the rhythm's **period**, e.g. 24 hours. Rhythms that continue in the absence of external cues are said to be **endogenous** (internal). Those that are direct responses to the environment and do not persist when conditions are kept constant are called **exogenous**. In most cases, the rhythm is the expression of both the internal (endogenous) timing mechanism (called the biological clock) and the environmental (exogenous) cue that synchronizes it. Biological rhythms are adaptive in that they ensure the appropriate behavior occurs at the appropriate time.

Term	Definition
Endogenous	An internally controlled response
Exogenous	An externally controlled response
Circadian (daily)	A rhythm that cycles over an approximately 24 hour period.

Term	Definition
Circatidal (tidal)	A rhythm that matches the movement of the tides and has a period of ~12.5 hours.
Circalunar (lunar)	A rhythm that cycles over an approximately 29.5 day period.
Circannual (annual)	An annual rhythm (one year)

Rhythm: circadian (daily)
Period: ~ 24 h
Example: Crickets are generally active during the night (nocturnal) when they forage in leaf litter or trees. Being active at night makes them less vulnerable to daytime predators.

CIRCADIAN: cricket
Joseph Berger; Bugwood.org

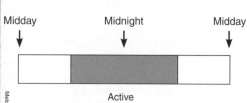
Melodi2

Rhythm: circatidal (tidal)
Period: ~ 12.4 h (coincident with tidal flows)
Example: Fiddler crabs inhabit mangrove swamps and mud flats. Locomotion and feeding occur at low tide when algae and detritus are exposed and easily collected. During high tide the crabs hide in burrows in the mud, away from predators.

CIRCATIDAL: fiddler crab
NOAA

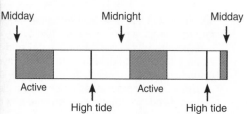

Rhythm: circalunar (lunar) / circannual
Period: ~ 29.5 days (a month) / one year
Example: In spring and summer during a new moon, grunions lay eggs above the high water mark, where the eggs are 'stranded' away from aquatic predators. At the next set of high tides the eggs hatch and the young return to the sea.

CIRCALUNAR: grunion
Eric Wittman CC 2.0

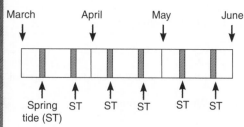

Rhythm: circannual (annual)
Period: ~ a year
Example: Arctic ground squirrels feed in the Arctic summer to store fat for winter hibernation. They hibernate from early August to late April when temperatures are low and food is scarce.

CIRCANNUAL: Arctic ground squirrel
Ianaré Sévi CC 3.0

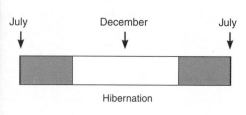

Rhythm: circannual (annual)
Period: ~ a year
Example: In many hoofed grazers (ungulates) including domestic livestock species, the reproductive cycle is timed so that young are born in spring when the weather is warmer and food is more plentiful.

CIRCANNUAL: bison

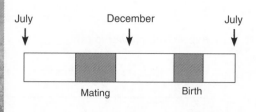

©2017 **BIOZONE** International
ISBN: 978-1-927309-65-0
Photocopying Prohibited

1. (a) Contrast endogenous and exogenous components of a biological rhythm: _____

    _____

    _____

    _____

   (b) Explain why endogenous and exogenous components are important in maintaining a biological rhythm: _____

    _____

    _____

    _____

    _____

2. For each of the following rhythms provide a definition and example and then describe the adaptive value of the rhythm:

   (a) Circadian rhythm: _____

    _____

    _____

    _____

   (b) Circatidal rhythm: _____

    _____

    _____

    _____

   (c) Circalunar rhythm: _____

    _____

    _____

    _____

    _____

   (d) Circannual rhythm: _____

    _____

    _____

    _____

3. For each of the examples below, describe an environmental cue that might be used to induce or maintain the activity:

   (a) Hedgehog's hibernation in winter: _____

   (b) Blackbird's foraging and social behavior during daylight: _____

   (c) Bat's activity of hunting for insects at night: _____

   (d) Coordinated flowering of plants in spring: _____

4. Suggest a way a scientist could discover if an organism's biological rhythm is endogenous or exogenous: _____

    _____

    _____

    _____

145 Human Biological Rhythms

Key Idea: Humans exhibit a number of periodic changes in behavior or physiology that are generated and maintained by a biological clock.

In humans, many physiological activities exhibit a circadian (below), circalunar, or circannual rhythm. These rhythms are endogenous, but are entrained by the environment.

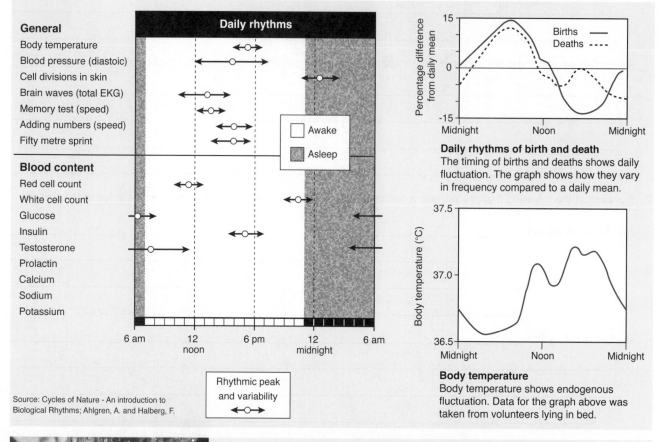

Daily rhythms of birth and death
The timing of births and deaths shows daily fluctuation. The graph shows how they vary in frequency compared to a daily mean.

Body temperature
Body temperature shows endogenous fluctuation. Data for the graph above was taken from volunteers lying in bed.

Source: Cycles of Nature - An introduction to Biological Rhythms; Ahlgren, A. and Halberg, F.

Rhythmic peak and variability

Rhythms in a long duration cave occupation

A woman spent four months isolated underground in a cave so that her biological rhythms could be studied in the absence of the normal day/night environmental cues. The air temperature inside the cave remained constant over this time. Her body temperature was measured 3 times a day for 4 months.

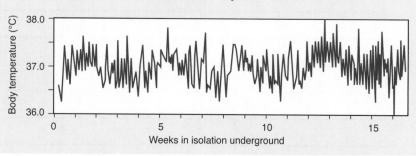

1. Study the graphs for 'daily rhythms of birth and death' and 'body temperature'. Is there any correlation (mutual relationship) between the time of day when most deaths occur and body temperature? Explain your answer:

_____

_____

2. Different cycles for the same biological variable can occur simultaneously in an organism. The graph above shows the body temperature of a woman who stayed underground for four months.

(a) What is represented by the sharp fluctuations in body temperature? _____

_____

(b) Explain why this experiment was carried out in a cave environment: _____

_____

WEB 145   CONNECT 166 AP2

DATA

146 Interpreting Actograms

Key Idea: Actograms are graphical records of an organism's activity and can be used to determine its activity patterns.
In a laboratory, the activity of an organism can be recorded continuously. The activity is often recorded as a bar on a line representing 24 hours. By placing the successive blocks of 24 hours under each other, a clearer picture of the pattern of activity can be seen. By keeping the environmental cues constant (e.g. constant dark) it is possible to see the length of time of the organism's biological clock runs for in the absence of environmental cues. This length of time is called the **free-running period**. A **phase shift** occurs when an organism is entrained to a new regime of environmental cues.

Making an actogram

Tape

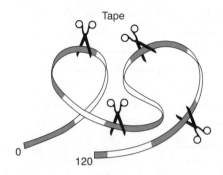

Raster

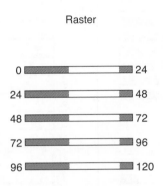

Double plotted raster

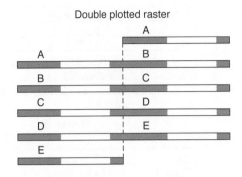

Imagine ticker tape rolling out of a machine. Every time an organism is active, a mark or bar is made on the tape. The tape is then cut up into lengths representing 24 hours.

The lengths are then laid out one under the other in order, forming a stack called a **raster**.

To maintain continuity between cuts, the raster is copied and pasted to the left but shifted down one line. This is called **double plotting** and is how most actograms are laid out.

Actogram of human activity

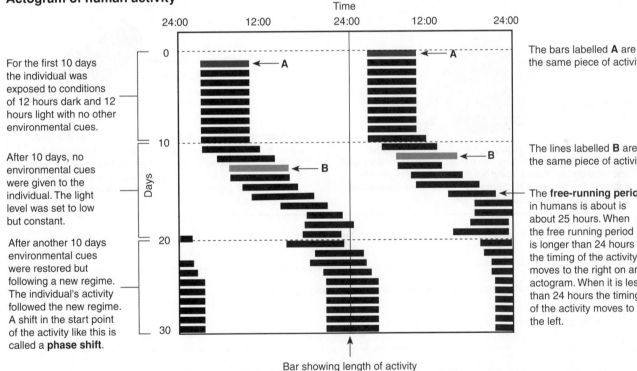

For the first 10 days the individual was exposed to conditions of 12 hours dark and 12 hours light with no other environmental cues.

After 10 days, no environmental cues were given to the individual. The light level was set to low but constant.

After another 10 days environmental cues were restored but following a new regime. The individual's activity followed the new regime. A shift in the start point of the activity like this is called a **phase shift**.

The bars labelled **A** are the same piece of activity.

The lines labelled **B** are the same piece of activity.

The **free-running period** in humans is about is about 25 hours. When the free running period is longer than 24 hours the timing of the activity moves to the right on an actogram. When it is less than 24 hours the timing of the activity moves to the left.

Bar showing length of activity

1. Why are actograms produced by double plotting the original raster? _____

_____

2. (a) What is a free-running period? _____

_____

(b) How would you tell if the free-running period was shorter or longer than 24 hours? _____

_____

_____

©2017 **BIOZONE** International
ISBN: 978-1-927309-65-0
Photocopying Prohibited

PRACTICES WEB

146

KNOW

147 Activity Patterns in Animals

Key Idea: A free-running endogenous rhythm usually has a slightly different period from the cyclic environmental variable that entrains it. The rhythm can be recorded as an actogram. In the absence of an external zietgeber (such as light) an endogenous rhythm will adopt a free-running period, which is usually slightly different from the cyclic environmental variable to which it is usually entrained. The activity patterns of two species were recorded in the absence of environmental cues to determine the free running period of the rhythm. The results are displayed below as **actograms**. On each actogram the first day of recording is only shown on the left (day two is outlined and shaded in blue).

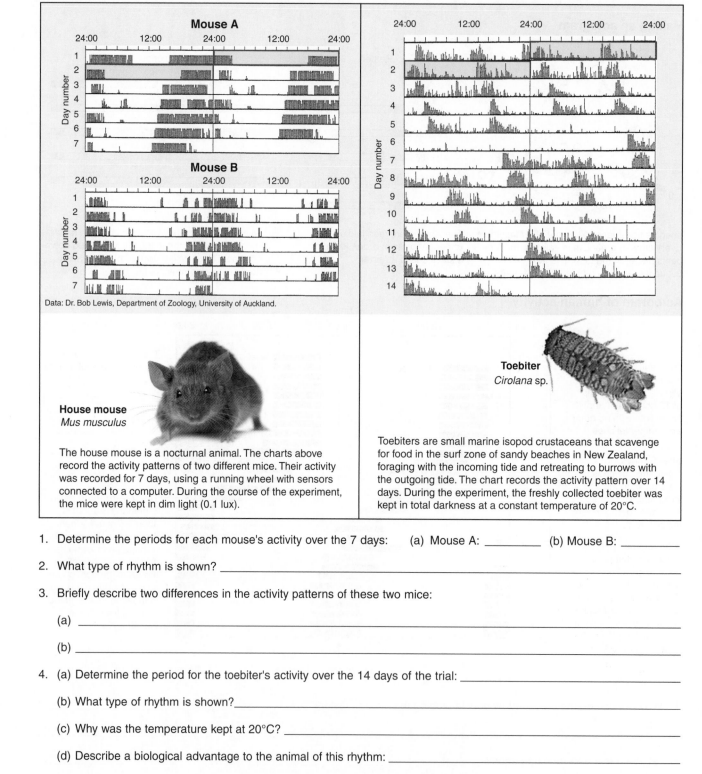

Data: Dr. Bob Lewis, Department of Zoology, University of Auckland.

House mouse
Mus musculus

The house mouse is a nocturnal animal. The charts above record the activity patterns of two different mice. Their activity was recorded for 7 days, using a running wheel with sensors connected to a computer. During the course of the experiment, the mice were kept in dim light (0.1 lux).

Toebiter
Cirolana sp.

Toebiters are small marine isopod crustaceans that scavenge for food in the surf zone of sandy beaches in New Zealand, foraging with the incoming tide and retreating to burrows with the outgoing tide. The chart records the activity pattern over 14 days. During the experiment, the freshly collected toebiter was kept in total darkness at a constant temperature of 20°C.

1. Determine the periods for each mouse's activity over the 7 days: (a) Mouse A: _____   (b) Mouse B: _____

2. What type of rhythm is shown? _____

3. Briefly describe two differences in the activity patterns of these two mice:

    (a) _____

    (b) _____

4. (a) Determine the period for the toebiter's activity over the 14 days of the trial: _____

    (b) What type of rhythm is shown?_____

    (c) Why was the temperature kept at 20°C? _____

    (d) Describe a biological advantage to the animal of this rhythm: _____

    _____

    (e) What cue on the shore would the animal use to synchronize its rhythm? _____

5. What evidence do you have that the rhythms shown above are endogenous? _____

_____

©2017 **BIOZONE** International
ISBN: 978-1-927309-65-0
Photocopying Prohibited

Cockroach activity

Cockroaches are nocturnal insects. The experiments below investigated the periodicity of their behavior under controlled conditions. The first experiment determined the **free-running period**. The results of the second experiment show the **entrainment** of the rhythm. Note the actograms below *are not* double plotted.

Free-running period

The charts on the right record the activity rhythm of a cockroach kept for 20 days in a running wheel actogram. There are activity records shown for each of the 20 days, with each day's record presented in succession down the page. The periods of activity are shown as grey rectangular blocks and periods of inactivity shown as no rectangle.

The onset of constant darkness on day 11 exposed the free-running period and produced a phase shift.

Light regime

This is the term used to describe the cycles of light and darkness. It is indicated by bars of 'light' and 'darkness' at the bottom of each table:

Days 1-10: The cockroach was in a 12 hour light / 12 hour dark cycle (**LD 12:12**)

Days 11-20: The cockroach was in constant darkness for these 10 days (**DD**)

Entrainment

Recall that the process by which the endogenous rhythm is synchronized to an environmental cue (or **zeitgeber**), such as a 12 hour light / 12 hour dark cycle, is termed **entrainment**. The chart on the right shows the activity record of a cockroach. It is being entrained to a new light cycle, which occurs nine hours earlier than the one it had been experiencing previously.

Entrainment usually has the following features:
- A phase shift for the start of the activity is gradual, without jumps.
- As the activity gets nearer the new lights-out signal, the daily phase shift is reduced.

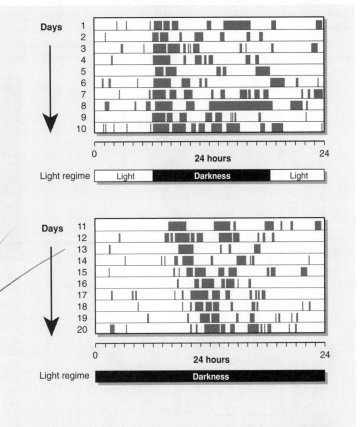

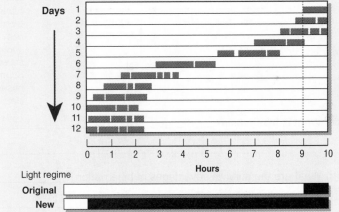

6. What type of rhythm is displayed in the first 10 days? _____

7. In what part of the light/darkness cycle was the cockroach most active? _____

8. Describe the activity pattern displayed by the cockroach: _____

9. Determine the free-running period (in hours) displayed between days 11-20: _____

10. Describe the effect of entrainment on the free-running period, and explain its adaptive value in a natural environment:

_____

_____

_____

_____

_____

©2017 **BIOZONE** International
ISBN: 978-1-927309-65-0
Photocopying Prohibited

148 Hibernation

Key Idea: Hibernation is a strategy that has evolved to conserve energy in times of extreme environmental conditions. It involves a rapid, drastic change in physiological processes. Many animals hibernate during winter to conserve energy while food supplies are limited. During **hibernation**, the animal enters a state of inactivity and reduced metabolic rate. The basal metabolic rate can drop to 2-4% of normal, conserving the body's reserves during the hibernation period. Body temperature, breathing rate, and heart rate are all significantly decreased during hibernation. The animal is also less responsive to environmental stimuli. Hibernation periods vary between species (days, weeks, or months). Some animals exhibit reduced metabolic activity during the summer months, this is called **estivation**.

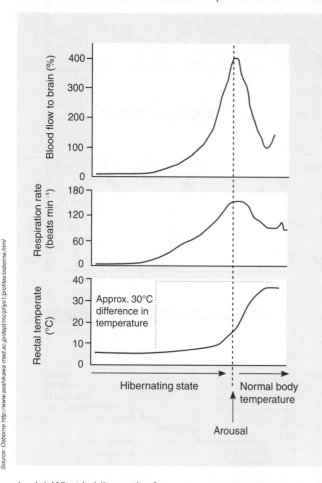

Animals prepare for hibernation at the onset of specific **environmental cues**. Common hibernation cues include a shortage of food, shorter daylight periods, and cooler temperatures. Many animals, such as this chipmunk (left), build up their energy stores by eating large quantities of food prior to hibernating. The excess energy is stored as fat deposits, which provide the energy to carry out metabolic activities during hibernation.

Obligate hibernators (e.g. rodents such as ground squirrels) drop their body temperature significantly, and cannot be awoken by external stimuli. However, they are periodically awoken by internal stimuli, so that maintenance functions can occur. **Facultative hibernators** (e.g. bats and bears) can be aroused by external stimuli and their period of inactivity is often described as **torpor**, rather than true hibernation. Arousal carries an energetic cost. Large animals, such as bears, do not drop their body temperature as much as smaller animals. Their body temperature drops only a few degrees because raising it from a very low temperature would take too long and would be too energy-expensive.

The data (left) shows metabolic activity and temperature in golden hamsters during hibernation. Note the large difference in the animal's temperature (~30°C) between hibernation and normal body temperature. Metabolic activity (blood flow to the brain and respiration rate) significantly decreases during hibernation, increases to a maximum during arousal, and tapers off once normal body temperature is achieved. The elevated metabolic rate observed during the arousal period speeds up arousal and rapidly clears waste products from the body.

Source: Osborne http://www.asahikawa-med.ac.jp/dept/mc/phys1/profiles/osborne.html

1. (a) What is hibernation? _____

    _____

    (b) What are the survival advantages of hibernation? _____

    _____

    _____

    (c) What are the common environmental cues triggering hibernation? ___

    _____

    _____

2. (a) What happens to the body temperature of the golden hamster during hibernation? _____

    _____

    (b) Why does this change in temperature occur? _____

    _____

    (c) Explain why blood flow to the brain and respiration rate may peak during arousal from hibernation:

    _____

    _____

©2017 **BIOZONE** International
ISBN: 978-1-927309-65-0
Photocopying Prohibited

149 The Components of Behavior

Key Idea: Animal behavior can be broadly divided into two categories, innate (inherited), and learned (from experience). Behavior in animals can be attributed to two components: **innate behavior**, which requires no prior experience, and **learned behavior**, which results from the experiences of the animal. Together they produce the total behavioral repertoire exhibited by the animal. Innate behaviors are adaptive and are performed for reasons related to fitness in particular environments (e.g. orientation away from light). Some, although not all, innate behaviors can be modified by learning, as occurs in habituation. Learning, which involves the modification of behavior by experience, is an important mechanism by which animals can modify their responses to behave in specific ways in particular situations.

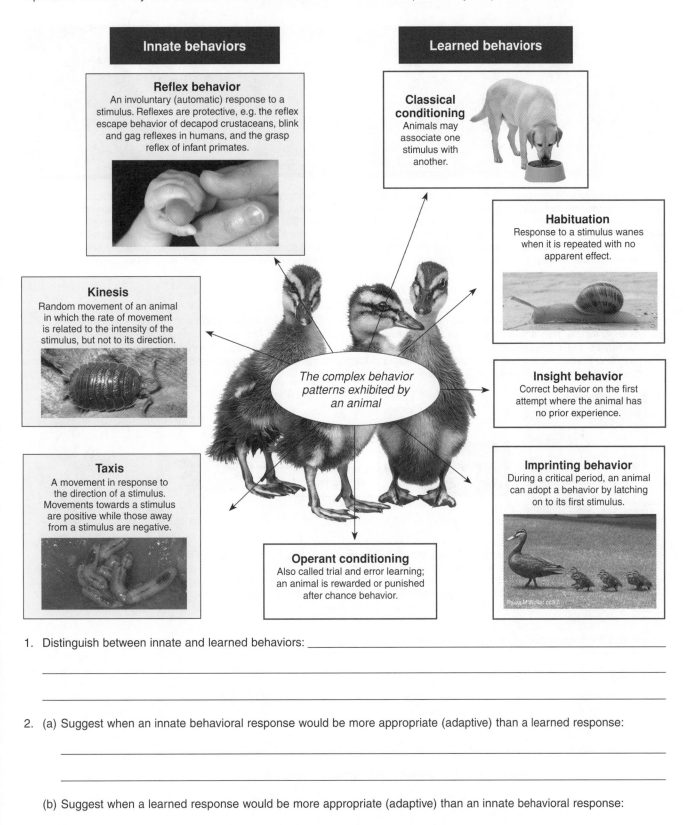

Innate behaviors

Reflex behavior
An involuntary (automatic) response to a stimulus. Reflexes are protective, e.g. the reflex escape behavior of decapod crustaceans, blink and gag reflexes in humans, and the grasp reflex of infant primates.

Kinesis
Random movement of an animal in which the rate of movement is related to the intensity of the stimulus, but not to its direction.

Taxis
A movement in response to the direction of a stimulus. Movements towards a stimulus are positive while those away from a stimulus are negative.

The complex behavior patterns exhibited by an animal

Operant conditioning
Also called trial and error learning; an animal is rewarded or punished after chance behavior.

Learned behaviors

Classical conditioning
Animals may associate one stimulus with another.

Habituation
Response to a stimulus wanes when it is repeated with no apparent effect.

Insight behavior
Correct behavior on the first attempt where the animal has no prior experience.

Imprinting behavior
During a critical period, an animal can adopt a behavior by latching on to its first stimulus.

Paula M Wicker cc3.0

1. Distinguish between innate and learned behaviors: _____

_____

_____

2. (a) Suggest when an innate behavioral response would be more appropriate (adaptive) than a learned response:

_____

_____

(b) Suggest when a learned response would be more appropriate (adaptive) than an innate behavioral response:

_____

_____

©2017 **BIOZONE** International
ISBN: 978-1-927309-65-0
Photocopying Prohibited

CONNECT
181
AP2

WEB
149

KNOW

150 Innate Behaviors

Key Idea: Innate behavior is a behavior that is carried out in response to stimuli and performed without prior experience. Innate behavior (often called instinct) is a pattern of behavior that is performed without any prior learning or experience. Innate behaviors are inherited without modification and have important roles in survival of young, courtship and mating, communication, and navigation. Many (but not all) innate behaviors follow an innate behavioral program called a **fixed-action pattern** (FAP), which is activated by a stimulus or **releaser** to direct some kind of behavioral response. Although 'pre-programmed', they are often complex and once begun, must run to completion.

Examples of innate behaviors

Newly hatched sea turtles innately head towards the brightest area, which is normally moonlight reflecting off the ocean. This takes them to the sea, reducing the risk of predation and overheating.

Most birds build nests that follow the same pattern within a species. Weaver birds all follow the same pattern of nest building, but experience does enhance the building technique.

Web building in spiders is innate and follows the same rules within species. Orb web spiders build large webs in open spaces. Jumping spiders leave a silk safety line behind them.

Courtship behaviors can be very complex but each species has its own specific behavior. Courtship dances in birds can be very elaborate and young birds may need to practice their dance.

Fixed action patterns

Some behaviors are so critical that they have become genetically fixed and the behavior can not be altered by learning or external stimuli. This kind of behavior is called a **fixed-action pattern** (FAP). FAPs are spontaneous, stereotyped (always the same), and indivisible. Once begun, a FAP runs to completion and is independent of learning. Many FAPs have been identified in animals. The one illustrated below relates to feeding in the southern black-backed gull (*Larus dominicanus*). Similar behavior occurs in many other gull species.

 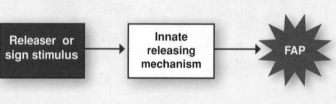 

The red dot on the bill of the adult gull acts as a releaser (sign stimulus) for a fixed action pattern in the chick (the chick pecks at the red dot). The pecking action acts as a releaser for a fixed action pattern in the adult gull, in which it regurgitates food (as above). The behavioral response to the sign stimulus is mediated via a hypothetical neural processing mechanism called the innate releasing mechanism. Thus the message of "I'm hungry, feed me" is communicated in a unchanging way, which is recognized by both parties.

1. What is the importance of innate behaviors to survival? _____

_____

_____

_____

_____

2. (a) Explain the role of releasers in behaviors involving fixed action patterns: _____

_____

(b) Name a releaser for a fixed action pattern in gulls and describe the behavior elicited: _____

_____

©2017 **BIOZONE** International
ISBN: 978-1-927309-65-0
Photocopying Prohibited

151 Kineses

Key Idea: Kineses are innate locomotory behaviors involving non-directional movements in response to external stimuli.

A **kinesis** (*pl.* kineses) is a non-directional response to a stimulus in which the speed of movement or the rate of turning is proportional to the stimulus intensity. Kineses are typical of many invertebrates (including protozoa) and do not involve orientation directly to the stimulus. Two main types can be identified. In an **orthokinesis**, the rate of **movement** is dependent on the stimulus intensity. In a **klinokinesis**, the rate of **turning** is related to the stimulus intensity.

Kinesis in woodlice

Woodlice are commonly found living in damp conditions under logs or bark. Many of the behavioral responses of woodlice are concerned with retaining moisture. Unlike most other terrestrial arthropods, they lack a waterproof cuticle, so water can diffuse through the exoskeleton, making them vulnerable to drying out. When exposed to low humidity, high temperatures, or high light levels, woodlice show a kinesis response to return them to their preferred, high humidity environment.

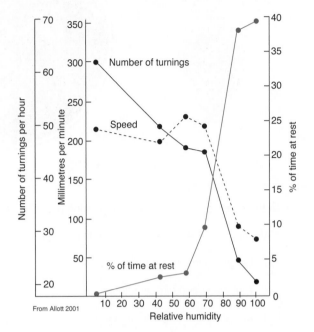

From Allott 2001

Investigating kinesis in woodlice

Experiment 1

To investigate the effect of a light-dark regime on the orthokinetic behavior of woodlice.

Method

A petri dish was laid out with 1 cm x 1 cm squares. The investigation was carried out at room temperature (about 21°C). A woodlouse was placed in the petri dish under constant light. The number of squares the woodlouse passed over in five minutes was recorded. This was repeated four times. The woodlouse was then placed in constant dark and the number of squares it passed over in five minutes recorded. Again, this was repeated four times. The results are shown below.

Results

Trial	Number of squares crossed	
	Light	Dark
1	122	15
2	206	68
3	103	57
4	70	59
Mean		

Experiment 2

To investigate the effect of a light-dark regime on the kilinokinetic behavior of woodlice.

Method

The woodlouse was again placed in the petri dish under constant light. The experiment was carried out at room temperature as in experiment 1. The number of turns the woodlouse performed in five minutes was recorded. This was repeated four times. The woodlouse was then placed in constant dark. Again the number of turns performed in five minutes was recorded. This was also carried out four times. The results are shown below.

Results

Trial	Number of turns	
	Light	Dark
1	80	10
2	165	20
3	110	122
4	90	55
Mean		

PRACTICES

CONNECT
149
AP2

WEB
151

KNOW

Kinesis in body lice

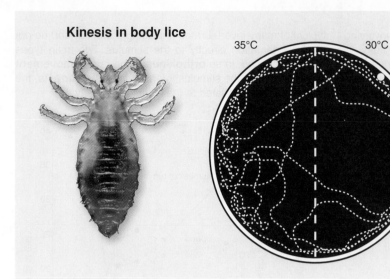

35°C   30°C

In a circular chamber, lice make relatively few turns at their preferred temperature of 30°C, but many random turns at 35°C. This response enables the lice to increase their chances of finding favorable conditions and remaining in them once found.

1. Use the graph on woodlice at the top of the previous page to answer the following questions:

   (a) At which relative humidities do the following occur:

   i.   Largest number of turnings per hour: _____

   ii. Highest speed of movement: _____

   iii. Largest percentage of time at rest: _____

   (b) Explain the significance of these movements: _____

   _____

   _____

   _____

   (c) What is the preferred range of relative humidity for the woodlice? _____

2. (a) Complete the results tables on the previous page by calculating the mean for each of the experiments.

   (b) Which regime (light or dark) does the woodlice appear to prefer? _____

   (c) Explain your reasoning: _____

   _____

   (d) Explain how increasing the number of turns or the speed of movement increases a woodlice's likelihood of survival when in a unfavorable environment.

   _____

   _____

   _____

3. (a) Identify the preferred temperature of a body louse: _____

   (b) The response of the body louse is a klinokinesis / orthokinesis (delete one)

   (c) Contrast the movements of the body louse when within and when outside its preferred temperature environment:

   _____

   _____

   _____

   _____

152 Taxes

Key Idea: A taxis is an innate locomotory behavior involving directional movements in response to external stimuli.

Taxes (*sing.* taxis) involve orientation and movement in response to a directional stimulus or a gradient in stimulus intensity. Taxes often involve moving the head until the sensory input from both sides is equal. Many taxes involve a simultaneous response to more than one stimulus, e.g.

fish orientate dorsal (back) side up in response to both light and gravity. As with plant tropisms, orientation responses are identified by the stimulus involved and whether the response is towards (positive) or away from (negative) the stimulus. Simple orientation responses are **innate** (genetically programmed). More complex orientation responses may involve learning (behavior modified based on experience).

When confronted with a vertical surface, snails will reorientate themselves so that they climb vertically upwards. The adaptive advantage of this may be to help the snail find food or shelter, or to avoid overly wet surfaces.

A

A flying male moth, encountering an odor (pheromone) trail left by a female, will turn and fly upwind until it reaches the female. This behavior increases the chances of the male moth mating and passing on its genes to the next generation.

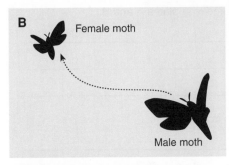
B
Female moth

Male moth

Male moths detect pheromones with their large antennae.

CSIRO cc3.0

Spiny lobsters (crayfish) will back into tight crevices so that their body is touching the crevice sides. The antennae may be extended out. This behavior gives the lobsters greater protection from predators.

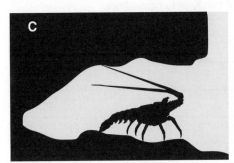
C

Spiny lobster in hole

At close range, mosquitoes use the temperature gradient generated by the body heat of a host to locate exposed flesh. This allows the female to find the blood needed for the development of eggs.

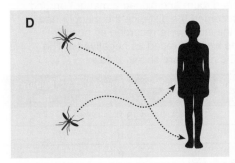
D

Blowfly maggots will turn and move rapidly away from a directional light source. Light usually indicates hot, dry areas and the maggots avoid predators and desiccation (drying out) by avoiding the light.

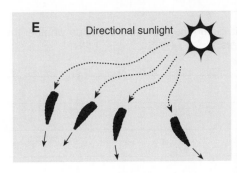
E
Directional sunlight

White fly larvae burrowing into soil.

CONNECT **169** AP2 CONNECT **149** AP2 WEB **152**

KNOW

1. Distinguish between a kinesis and a taxis: _____

_____

_____

2. Describe the adaptive value of simple orientation behaviors such as taxes: _____

_____

_____

3. For each example (**A-E**) on the previous page, describe the orientation response and whether it is positive or negative:

(a) **A:** _____ (d) **D:** _____

(b) **B:** _____ (e) **E:** _____

(c) **C:** _____

4. The diagrams on the right show the movement of nematodes on plates where a salt (NH_4Cl) was added (**A**) and on a plate where no NH_4Cl was added (**B**).

(a) Describe the movements of the nematodes in plates A and B:

_____

_____

_____

(b) Name the orientation behavior shown in plate A:

_____

(c) Describe an advantage of this kind of behavior to nematodes:

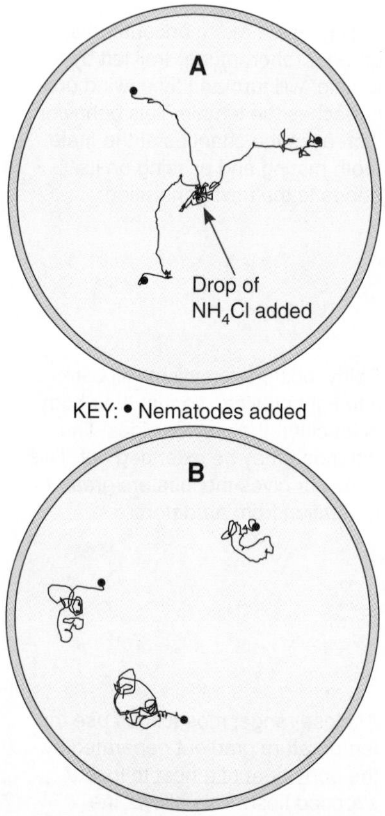

_____

_____

_____

_____

A

Drop of NH_4Cl added

KEY: • Nematodes added

B

S. Ward, Medical Research Council 1973

5. Some students carried out an investigation of the phototactic movements of maggots. They set up a lamp in a darkened room and placed a maggot on grid paper 10 cm from the lamp. They then recorded the distance the maggot had moved every 10 s. Movements towards the lamp were recorded as positive (+) while movements away from the lamp were recorded as negative (-). The investigation was repeated four times. The results are shown below:

Time (s)	Distance of maggot 1 from start point (cm)	Distance of maggot 2 from start point (cm)	Distance of maggot 3 from start point (cm)	Distance of maggot 4 from start point (cm)
10	-1.7	-3.7	-5.8	-3.0
20	-0.5	-0.0	+1.8	-1.5
30	+0.7	-0.8	-6.1	-0.2
40	+0.2	-1.4	+1.0	-1.0

Describe the movements of the maggots during the experiment. Include whether the maggots are positively or negatively phototactic and the rate of the movements.

_____

_____

_____

_____

©2017 **BIOZONE** International
ISBN: 978-1-927309-65-0
Photocopying Prohibited

153 Choice Chamber Investigations

Key Idea: Choice chambers are a simple way to investigate animal behaviors, including simple orientation behavior. Choice chambers are a simple way to investigate behavior in animals. A simple choice chamber consists of two distinct areas enclosing opposing environments, e.g. warm and cool, dry and humid, light and dark. Animals are placed in the middle of the chamber and given time to move to their preferred area before numbers in each chamber are counted.

Background

Students carried out two investigations on woodlice. The first was to determine woodlouse preference for light or dark. The second was to test preference for warm or cool environments.

Aim: investigation 1

To investigate if woodlice prefer a light or dark environment.

The method

A choice chamber was set up using two joined petri dishes, one painted black, the other left clear. The chamber was kept at room temperature (21°C). Ten woodlice were placed into the joining segment of the chamber and left for ten minutes to orientate themselves. The numbers of woodlice in each chamber were then recorded. The experiment was carried out a total of four times.

Aim: investigation 2

To investigate if woodlice prefer a warm or cool environment.

The method

A choice chamber was set up painted entirely black. One side was heated to 27°C by placing a heat pad underneath. The other side was kept cool at 14°C by placing a towel soaked in cool water around the chamber. Ten woodlice were placed into the joining segment of the chamber and left for ten minutes to orientate. The numbers of woodlice in each chamber were then recorded. The experiment was carried out a total of four times.

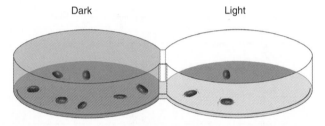

Dark   Light

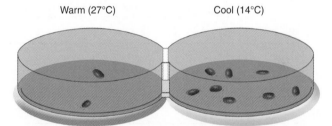

Warm (27°C)   Cool (14°C)

Results

	Number of woodlice in chamber	
Trial	Dark	Light
1	7	3
2	9	1
3	8	2
4	9	1

Results

	Number of woodlice in chamber	
Trial	Warm	Cool
1	2	8
2	3	7
3	2	8
4	1	9

1. (a) Write a hypothesis for investigation 1: _____

     _____

     _____

     (b) Write a hypothesis for investigation 2: _____

     _____

     _____

2. For each investigation, what would you expect if there was no difference in choice of environment by the woodlice?

     _____

     _____

3. How would you determine if the results for each investigation were significant? _____

     _____

4. Describe one way to improve investigation 1: _____

     _____

PRACTICES   PRACTICES   PRACTICES   CONNECT   WEB

   20 AP1   **153**

PRAC

Designing an investigation: fruit flies

Students wanted to investigate at which stage of ripeness bananas were the most attractive to fruit flies, *Drosophila melanogaster*.

They used a choice chamber made of two clear bottles end to end. Bananas of known age were used to determine the age of the banana that was most attractive. Bananas were purchased green and designated day 0. The age of the banana was determined from this date as they ripened.

Choice chamber made from soft drink bottles

5. Where will the bananas be place in the choice chamber? _____

_____

6. What range of ages of bananas would be suitable to investigate? _____

_____

7. What number of bananas (separate investigations) should be carried out? _____

8. Write a method that would allow you to determine the age of the banana that is most attractive to fruit flies:

_____

_____

_____

_____

_____

_____

_____

_____

_____

_____

_____

_____

_____

_____

_____

9. The results for a trial between 0 and 10 day old bananas are shown below:

Time (minutes)	Position in chamber (number of fruit flies)		
	End with 10 day old banana	Middle	End with 0 day old banana
1	21	18	21
10	45	3	12

(a) Which banana appeared the most attractive to the fruit flies? _____

(b) A student suggested that the sex of the flies might make a difference to their choice of banana ripeness. How could you test this?

_____

_____

10. As a group, discuss how the method and the design of the choice chamber could be improved. Draw and label your design and attach it to the page. HINT: Think about the accuracy of counting.

©2017 **BIOZONE** International
ISBN: 978-1-927309-65-0
Photocopying Prohibited

154 Migration Patterns

Key Idea: Migrations often involve very large distances and usually involve a return journey. They are initiated by the activity of internal clocks or timekeepers in response to environmental cues such as changes in daylength.

True migrations are those where animals travel from one well-defined region to another, for a specific purpose such as overwintering, breeding, or seeking food. Some mass movements of animals are not true migrations in that they do not involve a return journey and they are not governed by an internal biological clock. Such movements are best described as dispersals and are typical of species such as the 'migratory' locusts of north Africa and Australia.

Dispersal: one-way migration

Some migrations of animals involve a one-way movement. In such cases, the animal does not return to its original home range. This is typical of population dispersal. This often occurs to escape deteriorating habitats and to colonize new ones.

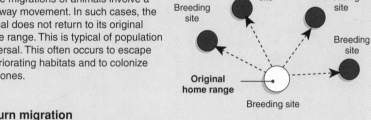

Dispersal: muskrat

Return migration

Animals that move to a winter feeding ground are making one leg of a return migration. The same animals return to their home range in the spring which is where they have their breeding sites. Sometimes they follow different routes on the return journey.

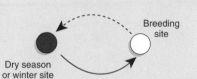

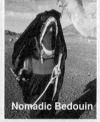
Return migration: caribou

Nomadic migration

Similar to one-way migration but individuals may breed at several locations during their lifetimes. These migrations are apparently directionless, with no set pattern. Each stopover point is a potential breeding site. There may also be temporary non-breeding stopovers for the winter or dry season.

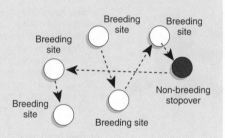

Nomadic Inuit   Nomadic Bedouin

Remigration circuits

In some populations, the return leg of a migration may have stopovers and may be completed by one or more subsequent generations. In addition to winter or dry season areas, there may be stops at feeding areas by juveniles or adults. Also included are closed circuits where animals die after breeding.

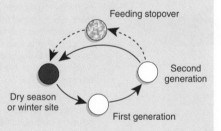

Remigration circuit: Pacific salmon

1. Giving an example, describe the conditions under which nomadic migration behavior might be necessary:

_____

_____

2. Which of the above forms of migration would lead to further dispersal of a population? Explain your answer: _____

_____

_____

3. Describe an environmental cue important in the regular migratory behavior of a named species: _____

_____

4. Explain the adaptive value of migratory behavior: _____

_____

_____

CONNECT
150
AP2

CONNECT
149
AP2

KNOW

155 Examples of Migration

Key Idea: Migrations are usually for the purpose of feeding or breeding. True migrations involve a return journey.

Migrations come in many forms. Some of the longer, more commonly noted ones are shown below. However, migrations do not have to be long distances or seasonal. Many marine animals migrate from the deep ocean to the surface daily to feed, this may only be a distance of a few hundred (vertical) meters or less.

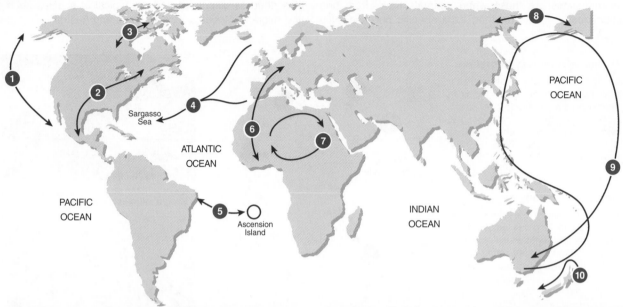

European eels migrate across the northern Atlantic ocean to spawn in the Sargasso Sea off the coast of Florida. The larvae that hatch from the eggs gradually drift back across to Europe, a migration that takes several years. Eventually they enter estuaries and move upriver where they feed, grow and mature.

3000 km

Caribou spend the winter feeding in the coniferous forests in central Canada. In the spring, they move north to the tundra of the Barren Lands, within the Arctic circle, a distance of some 1000 km. There they calves in the relative safety of the open tundra where food is plentiful.

750 to 1000 km

Migratory locusts are found in desert regions of northern and eastern Africa, the Middle East and Australia. Their migration is more strictly a dispersal in response to an expanding population with limited food. The lack of food triggers development of the voracious, migratory form.

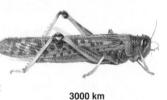

3000 km

The **European swift** is one of 140 bird species that follow one of Europe's migratory routes from northern Spain to Africa. Swifts breed throughout Europe, and migrate to south of the Sahara after breeding. Swifts feed on the wing and the onset of the migration is thought to be triggered by the lack of nutritious airborne insects.

3000 to 12,000 km

A number of **shearwater** species (including mutton birds) breed in Australia and islands around New Zealand, then migrate northwards with the onset of the southern winter to the north and northeast Pacific. The return journey across the eastern Pacific is assisted directly by the NE trade winds.

11,000 to 13,500 km

A number of **whale** species, including **humpback** and **grey whales**, follow an annual migration. In summer, they feed in the krill-rich waters of polar regions. In winter, they move closer to the equator to give birth to young conceived the previous year and to mate again. They seldom feed in transit.

7000 km

Green turtles migrate between coastal foraging areas and nesting grounds. They return from the coasts of South America to the beach of their own spawning on Ascension Island to lay eggs.

3000 km

Polar bears can cover distances of up to 1600 km walking across ice from Alaska, USA to set up winter dens across the Bering Strait in Siberia.

1600 km

Monarch butterflies have one of the longest of all insect migrations. Five or more generations are needed to complete one migration cycle. In North America, the insects overwinter in mass roosts in southern California or near Mexico City. In spring they migrate north, with some even reaching Canada by late summer, then return south for winter.

2000 to 4000 km

In New Zealand and elsewhere, **spiny lobsters** periodically make migrations of many hundreds of kilometers. The movement is predominantly against the prevailing current. It is thought to compensate for the long-term downstream movement of the population as planktonic larvae are swept in one direction by the ocean currents.

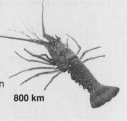

800 km

WEB 155
CONNECT 149 AP2
CONNECT 150 AP2

KNOW

©2017 **BIOZONE** International
ISBN: 978-1-927309-65-0
Photocopying Prohibited

The origins of migration

Migration is often an adaptive response to seasonal environments, allowing animals to exploit favorable conditions at different times of the year. Migrations probably evolved as a result of seasonal changes in distribution that offered benefits to fitness. However, very little evidence has been uncovered to shed light on how animals such as the Arctic tern came to migrate 70,000 km from one pole to the other and back every year. Some computer models based on the current distribution and migration of North American birds suggest the ancestral populations lived in North America all year round before evolving migrations that took them to the tropics during the Northern Hemisphere winter. The origins of oceanic crossings is still contested. It is possible that some species were migrating before the continents split and drifted apart, but this assumes that the same route has been used for millions of years.

Migrations are no doubt based in part on the search for food and shelter. As food becomes scarce in winter animals move to new areas in search of food, returning when conditions become more favorable.

1. Match up the ten numbered migration routes on the map on the previous page with each of the animals below:

(a) Migratory locust: _____ (f) Monarch butterfly: _____

(b) Caribou: _____ (g) European swift: _____

(c) Shearwater: _____ (h) Humpback whale: _____

(d) Polar bear: _____ (i) European eel: _____

(e) Green turtle: _____ (j) Spiny lobster: _____

2. Describe an adaptive advantage of migration for each of the organisms listed below:

(a) Monarch butterflies: _____

(b) Humpback whale: _____

(c) Spiny lobster: _____

(d) Caribou: _____

3. It has been suggested that the retreat of ice sheets during interglacials has been a major selection pressure in the evolution of long distance migratory behavior in many animals.

There have been a number of glacial periods over the last 2 million years, two of which occurred in the last 150,000 years. The cycle of glacial periods affected the geographical distribution of specific ecological regions, causing them to expand and contract, as well as moving them further away from the polar regions and then back again with each cycle. The speed of some of the temperature changes associated with the cycles may have been quite rapid: over tens of years, rather than hundreds. This must have most benefited those species that had a flexibility in their migratory strategy.

Explain how these events may have influenced the evolution of migratory behavior and migratory routes:

156 Migratory Navigation in Birds

Key Idea: Migratory birds use a wide range of environmental cues to navigate accurately and determine their destination. Navigation is the process by which an animal uses various cues to determine its position in reference to a goal. Migrating birds must know their flight direction and when they have reached their destination (goal). Cues include star and solar cues, landscape features, wind direction, polarized light, magnetic and gravitational field information, and smell.

Sun compass

Experiments have been carried out to investigate the existence of a sun compass and its importance for daytime migrations. Caged birds were placed in circular enclosures with four windows. Mirrors were used to alter the angle at which light entered the enclosures. At migration time, in natural conditions, these birds clearly showed a preferred flight direction (A). When mirrors bent the Sun's rays through 90°, the birds turned their preferred direction (B and C).

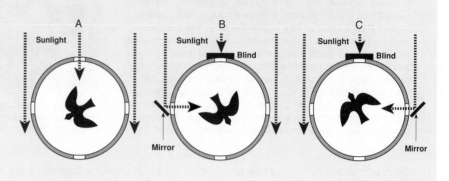

Magnetic compass

An experiment to investigate the existence of a magnetic compass in migratory birds used magnetic coils to mimic the Earth's magnetic field. The birds detect magnetic north, the direction of their spring migration. When the magnetic field was twisted so that north was in the east-southeast position, the birds kept their original path for the first two nights. By the third night, they had detected the change and altered their path accordingly.

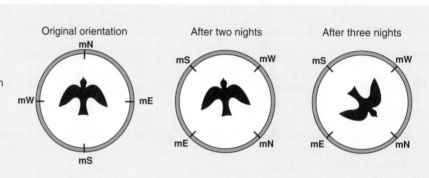

Star compass

An experiment to investigate the use of star positions in the night sky used an ink pad at the base of a cone of blotting paper. Nocturnal migrants flutter in their preferred direction of travel as the amount of ink shows. In a planetarium that projected the real sky, indigo buntings located the Pole Star and used it to find north, the direction of their spring migration. When the sky in the planetarium was rotated 90° counter-clockwise the birds altered their direction accordingly. Simulating a cloudy night, the obscured sky confused the birds.

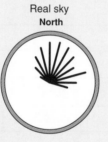 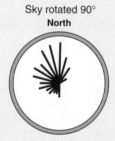 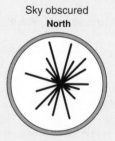

1. The experiments above investigated three possible compass mechanisms used by migratory birds to navigation. Summarize the results of each experiment, and state whether or not the experiment provided evidence that the birds were using the compass mechanism:

   (a) Sun compass: _____

   _____

   _____

   (b) Star compass: _____

   _____

   _____

   (c) Magnetic compass: _____

   _____

   _____

©2017 **BIOZONE** International
ISBN: 978-1-927309-65-0
Photocopying Prohibited

Navigation and migration in starlings and blackcaps

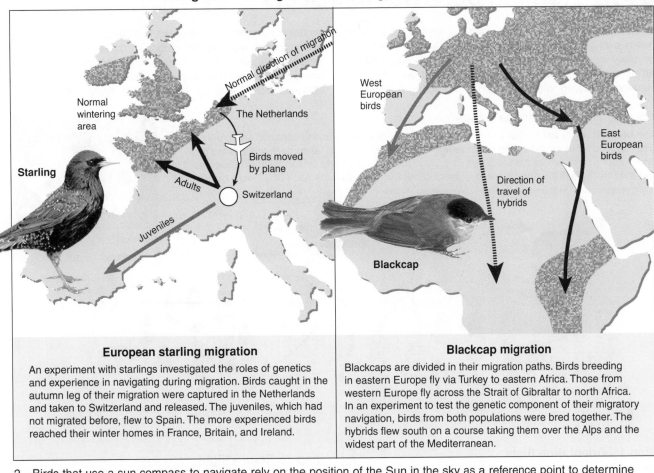

European starling migration

An experiment with starlings investigated the roles of genetics and experience in navigating during migration. Birds caught in the autumn leg of their migration were captured in the Netherlands and taken to Switzerland and released. The juveniles, which had not migrated before, flew to Spain. The more experienced birds reached their winter homes in France, Britain, and Ireland.

Blackcap migration

Blackcaps are divided in their migration paths. Birds breeding in eastern Europe fly via Turkey to eastern Africa. Those from western Europe fly across the Strait of Gibraltar to north Africa. In an experiment to test the genetic component of their migratory navigation, birds from both populations were bred together. The hybrids flew south on a course taking them over the Alps and the widest part of the Mediterranean.

2. Birds that use a sun compass to navigate rely on the position of the Sun in the sky as a reference point to determine north. Because the earth rotates on its axis once a day, the position of the Sun in the sky is constantly changing. Describe an essential mechanism that the birds must have in order to make use of this type of compass:

_____

_____

3. Study the information on the European starling migrations above and the answer the following:

(a) What direction did the juvenile starlings fly once they were relocated? _____

_____

(b) How is this different to the experienced birds? _____

_____

_____

(c) How might innate and learned behavior account for these differences? _____

_____

_____

_____

4. Study the information on the blackcap migrations above and the answer the following:

(a) What do you notice about the migration pattern of the hybrid blackcaps? _____

_____

_____

(b) What does this tell you about the origin of the migration pattern? _____

_____

157 Homing Behavior in Insects

Key Idea: Homing is the ability of an animal to return to its home site after being displaced and it involves navigation. In many insects, homing is important in increasing foraging efficiency because it reduces energy expenditure.

Homing (returning to a home site) is distinct from migration, although navigation is involved in both behaviors. Homing behavior often relies on the recognition of familiar landmarks, especially where the distances involved are relatively short. Navigation, often assisted by the use of trail pheromones, is also involved in the foraging behavior of many insects.

Homing and navigation in wasps

The **beewolf** (*Philanthus*) digs a nest in sand. It is a predator of bees and captures and paralyzes bees as food for its larvae during development. The paralyzed bee is taken back to the wasp's underground nest, where the wasp lays its eggs in the still living body. In a well-known experiment to test the homing behavior of this wasp, a scientist named Tinbergen, carried out a 2-step experiment.

(After Tinbergen, 1951. The Study of Instinct. Oxford University Press, London)

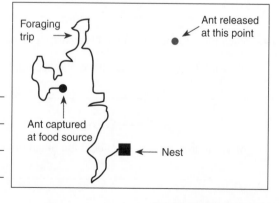

Beewolf
Philanthus triangulum

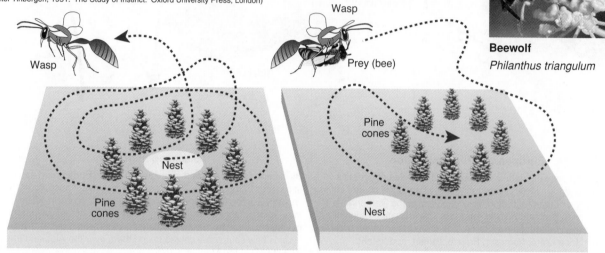

Step 1: Orientation flight

While a female wasp was in the burrow, Tinbergen placed a circle of pine cones around the nest entrance. When she emerged, the wasp reacted by carrying out a wavering orientation flight before flying off.

Step 2: Return flight

During her absence, the pine cones were moved away from the burrow leading to the nest. Returning to the nest with prey, the wasp orientated to the circle of pine cones, not the nest entrance.

Homing in ants

Cataglyphis desert ants use polarized light to navigate while foraging, often pausing and turning 360° to apparently note the position of the Sun and plane of light. When they discover a food source, they return directly to the nest as shown right. This ability to determine the direction to the nest reduces travel time when returning to the nest, making foraging more efficient.

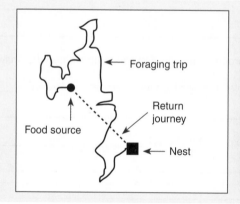

Cataglyphis ants collect a beetle

1. Why did moving the pine cones around the wasp nest result to the wasp being unable to find the nest?

_____

_____

2. (a) After a foraging trip, a *Cataglyphis* ant was displaced to another point some distance away. On the diagram shown right, draw in a line indicating the direction the ant traveled when released:

   (b) Explain why the ant moved in the direction you drew:

   _____

   _____

   _____

   _____

WEB
157
 CONNECT 149 AP2
 CONNECT 150 AP2
PRACTICES
PRACTICES

KNOW

©2017 **BIOZONE** International
ISBN: 978-1-927309-65-0
Photocopying Prohibited

158 Learned Behavior

Key Idea: Learning describes a relatively permanent modification of behavior that occurs as a result of practice or experience.

Learning is a critical process that affects the behavior of animals of all ages, across many taxa. Learning behaviors vary widely. The simplest are **habituation** and **imprinting**, when an animal learns to make a particular response only to

one type of animal or object. Like most behaviors, they are adaptive in that they enhance survival by ensuring appropriate responses in the given environment. More complex behaviors arise through **conditioning** and observational learning, such as imitation. Latent learning and insight are not readily demonstrated but have been shown experimentally in a range of species.

Filial (parent) imprinting

Filial imprinting is the process by which animals develop a social attachment. It differs from most other kinds of learning (including other types of imprinting), in that it normally can occur only at a specific time during an animal's life. This **critical period** is usually shortly after hatching (about 12 hours) and may last for several days. Ducks and geese have no innate ability to recognize *mother* or even their own species. They simply respond to, and identify as mother, the first object they encounter that has certain characteristics.

Sexual identity imprinting

Individuals learn to direct their sexual behavior at some stimulus objects, but not at others. Termed **sexual imprinting**, it may serve as a species identifying and species isolating mechanism. The mate preferences of birds have been shown to be imprinted according to the stimulus they were exposed to (other birds) during early rearing. Sexual imprinting generally involves longer periods of exposure to the stimulus than filial imprinting (*see left*).

Latent learning

Latent learning describes an association made without reinforcement and expressed later. Edward Tolman explored this concept in 1948 using rats in a maze. Rats given a reward (food) at the maze's end learned the maze quickly. Rats left to wander the maze apparently took much longer to reach the end. However when a reward was placed at the end of the maze this second group of rats completed the maze as quickly as the first. This showed they has learned the behavior required but had only demonstrated it when needed.

Habituation

Habituation is a very simple type of learning involving a loss of a response to a repeated stimulus when it fails to provide any form of reinforcement (reward or punishment). Habituation is different to fatigue, which involves loss of efficiency in a repeated activity, and arises as a result of the nature of sensory reception itself. An example of habituation is the waning response of a snail attempting to cross a platform that is being tapped at regular time intervals. At first, the snail retreats into its shell for a considerable period after each tap. As the tapping continues, the snail stays in its shell for a shorter duration, before resuming its travel.

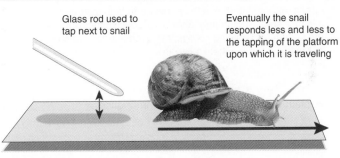

Glass rod used to tap next to snail

Eventually the snail responds less and less to the tapping of the platform upon which it is traveling

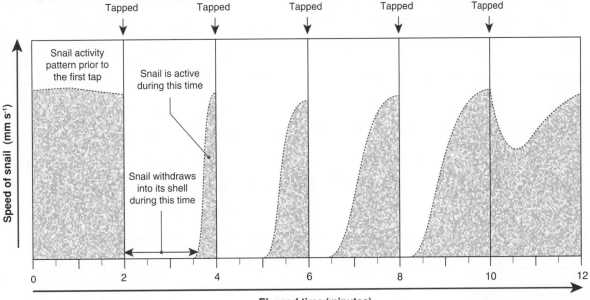

Operant conditioning

Operant conditioning describes situations where an animal learns to associate a particular behavior with a reward. The appearance of the reward is dependent on the appearance of the behavior. Burrhus Skinner studied operant conditioning using an apparatus he called a Skinner box (below). When an animal (usually a pigeon or rat) pushed a particular button it was rewarded with food. The animals learned to associate the pushing of the button with obtaining food (the reward). The behavior leads the animal to push the button in the first place is thought to be generated spontaneously (by accident or curiosity). This type of learning is also called instrumental learning and it is the predominant learning process in animals.

A Skinner box typically contains one or more buttons, which can be pressed to obtain a reward.

Food is delivered when the correct button is pushed.

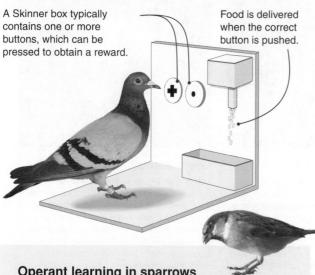

Operant learning in sparrows

Common house sparrows provide a good example of operant conditioning. They have learned to gain access to restaurants and cafés through automatic doors by triggering the motion sensors that control their opening. The birds will flutter in front of the sensor until the door opens, or perch on top of the sensor and lean over until it is triggered. Presumably, after accidentally triggering the sensor and gaining access, the birds learned which behaviors will bring them a reward.

Insight behavior

Insight behavior involves using reason to form conclusions or solve a new problem. It is not based on past experiences of a similar problem, but does involve linking together isolated experiences from different problems to reach an appropriate response. Insight learning is common in higher mammals (e.g. apes) and there is some evidence for its occurrence in other animals including dogs, pigeons and ravens.

Chimpanzees (Kohler 1917)

1 Chimps were presented with bananas hung out of reach. They were given various materials to use but not taught what to do with them.

2 After trying a number of times to jump up and reach the bananas, one chimp began to stack boxes under them. Although it took him a few times to stack the boxes correctly, he eventually reached the bananas.

There is still debate on just how much insight learning occurs in various animals, as it is difficult to demonstrate conclusively but it is generally accepted to involve:

▶ The sudden "appearance" of a solution - the "aha" moment.

▶ The solution appears after an impasse, a period of unsuccessful attempts.

▶ A new approach to or restructuring of the problem.

1. Explain the adaptive value of filial imprinting: _____

2. In what way is habituation adaptive? _____

3. Suggest when latent learning might be important to an animal's survival: _____

4. (a) Describe the basic features of operant conditioning: _____

   (b) Explain why operant conditioning is likely to be the predominant learning process in animals:

5. Explain why it is difficult to prove conclusively that an animal is using insight learning when solving a given problem:

©2017 **BIOZONE** International
ISBN: 978-1-927309-65-0
Photocopying Prohibited

159 Learning to Sing

Key Idea: Singing is instinctive, but the correct song must be practiced and learned.

Song birds use vocalizations (songs and calls) as a way to communicate, establish territories, and attract mates. The characteristics of a song may also be an indicator of fitness, as it has been shown that parasites and disease affect the song produced. While singing is instinctive, learning to sing the correct song is a learned behavior and, without it, a song bird is unlikely to gain a territory or a mate. Analyses of many bird species show that there are at least two major strategies for song development: (1) imitation of other birds, particularly adults of the same species, and (2) invention or improvisation. These strategies overlie the genetic template for the song learning process. The window during which a song can be learned varies between species. In some species, the inherited song pattern can be modified by learning only during early life. In others, the song is modified according to experience for at least another year, and some (e.g. blackbirds) continue to modify their songs throughout their entire life.

Learning to sing

The songs of different bird species vary but are generally characteristic of the species. The structure of bird song is studied using a technique called **acoustic spectroscopy**, a technique which produces a graphical representation of the sounds being made. This enables song patterns to be compared between individuals and has led to experimental work to establish how birds learn song and how much of the song is genetically determined.

Blackbirds modify their songs throughout their life

The sound spectrographs of **chaffinch** song (below) illustrate how the final song that is produced can be altered by exposure to the songs of same-species individuals during the first three months when the song is learned. The upper trace shows the song of a normal male, while the lower trace is the song of a male reared in isolation from the nest. The isolated male's song is the right pitch and relatively normal in length, but it is simpler and lacks the acoustic 'flourish' at the end (arrow), which is typical of the chaffinch's song.

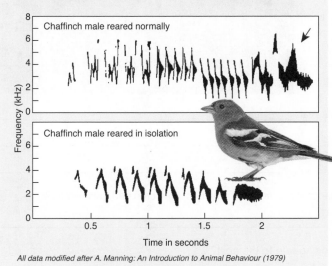

All data modified after A. Manning: An Introduction to Animal Behaviour (1979)

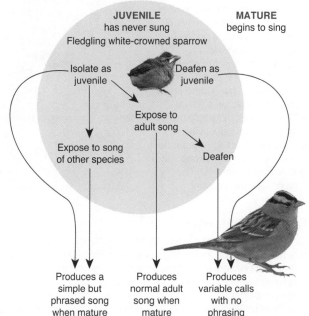

JUVENILE — has never sung
Fledgling white-crowned sparrow

MATURE — begins to sing

- Isolate as juvenile
- Deafen as juvenile
- Expose to adult song
- Expose to song of other species
- Deafen

Produces a simple but phrased song when mature

Produces normal adult song when mature

Produces variable calls with no phrasing

The diagram above summarizes experiments investigating song development in **white-crowned sparrows**. This small finch has a wide range on the Pacific Coast of America and birds from different regions have different **song dialects**. The experiments found:

▶ Isolated males will eventually sing similar and simplified versions of the normal song, regardless of which region they come from.

▶ Isolated males can be trained to sing their own dialect by playing them tape recorded songs of birds from their region.

▶ After 4 months of age, birds are unreceptive to further learning.

▶ A bird needs to be able to hear itself in order to produce the normal song. It requires auditory feedback to adjust the notes.

▶ Once birds have learned their normal adult song, they can continue to sing normally even if deafened.

1. Describe the evidence that the song of young chaffinches involves learning: _____

_____

_____

2. Discuss the possible evolutionary significance of modifying a basic (genetically determined) song pattern by learning:

_____

_____

_____

_____

CONNECT **168** AP2　WEB **159**

KNOW

160 Cooperative Behavior Improves Survival

Key Idea: By working together (directly or indirectly) members of a group increase each other's chances of survival. However the level of help depends on the level of relatedness.

Living in a group can improve the survival of individual group members, e.g. by improving foraging success or decreasing the chances of predation. Animals such as meerkats, ground squirrels, and prairie dogs decrease the chances of predation to others by having sentries that produce alarms calls when a predator approaches. Many animals live as family groups that help with foraging and raising the young.

Crowned fairy wrens

Purple crowned fairy wrens are found in northern Australia. They are cooperative breeders and maintain territories along rivers year round. Groups consist of a breeding pair and up to six helpers (earlier offspring). The graphs right show the effect of the number of helpers (A and B) and group size (C and D) on various aspects on survival.
A: The number of fledglings per nest (current productivity).
B: The total feeding rate per nest.
C: Breeder survival probability (future productivity). D: The feeding rate per breeder.

P Barden CC 4.0

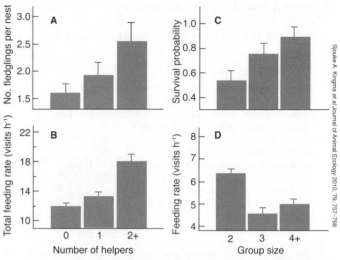

Sjouke A. Kingma *et al* Journal of Animal Ecology 2010, 79, 757–768

Sharing with friends

Cooperation does not have to be between family groups, or for direct actions such as feeding or defense. Sometimes cooperative activity occurs between unrelated individuals and involves sharing a resource with another in the knowledge that the act with be repaid at a later date. Common vampire bats feed on blood, usually of other mammals, by shaving off skin and lapping blood from the wound. Female bats live in colonies. Occasionally one bat may not feed successfully and will beg blood from a second bat, which the second bat may regurgitate. This sharing of blood between "friends" may be paid back at a later date. If the second bat refuses to share, the first bat often rebuffs any attempt by the second bat to beg a meal latter on. This is a "tit-for-tat" behavior.

Uwe Schmidt CC 4.0

1. Describe two ways in which living in a group may help survival: _____

_____

_____

2. (a) What is the effect of number of helpers on the number of fledglings per nest in purple crowned fairy wrens? _____

_____

   (b) What is the effect of number of helpers on the total feeding rate to the nest? _____

_____

3. (a) What is the effect of group size on the number of feeding visits breeders must make to the nest? _____

_____

   (b) What is the relationship between this and survival probability of breeders? Can you suggest why? _____

_____

_____

4. How does sharing a meal now help survival in the future for vampire bats? _____

_____

_____

©2017 **BIOZONE** International
ISBN: 978-1-927309-65-0
Photocopying Prohibited

161 Mutualism Involving Animals

Key Idea: Mutualistic relationships benefit both species involved. Often the benefit (to at least one party) is food.

Mutualism is a symbiotic relationship between two different species in which both interacting species benefit from the association. It can be contrasted with exploitation or **parasitism** in which one animal benefits while the other does not. Some mutualistic relationships are obligate, meaning that one (or both) species can not survive without its mutualistic partner. Other mutualistic relationships may not have the same level of dependency and the relationship is not necessary for the survival of either species. These associations are termed facultative mutualism.

Types of mutualistic relationships

Resource-resource relationships: One resource is traded for another (usually food or a nutrient)

Many reef building corals rely on a mutualism with algae in their tissues. The corals obtain some of their energy from the algae. The algae obtain a habitat and utilize the coral's nitrogenous waste and carbon dioxide.

Termites, which feed on wood, rely on a community of microbes in their gut to break down the cellulose in wood and produce the fatty acids the termites use for energy. The obligatory relationship provides food for both microbes and termites.

Service-resource relationships: A service is performed in exchange for a resource, e.g. food for protection

Some species of ants "farm" aphids by protecting the aphids from predation by ladybirds. In return the ants harvest the honeydew produced by the aphids.

Many large grazing mammals rely on oxpeckers to remove parasite infestations. The grazers provide food to the birds and the birds provide an anti-parasite service.

Service-service relationships: Both organisms provide a service to each other

True service-service mutualisms are very rare and there is usually some sort of resource component present.

Clownfish protect their home sea anemone by chasing away predators, e.g. butterfly fish. In return, the anemone protects the clownfish from its predators. However, the anemone's symbiotic algae also benefit from the nitrogen excreted by the clownfish.

Acacia ants nest within the thorns of the bullhorn acacia. In exchange for shelter, the ants protect acacias from attack by herbivores. There is a resource component though because the ants feed on the lipid rich bodies at the tips of the acacia leaflets.

PRACTICES

CONNECT	CONNECT	CONNECT	CONNECT	CONNECT	WEB
271 AP1	**223** AP2	**218** AP2	**84** AP2	**83** AP2	**161**

KNOW

Mutualistic relationships can be obligate or facultative

White-tailed deer

Pollen grains

Honeybee pollinating a purple crocus

In an **obligate mutualism**, neither species can survive without the other. There is a mutualistic relationship between many herbivores and the microbes in their gut, which enables cellulose to be digested. In ruminants, the rumen microflora break down the cellulose in forage and the ruminant obtains energy from the volatile fatty acids released by the microbial activity. The microbes benefit by having a stable growth environment and a food supply.

In a **facultative mutualism**, both species benefit from interacting with each other but can survive without the interaction. In many cases, a species may interact mutualistically with many similar species. For example, bees pollinate many different types of flower. The flower can use any type of bee as a pollinator and the bee can visit any type of flower to gather nectar.

1. (a) Define the term mutualism: _____

    _____

    _____

    (b) Distinguish between facultative and obligate mutualism: _____

    _____

    _____

2. The image shown right shows the *Riftia* tube worm, which lives near hydrothermal vents deep in the ocean. The worms have no digestive tract and rely on symbiotic bacteria for nutrition. In return, the bacteria are provided with the safe stable internal environment of the worm in which to live.

    (a) Is the relationship between the worm and the bacteria an obligate or facultative relationship?

    _____

    (b) Explain your answer: _____

    _____

    _____

3. Describe three broad classes of benefits that seem to be most common in mutualism:

    (a) _____

    (b) _____

    (c) _____

4. Service-service mutualisms, such as the clownfish-anemone relationship, appear to be very rare. What might suggest that there is a resource component to these relationships?

    _____

    _____

    _____

    _____

    _____

©2017 **BIOZONE** International
ISBN: 978-1-927309-65-0
Photocopying Prohibited

162 Mutualism Involving Plants

Key Idea: Plants form mutualistic relationships with animals, fungi, bacteria, and other plants.
Because plants are unable to move they must have effective strategies for reproduction and acquiring nutrients. Most plants form mutualistic relationships with fungi, which provide nutrients to the roots in exchange for carbohydrate from the plant. Plants have also coevolved with animals (usually insects), which may defend the plant or act as pollinators.

Nitrogen fixation in root nodules

Nitrogen fixation is a crucial part of the nitrogen cycle. Nitrogen is an abundant element, but biologically available nitrogen compounds are relatively scarce, so plants that are able to form a mutualistic relationship with bacteria to fix nitrogen have a nutritional advantage.

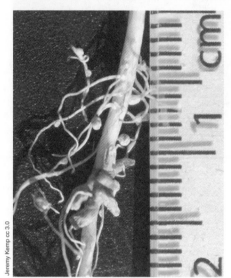

Plants in the legume family (e.g. peas, beans, and clover) and nitrogen fixing bacteria (e.g. *Rhizobium*) form a mutualistic relationship with considerable nutritional benefits to both parties.

Nitrogen fixing bacteria reduce nitrogen from the atmosphere to ammonium ions, combining them with organic acids to produce amino acids. The amino acids provide a nitrogen supply to the plant and the bacteria gain a supply of carbohydrate and a suitable environment in which to grow.

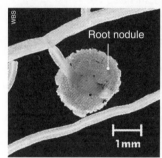

Root nodule

1mm

Nitrogen fixation in legumes occurs within root nodules, which are extensions of the root tissue formed in response to bacterial entry. The nodules provide the low oxygen environment necessary for nitrogen fixation. The presence of nodules allows plants to grow successfully even when soil nitrate is low.

Mycorrhizal associations

Mycorrhizae are formed by the mutualistic association between a fungus and the roots of a vascular plant. The fungus colonizes the plant roots, either intracellularly or extracellularly, forming the mutually beneficial mycorrhizal association.

Around 85% of vascular plant species have mycorrhizal associations and they are vitally important to plant health and forest ecology. The fungal mycelium provides a vast surface area, improving the plant's capacity to absorb minerals and water. In addition, the fungus can access phosphate ions, which are largely unavailable to the plant roots, and transfers them to the plant. In return, the plant provides the fungus with a supply of carbohydrate (produced by photosynthesis). Many conspicuous fungi in the forest, e.g. fly agaric (A), are ecotomycorrhizal. The mycorrhizal roots (B) are short and stubby with a film of fungal threads enveloping them.

1. Root nodules are a mutualistic relationship between a bacterium and a plant. Describe the benefits of the relationship to:

   (a) The plant: _____

   _____

   _____

   (b) The bacterium: _____

   _____

   _____

2. When might legumes have a clear competitive advantage over plants that cannot fix nitrogen? Explain your answer:

   _____

   _____

   _____

©2017 **BIOZONE** International
ISBN: 978-1-927309-65-0
Photocopying Prohibited

PRACTICES

CONNECT **271** AP1   CONNECT **247** AP2   CONNECT **223** AP2   CONNECT **84** AP2   CONNECT **83** AP2   WEB **162**

KNOW

Myrmecophytes

Myrmecophytes are plants that have a mutualistic relationship with ants. The plant usually provides either living spaces or feeding areas for the ants and in return the ants defend the plant from herbivorous insects and may even pollinate the plants. The relationship may be obligatory or facultative.

The ant species *Crematogaster borneensis* is found living only on plants of the *Macaranga* genus. These plants provide food bodies and domatia (living spaces) for the ants. The digestive tract of the ant is adapted to life on the host plant and workers cannot survive away from it.

Domatia are primarily found in stems, leaves, or spines. Acacia often produce swellings around the thorns that ants may occupy. The ant *Pseudomyrmex ferruginea* has an obligate mutualistic relationship with around five species of Central American acacia (*Vachellia* spp). The ants aggressively attack other insects that may land on the acacia and may also sting mammalian herbivores. The ants also prevent vines from growing on the acacia. The acacia provides the ant with food from nectar glands on its leaf stalks and areas on its leaf tips called Beltian bodies which are protein and lipid rich.

Vachellia cornigera showing enlarged thorns and light-colored Beltian bodies at the leaf tips.

3. Describe the mutualism between vascular plants and mycorrhizal fungi, including benefits to plant and fungus:

4. (a) Horticulturists frequently add a mycorrhizal inoculum when transplanting plants. Suggest why they would do this:

(b) Why would they not add phosphorus fertilizer when attempting to establish the mycorrhizae?

5. Describe the structures that a myrmecophyte produces to support ants:

6. (a) Describe the advantages for an ant living on a myrmecophyte:

(b) Describe the advantages to a myrmecophyte of harboring ants:

163 Pollination Relationships

Key Idea: Many angiosperms have a mutualistic relationship with their pollinators in which the plants achieve pollination by rewarding insects with food such as nectar or pollen. Mutualistic relationships involve exchanges between two species so that each species benefits in some way. The benefits are not always equal for each party, because each species acts in its own interests. Pollination relationships are a result of coevolution and are often highly specific.

▶ Pollination of flowers by insects is usually mutualistic. The insect benefits from the energy in the plant nectar or pollen it consumes. The plant benefits by having its gametes transferred to another plant. About 87.5% of all flowering plants are pollinated by animals, with the vast majority being insect pollinated.

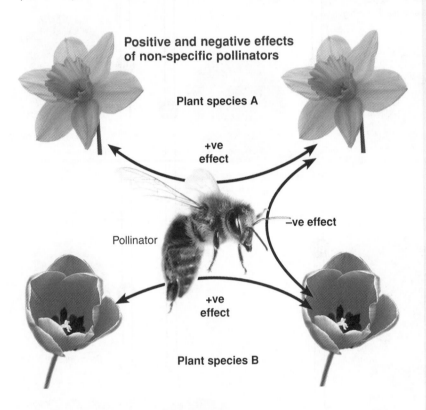

Positive and negative effects of non-specific pollinators

Plant species A

+ve effect

−ve effect

Pollinator

+ve effect

Plant species B

Orchid flowers (above) are highly variable in their structure and often highly specialized. They often have only one specific insect pollinator. This highly specific relationship is the result of reciprocal evolution between the plant and its animal pollinator (coevolution).

Magnolias are an ancient plant group with generalized flowers (above). They evolved before bees, and beetles are their main pollinators. Magnolias produce large amounts of pollen, some of which is food for the beetles.

▶ Most insect pollinators are generalists, meaning they do not form pollination relationships with specific plants. Honeybees, for example, pollinate many different kinds of plants. This can be of negative value to a particular plant, as the energy expended in producing pollen and nectar is wasted if the bee does not next fly to a plant of the same species. Plants of the same species often synchronous their flowering, this helps to ensure the insect will visit (and pollinate) flowers of the same species consecutively.

1. Describe the benefits of mutualism to both the flower and the pollinator: _____

_____

_____

2. (a) Identify one advantage and one disadvantage of a having specific pollinator: _____

_____

_____

(b) Identify one advantage and one disadvantage of having a generalist pollinator: _____

_____

_____

3. How might the presence of generalist pollinators provide a selection pressure for synchronous flowering? _____

_____

_____

CONNECT **271** AP1   CONNECT **223** AP2   WEB **163**

KNOW

164 Niche Differentiation

Key Idea: Interspecific competition is reduced when different species exploit slightly different feeding niches.

Competition is most intense between members of the same species because their habitat and resource requirements are identical. Interspecific competition is usually less intense, although many species exploit at least some of the same resources. Different species with similar ecological requirements may reduce direct competition by exploiting the resources within different microhabitats or by exploiting the same resources at different times of the day or year.

Catch insects in flight — **Sw**

Canopy

St Gleaning from foliage

Wt Extracts insects from bark

Lf Catch insects in flight

Secondary tree layer

Bt Gleaning from foliage

Understorey

Rf Catch insects in flight

Omnivorous

Ground **Ys**

Gt

Insectivorous

Reducing competition in a eucalypt forest

The diagram above illustrates how a layered forest structure provides the opportunities and resources for species with similar foraging niches to coexist. Different layers of the forest allow insectivorous birds to specialise in foraging at different heights and in different ways. The similar sized striated and brown thornbills feed at different heights, as do the leaden flycatcher and the rufous fantail. Adaptations reflect their feeding specialisations. The ground-dwelling yellow-throated scrubwren and the larger ground thrush have robust legs and feet, while the white-throated treecreeper has long toes and large curved claws, specialising in removing insects from the bark. The swifts are extremely agile fliers capable of catching insects on the wing.

Key to bird species

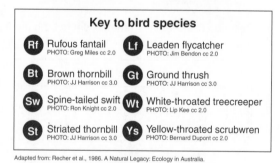

Rf Rufous fantail
PHOTO: Greg Miles cc 2.0

Lf Leaden flycatcher
PHOTO: Jim Bendon cc 2.0

Bt Brown thornbill
PHOTO: JJ Harrison cc 3.0

Gt Ground thrush
PHOTO: JJ Harrison cc 3.0

Sw Spine-tailed swift
PHOTO: Ron Knight cc 2.0

Wt White-throated treecreeper
PHOTO: Lip Kee cc 2.0

St Striated thornbill
PHOTO: JJ Harrison cc 3.0

Ys Yellow-throated scrubwren
PHOTO: Bernard Dupont cc 2.0

Adapted from: Recher et al., 1986. A Natural Legacy: Ecology in Australia.

1. How do the insectivorous birds in a layered forest avoid direct competition for the same resources?

_____

2. In forests where shrubs are absent or sparse, only the striated thornbill is present and in shrub habitats with few trees there are few striated thornbills and the brown thornbills are common. Suggest why this is the case:

_____

_____

_____

©2017 **BIOZONE** International
ISBN: 978-1-927309-65-0
Photocopying Prohibited

165 KEY TERMS AND IDEAS: Did You Get It?

1. (a) What is the name given to a plant growth response to directional light? _____

(b) What is the name given to a plant growth response to gravity? _____

(c) What plant hormone is principally responsible for the phototropic effect? _____

2. (a) What responses are being shown by the orchid in the photo (right):

_____

_____

(b) What is the stimulus involved? _____

_____

(c) How are these responses adaptive?

_____

_____

_____

Tangopaso CC 3.0

3. Cleaner shrimps are various species of shrimp that have the habit of picking parasites from the mouth and gills of fish that attend "cleaning stations" on tropical reefs.

(a) Identify this type of species interaction: _____

(b) Explain how each species benefits or is disadvantaged in the interaction: _____

_____

_____

4. Test your vocabulary by matching each term to its correct definition, as identified by its preceding letter code.

endogenous ....

exogenous ....

hibernation ....

innate behavior ....

kinesis ....

mutualism ....

taxis ....

tropism ....

A An orientation movement by an animal in response to a directional stimulus.

B A directional growth response in plants either towards (positive) or away (negative) from a stimulus.

C Originating within the organism itself, e.g. hormonal changes.

D Interaction between species in which both participants benefit.

E Non-directional response to a stimulus in which the speed or rate of turning is proportional to the stimulus intensity.

F A behavior that is not learned but has a genetic component and is already present in an animal at birth or hatching.

G Originating externally to the organism, i.e. in the external environment, e.g. light and dark.

H A physiological adaptation in animals to prolonged and predictable cold periods, characterized by a state of inactivity and depression of metabolic rate.

5. The activity of a species of fly was recorded under constant conditions in a lab and an actogram made:

Time (hours)   24   12

(a) Was the free-running period of the fly longer or shorter than 24 hours? _____

(b) How can you tell? _____

TEST

Key terms

acetylcholine
action potential
axon
cerebellum
cerebrum
colony
communication
courtship behavior
dendrites
depolarization
epinephrine
fight or flight response
flocking
herding
hindbrain
hyperpolarization
midbrain
migration
myelin sheath
Na+/K+ pump
neuron
neutrotransmiiter
nodes of Ranvier
norepinephrine
resting potential
schooling
Schwann cell
soma
summation
swarming behavior
synapse
synaptic integration
territory
threshold potential
voltage gated
channels

3.E.1 Individuals act on information and communicate it to others

Activity number

Essential knowledge

(a) Organisms exchange information with each other in response to internal changes and external cues, which can change behavior

☐ 1 Describe examples to show how the exchange of information in response to internal changes and external cues can change behavior. Examples include:
  • the fight or flight response
  • predator warnings and protection of young
  • plant-plant interactions as a result of herbivory.

166-169

(b) Communication occurs through several mechanisms

☐ 1 Describe the variety of signal behaviors or cues that produce changes in the behavior of other organisms and explain how these can result in differential reproductive success. Examples include:
  • responses to herbivory
  • territorial marking behavior in mammals
  • coloration in flowers

166-169
171

☐ 2 Describe the visual, tactile, audible, electrical, and chemical signals used to indicate dominance, find food, establish territory, and ensure reproductive success. Examples include:
  • bee dances
  • bird songs
  • territorial marking behavior in mammals
  • pack behavior
  • herd, flock, and schooling behavior
  • predator warnings
  • colony and swarming behavior in insects
  • coloration

168 169
171-177

(c) Responses to and communication of information are vital to natural selection and evolution

☐ 1 Understand that natural selection favors innate and learned behaviors that increase fitness (behavior is adaptive). use examples to show this including:
  • parent offspring interactions
  • migration patterns
  • courtship and mating behaviors
  • foraging in bees or other animals
  • avoidance behavior [also 2.E.3].

168
170-174

☐ 2 Use examples to explain how cooperative behavior increases individual fitness and population survival. Examples include:
  • herd, flock, and schooling behavior
  • predator warnings
  • colony and swarming behavior in insects [also 2.E.3].

173-177

3.E.2 Animals have nervous systems that detect signals, transmit and integrate information, and produce responses

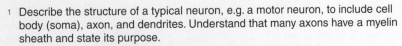

Activity number

Essential knowledge

(a) The neuron is the basic structure of the nervous system that reflects function

☐ 1 Describe the structure of a typical neuron, e.g. a motor neuron, to include cell body (soma), axon, and dendrites. Understand that many axons have a myelin sheath and state its purpose. **178 179**

☐ 2 Explain how a neuron's structure enables the detection, generation, transmission, and integration of signal information. **179 180**

☐ 3 Describe how gaps in the Scwhann cells and their myelin sheaths (nodes of Ranvier) affect the way an impulse travels along the neuron with reference to saltatory conduction. **179 180**

(b) Action potentials propagate impulses along neurons

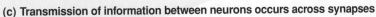

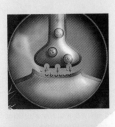

☐ 1 Describe how the membranes of neurons are polarized. Define resting potential and explain how it is generated. **180**

☐ 2 Describe how an action potential is generated and propagated. Include reference to Na^+ and K^+ gated channels, threshold, and depolarization. **180**

☐ 3 Explain the role of Na^+/K^+ pumps and ATP in maintaining membrane potential. **180**

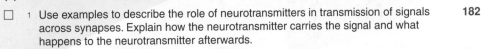

(c) Transmission of information between neurons occurs across synapses

☐ 1 Use examples to describe the role of neurotransmitters in transmission of signals across synapses. Explain how the neurotransmitter carries the signal and what happens to the neurotransmitter afterwards. **182**

☐ 2 Explain how transmission of information between neurons produces a response. Include reference to release of neurotransmitter from the presynaptic neuron and response of the postsynaptic cell (neuron, muscle, or gland) to the neurotransmitter. **178 181 182**

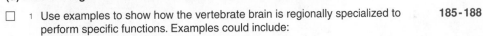

☐ 3 Use examples to show that the responses of a postsynaptic cell can be stimulatory or inhibitory. Recall the role of the nervous system in integration of information and understand how stimulatory and inhibitory inputs can sum to produce a final response. **182-184**

(d) Different regions of the vertebrate brain have different functions

☐ 1 Use examples to show how the vertebrate brain is regionally specialized to perform specific functions. Examples could include: **185-188**
- Vision (visual area of the cerebral cortex)
- Hearing (auditory region of the cerebral cortex)
- Muscle movement (primary motor cortex)
- Abstract thought and emotion (frontal lobe of the cerebrum)
- Neurohormone production (hypothalamus)
- Forebrain (cerebrum), midbrain, hindbrain (pons, medulla, cerebellum)

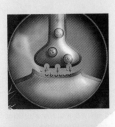

Malene Thyssen cc 3.0

166 Detecting Changing States

Key Idea: Sensory receptors allow the body to respond to a range of stimuli in the internal and external environments. A **stimulus** is any physical or chemical change in the environment capable of provoking a response in an organism. Organisms respond to stimuli in order to survive. Stimuli may be either external (outside the organism) or internal (within its body). Some of the sensory receptors that animals (including humans) use to detect stimuli are shown below. Sensory receptors respond only to specific stimuli, so the sense organs an animal has determines how it perceives the world.

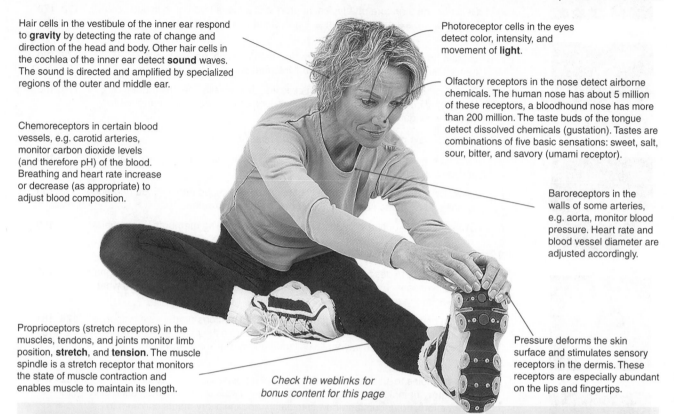

Hair cells in the vestibule of the inner ear respond to **gravity** by detecting the rate of change and direction of the head and body. Other hair cells in the cochlea of the inner ear detect **sound** waves. The sound is directed and amplified by specialized regions of the outer and middle ear.

Photoreceptor cells in the eyes detect color, intensity, and movement of **light**.

Olfactory receptors in the nose detect airborne chemicals. The human nose has about 5 million of these receptors, a bloodhound nose has more than 200 million. The taste buds of the tongue detect dissolved chemicals (gustation). Tastes are combinations of five basic sensations: sweet, salt, sour, bitter, and savory (umami receptor).

Chemoreceptors in certain blood vessels, e.g. carotid arteries, monitor carbon dioxide levels (and therefore pH) of the blood. Breathing and heart rate increase or decrease (as appropriate) to adjust blood composition.

Baroreceptors in the walls of some arteries, e.g. aorta, monitor blood pressure. Heart rate and blood vessel diameter are adjusted accordingly.

Proprioceptors (stretch receptors) in the muscles, tendons, and joints monitor limb position, **stretch**, and **tension**. The muscle spindle is a stretch receptor that monitors the state of muscle contraction and enables muscle to maintain its length.

Check the weblinks for bonus content for this page

Pressure deforms the skin surface and stimulates sensory receptors in the dermis. These receptors are especially abundant on the lips and fingertips.

Pain and temperature are detected by nerve endings in the skin. Deep tissue injury is sometimes felt on the skin as referred pain.

Humans rely heavily on hearing when learning to communicate; without it, speech and language development are more difficult.

The vibration receptors in the limbs of arthropods are sensitive to movement: either sound or vibration (from struggling prey).

The chemosensory Jacobson's organ in the roof of the mouth of reptiles (e.g. snakes) enables them to detect chemical stimuli.

Breathing and heart rates are regulated in response to sensory input from chemoreceptors.

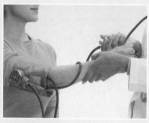

Baroreceptors and osmoreceptors act together to maintain blood pressure and volume.

Many insects, such as these ants, rely on chemical sense for location of food and communication.

Jacobson's organ is also present in mammals and is used to detect sexual receptivity in mates.

1. (a) What is a stimulus? _____

_____

(b) Why is the ability to be able to respond to stimuli important? _____

_____

_____

©2017 **BIOZONE** International
ISBN: 978-1-927309-65-0
Photocopying Prohibited

167 Plant Responses to Threats

Key Idea: Plants use physical mechanisms to protect themselves against herbivory. Some send chemical signals to other plants so they can employ protective measures.

Plants have evolved a number of mechanisms to protect them from herbivory (being grazed or browsed by animals). The most well known of these mechanisms are physical (e.g. thick cuticles, protective thorns) or involve specific nastic responses of plant parts. However, there is growing evidence that when a plant is being browsed, it is able to send chemical signals to nearby plants so that they can launch protective defenses of their own. These protective mechanisms improve the plant's chances of reproductive success.

Nastic responses to herbivory

▶ Nastic movements (nasties) are non-directional responses to stimuli, and include the folding of leaves in response to touch. This response is commonly observed in *Mimosa pudica*. *M. pudica* has long leaves composed of small leaflets. When a leaf is touched, it collapses and its leaflets fold together. Strong disturbances cause the entire leaf to droop from its base. This response takes only a few seconds and is caused by a rapid loss of turgor pressure from the cells at the bases of the leaves and leaflets.

▶ The message that the plant has been disturbed is passed quickly around the plant by electrical signals (changes in membrane potential) not by plant hormones (as occurs in tropisms). The response can be likened to the nerve impulses of animals, but it is much slower. After the disturbance is removed, turgor is restored to the cells, and the leaflets slowly return to their normal state.

▶ The adaptive value of leaf collapse nasties is most likely related to deterring browsers or reducing water loss during high winds.

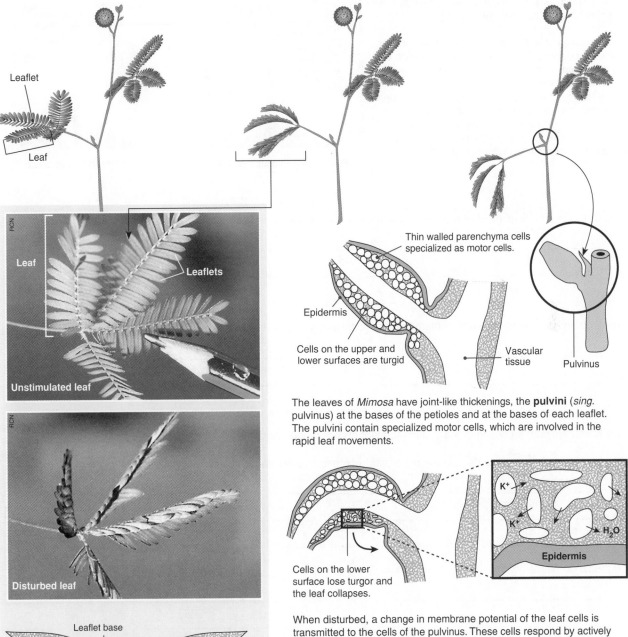

Leaflet

Leaf

Leaf **Leaflets**

Unstimulated leaf

Disturbed leaf

Leaflet base

Leaflet

Leaf axis

Thin walled parenchyma cells specialized as motor cells.

Epidermis

Cells on the upper and lower surfaces are turgid

Vascular tissue

Pulvinus

The leaves of *Mimosa* have joint-like thickenings, the **pulvini** (*sing.* pulvinus) at the bases of the petioles and at the bases of each leaflet. The pulvini contain specialized motor cells, which are involved in the rapid leaf movements.

K^+

K^+

H_2O

Epidermis

Cells on the lower surface lose turgor and the leaf collapses.

When disturbed, a change in membrane potential of the leaf cells is transmitted to the cells of the pulvinus. These cells respond by actively pumping potassium ions out of the cytoplasm (see inset above). Water follows osmotically and there is a sudden loss of turgor.

This mechanism also operates at the leaflet bases, except that the cells on the upper surface of the pulvinus lose turgor, and the individual leaflets fold up, rather than down (left).

CONNECT **81** AP1

CONNECT **122** AP2

CONNECT **40** AP2

WEB **167**

KNOW

Plants transmit warning signals through the air

Arthropods (e.g. aphids, caterpillars, and beetles) are common browsers of plants. The plants respond to browsing damage by activating defense mechanisms. These include the production of unpalatable chemicals, proteinase inhibitors, or metabolites involved in defense pathways.

Browsing may also cause plants to release volatile organic compounds (VOCs). Often these chemicals attract carnivorous arthropods to the plant and these prey upon the browsers, limiting the browse damage.

The VOCs can travel through the air and act as chemical signals to nearby plants, warning them of potential danger from browsers (right). Neighboring plants receiving the signal activate the appropriate defense related genes. Early activation of plant defense mechanisms can prevent an attack or limit the damage caused. These plants gain a competitive advantage over plants that have not received the warning.

The composition of VOCs released from a plant under attack varies depending on the herbivore browsing on them. For example, the composition of VOC released in response to damage caused by chewing beetle is different to the composition released when cell feeders such as mites are feeding from the plant. The specific VOC message enables neighboring plants to better coordinate their defense response and increases their chances of successfully repelling an attack.

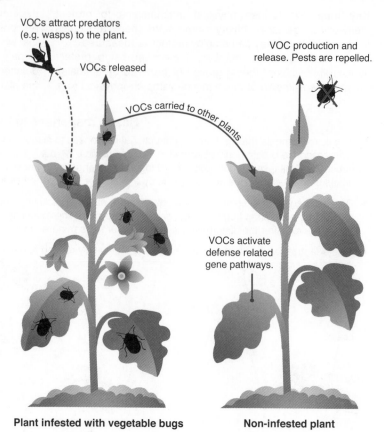

VOCs attract predators (e.g. wasps) to the plant.

VOCs released

VOC production and release. Pests are repelled.

VOCs carried to other plants

VOCs activate defense related gene pathways.

Plant infested with vegetable bugs   **Non-infested plant**

1. (a) Describe the basic mechanism behind the sudden leaf movements in *Mimosa*: _____

   _____

   _____

   _____

   (b) Explain how the movements of the *Mimosa pudica* might help its survival: _____

   _____

   _____

   _____

2. (a) Name the compounds involved in plant-to-plant herbivory signaling: _____

   (b) How do these compounds help protect a plant which is already under attack by browsers? _____

   _____

   _____

   _____

   (c) How do plants communicate a browser attack to each other? _____

   _____

   _____

   (d) How does this provide a competitive advantage to a plant not under attack? _____

   _____

   _____

168 Animal Communication

Key Idea: Communication is the transmission of (understood) information between individuals, usually of the same species. It is essential to species survival and reproductive success. Effective communication enables animals to avoid predators, coordinate foraging and hunting activity, maintain social behaviors, and attract mates. The messages are often highly ritualized (follow a fixed pattern) and therefore not easily misinterpreted. During conflict situations (e.g. asserting dominance) ritualization often prevents interactions escalating to a point where an individual is seriously injured. Messages can be passed between animals using a range of signals that may be visual, chemical, auditory, or tactile. The type of signal used depends on the activity pattern and habitat of the animal, e.g. sound carries well in dense forest.

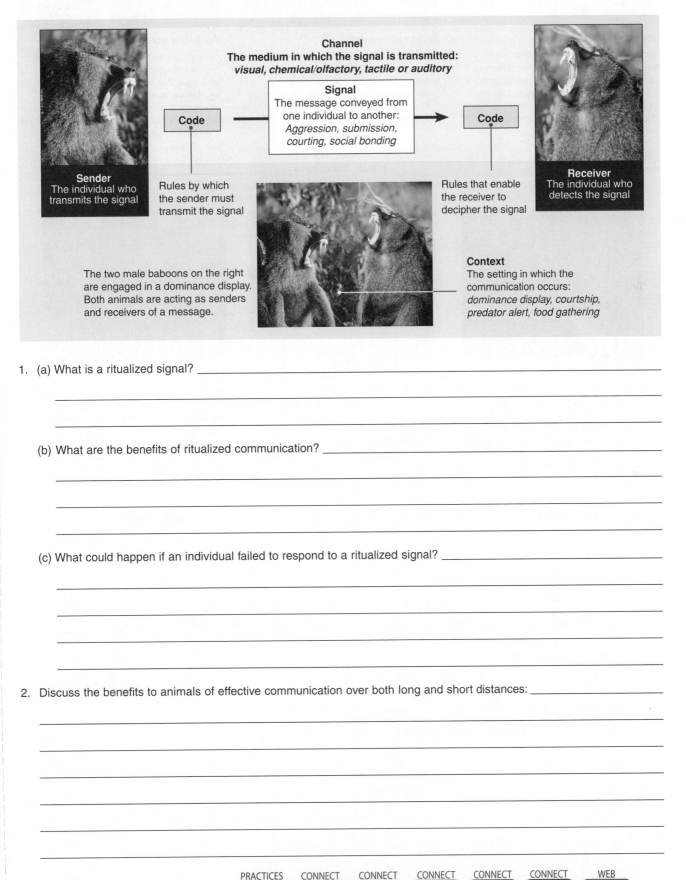

Channel
The medium in which the signal is transmitted:
visual, chemical/olfactory, tactile or auditory

Signal
The message conveyed from one individual to another:
Aggression, submission, courting, social bonding

Code

Code

Sender
The individual who transmits the signal

Rules by which the sender must transmit the signal

Rules that enable the receiver to decipher the signal

Receiver
The individual who detects the signal

The two male baboons on the right are engaged in a dominance display. Both animals are acting as senders and receivers of a message.

Context
The setting in which the communication occurs:
dominance display, courtship, predator alert, food gathering

1. (a) What is a ritualized signal? _____

_____

_____

(b) What are the benefits of ritualized communication? _____

_____

_____

_____

(c) What could happen if an individual failed to respond to a ritualized signal? _____

_____

_____

_____

_____

2. Discuss the benefits to animals of effective communication over both long and short distances: _____

_____

_____

_____

_____

_____

_____

PRACTICES

CONNECT **78** AP1

CONNECT **227** AP2

CONNECT **224** AP2

CONNECT **187** AP2

CONNECT **174** AP2

WEB **168**

KNOW

Olfactory messages

Some animals produce scents that are carried by the wind. Scents may advertize for a mate or warn neighboring competitors to keep out of a territory. In some cases, mammals use their urine and feces to mark territorial boundaries. Sniffing genitals is common among mammals.

Tactile messages

Touch may be part of a cooperative or an aggressive interaction. Grooming behavior between members of a primate group communicates social bonding. Vibrations sent along a web by a male spider signal to a potential mate not to eat him.

Auditory messages

Sound may be used to communicate over great distances. Birds keep rivals away and advertize for mates with song. Fin whales send messages over thousands of kilometers of ocean. Calls by mammals may attract mates, keep in touch with group members or warn away competitors.

Visual messages

Many animals convey information to other members of the species through body coverings and adornment, as well as through gestures and body language. Visual displays can show threat or submission, attract a mate, or exert control over a social group.

Bioluminescence

Many animals are bioluminescent (produce and emit light). The glow they produce can be used as a signal to others of their species, such as fireflies signaling to a mate. Some deep sea fish use bioluminescence to signal to other fish in the school.

Body position or facial expression

Social species with dominance hierarchies (e.g. wolves) use stereotyped expressions and body postures to avoid direct conflict with others in the group. The messages are well understood and rarely challenged and are crucial to maintenance of the hierarchy.

Attraction

Some animals produce stunning visual displays to attract a mate. The plumage of some birds can be extremely colorful and elaborate, such as in the peacock (above), bird of paradise, and lyrebird. Display of the plumage is often accompanied by specific body postures.

3. (a) Describe and explain the communication methods best suited to nocturnal animals in a forest habitat: _____

_____

_____

_____

(b) Describe and explain the communication methods best suited to solitary animals with large home ranges: _____

_____

_____

_____

4. Explain the role of dominant and submissive behaviors in animals with social hierarchies: _____

_____

_____

_____

_____

The fight or flight response

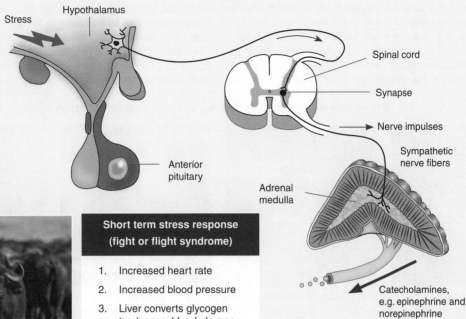

Stress

Hypothalamus

Spinal cord

Synapse

Nerve impulses

Sympathetic nerve fibers

Anterior pituitary

Adrenal medulla

Catecholamines, e.g. epinephrine and norepinephrine

Short term stress response (fight or flight syndrome)

1. Increased heart rate
2. Increased blood pressure
3. Liver converts glycogen to glucose; blood glucose levels increase
4. Dilation of bronchioles
5. Blood flow to gut and kidney reduced
   Blood flow to muscles and brain increased
6. Increased metabolic rate

A threat from another animal does not always cause an immediate fight or flight reaction. There may be a period of heightened awareness, during which each animal interprets behavioral signals and cues from the other before they decide to take action (fight or flight). The heightened awareness of one animal will be received by the rest of the group, potentially changing the behavior of the group as a whole (e.g. one individual's nervousness may unsettle the group to the point where a stampede occurs).

Internal changes or external cues can change the way an animal behaves in certain situations. A good example is the fight or flight response. When an animal is subjected to stress (e.g. being stalked by a predator) the way the animal reacts is controlled by complex hormonal and nervous interactions of the hypothalamus, and pituitary and adrenal glands. The stress response is triggered through sympathetic stimulation of the central medulla region of the adrenal glands. This stimulation causes the release of catecholamines (epinephrine and norepinephrine). These physiological changes occur as part of the short term stress response so animals operate at peak efficiency when endangered, competing, or whenever a high level of performance is required.

5. (a) What is the fight or flight response? _____

   _____

   _____

   (b) Describe some of the physiological changes that take place: _____

   _____

   _____

   _____

   _____

   (c) Describe how these changes prepare an animal to react to a threatening or potentially life threatening situation:

   _____

   _____

   _____

   _____

169 Pheromones

Key Idea: A pheromone is a chemical produced by an animal and released into the external environment where it affects the physiology or behavior of members of the same species. **Pheromones**, which are often sex attractants, are common amongst insects and mammals, and commonly relate to reproductive behavior. Many mammals, including canids and all members of the cat family, use scent marking to mark territories and advertize their readiness to mate. Other mammals, including rabbits, release a mammary pheromone that triggers nursing behavior in the young. Pheromones are also used as signaling molecules in social insects such as bees, wasps, and ants. They may be used to mark a scent trail to a food source or to signal alarm. Pheromones are widely used as baits to attract and trap insect pests.

Pheromones in animal communication and orientation

Pheromones produced by a honey bee queen and her daughters, the workers, maintain the social order of the colony. The pheromone is a blend of unsaturated fatty acids.

In mammals, pheromones are used to signal sexual receptivity and territory, or to synchronize group behavior. Pheromone detection relies on the vomeronasal organ (VNO), an area of receptor tissue in the nasal cavity. Mammals use a flehmen response, in which the upper lip is curled up, to better expose the VNO to the chemicals of interest.

Reptiles also use the VNO to detect chemicals. The flicking of a snake's tongue samples chemicals in the environment and delivers them to the VNO. This behavior is used to detect prey.

Communication in ants and other social insects occurs through pheromones. Foraging ants leave a trail along the ground that other ants will follow and reinforce until the food source is depleted. Ants also release alarm substances, which will send other ants in the vicinity into an attack frenzy. These signals dissipate rapidly if not reinforced.

The feathery antennae of male moths are specialized to detect the pheromone released by females. Males can detect concentrations as low as 2 ppm. They use wind direction to orientate, flying upwind to find the female. The sex attractant property of pheromones is used in traps, which are widely used to trap insect pests in orchards.

1. What is the significance of pheromones being species specific? _____
   _____

2. Giving examples, briefly describe the role of pheromones in three aspects of animal behavior:

   (a) _____
   _____

   (b) _____
   _____

   (c) _____
   _____

3. From what you know of pheromone activity, suggest how a pheromone trap would operate to control an insect pest:
   _____
   _____

©2017 **BIOZONE** International
ISBN: 978-1-927309-65-0
Photocopying Prohibited

KNOW

170 Courtship

Key Idea: Behaviors associated with breeding, such as courtship, are adaptations to ensure reproductive success. Many behaviors in animals, including territorial behavior, are associated with reproduction, reflecting the importance of this event in an individual's life cycle. Most animals breed during a specific season and show no reproductive behavior outside this time. During the breeding season, reproductive signals must be given and interpreted correctly or the chance for successful reproduction may be missed. The short time period that most sexually reproducing animals have in which to breed creates strong selective pressure for behavior that improves the chances of reproductive success (therefore fitness). Breeding pairs often establish territories (defended areas) to ensure reliable access to resources during breeding, while ritualized courtship behaviors reduce conflict between the sexes so that mating is achieved without injury.

What is courtship?

▶ Courtship refers to the behavior of animals just before, during, and just after mating. Courtship is a way for both male and female to evaluate the health, strength, and potential fitness of a possible mate.

▶ A potential mate may initially be attracted by a call (e.g. male frog calling). The caller (usually male) may then perform a more intricate display once the responder (usually female) arrives. In other cases the male's call and display may be the same performance.

▶ Sometimes the male may attract a mate by offering a gift of food. This is relatively common in insects, such as empid flies (right and below). Sometimes the male himself is unwittingly the "gift of food", such as in praying mantises in which the male is invariably eaten during mating. This behavior is also common in spiders.

Prey

Do females chose mates?

▶ Mate choice (or intersexual selection) occurs where members of one biological sex choose mates of the other sex to mate with. Where there is mate choice, one sex competes with same-sex members and the other sex chooses. This competition often involves elaborate rituals, calls, and displays to the choosing sex.

▶ Females usually have more invested in offspring so their mate choice is important and they are often the choosy sex. Female preference for certain features, e.g. eyes on a peacock's tail, is thought to be behind the elaborate structures and displays that have evolved in many species (e.g. peafowl, right).

Courtship is a often crucial part of breeding behavior

Female empid flies are aggressive hunters, so males have to be careful about how they approach them. Ritualized courtship behavior by the male helps him to be accepted by the female as a mate. The male's gift of food for the female pacifies her during mating and is a crucial component of mating success.

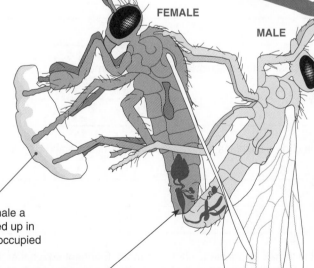
FEMALE
MALE

Male hangs on
The male empid fly grips on to a twig with its front legs during mating. It uses the other four legs to grip on to the female.

Courtship gift
The male gives the female a meal (an insect wrapped up in a cocoon) to keep her occupied while he mates.

Lock and key genitalia
The empid flies lock the tips of their abdomens together so that the male's sperm can enter the female. If the sperm were exposed to the air, they would dry out and die.

©2017 **BIOZONE** International
ISBN: 978-1-927309-65-0
Photocopying Prohibited

CONNECT **208** AP1 CONNECT **150** AP2 CONNECT **32** AP2 WEB **170**

KNOW

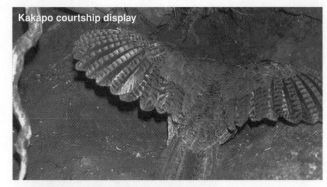

Kakapo courtship display

Albatross courtship ritual

In birds, song is an important mechanism for attracting a mate and proclaiming ownership of a territory. The song also acts as reproductive isolating mechanism, as differences between the songs of two species enables individuals to recognize their own species and mate only with them. Kakapo (a New Zealand ground parrot) are a lek species and males attempt to attract a mate to their lek (breeding territory) by producing a low frequency booming sound during the breeding season that can be heard over many kilometers of forest. When a female arrives the male begins a display in which he spreads his wings and rocks side to side (above).

One of the functions of courtship behavior is to synchronize the behaviors of the male and female so that mating can occur, and to override attack or escape behavior. Although courtship rituals may be complex, they are very stereotyped and not easily misinterpreted. Males display, usually through exaggerated physical posturing, and the females then select their mates. Courtship displays are species specific and may include ritualized behavior such as dancing, feeding, and nest-building. Many birds may form life long bonds (e.g. albatross above) and renew these every year by displaying to each other when they arrive at breeding grounds.

1. (a) Why might courtship behavior be necessary prior to mating? _____

   _____

   _____

   (b) Why is courtship behavior often ritualized, with stereotyped displays? _____

   _____

   _____

   _____

2. Describe two aspects of mating behavior in empid flies that help to ensure successful mating:

   (a) _____

   _____

   (b) _____

   _____

3. (a) Why is choosing the best mate particularly important for females? _____

   _____

   _____

   (b) Explain how female choice could lead to elaboration of structures and displays in males: _____

   _____

   _____

   _____

   (c) In some species, the female is unable to choose a mate. Elephant seal males fight for the right to mate with a female and defend a harem. Females arriving at the beach often try to avoid the harem, but with males being up to four times heavier this is difficult. How does this system ensure the offspring are likely to have the best genes?

   _____

   _____

   _____

©2017 **BIOZONE** International
ISBN: 978-1-927309-65-0
Photocopying Prohibited

171 Territories and Breeding Behavior

Key Idea: Territories are not always permanent and may be established only during the breeding season.

Territories (defended areas) are most often established in the breeding season, usually spring and summer. In autumn and winter, migratory animals leave their territories and return to winter grounds. Other animals that do not migrate may still only defend a territory during the breeding season, when the benefits of territory defense are higher. Establishing a territory uses energy and effort, but the benefit is the exclusive access to resources. During the breeding season males in particular spend time defending a territory with the goal of attracting a female and reproducing. In many cases, where territories are established purely for breeding purposes, the position of the territory is often the most important factor.

The speckled wood butterfly is found throughout northern Eurasia and Africa. During the breeding season, males have two breeding strategies. Dominant males defend a patch of sunlight in a wood, while others fly through the forest looking for unmated females. Studies have shown the males defending a territory have a greater chance of mating. This appears not to be because these males are more desirable *per se* but that they are more able to spot females flying through the sunlight than males with no sunlit patch.

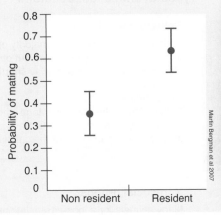

Martin Bergman et al 2007

Charlesjsharp CC 4.0

Lekking is a relatively common breeding behavior. Lekking areas (arenas) often contain numerous males. In most cases the more dominant males have leks in the center of the arena. The diagram right shows a schematic of a greater sage grouse lek arena. The most dominant alpha male (A) is found in the middle.

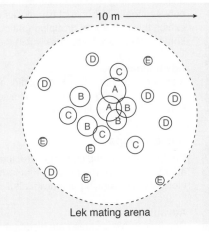

Lek mating arena

Topis are antelopes found on sub-Saharan grasslands. They establish leks during the mating season (March to May). Studies have shown that the closer the male is to the center of the lekking arena the larger the number of females that are mated with per day.

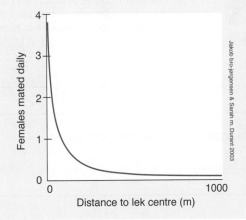

Jakob broı-jørgensen & Sarah m. Durant 2003

Snake3yes CC 2.0

1. Use examples to support the statement that the position of a territory can be important for reproductive success:

_____

_____

_____

_____

_____

PRACTICES   CONNECT   WEB

32 AP2   **171**

KNOW

172 Animal Associations and Sociality

Key Idea: The degree of sociality shown by animals varies. Animals may be solitary, form loosely associated groups, or form complex groups with clear social structures.

Sociality refers to the tendency of animals to come together or associate with other animals of the same species in social groups. No wild animal lives completely alone, because at some stage in their lives all animals interact with others of their species to reproduce or compete for food or resources. However, the degree of social interaction is highly variable. Some animals are solitary (spend most of their time alone) others form groups without any social order, while others form close social groups with a complex structure.

Types of animal associations

North American brown bear

Solitary animals

Solitary animals spend the majority of their lives alone, often in defended territories. They may only seek out others of their species for breeding. Offspring are often driven away shortly after they become independent.

Advantages:
Solitary life is often an advantage when resources are scarce or scattered over a large area. Solitary animals include bears (left), various invertebrates, and many of the cat family, e.g. tigers.

Geese

Non-social groups

Many animals form loose associations but do not interact socially. Each animal is acting directly for its own benefit with little or no direct cooperation between them. Schools of fish, flocks of birds and some herding mammals exhibit this non-social grouping (although most herding mammals have some sort of social structure within the herd).

Advantages:
Non-social groups provide protection from predators by reducing the possibility of being preyed upon individually. There may also be energetic benefits during feeding and moving.

Chimpanzees

Social groups

Primates form complex social structures, which are usually based around a family group. Some animals that form social groups also form dominance hierarchies (a ranking of individuals).

Advantages:
Dominance hierarchies help distribute resources and maintain social structure. In some species (e.g. ants and bees), members of the group are divided into castes with specialized roles. Some produce offspring or help raise young, others may be workers or help with defense of the colony.

1. What is sociality? _____

_____

2. Briefly describe the features of the three main types of animal associations:

(a) Solitary animals: _____

_____

_____

(b) Non-social groupings: _____

_____

_____

(c) Social groups: _____

_____

_____

©2017 **BIOZONE** International
ISBN: 978-1-927309-65-0
Photocopying Prohibited

173 Cooperation and Survival

Key Idea: Cooperative behavior involves two or more individuals working together to achieve a common goal. It increases the probability of survival for all individuals involved. **Cooperative behavior** involves two or more individuals working together to achieve a common goal (e.g. defense, food acquisition, or rearing young). Examples of cooperation include hunting as a team (e.g. hunts by wolves, lions, and chimpanzees), responding to the actions of others with the same goal (e.g. consensus decisions to migrate), or acting to benefit others (e.g. mobbing in small birds). Cooperation occurs most commonly between members of the same species and is adaptive in that it generally increases the survival probability for all those that cooperate. **Altruism** is an extreme form of cooperative behavior in which one individual disadvantages itself for the benefit of another. Altruism is often seen in highly social animal groups. Most often the individual who is disadvantaged receives some non-material benefit (e.g. increased probability of passing on genes).

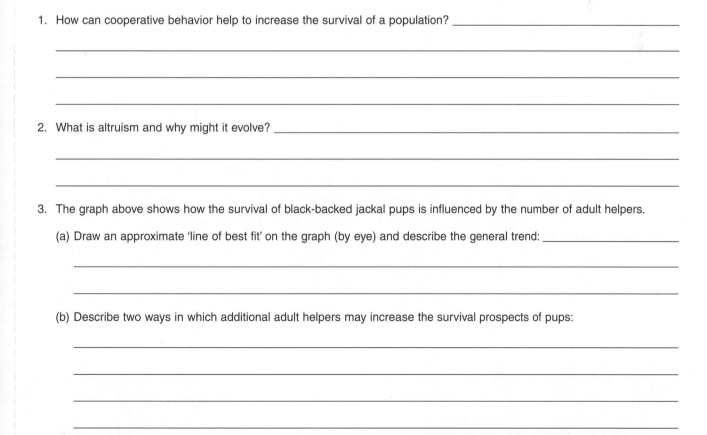

Cooperation in the black-backed jackal

Black-backed jackal (*Canis mesomelas*)

Black-backed jackals live in the brushland of Africa. Monogamous pairs (single male and female parents) hunt cooperatively, share food and defend territories. Offspring from the previous year's litter frequently help rear their siblings by regurgitating food for the lactating mother and for the pups themselves. The pup survival results of 15 separate jackal groups are shown in the graph on the right.

The effect of the number of adults in the family on pup survival in black-backed jackals

SOURCE: Drickamer & Vessey, Animal behavior (3rd Ed) PWS, 1992

1. How can cooperative behavior help to increase the survival of a population? _____

_____

_____

_____

2. What is altruism and why might it evolve? _____

_____

_____

3. The graph above shows how the survival of black-backed jackal pups is influenced by the number of adult helpers.

(a) Draw an approximate 'line of best fit' on the graph (by eye) and describe the general trend: _____

_____

_____

(b) Describe two ways in which additional adult helpers may increase the survival prospects of pups:

_____

_____

_____

_____

©2017 **BIOZONE** International
ISBN: 978-1-927309-65-0
Photocopying Prohibited

CONNECT
15
AP1

CONNECT
161
AP2

CONNECT
160
AP2

WEB
173

KNOW

Honeybees: The ultimate in unselfish behavior?

Are honeybees altruistic?

Each female worker in the colony:

▶ Sacrifices her life to defend the colony against danger.

▶ Produces no eggs.

▶ Raises the young of the queen.

Workers: female diploid
The queen's daughters will share identical genes from the father and will share half the genes from the queen.

Todd Huffman Lattice cc2.0

Drone: male haploid

Kin selection explains the behavior

Kin selection is altruistic behavior towards relatives (i.e. animals help more often when more closely related).

▶ Honeybee males (drones) are haploid and females are diploid.

▶ Workers therefore all have the same male genes and half the queen's genes.

▶ Workers are more closely related to each other than they would be to their own daughters.

▶ Therefore care-giving behavior of sisters will increase faster than genes promoting investment in offspring.

Honeypot ants

Honeypot ants of central Australia have a special group of workers called 'repletes'. These never leave the nest, but stay in underground galleries where they serve as vessels for storing a rich food supply. Regular workers that have been foraging for honeydew and nectar return to the nest where they regurgitate food from their crops to feed the replete. The replete will continue to accept these offerings until its abdomen has swollen to the size of a pea (normally it is the size of a grain of rice). The repletes become so swollen that their movements are restricted to clinging to the gallery ceiling where many hundreds of them hang in a row. When the dry season arrives and food supplies become scarce, workers return to the repletes, coaxing them to regurgitate droplets of honey.

Greg Hume at en.wikipedia CC 2.5

4. (a) How does kin selection account for the evolution of apparently altruistic behavior? _____

_____

_____

_____

_____

(b) Do you think such behavior is truly self sacrificing? Explain: _____

_____

_____

_____

_____

5. How are honeypot ant repletes an extreme form of cooperation? _____

_____

_____

_____

_____

_____

_____

©2017 **BIOZONE** International
ISBN: 978-1-927309-65-0
Photocopying Prohibited

174 Honeybee Communication

Key Idea: Honeybees use specific dances to communicate the direction and distance to food or water sources. Honeybees navigate using a sun compass, so honeybees communicate the direction and distance to food relative to the current position of the sun. In the waggle dance, they adjust their dance to account for the changing direction of the sun.

The waggle dance

Bees communicate the direction and distance of the food source through the waggle dance (below). If food is located directly in line with the sun, the communicator (bee in the blue circle) demonstrates it by running directly up the comb. To direct bees to food located either side of the sun, the bee introduces the corresponding angle to the right or left of the upward direction into the dance. Bees adjust the angles of their dance to account the changing direction of the sun throughout the day. This means directions to the food source are still correct even though the sun has changed positions.

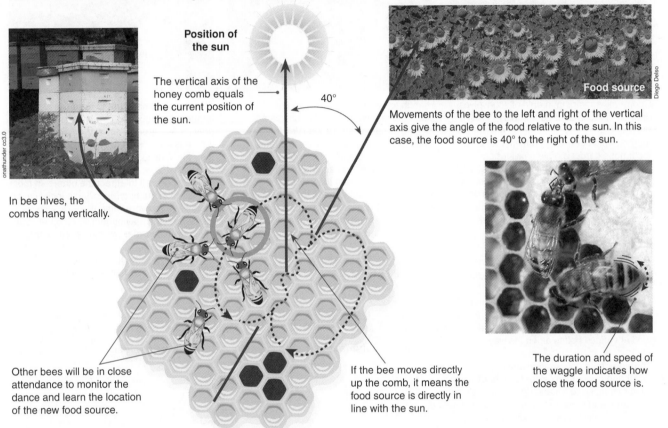

In bee hives, the combs hang vertically.

Position of the sun

The vertical axis of the honey comb equals the current position of the sun.

40°

Food source

Movements of the bee to the left and right of the vertical axis give the angle of the food relative to the sun. In this case, the food source is 40° to the right of the sun.

The duration and speed of the waggle indicates how close the food source is.

Other bees will be in close attendance to monitor the dance and learn the location of the new food source.

If the bee moves directly up the comb, it means the food source is directly in line with the sun.

The round dance

If the food source is very close (less than 50 m) the honeybee will perform a round dance. The honeybee's round dance stimulates other workers to leave the hive and search within 50 m for a food source (below).

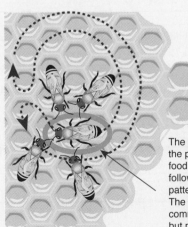

The bee communicating the presence of the food source (blue oval) follows a circular pattern (dotted line). The dance communicates distance but not direction.

1. What environmental reference is used by honey bees to orientate for navigation?

   _____

2. How does a bee communicate the proximity of a food source in the waggle dance?

   _____

   _____

   _____

3. How does the bee compensate for the time it takes between finding the food and delivering its message to the hive?

   _____

   _____

   _____

   _____

4. When is the round dance used? _____

   _____

PRACTICES   WEB

174

KNOW

175 Cooperative Foraging

Key Idea: Cooperative behavior in gathering food increases the chances of foraging success and improves efficiencies. Cooperating to gather food can be much more efficient than finding it alone. It increases the chances of finding food or capturing prey. Cooperative hunting will evolve in a species if there is a sustained benefit to the participants, the benefit for a single hunter is less than that of the benefit of hunting in a group, and cooperation within the group is guaranteed.

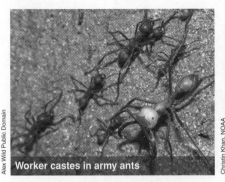
Worker castes in army ants

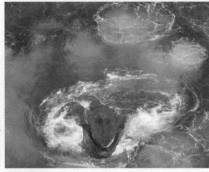

Cooperative foraging in ants often involves division of labor. Leaf-cutter ants harvest parts of leaves and use them to cultivate a fungus, which they eat. Workers that tend the fungus gardens have smaller heads than the foragers, which cut and transport the leaves. Similarly, army ants have several distinct worker castes. The smaller castes collect small prey, and larger porter ants collect larger prey. The largest workers defend the nest.

Humpback whales (above) use a cooperative technique called bubble net feeding to catch fish. The whales form a circle around a school of fish, and blow bubbles of air from their blowholes. The bubbles enclose the fish and prevent them escaping. The whales then simultaneously swim upwards and feed off the fish. Up to 60 individuals may be involved, so a high degree of learning and communication is required for this technique to be successful.

The mountain caracara in Peru (above) forages in groups of three or four, looking for prey hidden around rocks. Working together, the birds are able to overturn rocks far bigger than any individual could move. If a bird finds a rock that is worth turning over, it produces a high pitched call to attract the others. In most cases, only one bird (usually the initial caller) benefits from overturning the rock. However, the other birds may benefit when other rocks are overturned later (reciprocal altruism).

Army ants foraging

Two species of army ant have quite different raiding patterns (right):
The columns of *Eciton hamatum* go in many directions, whereas the swarm-raider *Eciton burchelli* forms a broad front. Both species cache food at various points along the way.

Through group cooperation, the tiny ants are able to subdue prey much larger than themselves, even managing to kill and devour animals such as lizards and small mammals. This would not be possible if they hunted as individuals.

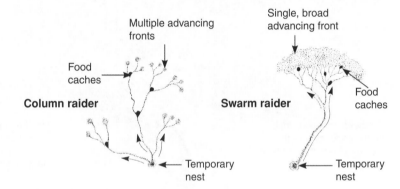

Multiple advancing fronts

Food caches

Column raider

Temporary nest

Single, broad advancing front

Food caches

Swarm raider

Temporary nest

1. What are the advantages of cooperative food gathering?_____

_____

2. What conditions favour group cooperation in food gathering? _____

_____

_____

3. Describe how the division of roles within ants increase the colony's success in obtaining food: _____

_____

_____

_____

4. (a) Describe the advantages of reciprocal altruism in mountain caracara: _____

_____

(b) Suggest why this a successful strategy even when birds do benefit all the time: _____

_____

_____

©2017 **BIOZONE** International
ISBN: 978-1-927309-65-0
Photocopying Prohibited

176 Cooperative Defense

Key Idea: Working together in defense decreases individual risk and increases the chances of a successful defense. Group defense is a key strategy for survival in social or herding mammals. Forming groups during an attack by a predator decreases the chances of being singled out, while increasing the chances of a successful defense.

Group defense in musk oxen

In the Siberian steppes, which are extensive grasslands, musk oxen must find novel ways to protect themselves from predators. There is often no natural cover, so they must make their own barrier in the form of a defensive circle. When wolves (their most common predator) attack, they shield the young inside the circle. Lone animals have little chance of surviving an attack as wolves hunt in packs.

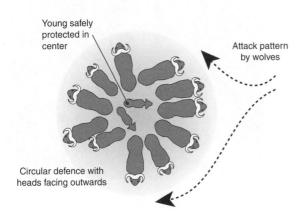

Young safely protected in center

Attack pattern by wolves

Circular defence with heads facing outwards

Red colobus monkey defense

Red colobus monkeys are a common target during chimpanzee hunts. They counter these attacks by fleeing (especially females with young), hiding, or mounting a group defense. The group defense is usually the job of the males and the more defenders there are, the greater the likelihood of the defense being successful.

A.Tamari

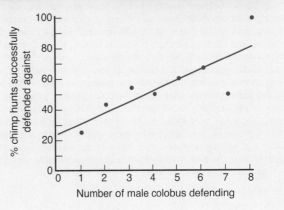

% chimp hunts successfully defended against

Number of male colobus defending

1. Describe two benefits of cooperative defense: _____
_____
_____
_____

2. How many colobus males are needed to effectively guarantee a successful defense against chimpanzees?

_____

3. Sheep need to spend most of their day feeding on grass. They form mobs both naturally in the wild as well as on farms.

   (a) Explain why sheep form mobs: _____
   _____
   _____

   (b) Explain how this might enhance an individual sheep's ability to feed: _____
   _____
   _____

©2017 **BIOZONE** International
ISBN: 978-1-927309-65-0
Photocopying Prohibited

CONNECT
160
AP2

WEB
176

KNOW

177 Cooperative Attack

Key Idea: Working together in attack can help increase the chance of success especially if specific roles are allocated between the attacking members.

Group attack is often used for hunting for food, but may be used by some species for raiding nests or territories in order to gain access to new resources (e.g. space or workers). Group attacks may be highly organized, with individuals taking specific roles.

Lionesses hunt as a coordinated group. Several lionesses hide downwind of the prey, while others circle upwind and stampede the prey towards the lionesses in wait. Group cooperation reduces the risk of injury and increases the chance of a kill. Only 15% of hunts by a solitary lioness are successful. Those hunting in a group are successful 40% of the time.

Many ant species, e.g. **slavemaker ants** (above left), raid other ant nests (called slave-raiding), killing workers and capturing grubs. The grubs are carried back to the home nest where they grow and tend the slavemaker ants' own young. Sometimes, however, the slaves rebel and can destroy the slavemaker nest. In his book **Journey of Ants** Edward O. Wilson, the world's leading ant expert, noted (not wholly tongue-in-cheek) that with ants *"their foreign policy can be summed up as follows: restless aggression, territorial conquest, and genocidal annihilation of neighboring colonies wherever possible. If ants had nuclear weapons, they would probably end the world in a week."*

The Gombe Chimpanzee War

Group attacks between members of the same species and even the same social groups do occur. They usually involve disputes over resources or territory, but may be due simply to rifts in social groups. One of the most well recorded and startling examples of group fighting is the Gombe Chimpanzee War. Observed by Jane Goodall, the violence began in 1974, after a split in a group of chimpanzees in the Gombe Stream National Park, in Tanzania. The group divided into two, the Kasakela in the northern part of the former territory and the Kahama in the south. Over the course of four years, the Kasakela systematically destroyed the Kahama, killing all six males and one female and kidnapping three more females. The Kasakela then took over the Kahama territory. However, ironically, the territorial gains made by the Kasakela were quickly lost as their new territory bordered a larger more powerful group of chimpanzees, the Kalande. After a few violent skirmishes along this border, the Kasakela were pushed back into their former territory.

1. (a) Suggest two reasons for cooperative attacks: _____

_____

_____

(b) Suggest why cooperative attacks are more likely to be successful than individual attacks:

_____

_____

_____

_____

_____

_____

2. Chimpanzees often hunt cooperatively. Use the data below to plot the relationship between hunting success and group size.

No. hunters	1	2	3	4	5	6
Hunt success (%)	13	29	49	72	75	42

©2017 **BIOZONE** International
ISBN: 978-1-927309-65-0
Photocopying Prohibited

178 Nervous Regulation in Vertebrates

Key Idea: The nervous and endocrine systems work together to maintain homeostasis. Neurons of the nervous system transmit information as nerve impulses to the central nervous system, which coordinates appropriate responses to stimuli. In humans, the nervous and endocrine (hormonal) systems work together to regulate the internal environment and maintain homeostasis in a fluctuating environment. The

nervous system includes cells called **neurons** (nerve cells) which are specialized to transmit information in the form of electrochemical impulses (action potentials). The nervous system is a signaling network with branches carrying information directly to and from specific target tissues. Impulses can be transmitted over considerable distances and the response is very precise and rapid.

Coordination by the nervous system

The vertebrate nervous system consists of the **central nervous system** (brain and spinal cord), and the nerves and receptors outside it (**peripheral nervous system**). Sensory input to receptors comes via stimuli. Information about the effect of a response is provided by feedback mechanisms so that the system can be readjusted. The basic organization of the nervous system can be simplified into a few key components: the sensory receptors, a central nervous system processing point, and the effectors, which bring about the response (below).

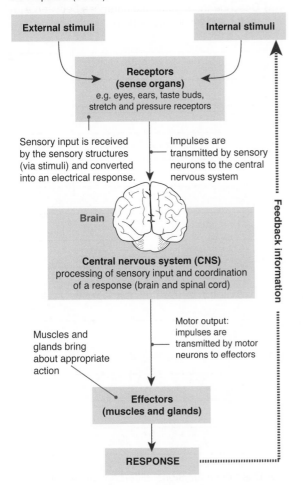

External stimuli · **Internal stimuli**

Receptors (sense organs)
e.g. eyes, ears, taste buds, stretch and pressure receptors

Sensory input is received by the sensory structures (via stimuli) and converted into an electrical response.

Impulses are transmitted by sensory neurons to the central nervous system

Brain

Central nervous system (CNS)
processing of sensory input and coordination of a response (brain and spinal cord)

Muscles and glands bring about appropriate action

Motor output: impulses are transmitted by motor neurons to effectors

Effectors (muscles and glands)

RESPONSE

Feedback information

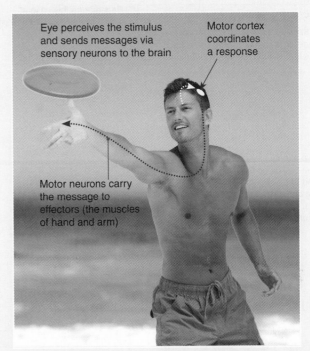

Eye perceives the stimulus and sends messages via sensory neurons to the brain

Motor cortex coordinates a response

Motor neurons carry the message to effectors (the muscles of hand and arm)

In the example above, the frisbee's approach is perceived by the eye. The motor cortex of the brain integrates the sensory message. Coordination of hand and body orientation is brought about through motor neurons to the muscles.

Comparison of nervous and hormonal control

	Nervous control	Hormonal control
Communication	Impulses across synapses	Hormones in the blood
Speed	Very rapid (within a few milliseconds)	Relatively slow (over minutes, hours, or longer)
Duration	Short term and reversible	Longer lasting effects
Target pathway	Specific (through nerves) to specific cells	Hormones broadcast to target cells everywhere
Action	Causes glands to secrete or muscles to contract	Causes changes in metabolic activity

1. Identify the three basic components of a nervous system and describe their role:

   (a) _____

   (b) _____

   (c) _____

2. Comment on the significance of the differences between the speed and duration of nervous and hormonal controls:

   _____

   _____

   _____

179 Neurons

Key Idea: Neurons conduct electrical impulses from sensory receptors along axons to other neurons or to effector cells.

Neurons (nerve cells) are electrically excitable cells that transmit nerve impulses. Neurons have a recognizable structure with a cell body (soma) and long processes (dendrites and axons). Most neurons in the peripheral nervous system (nerves outside the brain and spinal cord) are also supported by a fatty insulating sheath of myelin, which increases the speed of impulse conduction. Information, in the form of electrochemical impulses, is transmitted along neurons from receptors to a coordination center and then to effectors. The speed of impulse conduction depends primarily on the axon diameter and whether or not the axon is myelinated.

Sensory neuron

Transmit impulses from sensory receptors to the central nervous system (CNS), i.e. brain or spinal cord.

Receptor (in this case a pressure receptor in the skin).

Axon surrounded by myelin sheath, which acts as an insulator, increasing the speed of the nerve impulse.

Two axonal branches, one central (to the CNS) and one peripheral (to the sensory receptor). The axons of sensory neurons tend to be short.

Node of Ranvier, a gap in the myelin sheath.

Impulse direction

Soma (cell body) contains the nucleus and other organelles to keep the neuron functioning.

Axon terminals of all neurons have synaptic knobs or end buttons.

Motor neuron

Transmit impulses from the central nervous system to effectors (muscles or glands).

Dendrites are thin processes from the cell body that receive stimuli.

Soma containing the organelles to keep the neuron alive and functioning.

Axon hillock region where nerve impulse is generated.

Axon: A long extension of the cell transmits the nerve impulse to another neuron or to an effector (e.g. muscle). In the peripheral nervous system, motor axons are myelinated.

Node of Ranvier

Myelin sheath

Impulse direction

The axon branches of motor neurons have synaptic knobs at each end. These release **neurotransmitters**, chemicals that transmit the impulse between neurons or between a neuron and a muscle cell.

1. Describe the basic structure of a neuron:

_____

_____

_____

_____

_____

2. (a) Describe the structural differences between a motor and a sensory neuron:

_____

_____

_____

_____

_____

_____

_____

_____

_____

_____

_____

_____

_____

_____

_____

_____

(b) Describe a functional difference between a motor and a sensory neuron:

_____

_____

_____

_____

_____

_____

_____

WEB
179

CONNECT
27
AP2

PRACTICES

KNOW

©2017 **BIOZONE** International
ISBN: 978-1-927309-65-0
Photocopying Prohibited

- Where conduction speed is important, the axons of neurons are sheathed within a lipid-rich substance called **myelin**.

- Myelin is produced by oligodendrocytes in the central nervous system (CNS) and by Schwann cells in the peripheral nervous system (PNS).

- At intervals along the axons of myelinated neurons, there are gaps between neighboring Schwann cells and their sheaths called **nodes of Ranvier**. Myelin acts as an insulator, increasing the speed at which nerve impulses travel because it forces the impulse to "jump" from one uninsulated region to the next.

- Some invertebrates, e.g. earthworms and squid, increase conduction speed by increasing axon diameter.

- **Non-myelinated axons** are relatively more common in the CNS where the distances traveled are less than in the PNS. Here, the axons are encased within the cytoplasmic extensions of oligodendrocytes or Schwann cells, rather than within a myelin sheath.

- Impulses travel more slowly because the nerve impulse is propagated along the entire axon membrane, rather than jumping from node to node as occurs in myelinated neurons.

Myelinated neurons
Diameter: 1-25 µm
Conduction speed: 6-120 ms^{-1}

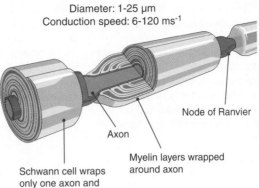

Node of Ranvier

Axon

Myelin layers wrapped around axon

Schwann cell wraps only one axon and produces myelin

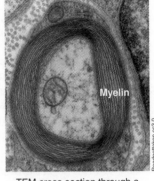

Myelin

TEM cross section through a myelinated axon

Non-myelinated neurons
Diameter: <1 µm
Conduction speed: 0.2-0.5 ms^{-1}

Cytoplasmic extensions

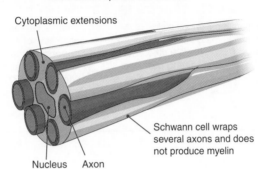

Schwann cell wraps several axons and does not produce myelin

Nucleus Axon

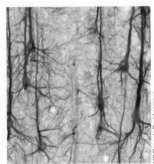

Unmyelinated pyramidal neurons of the cerebral cortex

3. (a) What do neurons do? _____

_____

(b) How does this differ from supporting cells (e.g. Schwann cells)? _____

_____

_____

4. What is the purpose of the synaptic knobs at axon terminals? _____

_____

_____

5. (a) What is the function of myelination in neurons? _____

_____

(b) What cell type produces the myelin sheath in the peripheral nervous system? _____

(c) Explain how an action potential travels in a myelinated neuron: _____

_____

_____

(d) How does this differ from its travel in a non-myelinated neuron? _____

_____

_____

(e) Why do motor neurons outside the CNS tend to be myelinated? _____

_____

©2017 **BIOZONE** International
ISBN: 978-1-927309-65-0

180 Transmission of Nerve Impulses

Key Idea: A nerve impulse occurs in response to a stimulus and involves the transmission of a membrane depolarization along the axon of a neuron.

The plasma membranes of cells, including neurons, contain **sodium-potassium ion pumps** which actively pump sodium ions (Na^+) out of the cell and potassium ions (K^+) into the cell. The action of these ion pumps in neurons creates a separation of charge (a potential difference or voltage) either side of the membrane and makes the cells **electrically excitable**. It

is this property that enables neurons to transmit electrical impulses. The **resting state** of a neuron, with a net negative charge inside, is maintained by the sodium-potassium pumps, which actively move two K^+ into the neuron for every three Na^+ moved out (below left). When a nerve is stimulated, a brief increase in membrane permeability to Na^+ temporarily reverses the membrane polarity (a **depolarization**). After the nerve impulse passes, the sodium-potassium pump restores the resting potential.

The resting neuron

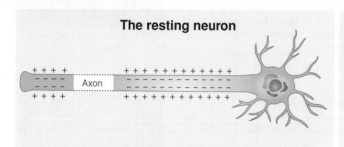

When a neuron is not transmitting an impulse, the inside of the cell is negatively charged relative to the outside and the cell is said to be electrically polarized. The potential difference (voltage) across the membrane is called the **resting potential**. For most nerve cells this is about -70 mV. Nerve transmission is possible because this membrane potential exists.

The nerve impulse

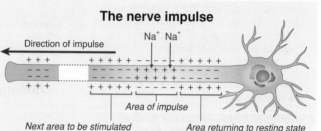

When a neuron is stimulated, the distribution of charges on each side of the membrane briefly reverses. This process of **depolarization** causes a burst of electrical activity to pass along the axon of the neuron as an **action potential**. As the charge reversal reaches one region, local currents depolarize the next region and the impulse spreads along the axon.

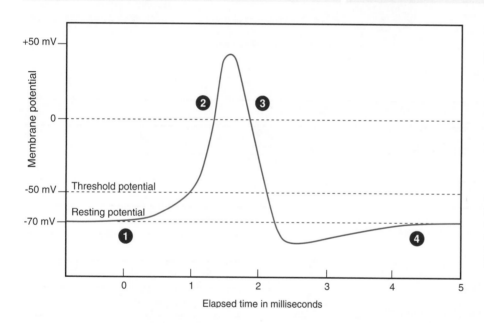

The depolarization in an axon can be shown as a change in membrane potential (in millivolts). A stimulus must be strong enough to reach the **threshold potential** before an action potential is generated. This is the voltage at which the depolarization of the membrane becomes unstoppable.

The action potential is **all or nothing** in its generation and because of this, impulses (once generated) always reach threshold and move along the axon without attenuation. The resting potential is restored by the movement of potassium ions (K^+) out of the cell. During this **refractory period**, the nerve cannot respond, so nerve impulses are discrete.

Voltage-gated ion channels and the course of an action potential

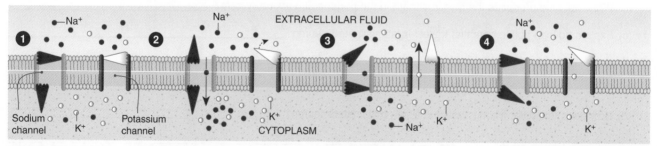

Resting state:
Voltage activated Na^+ and K^+ channels are closed. Negative interior is maintained by the Na^+/K^+ pump.

Depolarization:
Voltage activated Na^+ channels open and there is a rapid influx of Na^+ ions. The interior of the neuron becomes positive relative to the outside.

Repolarization:
Voltage activated Na^+ channels close and the K^+ channels open; K^+ moves out of the cell, restoring the negative charge to the cell interior.

Returning to resting state:
Voltage activated Na^+ and K^+ channels close and the Na^+/K^+ pump restores the original balance of ions, returning the neuron to its resting state.

©2017 **BIOZONE** International
ISBN: 978-1-927309-65-0
Photocopying Prohibited

Axon myelination is a feature of vertebrate nervous systems and it enables them to achieve very rapid speeds of nerve conduction. Myelinated neurons conduct impulses by **saltatory conduction**, a term that describes how the impulse jumps along the fiber. In a myelinated neuron, **action potentials are generated only at the nodes**, which is where the voltage gated channels occur. The axon is insulated so the action potential at one node is sufficient to trigger an action potential in the next node and the impulse jumps along the fiber. Contrast this with a non-myelinated neuron in which voltage-gated channels occur along the entire length of the axon.

As well as increasing the speed of conduction, the myelin sheath reduces energy expenditure because the area over which depolarization occurs is less (and therefore also the number of sodium and potassium ions that need to be pumped to restore the resting potential).

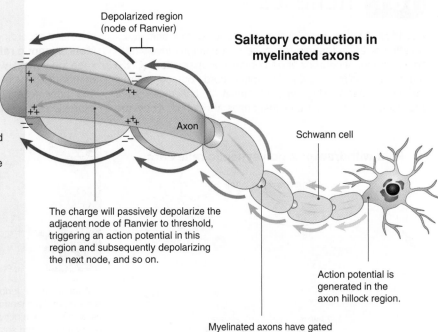

Depolarized region (node of Ranvier)

Saltatory conduction in myelinated axons

Axon

Schwann cell

The charge will passively depolarize the adjacent node of Ranvier to threshold, triggering an action potential in this region and subsequently depolarizing the next node, and so on.

Action potential is generated in the axon hillock region.

Myelinated axons have gated channels only at their nodes.

1. What is an action potential? _____
_____

2. Describe the movement of voltage-gated channels and ions associated with:

   (a) Depolarization of the neuron: _____

   (b) Repolarization of the neuron: _____

3. Summarize the sequence of events in a neuron when it receives a stimulus sufficient to reach threshold:

   (i): _____
   _____

   (ii): _____
   _____

   (iii): _____
   _____

   (iv): _____
   _____

4. (a) Explain why the nerve impulse in a myelinated neuron jumps along the axon from node to node:

   _____

   _____

   (b) How does myelination reduce the energetic costs of impulse conduction? _____

   _____

   _____

5. How is the resting potential restored in a neuron after an action potential has passed?_____

   _____

6. Explain how the refractory period influences the direction in which an impulse will travel: _____

   _____

©2017 **BIOZONE** International
ISBN: 978-1-927309-65-0
Photocopying Prohibited

181 Reflexes

Key Idea: A reflex is an involuntary response to a stimulus.
A **reflex** is an automatic response to a stimulus involving a small number of neurons and a central nervous system (CNS) processing point (usually the spinal cord, but sometimes the brain stem). This type of circuit is called a **reflex arc**. Reflexes permit rapid responses to stimuli. They are classified according to the number of CNS synapses involved. **Monosynaptic reflexes** involve only one CNS synapse (e.g. knee jerk reflex), whereas **polysynaptic reflexes** involve two or more (e.g. pain withdrawal reflex). Both are spinal reflexes. The pupil reflex (opening and closure of the pupil) is an example of a cranial reflex.

Pain withdrawal: A polysynaptic reflex arc

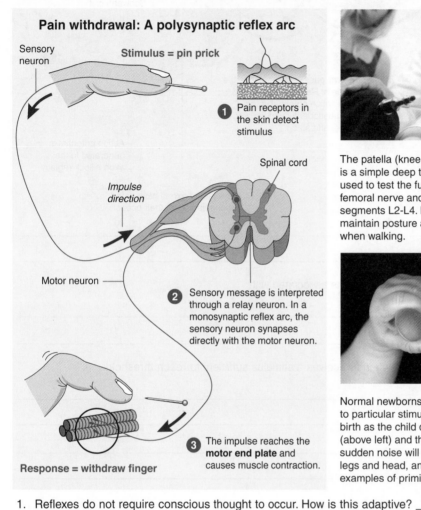

Sensory neuron

Stimulus = pin prick

1 Pain receptors in the skin detect stimulus

Spinal cord

Impulse direction

Motor neuron

2 Sensory message is interpreted through a relay neuron. In a monosynaptic reflex arc, the sensory neuron synapses directly with the motor neuron.

3 The impulse reaches the **motor end plate** and causes muscle contraction.

Response = withdraw finger

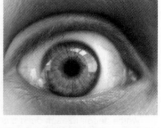

The patella (knee jerk) reflex is a simple deep tendon reflex used to test the function of the femoral nerve and spinal cord segments L2-L4. It helps to maintain posture and balance when walking.

The pupillary light reflex refers to the rapid expansion or contraction of the pupils in response to the intensity of light falling on the retina. It is a polysynaptic cranial reflex and can be used to test for brain death.

Normal newborns exhibit a number of primitive reflexes in response to particular stimuli. These reflexes disappear within a few months of birth as the child develops. Primitive reflexes include the grasp reflex (above left) and the startle or Moro reflex (above right) in which a sudden noise will cause the infant to throw out its arms, extend the legs and head, and cry. The rooting and sucking reflexes are further examples of primitive reflexes.

1. Reflexes do not require conscious thought to occur. How is this adaptive? _____

_____

_____

2. Distinguish between a spinal reflex and a cranial reflex and give an example of each: _____

_____

_____

_____

3. (a) Describe the difference between a monosynaptic and a polysynaptic reflex arc: _____

_____

_____

_____

(b) Which would produce the most rapid response, given similar length sensory and motor pathways? Explain: _____

_____

_____

WEB
181

CONNECT
118
AP2

PRACTICES

PRACTICES

KNOW

©2017 **BIOZONE** International
ISBN: 978-1-927309-65-0
Photocopying Prohibited

182 Chemical Synapses

Key Idea: Synapses are junctions between neurons or between neurons and effector cells (e.g. muscle).

Action potentials are transmitted across junctions called **synapses**. Synapses can occur between two neurons or between a neuron and an effector cell (e.g. muscle). Chemical synapses are the most widespread type of synapse in nervous systems. In these, the axon terminal is a swollen knob, and a gap (the synaptic cleft) separates it from the receiving cell. The synaptic knobs are filled with tiny vesicles of neurotransmitter, which diffuses across the synaptic

cleft and causes an electrical response in the receiving (post-synaptic) cell. This response may be a depolarization (making an action potential more likely) or a hyperpolarization (making an action potential less likely). However transmission at chemical synapses is always unidirectional. Synapses are classified as cholinergic or adrenergic according to the neurotransmitter involved. The junction between a neuron and a muscle cell is a specialized cholinergic synapse called the motor end plate but it is functionally no different. The diagram below depicts an excitatory synapse.

The structure of an excitatory chemical synapse

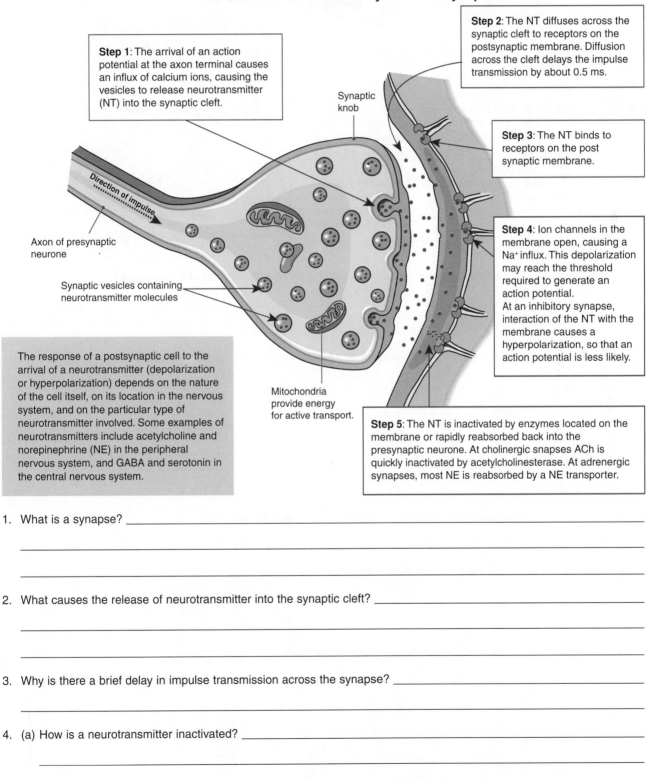

Step 1: The arrival of an action potential at the axon terminal causes an influx of calcium ions, causing the vesicles to release neurotransmitter (NT) into the synaptic cleft.

Synaptic knob

Step 2: The NT diffuses across the synaptic cleft to receptors on the postsynaptic membrane. Diffusion across the cleft delays the impulse transmission by about 0.5 ms.

Step 3: The NT binds to receptors on the post synaptic membrane.

Direction of impulse

Axon of presynaptic neurone

Synaptic vesicles containing neurotransmitter molecules

Step 4: Ion channels in the membrane open, causing a Na⁺ influx. This depolarization may reach the threshold required to generate an action potential.
At an inhibitory synapse, interaction of the NT with the membrane causes a hyperpolarization, so that an action potential is less likely.

Mitochondria provide energy for active transport.

The response of a postsynaptic cell to the arrival of a neurotransmitter (depolarization or hyperpolarization) depends on the nature of the cell itself, on its location in the nervous system, and on the particular type of neurotransmitter involved. Some examples of neurotransmitters include acetylcholine and norepinephrine (NE) in the peripheral nervous system, and GABA and serotonin in the central nervous system.

Step 5: The NT is inactivated by enzymes located on the membrane or rapidly reabsorbed back into the presynaptic neurone. At cholinergic snapses ACh is quickly inactivated by acetylcholinesterase. At adrenergic synapses, most NE is reabsorbed by a NE transporter.

1. What is a synapse? _____

_____

_____

2. What causes the release of neurotransmitter into the synaptic cleft? _____

_____

_____

3. Why is there a brief delay in impulse transmission across the synapse? _____

_____

4. (a) How is a neurotransmitter inactivated? _____

_____

_____

PRACTICES

PRACTICES

CONNECT **82** AP1

CONNECT **80** AP1

CONNECT **118** AP2

WEB **182**

KNOW

Acetylcholine and norepinephrine are neurotransmitters

Neurotransmitters (NT) are chemicals that transmit signals between neurons. There are many different types and their effects (excitatory or inhibitory) depend on the NT and properties of the cell receiving them. Excitatory NTs increase the chance of an action potential being propagated, whereas inhibitory NTs have the opposite effect. The table below summarizes the properties of two common NTs, acetylcholine (ACh) and norepinephrine (NE).

	Acetylcholine	Norepinephrine
Structure		
Synapse	Cholinergic	Adrenergic
Effect	Excitatory/inhibitory	Excitatory
Targets	The NT in all autonomic ganglia and the only NT in the somatic nervous system. Skeletal muscle, brain, many organs.	Body wide. The primary NT in the sympathetic nervous system.
Action	Wide range of effects, including stimulation of muscles, memory, alertness, arousal, learning, and control of many organs.	Alerts the nervous system for fight or flight response, e.g. increases heart rate and blood pressure.

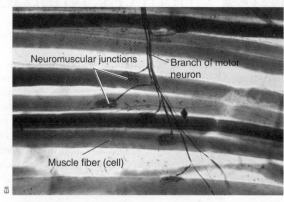

Neuromuscular junctions — Branch of motor neuron

Muscle fiber (cell)

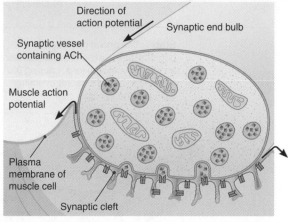

Direction of action potential — Synaptic end bulb

Synaptic vessel containing ACh

Muscle action potential

Plasma membrane of muscle cell

Synaptic cleft

Diagrammatic representation of part of a motor end plate.

(b) Why do you think it is important for the neurotransmitter to be inactivated soon after its release? _____

_____

_____

(c) Why is transmission at chemical synapses unidirectional and what is the significance of this? _____

_____

_____

_____

5. In what way is the motor end plate like any other cholinergic synapse? _____

_____

6. (a) What is the response of the postsynaptic membrane at an excitatory synapse?_____

(b) What is the response of the postsynaptic membrane at an inhibitory synapse? _____

(c) What factors determine the response of the postsynaptic cell? _____

_____

_____

7. Contrast the effect and roles of acetylcholine and norepinephrine as neurotransmitters: _____

_____

_____

_____

_____

_____

©2017 **BIOZONE** International
ISBN: 978-1-927309-65-0
Photocopying Prohibited

183 Integration at Synapses

Key Idea: Synapses play a pivotal role in the ability of the nervous system to respond appropriately to stimulation and to adapt to change by integrating all inputs.

The nature of synaptic transmission in the nervous system allows the **integration** (interpretation and coordination) of inputs from many sources. These inputs can be excitatory (causing depolarization) or inhibitory (making an action potential less likely). It is the sum of all excitatory and inhibitory inputs that leads to the final response in a post-synaptic cell. Synaptic integration is behind all the various responses we have to stimuli. It is also the most probable mechanism by which learning and memory are achieved.

Summation at synapses

Graded postsynaptic responses may sum to produce an action potential.

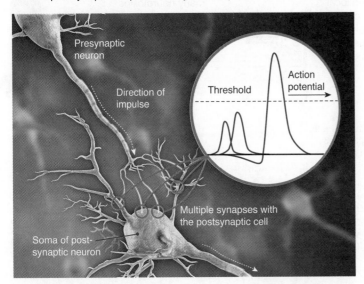

Nerve transmission across chemical synapses has several advantages, despite the delay caused by neurotransmitter diffusion. Chemical synapses transmit impulses in one direction to a precise location and, because they rely on a limited supply of neurotransmitter, they are subject to fatigue (inability to respond to repeated stimulation). This protects the system against overstimulation.

Synapses also act as centers for the **integration of inputs** from many sources. The response of a postsynaptic cell is often not strong enough on its own to generate an action potential. However, because the strength of the response is related to the amount of neurotransmitter released, subthreshold responses can sum together to produce a response in the post-synaptic cell. This additive effect is called **summation**. Summation can be temporal or spatial (right).

1 Temporal summation

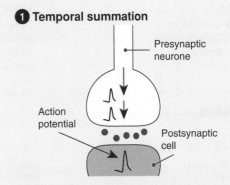

Several impulses may arrive at the synapse in quick succession from a single axon. The individual responses are so close in time that they sum to reach threshold and produce an action potential in the postsynaptic neuron.

2 Spatial summation

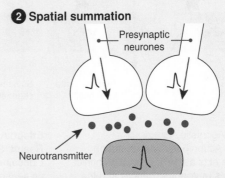

Impulses from spatially separated axon terminals may arrive simultaneously at different regions of the same postsynaptic neuron. The responses from the different places sum to produce an action potential.

1. Explain the purpose of nervous system integration: _____

_____

_____

2. Describe two advantages of chemical synapses:

(a) _____

_____

(b) _____

_____

3. (a) Explain what is meant by summation: _____

_____

(b) In simple terms, distinguish between temporal and spatial summation: _____

_____

_____

PRACTICES   PRACTICES   WEB

183

KNOW

184 Drugs at Synapses

Key Idea: Drugs may increase or decrease the effect of neurotransmitters at synapses.

Drugs may act at synapses either mimicking or blocking the usual effect of a neurotransmitter (whether it be excitatory or inhibitory). Drugs that increase the usual effect of a neurotransmitter are called **agonists** while those that decrease their effect are called **antagonists**. Many recreational and therapeutic drugs work through their action at synapses, controlling the response of the receiving cell to incoming action potentials.

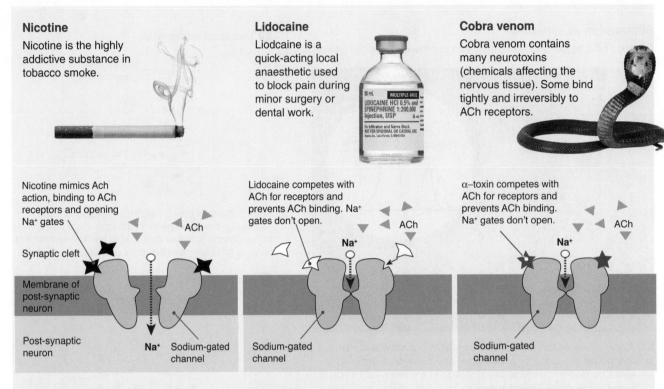

Nicotine

Nicotine is the highly addictive substance in tobacco smoke.

Nicotine mimics Ach action, binding to ACh receptors and opening Na⁺ gates

Synaptic cleft

Membrane of post-synaptic neuron

Post-synaptic neuron

Na⁺ Sodium-gated channel

Effect: Agonistic
Result: Action potential generation
Nicotine acts as an agonist at nicotinic synapses (autonomic ganglia and the motor end plate). It binds to and activates ACh receptors on the postsynaptic membrane (e.g. of a muscle cell). This opens sodium gates, leading to a sodium influx and membrane depolarization.

Lidocaine

Liodcaine is a quick-acting local anaesthetic used to block pain during minor surgery or dental work.

Lidocaine competes with ACh for receptors and prevents ACh binding. Na⁺ gates don't open.

Na⁺

Sodium-gated channel

Effect: Antagonistic
Result: Sensory inhibition
Lidocaine binds to the ACh receptors on sensory neurons and prevents ACh binding. No depolarization occurs, so no action potential is generated on the post-synaptic neuron. Pain signals are not generated.

Cobra venom

Cobra venom contains many neurotoxins (chemicals affecting the nervous tissue). Some bind tightly and irreversibly to ACh receptors.

α–toxin competes with ACh for receptors and prevents ACh binding. Na⁺ gates don't open.

Na⁺

Sodium-gated channel

Effect: Antagonistic
Result: Muscular paralysis
Toxins in cobra venom bind to ACh receptors and prevent ACh binding to receptors on the plasma membrane of muscle cells. As a result, sodium channels remain closed and no action potentials are produced. They can cause muscular paralysis, respiratory failure, and death.

1. Explain the difference between an agonistic and antagonistic drug: _____

_____

_____

2. Nicotine and cobra venom both bind to acetylcholine receptors. Explain why their effects are different: _____

_____

_____

_____

_____

3. Explain why lidocaine is used as a local anesthetic: _____

_____

_____

_____

©2017 **BIOZONE** International
ISBN: 978-1-927309-65-0
Photocopying Prohibited

185 The Structure of the Human Brain

Key Idea: The human brain is a highly complex organ that acts as the body's control center. It has a highly developed cerebrum relative to other animals.

The brain is the body's control center and one of its largest organs with a mass of ~1300 g (about 2% of the body's total mass). It lacks fuel stores so depends a constant supply of glucose, consuming about 120 g daily (about 60% of the glucose used by the body at rest). A constant flow of sensory information is received by the brain but it only responds to what is important at the time. Some responses are simple and rapid (e.g. cranial reflexes), whereas others may require many levels of processing. The brain is protected by the skull, the meninges (the membranes covering the brain and spinal cord), and the cerebrospinal fluid (CSF), which circulates throughout the brain's ventricles. The CSF acts as a shock absorber providing protection if the head is jolted or hit.

Primary structural regions of the brain

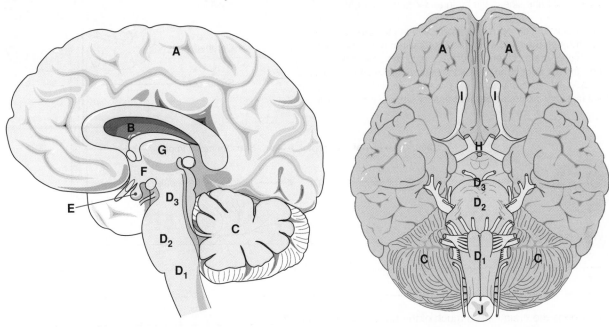

Section through lateral (side) view

Inferior (bottom) view

	Structure	Function
A	Cerebrum	Divided into left and right cerebral hemispheres. It contains sensory, motor, and association areas, and is involved in memory, emotion, language, reasoning, and sensory processing.
B	Ventricles	Cavities containing the CSF, which absorbs shocks and delivers nutritive substances.
C	Cerebellum	Coordinates body movements, posture, and balance.
D	Brainstem	Relay center for impulses between the rest of the brain and the spinal cord. Controls breathing, heartbeat, and the coughing and vomiting reflexes. D_1: Medulla oblongata, D_2: Pons, D_3: Midbrain.
E	Pituitary gland	An endocrine gland often called "the master gland" as it controls the actions of many other glands.
F	Hypothalamus	Controls the autonomic nervous system and links nervous and endocrine systems. Regulates appetite, thirst, body temperature, and sleep.
G	Thalamus	The main relay center for all sensory messages that enter the brain, before they are transmitted to the cerebrum.
H	Optic chiasm and optic nerves	The region immediately below the hypothalamus where the optic nerves from the retina of the eye cross. Conveys information about the visual fields.
I	Olfactory bulb and tract	Transmits information about airborne chemicals from the nose to the brain (the sense of smell). The olfactory tract comprises axons connecting the olfactory bulb to several regions in the brain.
J	Spinal cord	A long cord of nervous tissue extending from the brain down the spine. It carries impulses between the brain and the rest of the body and processes simple spinal reflexes.

PRACTICES   WEB

185

KNOW

Pathways for the somatosensory system

The somatosensory (somatic-sensory) system is a system of sensory neurons and pathways that monitor and respond to changes within or at the surface of the body. Messages are relayed to the cerebral cortex, where specific areas are responsible for specific senses (see next activity).

The brain is divided into left and right hemispheres. Any sensory system going to the cerebral cortex must cross over at some point, because the cerebral cortex operates on a contralateral (opposite side) basis. Thus, nerve fibers in the left or right halves of the spinal cord cross over in the medulla (lower brain stem) so that signals from sensory neurons on the body's left hand side are processed by the right hand side of the brain and vice versa (right).

Note that voluntary nerves are controlled in the cerebral cortex and pass through the cerebellum. The cerebellum receives information from higher brain centers about what muscles *should* be doing and from the peripheral nervous system about what the muscles *are* doing. It has a critical role in providing corrective feedback to minimize any discrepancies and ensure smooth motor activity.

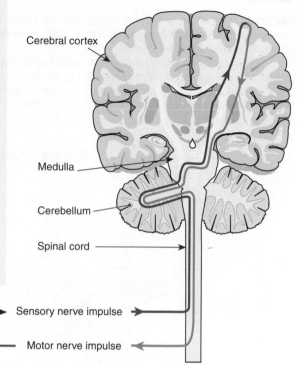

Cerebral cortex

Medulla

Cerebellum

Spinal cord

Stimulus → Receptor → Sensory nerve impulse →

Response ← Effector ← Motor nerve impulse ←

1. What is the main function of the brain? _____

_____

_____

2. Why does the brain use so much of the body's energy supply? _____

_____

3. Explain how the brain is protected against physical damage: _____

_____

_____

4. Suggest why damage to the brainstem could cause serious impairment or even death: _____

_____

_____

_____

5. (a) Describe the general stimulus-response pathway in the somatosensory system: _____

_____

_____

_____

(b) Explain why nerve impulses for muscle control travel through the cerebellum: _____

_____

_____

6. Suggest which region of the brain is important in the perfection of motor skills and why? _____

_____

_____

©2017 **BIOZONE** International
ISBN: 978-1-927309-65-0
Photocopying Prohibited

186 Functions of the Cerebrum

Key Idea: The cerebrum is the largest region in the brain. It is involved in the control of voluntary actions.

The human brain has a large, well developed cerebrum divided into two hemispheres. It has prominent folds (**gyri**) and grooves (**sulci**). Each cerebral hemisphere has an outer region of grey matter and an inner region of white matter, and is divided into four lobes by deep sulci or fissures. These lobes are the temporal, frontal, occipital, and parietal lobes. The cerebrum provides us with the ability to write, speak, calculate, plan, and produce new ideas.

Functional regions in the cerebrum

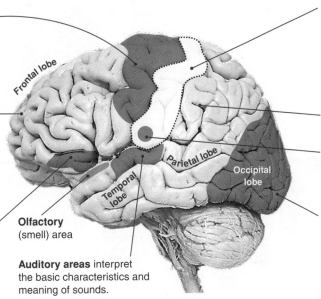

Primary motor area controls voluntary muscle movement. Stimulation of a point one side of the motor area results in muscular contraction on the opposite side of the body.

Frontal lobe includes the primary motor cortex but also responsible for emotion, abstract thought, problem solving, and memory.

Language areas: The motor speech area (Broca's area) is concerned with speech production. The sensory speech area (Wernicke's area) is concerned with speech recognition and coherence.

Olfactory (smell) area

Auditory areas interpret the basic characteristics and meaning of sounds.

Primary somatosensory area receives sensations from receptors in the skin, muscles and viscera, allowing recognition of pain, temperature, or touch. The size of the sensory region for different body parts depends on the number of receptors in that particular body part.

Sensory association area gives meaning to the sensations.

Primary gustatory area interprets sensations related to taste.

Visual areas within the occipital lobe receive, interpret, and evaluate visual stimuli. In vision, each eye views both sides of the visual field but the brain receives impulses from left and right visual fields separately. The visual cortex combines the images into a single impression or perception of the image.

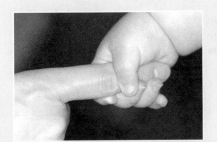

Touch is interpreted in the primary somatosensory area. The fingertips and the lips have a relatively large amount of area devoted to them.

Humans rely heavily on vision. The importance of this special sense in humans is indicated by the large occipital region of the brain.

The olfactory tract connects the olfactory bulb (innervated from the nasal epithelium) with the cerebral hemispheres where olfactory information is interpreted.

1. What is the function of the primary somatosensory area? _____

2. What is the function of the primary motor area? _____

3. For each of the following bodily functions, identify the region(s) of the brain involved in its control:

   (a) Visual processing: _____

   (b) Taste: _____

   (c) Language: _____

   (d) Memory: _____

PRACTICES   WEB

186

KNOW

The cerebral cortex

▶ The surface area of the cerebral cortex is increased by the gyri and sulci.

▶ The outer part of the cerebrum is the cerebral cortex, comprising a layer of grey matter around 3-4 mm thick.

▶ Most the brain's neural activity occurs in the cerebral cortex. It also contains 10% of the brain's neurons.

▶ The cerebral cortex has three primary activities; sensory, motor, and associative. The associative area of the cerebral cortex is the site of higher mental functions. In humans makes up 95% of the cerebral cortex.

Motor and sensory cortex

The **primary somatosensory cortex** and **primary motor cortex** (below) are regions of the cerebral cortex involved in sensing and responding respectively. Different parts of these areas receive signals from and control different areas of the body. Parts of the body with many nerve endings (e.g. the hands) take up greater areas of these regions in the brain than parts of the body with only a few nerve endings.

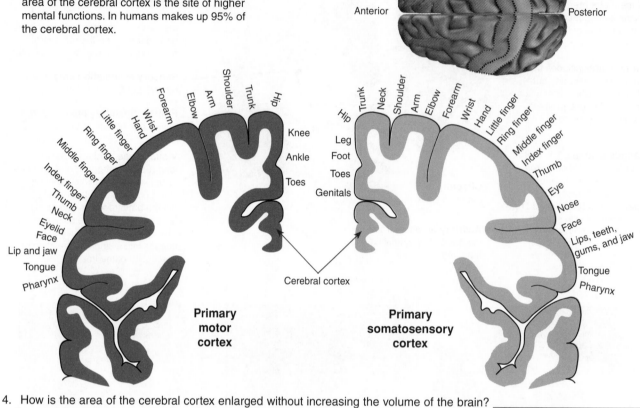

4. How is the area of the cerebral cortex enlarged without increasing the volume of the brain? _____

_____

_____

_____

5. (a) What are the three primary activities of the cerebral cortex? _____

_____

(b) Suggest why the associative area of the brain is so large in humans: _____

_____

_____

_____

6. (a) Why do you think the occipital lobe in a human brain so large? _____

_____

_____

(b) Contrast the size of the olfactory lobe and suggest a reason for the size difference: _____

_____

_____

©2017 **BIOZONE** International
ISBN: 978-1-927309-65-0
Photocopying Prohibited

187 The Hypothalamus and Neurosecretion

Key Idea: Neurohormones are hormones produced and released by specialized neurosecretory cells.

The hypothalamus is central to many homeostatic processes, linking the nervous and endocrine systems via the pituitary. It monitors hormone levels and indirectly regulates many functions, including body temperature, food and fluid intake,

and sleep. The hypothalamus synthesizes and secretes a variety of signaling molecules called **neurohormones**. These are produced and released by **neurosecretory cells** (also called neuroendocrine cells). These are specialized cells that function as both nerve cells and endocrine cells, releasing hormones in response to neural input.

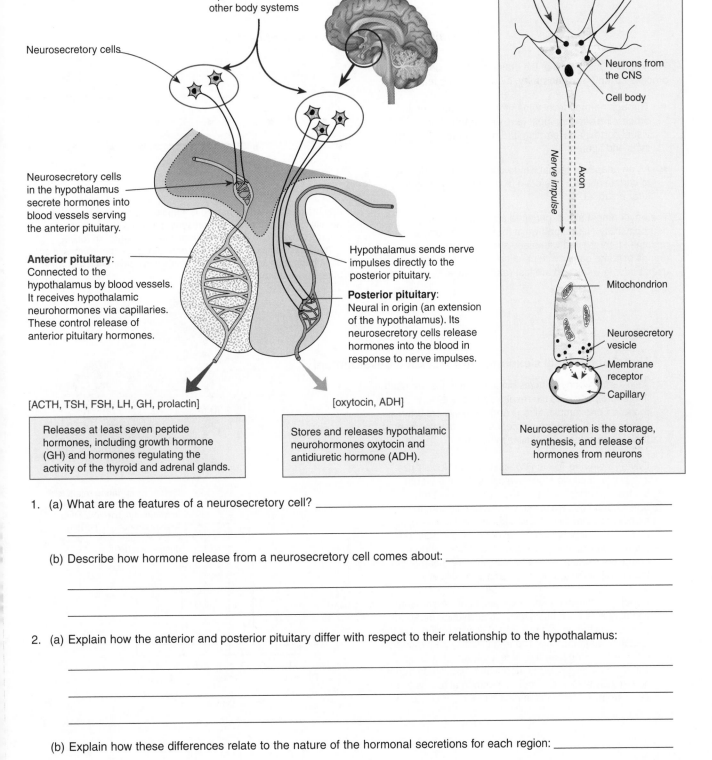

Input from brain and other body systems

Neurosecretory cells

Neurosecretory cells in the hypothalamus secrete hormones into blood vessels serving the anterior pituitary.

Anterior pituitary:
Connected to the hypothalamus by blood vessels. It receives hypothalamic neurohormones via capillaries. These control release of anterior pituitary hormones.

Hypothalamus sends nerve impulses directly to the posterior pituitary.

Posterior pituitary:
Neural in origin (an extension of the hypothalamus). Its neurosecretory cells release hormones into the blood in response to nerve impulses.

Neurons from the CNS

Cell body

Nerve impulse

Axon

Mitochondrion

Neurosecretory vesicle

Membrane receptor

Capillary

[ACTH, TSH, FSH, LH, GH, prolactin]

Releases at least seven peptide hormones, including growth hormone (GH) and hormones regulating the activity of the thyroid and adrenal glands.

[oxytocin, ADH]

Stores and releases hypothalamic neurohormones oxytocin and antidiuretic hormone (ADH).

Neurosecretion is the storage, synthesis, and release of hormones from neurons

1. (a) What are the features of a neurosecretory cell? _____

_____

(b) Describe how hormone release from a neurosecretory cell comes about: _____

_____

_____

2. (a) Explain how the anterior and posterior pituitary differ with respect to their relationship to the hypothalamus:

_____

_____

_____

(b) Explain how these differences relate to the nature of the hormonal secretions for each region: _____

_____

_____

_____

©2017 **BIOZONE** International
ISBN: 978-1-927309-65-0
Photocopying Prohibited

CONNECT   WEB
40   187
AP2

KNOW

188 Investigating Brain Function

Key Idea: If neurons and neural pathways are not used they are lost. If the lack of use occurs during a critical period during development the pathway will never be regained.

The brain is a complex organ containing around 90 billion neurons. To function correctly, the neurons must be able to make connections with one another, so that messages can be sent and received. Theories suggest that the acquisition of new information (learning) creates new neural connections. The more frequent the learning event, the more numerous the connections. This makes it easier to remember and retrieve the information in the future. If neurons and neural pathways are not used they are removed in a process call neural pruning. A similar process occurs during the development of the visual pathway in mammals. Cells that are stimulated by visual input strengthen their connections, whereas those that are not stimulated are lost. Faulty visual development in the young can occur if vision is impaired during critical periods of development.

The brain can reorganize neural connections

The neural organization of the brain constantly changes, a process called **neuroplasticity**. It occurs in two main ways:

▶ Through normal development of the immature brain processing sensory input (e.g. development of the visual cortex) and learning and memory (e.g. learning facts and figures).

▶ As an adaptive mechanism to compensate for damage to certain areas of the brain (e.g. relearning to walk after a stroke).

The environment and experience play a big role in neuroplasticity. Neural pathways that are commonly used become strengthened with many synaptic connections (things become easier with practise) while pathways not used become weak and are lost (use it or lose it).

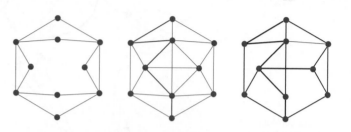

The brain of a newborn human is immature (above left) and there are still many connections (represented as lines above) between neurons (blue dots) that can be made. At 3-4 years old (above center) the maximum number of connections are made while in an adult brain (above right) unused connections have been lost while others have been strengthened.

A newborn kitten's eyes open around 9-14 days after birth.

Hubel and Wiesel's experiments on visual development

Animals with body systems similar to humans are used in research because the results can be applied to human models. For example, kittens and monkeys are often used in visual experiments because the development of their visual systems is similar to humans.

David Hubel and Torsten Wiesel won the Nobel prize in 1981 for their work on development of the **visual cortex** in mammals. The visual cortex is the area of the brain that processes visual stimuli. Specifically, they established that a critical window exists for developing a mature, fully functioning visual cortex. If the brain does not receive visual stimuli during the critical window, neural connections are not made, and development of the neural pathways does not proceed normally.

Hubel and Wiesel carried out a series of experiments on infant kittens and monkeys. They stitched (sutured) one eyelid shut so that the animal was deprived of visual stimuli through that eye. The age at which the procedure occurred and the length of time it remained sutured varied. The effect of visual deprivation was measured by neural activity. Some of their experimental results from 1970 are presented on the next page.

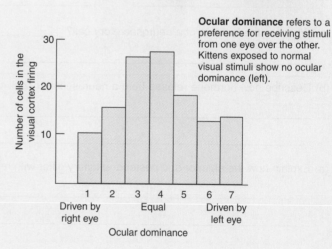

Ocular dominance refers to a preference for receiving stimuli from one eye over the other. Kittens exposed to normal visual stimuli show no ocular dominance (left).

1. (a) What is neural plasticity? _____

_____

_____

©2017 **BIOZONE** International
ISBN: 978-1-927309-65-0
Photocopying Prohibited

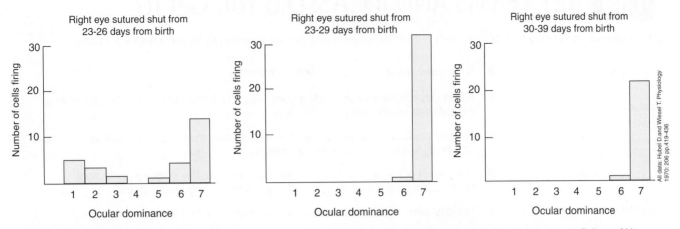

Right eye sutured shut from 23-26 days from birth

Right eye sutured shut from 23-29 days from birth

Right eye sutured shut from 30-39 days from birth

All data: Hubel D.and Wiesel T. Physiology 1970: 206 pp.419-436

Hubel and Wiesel's experiments showed that kittens have a critical window around 4-8 weeks of age for visual development. Failure of kittens to receive stimuli during this period results in pronounced ocular dominance and limited visual responses. The effects of visual deprivation were most obvious in kittens deprived of all visual stimulation from birth, but less obvious in kittens that received some visual stimulus and in older kittens and adult cats. The visual cortex in visually deprived kittens developed physiological abnormalities and atrophied (shrunk).

(b) Compare the number of neural connections between a newborn, a child 3-4 years old, and an adult: _____

_____

_____

_____

(c) Explain the differences between them: _____

_____

_____

_____

_____

2. What evidence is there that kittens exposed to normal visual stimuli show no ocular dominance? _____

_____

_____

_____

3. Explain the evidence supporting the importance of critical windows in brain development: _____

_____

_____

_____

_____

_____

_____

4. Identify the stage at which the critical period of visual development occurs in kittens: _____

_____

5. Why did Hubel and Wiesel use kittens in their experiments? _____

_____

_____

_____

189 KEY TERMS AND IDEAS: Did You Get It?

1. Test your vocabulary by matching each term to its correct definition, as identified by its preceding letter code.

acetylcholine _____

cerebellum _____

cerebrum _____

communication _____

courtship behavior _____

fight or flight response _____

myelin _____

neuron _____

synapse _____

territory _____

voltage gated channels _____

A A cell specialized to transmit electrical impulses.

B A lipid rich substance surrounding some neurons. It acts as an insulator, increasing the speed at which nerve impulses travel.

C The transmission of (understood) information between individuals of the same species. Can be by vocal, visual, or olfactory means.

D This region of the brain consists of two hemispheres. It is involved in memory, emotion, language, reasoning, sensory processing, and motor control.

E The gap between neighboring neurons or between a neuron and an effector.

F The part of the brain that coordinates and regulates body movements, posture, and balance.

G A defined area, used by an animal for a specific purpose, delineated in some way (e.g. by scent) and defended against individuals of the same species.

H A class of transmembrane proteins that form ion channels that are activated by changes in the electrical membrane potential near the channel.

I A neurotransmitter that acts at cholinergic synapses.

J A physiological reaction that occurs to ready the body to respond to a perceived harmful event.

K Behavior that acts as a prelude to mating and reproduction. It may involve a suite of visual, auditory, chemical, and postural cues.

2. (a) Label the components of the neuron pictured on the right:

A: _____

B: _____

C: _____

D: _____

E: _____

F: _____

(b) Is this neuron myelinated or unmyelinated? (delete one)

(c) Explain your answer: _____

_____

(d) In what form do electrical signals travel in this cell? _____

_____

3. Orcas (killer whales) hunt seal on icebergs by swimming towards the iceberg in a group at high speed before ducking under the ice causing a large wave to wash over the iceberg, knocking the seal into the sea where it can be captured.

(a) Identify the type of behavior displayed by the orcas: _____

_____

(b) Explain the advantage of this behavior: _____

_____

_____

_____

©2017 **BIOZONE** International
ISBN: 978-1-927309-65-0
Photocopying Prohibited

TEST

190 Synoptic Questions

Questions 1-3 relate to the photo of a plant leaf below.

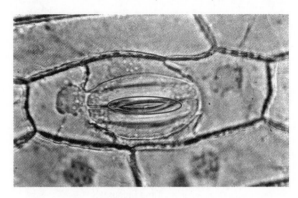

1. The structure circled is a stoma. They are most likely to be open during the day and closed at night:

   (a) True

   (b) False

2. When the guard cells surrounding the stoma take up water this causes them to:

   (a) Buckle outward, opening the stomatal pore.

   (b) Buckle outward, closing the stomatal pore.

   (c) Sag inward, opening the stomatal pore.

   (d) Sag inward, closing the stomatal pore.

3. Identify the ion responsible for bringing about the changes in turgor of the guard cell:

   (a) Calcium

   (b) Sodium

   (c) Potassium

   (d) Magnesium

Questions 4 and 5 relate to the photo below.

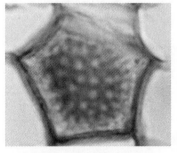

4. What does the image show?

   (a) Xylem

   (b) Companion cell

   (c) Vessel element

   (d) Sieve tube

5. Identify the correct statement for this image:

   (a) This tissue is dead and transports water

   (b) This tissue is alive and transports water

   (c) This tissue is dead and transports sugar

   (d) This tissue is alive and transports sugar

6. Which describes the **correct** order for processing food?

   (a) Egestion, digestion, absorption, ingestion.

   (b) Egestion, absorption, digestion, ingestion.

   (c) Ingestion, digestion, absorption, egestion.

   (d) Ingestion, absorption, digestion, egestion.

7. Oxygen extraction from the environment is easier for a terrestrial mammal than a fish because:

   (a) The percentage of oxygen in air is higher than in water.

   (b) Oxygen diffuses more readily across the surface of lungs than the surface of gills.

   (c) Gills have a much smaller surface area than lungs for the same size animal.

   (d) Countercurrent flow across the gills of fish is inefficient.

8. The main nitrogenous waste excreted by birds is:

   (a) Ammonia

   (b) Urea

   (c) Uric acid

   (d) Nitrate

9. Which cell types are responsible for initiating a secondary immune response?

   (a) Dendritic cells

   (b) Memory cells

   (c) Neutrophils

   (d) Macrophages

10. The response shown by the seedlings below is

    _____, and is caused by _____.

    (a) Phototropism, auxin.

    (b) Apical dominance, auxin.

    (c) Photoperiodism, phytochrome.

    (d) Chemotropism, minerals.

11. Habituation is:

    (a) When an individual develops social attachment to another individual.

    (b) When an individual's response becomes more predictable over time.

    (c) When an individual decreases or ceases its response to a stimulus.

    (d) None of the above.

12. Identify the **incorrect** statement about reflexes:

    (a) A reflex requires no conscious thought to occur.

    (b) Reflexes can be monosynaptic or polysynaptic.

    (c) The knee jerk response is an example of a reflex.

    (d) A reflex involves some conscious thought.

PRACTICES

TEST

13. Students carried out an experiment under controlled conditions (20°C and indirect lighting) to determine if transpiration rates varied between two plant species. One of the plant species is adapted for arid (dry) environments, the other is not. The results are plotted, right.

(a) Compare the transpiration rates between the two species:

_____

_____

_____

_____

(b) Identify the arid adapted species: _____

(c) Why did you choose this species? _____

_____

Water loss over time for two plant species

Total water loss (mL 100 g⁻¹ tissue) vs Time (minutes)

Species 1

Species 2

(d) Suggest some structural features that could account for differences in transpiration rates between the two species:

_____

_____

_____

_____

_____

14. The plot right shows the change in heart rate of a submerged harbor seal. The vertical broken lines mark the beginning and end of a dive:

(a) What happens to the seal's heart rate during a dive? _____

_____

(b) What is it about this plot that tells you the change in heart rate is a reflex response to the dive and not the result of a gradual depletion of oxygen:

_____

_____

Heart rate of a harbor seal

Heart rate (beats per minute) vs Time (minutes)

(c) During the dive, only the vital organs receive oxygen and blood is diverted from the muscles and gut. What do you think determines the duration of the dive? Explain:

_____

_____

15. A student investigated the effect of increasing seawater dilution on the cumulative weight gain in common rock crabs in different seawater dilutions. The results are plotted right.

(a) Explain the differences in crab weight gain: _____

_____

_____

(b) What does this experiment suggest about the osmoregulatory ability of this crab species?

_____

_____

Weight gain in crabs at two seawater dilutions

Weight (mg) vs Time (minutes)

Weight gain 50% seawater

Weight gain 75% seawater

©2017 **BIOZONE** International
ISBN: 978-1-927309-65-0
Photocopying Prohibited

16. In 1998, Dr Andrew Wakefield and his colleagues published a paper linking the measles, mumps, and rubella vaccine (MMR) to an increase in autism rates.

Dr Wakefield's paper has since been retracted by the journal in which it was published as it was found to be fraudulent and flawed in several aspects, e.g. sample size of only 12, with no control group.

Since the publication of Wakefield's paper, 20 large scale epidemiologic studies into MMR and autism have been carried out in several countries. **All have shown that the MMR vaccine does not cause autism**. However, the damage has been done, and health authorities must now convince the public that the vaccine is safe.

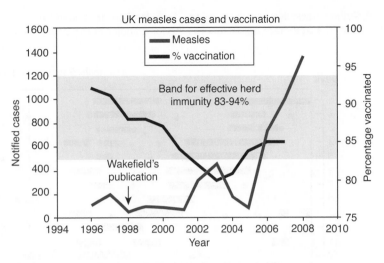

UK measles cases and vaccination

Band for effective herd immunity 83-94%

Wakefield's publication

The graph above shows the number of measles cases in the UK, together with percentage vaccination, 1994-2008.

(a) What happened to MMR vaccination rates after the publication of Wakefield's study? _____

_____

_____

(b) What is the trend in measles cases in the UK since 2006? _____

_____

(c) Give a likely explanation for this trend: _____

_____

_____

_____

17. Top professional endurance cyclists may have resting heart rates below 30 beats per minute (bpm). In contrast, most people have a resting heart rate of 60-70 bpm.

Five times Tour de France winner Miguel Indurain (right) had a resting heart rate of 28 bpm and could transport 7 liters of oxygen around his body per minute, almost twice as much as an ordinary person. Not only was his heart efficient, but his lung capacity (the volume of gas the lungs can hold) was 1.8 liters more than an average person.

(a) During a mountain climb, Indurain's heart rate could increase to 190 bpm. Calculate the percentage increase in heart rate relative to his resting heart rate of 28 bpm:

_____

(b) Explain why heart rate needs to increase during exercise: _____

_____

_____

_____

Miguel Indurain

Eric Houdas CC SA3.0

(c) How would having a larger than normal lung capacity give a professional athlete an advantage? _____

_____

_____

(d) Some athletes have used a banned, naturally occurring hormone, called erythropoietin which boosts red blood cell production. Why this would provide an unfair advantage to an athlete using it?

_____

_____

18. Fiddler crabs live on tidal mudflats. They feed on detritus in the mud at low tide and return to their burrow at high tide to rest. The actogram below records the for a fiddler crab's activity in a laboratory. The crab was kept in constant light.

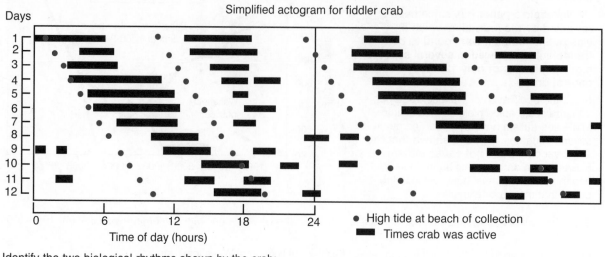

Simplified actogram for fiddler crab

• High tide at beach of collection
▬ Times crab was active

(a) Identify the two biological rhythms shown by the crab: _____

(b) Identify the origin of the biological rhythms: _____

(c) Explain the pattern seen in the actogram. Predict the continued activity of the crab in the lab in relation to the high tide at the beach of collection:

_____

_____

_____

_____

_____

_____

_____

19. (a) The graph below shows a recording of the changes in membrane potential in an axon during transmission of an action potential. Match each stage (A-E) to the correct summary provided below.

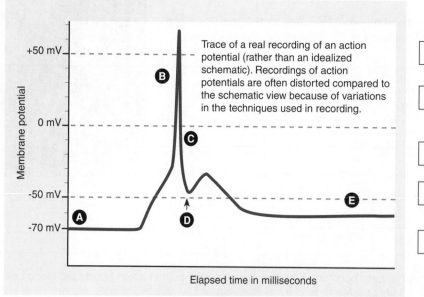

Trace of a real recording of an action potential (rather than an idealized schematic). Recordings of action potentials are often distorted compared to the schematic view because of variations in the techniques used in recording.

☐ Membrane depolarization (due to rapid Na⁺ entry across the axon membrane).

☐ Hyperpolarization (an overshoot caused by the delay in closing of the K⁺ channels).

☐ Return to resting potential after the stimulus has passed.

☐ Repolarization as the Na⁺ channels close and slower K⁺ channels begin to open.

☐ The membrane's resting potential.

(b) What is the resting potential of the axon? _____

(c) What is the maximum voltage reached by the action potential? _____

©2017 **BIOZONE** International
ISBN: 978-1-927309-65-0
Photocopying Prohibited

Populations and Communities

Key terms

abiotic factor

abundance

biotic factor

commensalism

community

competition

cooperation

demography

density

density-dependent factor

density-independent factor

distribution

ecological niche

ecosystem

exponential growth

fecundity

habitat

logistic growth

mortality

mutualism

natality

population

predation

resilience

species composition

species diversity

symbiosis

2.D.1 Biological systems are affected by biotic and abiotic interactions involving exchange of matter and free energy

Essential knowledge

Activity number

(a) Cell activities are affected by interactions with biotic and abiotic factors

☐ 1 Recall how the activities of cells are affected by their interactions with other cells and with the physical environment.

18 23

(b) Organism activities are affected by interactions with biotic and abiotic factors

☐ 1 Explain how nutrient and water availability, temperature, salinity and pH can determine the activities of organisms.

193 217
220-222

(c) Stability is affected by interactions with biotic and abiotic factors

☐ 1 Use examples to illustrate how abiotic and biotic factors can influence the stability of populations, communities, and ecosystems. Examples include:
- changes in distribution in response to water and nutrient availability
- population density regulated by availability of nesting materials and sites and by energy and matter transfers through food chains and webs
- species diversity determined by number of available niches

193 217
224 225
227 229

Butterfly austral cc 3.0

4.A.5 Communities are composed of populations of organisms that interact in complex ways

Essential knowledge

Activity number

(a) Community structure is described in terms of species composition and diversity

☐ 1 Using examples, explain how the structure of communities can be quantified and described by the species composition and the species diversity.

191-194
197-203

(b) Models are used to investigate and illustrate population interactions within, and environmental impacts on, a community

☐ 1 Use examples to explain how mathematical or computer models can be used to illustrate and investigate population interactions within a community. Examples include models of population growth under competition and predator-prey models and the stability of population cycles. *[The content for this point in its entirety is shared with the final chapter, 'The Diversity and Stability of Ecosystems'.]*

226 229

(c) Mathematical models and graphical representations are use to illustrate species interactions and patterns of population growth

☐ 1 Explain how unconstrained reproduction results in exponential population growth.

207-210

☐ 2 Explain how a population can produce a density of individuals exceeding the carrying capacity of the ecosystem (the system's resource availability). Show this on a mathematical or graphical representation of population growth.

207-210

☐ 3 Distinguish between density dependent and density-independent factors. Explain how density-dependent factors increasingly limit a population's growth as it approaches carrying capacity and result in logistic growth.

208 210

☐ 4 Using examples, show how demographics data with respect to age distributions and fecundity (fertility) can be applied to human populations.

211-214

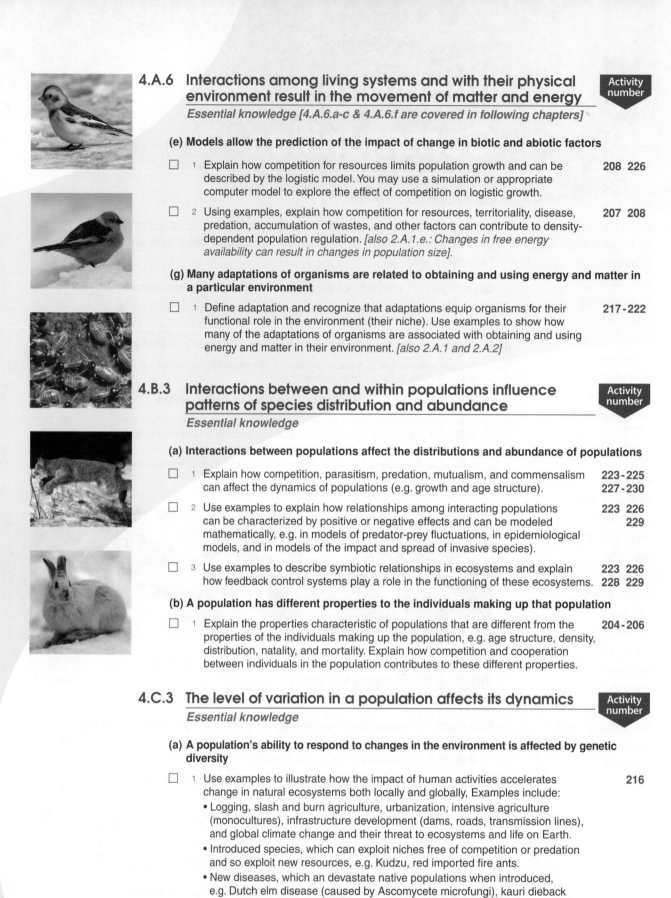

4.A.6 Interactions among living systems and with their physical environment result in the movement of matter and energy

Activity number

Essential knowledge [4.A.6.a-c & 4.A.6.f are covered in following chapters]

(e) Models allow the prediction of the impact of change in biotic and abiotic factors

- ☐ 1 Explain how competition for resources limits population growth and can be described by the logistic model. You may use a simulation or appropriate computer model to explore the effect of competition on logistic growth. — **208 226**
- ☐ 2 Using examples, explain how competition for resources, territoriality, disease, predation, accumulation of wastes, and other factors can contribute to density-dependent population regulation. *[also 2.A.1.e.: Changes in free energy availability can result in changes in population size].* — **207 208**

(g) Many adaptations of organisms are related to obtaining and using energy and matter in a particular environment

- ☐ 1 Define adaptation and recognize that adaptations equip organisms for their functional role in the environment (their niche). Use examples to show how many of the adaptations of organisms are associated with obtaining and using energy and matter in their environment. *[also 2.A.1 and 2.A.2]* — **217-222**

4.B.3 Interactions between and within populations influence patterns of species distribution and abundance

Activity number

Essential knowledge

(a) Interactions between populations affect the distributions and abundance of populations

- ☐ 1 Explain how competition, parasitism, predation, mutualism, and commensalism can affect the dynamics of populations (e.g. growth and age structure). — **223-225 227-230**
- ☐ 2 Use examples to explain how relationships among interacting populations can be characterized by positive or negative effects and can be modeled mathematically, e.g. in models of predator-prey fluctuations, in epidemiological models, and in models of the impact and spread of invasive species). — **223 226 229**
- ☐ 3 Use examples to describe symbiotic relationships in ecosystems and explain how feedback control systems play a role in the functioning of these ecosystems. — **223 226 228 229**

(b) A population has different properties to the individuals making up that population

- ☐ 1 Explain the properties characteristic of populations that are different from the properties of the individuals making up the population, e.g. age structure, density, distribution, natality, and mortality. Explain how competition and cooperation between individuals in the population contributes to these different properties. — **204-206**

4.C.3 The level of variation in a population affects its dynamics

Activity number

Essential knowledge

(a) A population's ability to respond to changes in the environment is affected by genetic diversity

- ☐ 1 Use examples to illustrate how the impact of human activities accelerates change in natural ecosystems both locally and globally, Examples include: — **216**
  - Logging, slash and burn agriculture, urbanization, intensive agriculture (monocultures), infrastructure development (dams, roads, transmission lines), and global climate change and their threat to ecosystems and life on Earth.
  - Introduced species, which can exploit niches free of competition or predation and so exploit new resources, e.g. Kudzu, red imported fire ants.
  - New diseases, which an devastate native populations when introduced, e.g. Dutch elm disease (caused by Ascomycete microfungi), kauri dieback and potato blight (both caused by species of the water mold *Phytophthora*), smallpox epidemics in Native American populations (including in Mexico).

(b) Genetic diversity allows individuals in a population to respond differently to the same changes in environmental conditions

- ☐ 1 Describe how individuals in a population respond differently to the same change in environment. Explain the consequences of this variability in response. — **216**

191 Components of an Ecosystem

Key Idea: An ecosystem consists of all the organisms living in a particular area and their physical environment.

An **ecosystem** is a community of living organisms and the physical (non-living) components of their environment. The community (living component of the ecosystem) is in turn made up of a number of **populations**, these being organisms of the same species living in the same geographical area. The structure and function of an ecosystem is determined by the physical (abiotic) and the living (biotic) factors, which determine species distribution and survival.

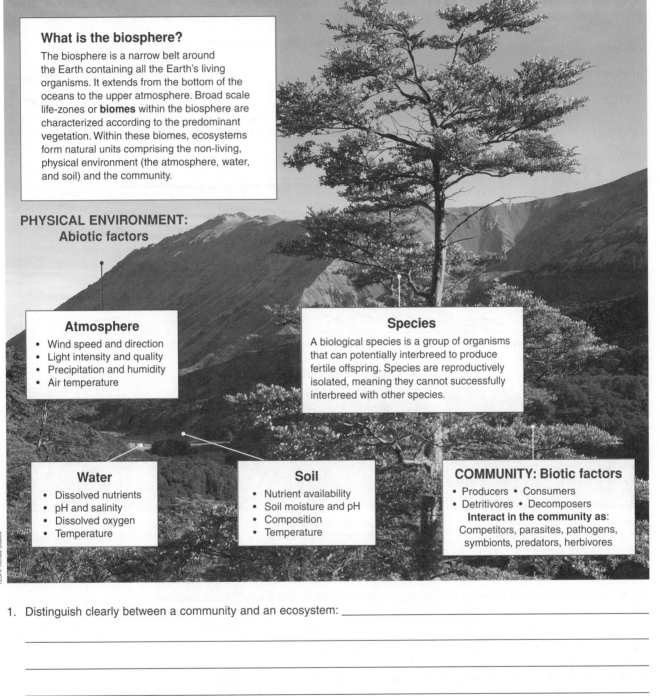

What is the biosphere?

The biosphere is a narrow belt around the Earth containing all the Earth's living organisms. It extends from the bottom of the oceans to the upper atmosphere. Broad scale life-zones or **biomes** within the biosphere are characterized according to the predominant vegetation. Within these biomes, ecosystems form natural units comprising the non-living, physical environment (the atmosphere, water, and soil) and the community.

PHYSICAL ENVIRONMENT:
Abiotic factors

Atmosphere
• Wind speed and direction
• Light intensity and quality
• Precipitation and humidity
• Air temperature

Species
A biological species is a group of organisms that can potentially interbreed to produce fertile offspring. Species are reproductively isolated, meaning they cannot successfully interbreed with other species.

Water
• Dissolved nutrients
• pH and salinity
• Dissolved oxygen
• Temperature

Soil
• Nutrient availability
• Soil moisture and pH
• Composition
• Temperature

COMMUNITY: Biotic factors
• Producers • Consumers
• Detritivores • Decomposers
  Interact in the community as:
  Competitors, parasites, pathogens, symbionts, predators, herbivores

123RF Tomas Sobek

1. Distinguish clearly between a community and an ecosystem: _____

_____

_____

_____

2. Distinguish between biotic and abiotic factors: _____

_____

_____

3. Use one or more of the following terms to describe each of the features of a beech community listed below:
   Terms: *population, community, ecosystem, physical factor.*

   (a) All the beech trees present: _____   (c) All the organisms present: _____

   (b) The entire forest: _____   (d) The humidity: _____

CONNECT **238** AP2   CONNECT **233** AP2   WEB **191**

KNOW

192 Habitat

Key Idea: The environment in which an organism lives is its habitat. The habitat may not be homogeneous in its quality.

The environment in which an organism (or species) lives (including all the physical and biotic factors) is its **habitat**. Within any habitat, each species has a range of tolerance to variations in its environment. Within the population, individuals will have slightly different tolerance ranges based on small differences in genetic make-up, age, and health. The wider an organism's tolerance range for any one factor (e.g. temperature) the more likely it is that the organism will survive variations in that factor. For the same reasons, species with a wider tolerance range are likely to be more widely distributed. Organisms have a narrower **optimum range** within which they function best. This may vary seasonally or during development. Organisms are usually most abundant where the abiotic factors are closest to the optimum range.

Habitat occupation and tolerance range

Examples of abiotic factors influencing available habitat:

Too acidic ←——————————— pH ———————————→ Too alkaline

Too cold ←——————————— Temperature ———————————→ Too hot

The law of tolerances states that "*for each abiotic factor, a species population (or organism) has a tolerance range within which it can survive. Toward the extremes of this range, that abiotic factor tends to limit the organism's ability to survive*".

←——————————————— Tolerance range ———————————————→

| Zone of death or avoidance | Zone of physiological stress | ←——— Optimum range ———→ | Zone of physiological stress | Zone of death or avoidance |

Number of organisms (y-axis)

Unavailable habitat   Marginal habitat   Preferred habitat   Marginal habitat   Unavailable habitat

The scale of available habitats

A habitat may be vast and relatively homogeneous for the most part, as is the open ocean. Barracuda (above) occur around reefs and in the open ocean where they are aggressive predators.

For non-motile organisms, such as the fungus pictured above, a suitable habitat may be defined by the particular environment in a relatively small area, such as on this decaying log.

For microbial organisms, such as the bacteria and protozoans of the ruminant gut, the habitat is defined by the chemical environment within the rumen (R) of the host animal, in this case, a cow.

1. What is the relationship between an organism's tolerance range and the habitat it occupies? _____

_____

_____

2. (a) In the diagram above, in which range is most of the population found? Explain your answer: _____

_____

_____

(b) What are the greatest constraints on an organism's growth and reproduction within this range? _____

_____

3. Describe some probable stresses on an organism forced into a marginal habitat: _____

_____

_____

©2017 **BIOZONE** International
ISBN: 978-1-927309-65-0
Photocopying Prohibited

193 Physical Factors and Gradients

Key Idea: Gradients in the physical environment influence the range of physical conditions and may create microhabitats. Gradients in abiotic factors are found in all environments. They create microhabitats and microclimates (see definitions below) within a larger area and influence patterns of species distribution. Organisms can exploit the microclimates produced by physical gradients and so occupy apparently inhospitable environments, e.g. frogs living in deserts.

A desert environment

Deserts experience extremes in temperature and humidity, but they are not uniform with respect to these factors. The diagram below gives hypothetical values for temperature and humidity for typical microclimates in a desert environment at midday.

> **Microclimate**: The climate of a very small or restricted area.
>
> **Microhabitat**: A habitat of limited extent, which differs in its characteristics from the surrounding more extensive habitat.

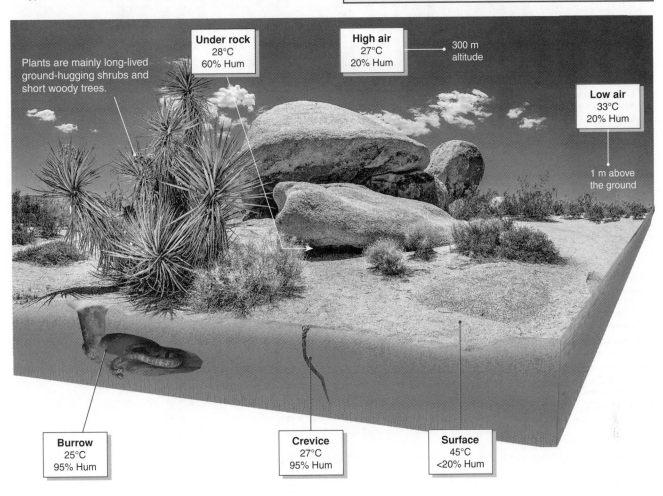

Plants are mainly long-lived ground-hugging shrubs and short woody trees.

Under rock
28°C
60% Hum

High air
27°C
20% Hum

300 m altitude

Low air
33°C
20% Hum

1 m above the ground

Burrow
25°C
95% Hum

Crevice
27°C
95% Hum

Surface
45°C
<20% Hum

1. (a) Study the diagram above and describe the general microhabitats where humidity is highest: _____

_____

_____

   (b) How do these microhabitats enable land animals to survive the extreme high temperatures of midday: _____

_____

_____

2. Desert surfaces not only receive more solar radiation than humid regions, but they lose much more heat at night, so night-time temperatures can be very low. How does this advantage both the plants and animals living there?

_____

_____

3. Suggest why there are relatively few large mammals in deserts: _____

_____

_____

CONNECT **222** AP2 CONNECT **221** AP2 CONNECT **220** AP2 CONNECT **194** AP2 WEB **193**

KNOW

A forest environment

Environmental gradients can arise as a result of vertical distance from the ground. In a forest, light quantity and quality, wind speed, humidity, and temperature change gradually from the canopy to the forest floor. These changes give rise to a layered (stratified) community in which different plant and animal species occupy different vertical positions in the forest according to their tolerances.

Light: 70%
Wind: 15 km h^{-1}
Humid: 67%

Canopy

Light: 50%
Wind: 12 km h^{-1}
Humid: 75%

Light: 12%
Wind: 9 km h^{-1}
Humid: 80%

Light: 6%
Wind: 5 km h^{-1}
Humid: 85%

A **datalogger** fitted with suitable probes was used to gather data on wind speed (**Wind**), humidity (**Humid**), and light intensity (**Light**) for each layer (left). Light intensity is given as a percentage of full sunlight.

Light: 1%
Wind: 3 km h^{-1}
Humid: 90%

Light: 0%
Wind: 0 km h^{-1}
Humid: 98%

Leaf litter

Tropical rainforests are complex communities with a vertical structure which divides the vegetation into layers. This pattern of vertical layering is called **stratification**.

4. Describe the general trend from the canopy to the leaf litter for each of the following:

    (a) Light intensity: _____

    (b) Wind speed: _____

    (c) Humidity: _____

5. Explain why each of these factors changes as the distance from the canopy increases:

    (a) Light intensity: _____

    _____

    (b) Wind speed: _____

    _____

    (c) Humidity: _____

    _____

6. What other feature of light, other than intensity, will also change with distance from the canopy and why?

    _____

    _____

©2017 **BIOZONE** International
ISBN: 978-1-927309-65-0
Photocopying Prohibited

194 Describing Community Structure

Key Idea: Descriptions of community structure include species diversity and composition.

Because the species in a community are so varied it is necessary that several measures be used in their description to prevent skewed interpretation of the community structure. These measures include the diversity of species, their density and composition, and abiotic factors. For example, a community may comprise 30 species, but one of those species may be far more common than the others. Moreover, the most common species numerically may actually be very small physically, and have the least biomass of all the other species combined.

Species diversity

Species diversity is a measure of the number of different types of species and their number in a community. When referring to species diversity it is useful to quantify both **species richness** and **species evenness**.

One measure of biodiversity is to simply count all the species present (the species richness). Species richness (**S**) is directly related to the number of species in a sampled area. It is a crude measure of the homogeneity of a community but it does not give any information about the relative abundance of particular species and so is relatively meaningless by itself. Thus a sample area with 500 daisies and 3 dandelions has the same species richness as a sample area with 200 daisies and 300 dandelions.

Species evenness is a measure of the proportion of each species in an area (the relative abundance). Species evenness is highest when the proportions of all species are the same and decreases as the proportions of species become less similar.

High species richness: 23 species

Low species richness 3 species

Table 1

Sample of freshwater invertebrates in a stream			
Common name	Site 1 (no. m^{-2})	Site 2 (no. m^{-2})	Site 3 (no. m^{-2})
Freshwater shrimp	20	67	5
Freshwater mite	15	0	0
Flat mayfly	21	23	0
Bighead stonefly	18	12	2
Blackfly	40	78	100
Bloodworm	22	21	43

Data for species richness and species evenness can be obtained by sampling, e.g. using quadrats. In table 1 above, three sites in a stream were sampled using quadrats and the species and number of individuals per m^2 recorded for each site. Using Site 1 as an example, species richness is 6, because S = number of species. In terms of species evenness, site 1 > site 2 > site 3.

Table 2

Plant species	Biomass (g)	Composition
A	105	53%
B	37	19%
C	56	28%

Species composition

Species composition refers to the proportions of plant species in relation to the total in a given area. It is usually expressed as a percentage and calculated as percentage cover, density, or biomass. Table 2, right, shows that in a community of three plants, species A comprises 53% of the community based on biomass. However measurements of density and cover should also be taken, to more accurately reflect the community composition.

1. Distinguish between species richness and species evenness: _____

_____

_____

2. (a) Why should several measures be included when describing community structure and composition? _____

_____

_____

   (b) In the plant community described in table 2, why should measurements of density and cover also be taken?

_____

3. Study the data on the freshwater invertebrates at the top of the page:

   (a) Which site has the lowest species diversity? _____

   (b) Which is the most abundant species in the stream? _____

   (c) Is there a trend in the data? Suggest what could be causing this in the stream: _____

_____

PRACTICES WEB

KNOW

Characteristics of a community

Dominance. A community is often dominated numerically by a particular species (e.g. coastal redwoods) or type of organism (e.g. corals). The community is often named after this species or type, e.g. redwood forest, broadleaf forest, coral reef.

Stratification: Forest communities possess a number of layers, e.g. ground layer, shrub layer, canopy. This produces numerous different environments within the one vertical space so that a description of the community must account for the vertical as well as horizontal distribution of species.

Stability: Some communities are successional. They modify the environment but will last for only a short time before other organisms move in and replace them.

Productivity: The rate of biomass production is a useful figure in describing the characteristics of a community (e.g. some communities, such as estuaries, have very high primary and secondary net productivity compared with others).

Edge effect: An ecosystem may take in many smaller habitats that inevitably overlap, producing edges or marginal zones. Species diversity is often higher here than elsewhere and will not represent any one single habitat.

Periodicity: Even in established, stable communities, community composition may change over time as a result of seasonal variations (e.g. annual plants, seasonal migrants). The timing of a sampling program may affect the information obtained about the community composition.

4. Explain the importance of each of the following in describing a community:

   (a) Dominance: _____

   _____

   _____

   (b) Stability: _____

   _____

   _____

   (c) Edge effect: _____

   _____

   _____

5. How might stratification influence species diversity in a forest community? _____

   _____

   _____

©2017 **BIOZONE** International
ISBN: 978-1-927309-65-0
Photocopying Prohibited

195 Features of Populations

Key Idea: Populations exhibit attributes, such as density, age structure, and mortality, that are not shown by individuals.

A **population** refers to all the organisms of the same species in a particular area. Biologists are often interested in population size (abundance) and density, both of which reflect the carrying capacity of the environment, i.e. how many organisms the environment can support. Populations have structure, with particular ratios of different ages and sexes. Analysis of age structure can show if the population is declining or increasing in size. We can also look at the distribution of organisms in their environment and so determine what aspects of the habitat are favored over others. One way to retrieve information about populations is to sample them. Sampling involves collecting data, directly or indirectly, about features of the population from samples (since populations are usually too large to examine in total).

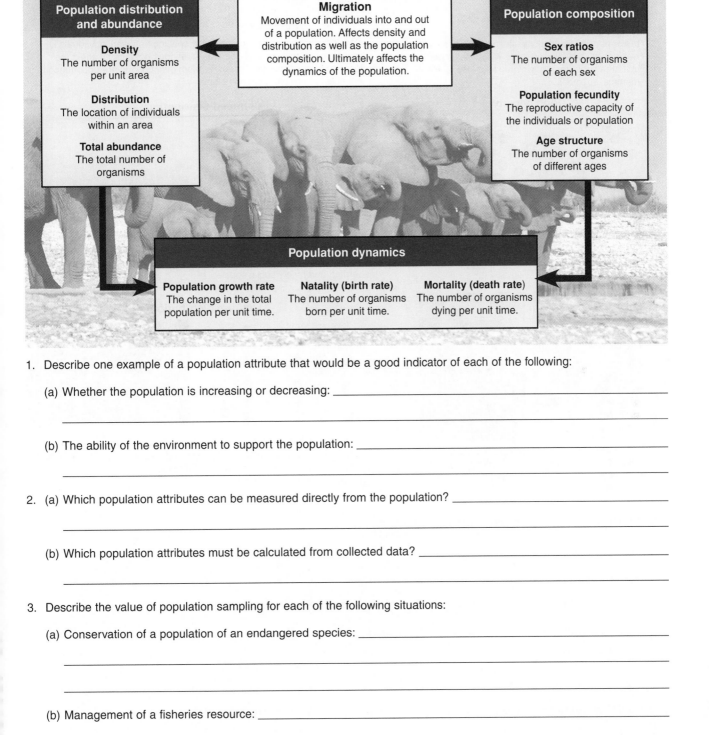

Population distribution and abundance

Density
The number of organisms per unit area

Distribution
The location of individuals within an area

Total abundance
The total number of organisms

Migration
Movement of individuals into and out of a population. Affects density and distribution as well as the population composition. Ultimately affects the dynamics of the population.

Population composition

Sex ratios
The number of organisms of each sex

Population fecundity
The reproductive capacity of the individuals or population

Age structure
The number of organisms of different ages

Population dynamics

Population growth rate
The change in the total population per unit time.

Natality (birth rate)
The number of organisms born per unit time.

Mortality (death rate)
The number of organisms dying per unit time.

1. Describe one example of a population attribute that would be a good indicator of each of the following:

   (a) Whether the population is increasing or decreasing: _____

   _____

   (b) The ability of the environment to support the population: _____

   _____

2. (a) Which population attributes can be measured directly from the population? _____

   _____

   (b) Which population attributes must be calculated from collected data? _____

   _____

3. Describe the value of population sampling for each of the following situations:

   (a) Conservation of a population of an endangered species: _____

   _____

   _____

   (b) Management of a fisheries resource: _____

   _____

   _____

   _____

©2017 **BIOZONE** International
ISBN: 978-1-927309-65-0
Photocopying Prohibited

CONNECT **229** AP2   CONNECT **216** AP2   WEB **195**

KNOW

196 Density and Distribution

Key Idea: Population density refers to the number of organisms of one species in a specified area. Distribution describes how the organisms are distributed relative to each other. Population density is the number of individuals per unit area or volume. The three basic distribution patterns (random, clumped, and uniform) are illustrated below using circles to represent individuals in a population. Distribution patterns are usually determined by resource distribution and/or behavior.

Low density

In low density populations, individuals are spaced well apart. There are only a few individuals per unit area or volume (e.g. highly territorial, solitary mammal species).

High density

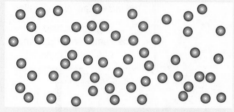

In high density populations, individuals are crowded together. There are many individuals per unit area or volume (e.g. colonial organisms, such as many corals).

Tigers are solitary animals, found at low densities.

Termites form well organized, high density colonies.

Random distribution

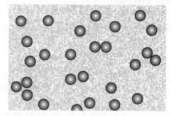 

Random distributions occur when the spacing between individuals is irregular. The presence of one individual does not directly affect the location of any other individual. Random distributions are uncommon in animals but are often seen in plants.

Clumped distribution

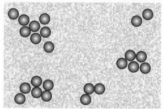 

Clumped distributions occur when individuals are grouped in patches (sometimes around a resource). The presence of one individual increases the probability of finding another close by. Such distributions occur in herding and highly social species.

Uniform distribution

Regular distribution patterns occur when individuals are evenly spaced within the area. The presence of one individual decreases the probability of finding another individual immediately adjacent. The gannets illustrated above are also at a high density.

1. (a) Distinguish between density and distribution: _____

_____

_____

(b) Why are both these measures necessary to describe the characteristics of a population? _____

_____

_____

_____

2. Describe an example of an organism that typically shows each of the following types of distribution pattern:

(a) Clumped: _____

(b) Random (more or less): _____

(c) Uniform (more or less): _____

©2017 **BIOZONE** International
ISBN: 978-1-927309-65-0
Photocopying Prohibited

197 Measuring Distribution and Abundance

Key Idea: Random sampling using an appropriate technique provides unbiased information about the distribution and abundance of species in a community.

Most practical exercises in ecology involve collecting data about the distribution and abundance of one or more species in a community. Most studies also measure the physical factors in the environment as these may help to explain the patterns of distribution and abundance observed. The use of **random sampling** methods, in which every possible sample of a given size the same chance of selection, provides unbiased data. As long as the sample size is large enough and the sampling technique is appropriate to the community being studied, sample data enables us to make inferences about aspects of the whole population.

Distribution and abundance

Ecological sampling collects data about where organisms are found and how they are distributed in the environment. This information can be used to determine the health and viability of a population and its ecosystem. When investigating populations it is useful to quantify:

▶ Species **distribution** (where the species are located)

▶ Species **abundance** (how many of a species there are in an area)

The methods used to sample communities and their constituent populations must be appropriate to the ecosystem being investigated. Communities in which the populations are at low density and have a random or clumped distribution will require a different sampling strategy to those where the populations are uniformly distributed and at higher density. There are many sampling options (below), each with advantages and drawbacks for particular communities.

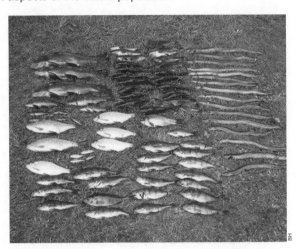

Sampling designs and techniques

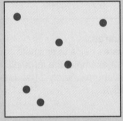

Random

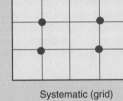

Systematic (grid)

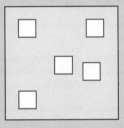

Point sampling

Individual points are chosen (using a grid reference or random numbers applied to a map grid) and the organisms are sampled at those points. Point sampling is most often used to collect data about vegetation distribution. It is time efficient and good for determining species abundance and community composition, but organisms in low abundance may be missed.

Area sampling using quadrats

A quadrat is a sampling tool that provides a known unit area of sample (e.g. 0.5 m^2). Quadrats are placed randomly or in a grid pattern on the sample area. The presence and abundance of organisms in these squares is noted. Quadrat sampling is appropriate for plants and slow moving animals and can be used to evaluate community composition.

First sample: marked

Second sample: proportion recapture

Mark and recapture sampling

Animals are captured, marked, and released. After a suitable time, the population is resampled. The number of marked animals recaptured in a second sample is recorded as a proportion of the total. Mark and recapture is useful for highly mobile species which are otherwise difficult to record. However, it is time consuming to do well.

Line transects

A tape marks a line and species occurring at regular points along the line are recorded. Lines can be chosen randomly (above) or may follow an environmental gradient. Line transects have little impact on the environment and are good for Assessing the presence/absence of plant species. However, rare species may be missed.

Belt transects

A measured strip is located across the study area and quadrats are used to sample the plants or animals at regular intervals along the belt. Belt transects provide information on abundance and distribution as well as presence/absence. Depending on the width of the belt and length of the transect, they can be time consuming.

PRACTICES

WEB

197

KNOW

Sampling plants

Plants can be sampled by transects and quadrat in the same way as animals. However plants, especially ground cover plants, tend to spread and often overlap with other plants, making it difficult to count individuals. To overcome this, an estimate of percentage cover of the plant can be used. Because plant leaves overlap, it is possible that the summed estimates of percentage cover can add to more than 100%.

Estimating the percentage cover from a quadrat can be rather subjective (is it 45% or 50%?) and the larger the quadrat the more subjective it can be. To retain objectivity, the quadrat can be divided into smaller squares and the number of squares covering a plant counted. Another way of estimating percentage cover is using a point quadrat.

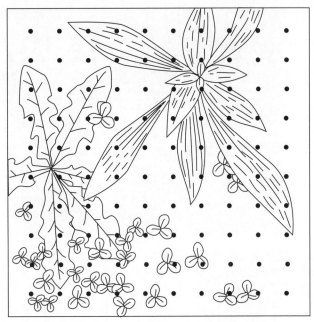

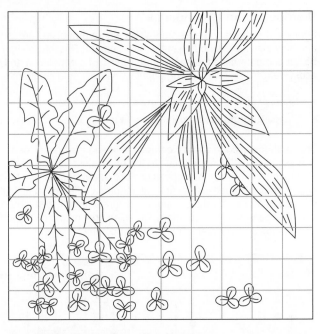

Point quadrat

When using a point quadrat a pin to placed vertically downward at each point and the plants it touches are recorded. The percentage cover is then calculated by the formula:

Percentage cover = (Hits (touches) ÷ total sampling points) x 100

Quadrat divided into smaller squares.

Dividing the quadrat into smaller squares helps to estimate percentage cover more accurately. For example, the large plant in the top right of the quadrat above fills approximately 21 squares out of 100 so covers 21% of the quadrat.

1. Distinguish between distribution and abundance: _____

_____

_____

2. Name a sampling technique that would be appropriate for determining:

    (a) Percentage cover of a plant species in pasture: _____

    _____

    (b) Change in community composition from low to high altitude on a mountain: _____

    _____

    (c) Association of plant species with particular soil types in a nature reserve: _____

    _____

3. Why is it common practice to also collect information about the physical environment when sampling populations?

    _____

    _____

    _____

4. Use the **point quadrat** to calculate the percentage cover of the large plant in the top right corner of the quadrat. How does this estimate compare with the estimate obtained by dividing the quadrat into smaller squares?

    _____

198 Quadrat Sampling

Key Idea: Quadrat sampling involves a series of random placements of a frame of known size over an area of habitat to assess the abundance or diversity of organisms.

Quadrat sampling is a method by which organisms in a certain proportion (sample) of the habitat are counted directly. It is used when the organisms are too numerous to count in total. It can be used to estimate population abundance (number), density, frequency of occurrence, and distribution. Quadrats may be used without a transect when studying a relatively uniform habitat. In this case, the quadrat positions are chosen randomly using a random number table. The general procedure is to count all the individuals (or estimate their percentage cover) in a number of quadrats of known size and to use this information to work out the abundance or percentage cover value for the whole area.

$$\text{Estimated density} = \frac{\text{Total number of individuals counted}}{\text{(Number of quadrats X area of each quadrat)}}$$

Guidelines for quadrat use:

1. The **area of each quadrat** must be known. Quadrats should be the same shape, but not necessarily square.

2. **Enough quadrat samples** must be taken to provide results that are representative of the total population.

3. The **population of each quadrat** must be known. Species must be distinguishable from each other, even if they have to be identified at a later date. It has to be decided beforehand what the count procedure will be and how organisms over the quadrat boundary will be counted.

4. The size of the quadrat should be appropriate to the organisms and habitat, e.g. a large size quadrat for trees.

5. The quadrats must be **representative of the whole area.** This is usually achieved by **random sampling** (right).

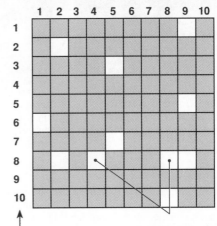

The area to be sampled is divided up into a grid pattern with indexed coordinates.

Quadrats are applied to the predetermined grid on a random basis. This can be achieved by using a random number table.

Sampling a centipede population

A researcher by the name of Lloyd (1967) sampled centipedes in Wytham Woods, near Oxford in England. A total of 37 hexagon–shaped quadrats were used, each with a diameter of 30 cm (see diagram on right). These were arranged in a pattern so that they were all touching each other. Use the data in the diagram to answer the following questions.

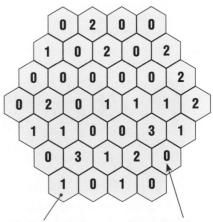

1. Determine the density of centipedes per square meter (remember that each quadrat is 0.08 square meters in area):

_____

2. Looking at the data for individual quadrats, describe in general terms the distribution of the centipedes in the sample area:

_____

Each quadrat was a hexagon with a diameter of 30 cm and an area of 0.08 m².

The number in each hexagon indicates how many centipedes were caught in that quadrat.

3. In this study, the quadrats were adjacent and all were sampled. Working in small groups, take a sample of 7 of the quadrats in the pattern pictured and estimate the centipede density. How did you choose your sample? How did your estimate compare to the value in question 1?

_____

_____

_____

_____

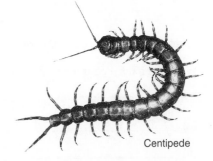

Centipede

PRACTICES

WEB

198

DATA

199 Quadrat-Based Estimates

Key Idea: The size and number of quadrats used to sample a community must be sufficient to be representative of that community without taking an excessively long time to use.
The simplest description of a community is a list of the species present. This does not provide information about the relative abundance of the species, although this can be estimated using abundance scales (e.g. ACFOR). Quadrats can provide quantitative information about a community. The size of the quadrat and the number of samples taken must represent the community as fairly as possible.

What size quadrat?

Quadrats are usually square, and cover 0.25 m² (0.5 m x 0.5 m) or 1 m², but they can be of any size or shape, even a single point. The quadrats used to sample plant communities are often 0.25 m². This size is ideal for low-growing vegetation, but quadrat size needs to be adjusted to habitat type. The quadrat must be large enough to be representative of the community, but not so large as to take a very long time to use.

A quadrat covering an area of 0.25 m² is suitable for most low growing plant communities, such as this alpine meadow, fields, and grasslands.

Larger quadrats (e.g.1m²) are needed for communities with shrubs and trees. Quadrats as large as 4 m x 4 m may be needed in woodlands.

Small quadrats (0.01 m² or 100 mm x 100 mm) are appropriate for lichens and mosses on rock faces and tree trunks.

How many quadrats?

As well as deciding on a suitable quadrat size, the other consideration is how many quadrats to take (the sample size). In species-poor or very homogeneous habitats, a small number of quadrats will be sufficient. In species-rich or heterogeneous habitats, more quadrats will be needed to ensure that all species are represented adequately.

Determining the number of quadrats needed

- Plot the cumulative number of species recorded (on the y axis) against the number of quadrats already taken (on the x axis).

- The point at which the curve levels off indicates the suitable number of quadrats required.

Fewer quadrats are needed in species-poor or very uniform habitats, such as this bluebell woodland.

Describing vegetation

Density (number of individuals per unit area) is a useful measure of abundance for animal populations, but can be problematic in plant communities where it can be difficult to determine where one plant ends and another begins. For this reason, plant abundance is often assessed using **percentage cover**. Here, the percentage of each quadrat covered by each species is recorded, either as a numerical value or using an abundance scale such as the ACFOR scale.

The ACFOR abundance scale

A = Abundant (30% +)
C = Common (20-29%)
F = Frequent (10-19%)
O = Occasional (5-9%)
R = Rare (1-4%)

The ACFOR scale could be used to assess the abundance of species in this wildflower meadow. Abundance scales are subjective, but it is not difficult to determine which abundance category each species falls into.

1. Describe one difference between the methods used to assess species abundance in plant and in animal communities:

   _____

2. What is the main consideration when determining appropriate quadrat size? _____

3. What is the main consideration when determining number of quadrats? _____

4. Explain two main disadvantages of using the ACFOR abundance scale to record information about a plant community:

   (a) _____

   (b) _____

©2017 **BIOZONE** International
ISBN: 978-1-927309-65-0
Photocopying Prohibited

200 Transect Sampling

Key Idea: Transect sampling is useful for providing information about species distribution along an environmental gradient.
A **transect** is a line placed across a community of organisms. Transects provide information on the distribution of species in the community. They are particularly valuable when the transect records community composition along an **environmental gradient** (e.g. up a mountain or across a seashore). The usual practice for small transects is to stretch a string between two markers. The string is marked off in measured distance intervals and the species at each marked point are noted. The sampling points along the transect may also be used for the siting of quadrats, so that changes in density and community composition can be recorded. Belt transects are essentially a form of continuous quadrat sampling. They provide more information on community composition but can be difficult to carry out. Some transects provide information on the vertical, as well as horizontal, distribution of species (e.g. tree canopies in a forest).

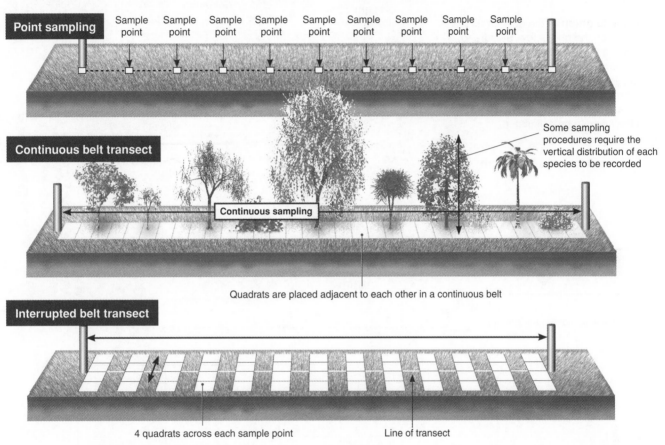

1. Belt transect sampling uses quadrats placed along a line at marked intervals. In contrast, point sampling transects record only the species that are touched or covered by the line at the marked points.

    (a) Describe one disadvantage of belt transects: _____

    _____

    (b) Why might line transects give an unrealistic sample of the community in question? _____

    _____

    _____

    (c) How do belt transects overcome this problem? _____

    _____

    (d) When would it not be appropriate to use transects to sample a community? _____

    _____

2. How could you test whether or not a transect sampling interval was sufficient to accurately sample a community?

    _____

    _____

    _____

©2017 **BIOZONE** International
ISBN: 978-1-927309-65-0
Photocopying Prohibited

PRACTICES

WEB

200

DATA

A **kite graph** is a good way to show the distribution of organisms sampled using a belt transect. Data may be expressed as abundance or percentage cover along an environmental gradient. Several species can be shown together on the same plot so that the distributions can be easily compared.

3. The data on the right were collected from a rocky shore field trip. Four common species of barnacle were sampled in a continuous belt transect from the low water mark, to a height of 10 m above that level. The number of each of the four species in a 1 m² quadrat was recorded.

Plot a **kite graph** of the data for all four species on the grid below. Be sure to choose a scale that takes account of the maximum number found at any one point and allows you to include all the species on the one plot. Include the scale on the diagram so that the number at each point on the kite can be calculated.

An example of a kite graph

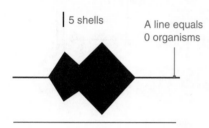

| | 5 shells |
A line equals
0 organisms

1   10
Distance above water line (m)

Field data notebook

Numbers of barnacles (4 common species) showing distribution on a rocky shore

	Barnacle species			
Height above low water (m)	Plicate barnacle	Columnar barnacle	Brown barnacle	Sheet barnacle
0	0	0	0	65
1	10	0	0	12
2	32	0	0	0
3	55	0	0	0
4	100	18	0	0
5	50	124	0	0
6	30	69	2	0
7	0	40	11	0
8	0	0	47	0
9	0	0	59	0
10	0	0	65	0

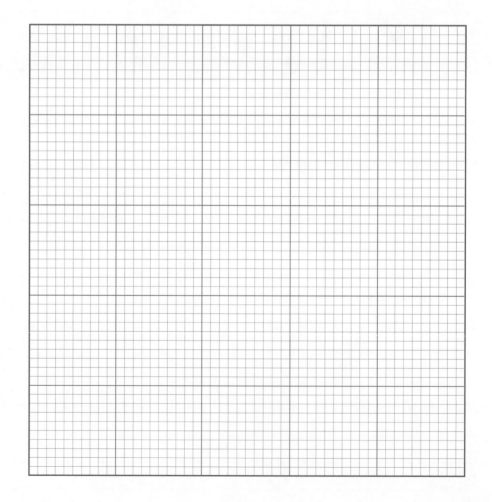

©2017 **BIOZONE** International
ISBN: 978-1-927309-65-0
Photocopying Prohibited

201 Sampling Techniques and Population Estimates

Key Idea: Population estimates made from samples may vary depending on the sampling technique, the number of samples, and where the data was collected.

The diagram (following page) represents an area of wasteland that has been invaded by various weeds. The distribution of five weed species is shown (simply noted as plants 1 to 5). This exercise is designed to show you how different sampling techniques and the way those techniques are applied can give different pictures of the make up of the community being studied. This exercise can be carried out in groups.

Setting up the sampling grid

1. **Mark out a grid pattern**
   Use a ruler to mark out 3 cm intervals along each side of the sampling area on the next page (area of quadrat = 0.03 x 0.03 m). **Draw lines** between these marks to create a 6 x 6 grid pattern (total area = 0.18 x 0.18 m). This will provide a total of 36 quadrats that can be investigated.

2. **Number the axes of the grid**
   Only a small proportion of the possible quadrat positions will be sampled. It is necessary to select the quadrats in a random manner. It is not sufficient to simply guess or choose your own on a 'gut feeling'. The best way to choose the quadrats randomly is to create a numbering system for the grid pattern and then select the quadrats from a random number table. Starting at the top left hand corner, **number the columns and rows from 1 to 6 on each axis**.

Quadrat sampling (counts and % cover)

3. **Choose quadrats randomly**
   To select the required number of quadrats randomly, use random numbers from a random number table. The random numbers are used as an index to the grid coordinates. Choose 6 quadrats from the total of 36 using table of random numbers provided at the bottom of the next page. Make a note of which column of random numbers you choose. Each member of your group should choose a different set of random numbers (i.e. different column: A–D) so that you can compare the effectiveness of the sampling method.

   Column of random numbers chosen: _____

   NOTE: Highlight the boundary of each selected quadrat with colored pen/highlighter.

4a. **Decide on the counting criteria**
   Before you count the individuals of each species, the criteria for counting need to be established. You must decide before sampling begins as to what to do about individuals that are only partly inside the quadrat. Possible answers include:

   (a) Only counting individuals that are completely inside the quadrat.

   (b) Only counting individuals with a certain part of the body (e.g. the main stem and root mass) inside the quadrat.

   (c) Allowing for 'half individuals' (e.g. 3.5 plants).

   (d) Counting an individual that is inside the quadrat by half or more as one complete individual.

   Discuss the merits and problems of the suggestions above with other members of the class (or group). You may even have counting criteria of your own. Think about other factors that could cause problems with your counting.

4b. **Carry out the sampling**
   Examine each selected quadrat and **count the number of individuals** of each species present. Record your data in the spaces provided in the quadrat count table (Table 1).

5. **Percentage cover**
   Carefully examine each selected quadrat again and **estimate the percentage cover** of each species present. Record your data in the spaces provided in the percentage cover table (Table 2) (to help you, the area of each plant type is given in the table. The quadrat area in 9 cm²).

Transect sampling

6. **Line transect**
   Draw **6 transect lines** across the sampling area from side to side or top to bottom. Use the random number table to work out where the transect line should go. For example using the numbers in column A (2,2) the first line should start in the middle of column 2 in the top row and go across the grid to the middle of column 2 in the bottom row.

   Mark the line every 1 cm. **Count** and record every plant the line touches at every 1 cm mark. Record the results in the line transect table (Table 3).

Table 1: Quadrat count

Coordinates for each quadrat	Plant 1	Plant 2	Plant 3	Plant 4	Plant 5
1:					
2:					
3:					
4:					
5:					
6:					
TOTAL					

Table 2: Percentage cover

Coordinates for each quadrat	2.59 cm² Plant 1	2.27 cm² Plant 2	0.79 cm² Plant 3	0.13 cm² Plant 4	0.5 cm² Plant 5
1:					
2:					
3:					
4:					
5:					
6:					
MEAN %					

©2017 **BIOZONE** International
ISBN: 978-1-927309-65-0
Photocopying Prohibited

PRACTICES PRACTICES PRACTICES ? PRACTICES WEB 201

KNOW

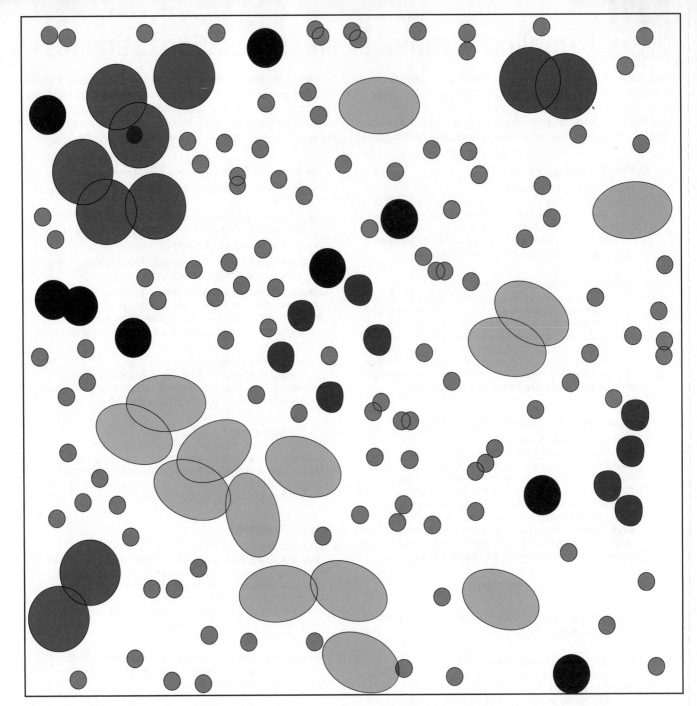

Table 3: Line transect

	Tally plant species	Total number	Relative abundance / % (plant X ÷ total plants x 100)
Plant 1			
Plant 2			
Plant 3			
Plant 4			
Plant 5			

Table of random numbers

A	B	C	D
2 2	3 1	6 2	2 2
3 2	1 5	6 3	4 3
3 1	5 6	3 6	6 4
4 6	3 6	1 3	4 5
4 3	4 2	4 5	3 5
5 6	1 4	3 1	1 4

The table above has been adapted from a table of random numbers from a statistics book. Use this table to select quadrats randomly from the grid above. Choose one of the columns (A to D) and use the numbers in that column as an index to the grid. The first digit refers to the row number and the second digit refers to the column number. To locate each of the 6 quadrats, find where the row and column intersect, as shown below:

Example: 5 2 refers to the 5th row and the 2nd column

©2017 **BIOZONE** International
ISBN: 978-1-927309-65-0
Photocopying Prohibited

1. (a) Use the combined data TOTALS (Table 1) for the sampled quadrats to estimate the average density for each species by using the formula:

$$\text{Density} = \frac{\text{Total number in all quadrats sampled}}{\text{Number of quadrats sampled} \ \times \ \text{area of a quadrat}}$$

Remember that a total of 6 quadrats are sampled and each has an area of 0.0009 m². The density should be expressed as the number of individuals per square metre (no. m⁻²).

Plant 1: _____ Plant 2: _____ Plant 3: _____ Plant 4: _____ Plant 5: _____

(b) Use the direct count data from your quadrats (Table 1) to calculate a relative abundance of the plants (abundance as a percentage of the community). Use the formula:

$$\text{Relative abundance} = \frac{\text{Number of plant } X \text{ in all quadrats sampled}}{\text{Total number of all plants counted}} \times 100$$

Plant 1: _____ Plant 2: _____ Plant 3: _____ Plant 4: _____ Plant 5: _____

2. Use the line transect data (Table 3) to calculate relative abundance as in question 1 (b). Transfer your result here:

Plant 1: _____ Plant 2: _____ Plant 3: _____ Plant 4: _____ Plant 5: _____

3. (a) Carry out a direct count of all 5 plant species for the whole sample area (all 36 quadrats). Apply the data from your direct count to the equation given in (1) above to calculate the actual population density (remember that the number of quadrats in this case = 36):

Plant 1: _____ Plant 2: _____ Plant 3: _____ Plant 4: _____ Plant 5: _____

(b) Compare your estimated population density to the actual population density for each species:

_____

_____

_____

(c) Now calculate the actual relative abundance of the plants as in question 1(b):

Plant 1: _____ Plant 2: _____ Plant 3: _____ Plant 4: _____ Plant 5: _____

(d) Compare your estimated population abundance calculations to the actual population abundance for each species:

_____

_____

_____

4. Comment on how the picture of the community provided by the different sampling methods compare. Compare your results with people in your class who have used different coordinates for their quadrats. What picture of the community was provided by their results?

_____

_____

_____

_____

_____

_____

_____

_____

_____

202 Using the Chi-Squared Test in Ecology

Key Idea: The chi-squared test for goodness of fit is used to compare sets of categorical data and evaluate if differences between them are statistically significant or due to chance. The **chi-squared test** (χ^2) is used when you are working with frequencies (counts) rather than measurements. It is a simple test to perform but the data must meet the requirements of the test. Firstly, it can only be used for data that are raw counts (not measurements or transformed data such as percentages).

Secondly, it is used to compare an experimental result with an expected theoretical outcome (e.g. an expected Mendelian ratio or a theoretical value indicating "no difference" between groups in a response such as habitat preference). Thirdly, it is not a valid test when sample sizes are small (<20). Like all statistical tests, it aims to test the null hypothesis. The following exercise is a worked example using the chi-squared test for goodness of fit to test habitat preference.

Using χ^2 in ecology

Pneumatophores

In an investigation of the ecological niche of the mangrove, *Avicennia marina var. resinifera*, the density of pneumatophores was measured in regions with different substrate. The mangrove trees were selected from four different areas: mostly sand, some sand, mostly mud, and some mud. Note that the variable, substrate type, is categorical in this case. Quadrats (1 m by 1 m) were placed around a large number of trees in each of these four areas and the numbers of pneumatophores were counted. Chi-squared was used to compare the observed results for pneumatophore density (as follows) to an expected outcome of no difference in density between substrates.

Mangrove pneumatophore density in different substrate areas			
Mostly sand	85	Mostly mud	130
Some sand	102	Some mud	123

Using χ^2, the probability of this result being consistent with the expected result could be tested. Worked example as follows:

Step 1: Calculate the expected value (E)
In this case, this is the sum of the observed values divided by the number of categories.

$$\frac{440}{4} = 110$$

Step 2: Calculate O – E
The difference between the observed and expected values is calculated as a measure of the deviation from a predicted result. Since some deviations are negative, they are all squared to give positive values. This step is usually performed as part of a tabulation (right, darker blue column).

Category	O	E	O - E	(O - E)²	$\frac{(O - E)^2}{E}$
Mostly sand	85	110	-25	625	5.68
Some sand	102	110	-8	64	0.58
Mostly mud	130	110	20	400	3.64
Some mud	123	110	13	169	1.54

Total = 440 $\chi^2 \longrightarrow$ $\Sigma = 11.44$

Step 3: Calculate the value of χ^2

$$\chi^2 = \sum \frac{(O - E)^2}{E}$$

Where: O = the observed result
E = the expected result
Σ = sum of

The calculated χ^2 value is given at the bottom right of the last column in the tabulation.

Step 5a: Using the χ^2 table
On the χ^2 table (part reproduced in Table 1 below) with 3 degrees of freedom, the calculated value for χ^2 of 11.44 corresponds to a probability of between 0.01 and 0.001 (see arrow). *This means that by chance alone a χ^2 value of 11.44 could be expected between 1% and 0.1% of the time.*

Step 4: Calculating degrees of freedom
The probability that any particular χ^2 value could be exceeded by chance depends on the number of degrees of freedom.
This is simply **one less than the total number of categories** (the number that could vary independently without affecting the last value). *In this case: 4–1 = 3.*

Step 5b: Using the χ^2 table
The probability of between 0.1 and 0.01 is lower than the 0.05 value (generally regarded as significant). The null hypothesis can be rejected and we can state that the observed results differ significantly from the expected (at P = 0.05).

Table 1: Critical values of χ^2 at different levels of probability. By convention, the critical probability for rejecting the null hypothesis (H_0) is 5%. If the test statistic is less than the tabulated critical value for P = 0.05 we cannot reject H_0 and the result is not significant. If the test statistic is greater than the critical value for P = 0.05 we reject H_0 in favor of the alternative hypothesis.

Degrees of freedom	Level of probability (P)									
	0.98	0.95	0.80	0.50	0.20	0.10	0.05	0.02	0.01	0.001
1	0.001	0.004	0.064	0.455	1.64	2.71	3.84	5.41	6.64	χ^2 10.83
2	0.040	0.103	0.466	1.386	3.22	4.61	5.99	7.82	9.21	13.82
3	0.185	0.352	1.005	2.366	4.64	6.25	7.82	9.84	11.35	16.27
4	0.429	0.711	1.649	3.357	5.99	7.78	9.49	11.67	13.28	18.47
5	0.752	0.145	2.343	4.351	7.29	9.24	11.07	13.39	15.09	20.52

$\longleftarrow$ **Do not reject H_0**   **Reject H_0** $\longrightarrow$

©2017 **BIOZONE** International
ISBN: 978-1-927309-65-0
Photocopying Prohibited

203 Investigating Distribution and Abundance

Key Idea: Sampling populations *in-situ* can reveal patterns of distribution, which can be attributed to habitat preference. These investigations are common in ecological studies.

Use this activity to practice analyzing data from a field study in which the aim was to identify and describe an existing pattern of species distribution.

The aim

To investigate the effect of fallen tree logs on millipede distribution in a forest.

Background

Millipedes consume decaying vegetation, and live in the moist conditions beneath logs and in the leaf litter of forest floors. The moist environment protects them from drying out as their cuticle is not a barrier to water loss.

Experimental method

The distribution of millipede populations in relation to fallen tree logs was investigated in a small forest reserve. Six logs of similar size were chosen from similar but separate regions of the forest. Logs with the same or similar surrounding environment (e.g. leaf litter depth, moisture levels) were selected.

For each log, eight samples of leaf litter at varying distances from the fallen tree log were taken using 30 cm² quadrats. Samples were taken from two transects, one each side of the log. The sample distances were: directly below the log (0 m), 1.5 m, 2.5 m, and 3.5 m from the log. It was assumed that the conditions on each side of the log would be essentially the same. The leaf litter was placed in Tullgren funnels and the invertebrates extracted. The number of millipedes in each sample was counted. The raw data are shown below.

Giant millipede, *Narceus americanus*

Jud McCranie CC 3.0

Experimental Setup

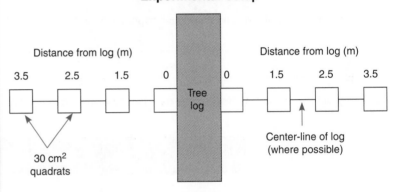

Distance from log (m): 3.5 2.5 1.5 0 | Tree log | 0 1.5 2.5 3.5

30 cm² quadrats

Center-line of log (where possible)

Environmental conditions on either side of the log were assumed to be equal.

Raw data for tree log and millipede investigation

Distance from log (m)

Tree log	Transect	0	1.5	2.5	3.5
1	1	12	11	3	2
	2	10	12	2	1
2	1	8	3	4	4
	2	9	5	2	1
3	1	14	6	3	3
	2	3	8	7	2

Distance from log (m)

Tree log	Transect	0	1.5	2.5	3.5
4	1	2	4	1	6
	2	4	5	2	2
5	1	12	10	16	10
	2	6	3	2	5
6	1	10	9	7	2
	2	11	11	8	1

PRACTICES   CONNECT

20 AP1

DATA

1. Complete the table below using the raw data on the previous page. Combine the values for each transect:

Tree log	Distance from log (m)			
TOTAL				

2. Explain why a chi-squared test is an appropriate statistical analysis for this data: _____

_____

_____

_____

3. State the null hypothesis and alternative hypothesis for the statistical test: _____

_____

_____

4. Carry out the chi-squared test on the data by completing the table below. Combine the values of all the tree logs.

Distance (m)	O	E	O-E	$(O-E)^2$	$\dfrac{(O-E)^2}{E}$
				$\Sigma \dfrac{(O-E)^2}{E}$	

5. Use the critical values table on page 298 to decide if the null hypothesis should be rejected or not rejected:

_____

_____

6. Discuss your findings in relation to how millipedes live. Include the validity of the findings and any biological ideas relevant to the findings in your discussion:

_____

_____

_____

_____

_____

_____

_____

_____

_____

©2017 **BIOZONE** International
ISBN: 978-1-927309-65-0
Photocopying Prohibited

204 Population Age Structure

Key Idea: Population age structure can be represented as a pyramid with the youngest individuals forming the base.

The **age structure** of a population refers to the relative proportion of individuals in each age group in the population. The age structure of populations can be categorized according to specific age groups (e.g. years), or by other measures such as life stage (egg, larvae, pupae) or size class (height or diameter in plants). Population growth is strongly influenced by age structure. Population age structures are represented as pyramids, in which the proportions of individuals in each age/size class are plotted with the youngest individuals at the pyramid's base. The number of individuals moving from one age class to the next influences the age structure of the population from year to year. The loss of an age class (e.g. through over-harvesting) can profoundly affect a population's ability to sustain itself.

Age structures in animal populations

Population growth is strongly influenced by age structure. A population with a high proportion of reproductive and pre-reproductive aged individuals has a much greater potential for population growth than one that is dominated by older individuals. These theoretical age pyramids show how growing populations are characterized by a high ratio of young (white bar) to adult age classes (blue bars). Aging populations with poor production are typically dominated by older individuals.

4 / 8 / 12 / 76	76 young : 24 adults **Rapidly growing population**

Example: Virginia opposum

4 / 8 / 24 / 64	64 young : 36 adults **Normal growth**

Example: White tailed deer

4 / 8 / 12 / 24 / 48	48 young : 52 adults **Poor production (aging)**

Example: Serval (locally at risk)

4 / 6 / 12 / 16 / 16 / 20 / 24	24 young : 76 adults **Very poor production**

Example: Kakapo (endangered ground parrot)

Age structures in human populations

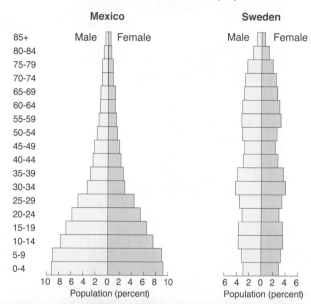

Mexico   **Sweden**

Age classes: 85+, 80-84, 75-79, 70-74, 65-69, 60-64, 55-59, 50-54, 45-49, 40-44, 35-39, 30-34, 25-29, 20-24, 15-19, 10-14, 5-9, 0-4

Male | Female

Population (percent)

Extended family: Nigeria

Most of the growth in human populations in recent years has occurred in the developing countries in Africa, Asia, and Central and South America. This is reflected in their age structure; a large proportion of the population comprises individuals younger than 15 years (age pyramid above, left). Even if each has fewer children, the population will continue to increase for many years. The stable age structure of Sweden is shown for comparison

---

1. For the theoretical age pyramids above left:

   (a) Determine the approximate ratio of young to adults in a rapidly increasing population: _____

   (b) Why are changes in population age structure alone not necessarily a reliable predictor of population trends?

   _____

   _____

2. Why is the population of Mexico likely to continue to increase rapidly even if the rate of population growth slows?

   _____

   _____

Managed fisheries

Analysis of a population's age structure can help in its management because it can indicate where most of the mortality occurs and whether or not reproductive individuals are being replaced. The age structure of plant and animal populations can be assessed through an analysis of size which is often related to age in a predictable way. The graphs below show the age structure of a hypothetical fish population under different fishing pressures. The age structure of the population is determined by analyzing the fish catch to determine the frequency of fish in each size (age) class.

Heavy fishing

Percentage of catch vs Age (years)

Moderate fishing

Percentage of catch vs Age (years)

Light fishing

Percentage of catch vs Age (years)

Thatch palm populations on Lord Howe Island

Lord Howe Island is a narrow sliver of land 770 km northeast of Sydney, Australia. The age structure of populations of the thatch palm *Howea forsteriana* was determined at three locations on the island: the golf course, Gray Face, and Far Flats. The height of the stem was used as an indication of age. The differences in age structure between the three sites are due mainly to the extent of grazing at each site.

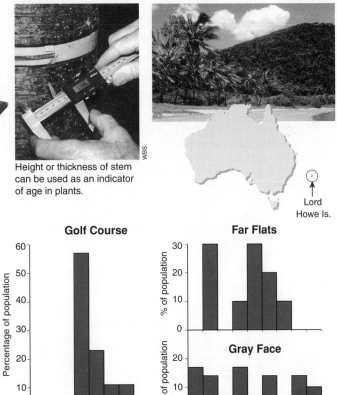

Height or thickness of stem can be used as an indicator of age in plants.

Lord Howe Is.

Golf Course

Percentage of population vs Stem height (m)

Far Flats

% of population vs Stem height (m)

Gray Face

% of population vs Stem height (m)

3. For the managed fish populations above left:

   (a) What general factor changes the population age structure? _____

   (b) How does the age structure change when the fishing pressure increases from light to heavy? _____

   _____

4. Determine the most common age class for each of the above fish populations under the following fishing pressures:

   (a) Heavy: _____ (b) Moderate: _____ (c) Light: _____

5. Determine which of the three sites sampled on Lord Howe Island (above, right), best reflects the age structure of:

   (a) An ungrazed population: _____

   Reason for your answer: _____

   (b) A heavily grazed and mown population: _____

   Reason for your answer: _____

6. What are the likely long term prospects for the population at the golf course? _____

   _____

7. Describe a potential problem with using size to estimate age: _____

   _____

8. Why is it important to know the age structure of a harvested population? _____

   _____

©2017 **BIOZONE** International
ISBN: 978-1-927309-65-0
Photocopying Prohibited

205 Life Tables and Survivorship

Key Idea: Life tables summarize the mortality for a population. They can provide information about life span and age structure.

Life tables, such as those shown below, provide a summary of mortality for a population (usually for a group of individuals of the same age or **cohort**). The basic data are just the number of individuals remaining alive at successive sampling times (the **survivorship** or l_x). Life tables are an important tool when analyzing changes in populations over time. They can tell us the ages at which most mortality occurs in a population and can also provide information about life span and population age structure. Biologists use the l_x column of a basic life table to derive a survivorship curve. Survivorship curves are standardized as the number of survivors per 1000 (or 1) individuals so that populations of different types can be easily compared.

Life table and survivorship curve for a population of the barnacle *Balanus*

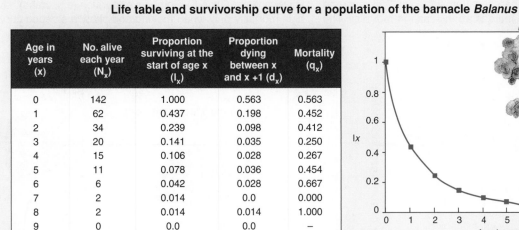

Age in years (x)	No. alive each year (N_x)	Proportion surviving at the start of age x (l_x)	Proportion dying between x and x +1 (d_x)	Mortality (q_x)
0	142	1.000	0.563	0.563
1	62	0.437	0.198	0.452
2	34	0.239	0.098	0.412
3	20	0.141	0.035	0.250
4	15	0.106	0.028	0.267
5	11	0.078	0.036	0.454
6	6	0.042	0.028	0.667
7	2	0.014	0.0	0.000
8	2	0.014	0.014	1.000
9	0	0.0	0.0	–

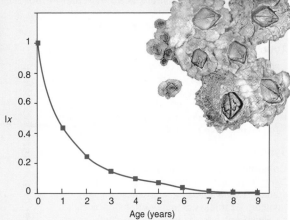

Life table for female elk, Northern Yellowstone

x	l_x	d_x	q_x
0	1000	323	0.323
1	677	13	0.019
2	664	2	0.003
3	662	2	0.003
4	660	4	0.006
5	656	4	0.006
6	652	9	0.014
7	643	3	0.005
8	640	3	0.005
9	637	9	0.014
10	628	7	0.001
11	621	12	0.019
12	609	13	0.021
13	596	41	0.069
14	555	34	0.061
15	521	20	0.038
16	501	59	0.118
17	442	75	0.170
18	367	93	0.253
19	274	82	0.299
20	192	57	0.297
21+	135	135	1.000

Survivorship curve for female elk of Northern Yellowstone National Park

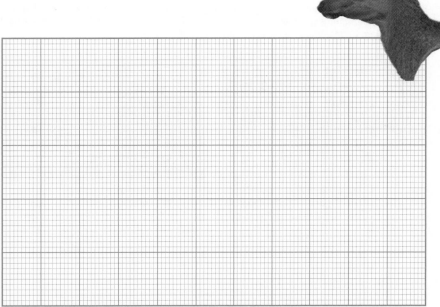

1. (a) In the example of the barnacle *Balanus* above, state when most of the population dies: _____

   (b) Identify the type of survivorship curve is represented by these data (see next activity): _____

2. (a) Using the grid, plot a survivorship curve for elk hinds (above) based on the life table data provided:

   (b) Describe the survivorship curve for these large mammals: _____

   _____

   _____

PRACTICES   PRACTICES   PRACTICES   WEB

     205

DATA

206 Survivorship Curves

Key Idea: Survivorship curves illustrate change in mortality with age. The shape reflects an organism's life history strategy. The survivorship curve depicts age-specific mortality. It is obtained by plotting the number of individuals of a particular cohort against time. Survivorship curves are standardized to start at 1000 and, as the population ages, the number of survivors progressively declines. The shape of a survivorship curve thus shows graphically at which life stages the highest mortality occurs. Survivorship curves in many populations fall into one of three hypothetical patterns (below). Wherever the curve becomes steep, there is an increase in mortality. The convex Type I curve is typical of populations whose individuals tend to live out their physiological life span. Such populations usually produce fewer young and show some degree of parental care. Organisms that suffer high losses of the early life stages (a Type III curve) compensate by producing vast numbers of offspring. These curves are conceptual models only, against which real life curves can be compared. Many species exhibit a mix of two of the three basic types. Some birds have a high chick mortality (Type III) but adult mortality is fairly constant (Type II). Some invertebrates (e.g. crabs) have high mortality only when molting and show a stepped curve.

Hypothetical survivorship curves

Type I

Late loss survivorship curve
Mortality (death rate) is very low in the infant and juvenile years, and throughout most of adult life. Mortality increases rapidly in old age. **Examples**: Humans (in developed countries) and many other large mammals (e.g. big cats, elephants).

Type II

Constant loss survivorship curve
Mortality is relatively constant through all life stages (no one age is more susceptible than another). **Examples**: Some invertebrates such as *Hydra*, some birds, some annual plants, some lizards, and many rodents.

Type III

Early loss survivorship curve
Mortality is very high during early life stages, followed by a low death rate for the few individuals reaching adulthood. **Examples**: Many fish (not mouth brooders), frogs, most plants, and most marine invertebrates (e.g. oysters, barnacles).

Graph of age specific survival

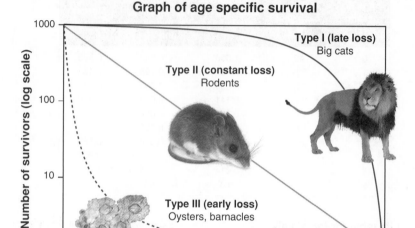

Three basic types of survivorship curves and representative organisms for each. The vertical axis may be scaled arithmetically or logarithmically.

Elephants have a close matriarchal society and a long period of parental care. Elephants are long-lived and females usually produce just one calf.

Rodents are well known for their large litters and prolific breeding capacity. Individuals are lost from the population at a more or less constant rate.

Despite vigilant parental care, many birds suffer high juvenile losses (Type III). For those surviving to adulthood, deaths occur at a constant rate.

1. Explain why human populations might not necessarily show a Type I curve: _____

   _____

2. Explain how organisms with a Type III survivorship compensate for the high mortality during early life stages:

   _____

3. Describe the features of a species with a Type I survivorship that aid in high juvenile survival:

   _____

4. Discuss the following statement: "There is no standard survivorship curve for a given species; the curve depicts the nature of a population at a particular time and place and under certain environmental conditions.":

   _____

   _____

   _____

WEB
206

CONNECT
32
AP2

PRACTICES

DATA

©2017 **BIOZONE** International
ISBN: 978-1-927309-65-0
Photocopying Prohibited

207 Factors Determining Population Growth

Key Idea: Population size increases through births and immigration and decreases through deaths and emigration. Populations are dynamic and the number of individuals in a population may fluctuate considerably over time. Populations gain individuals through births or immigration, and lose individuals through deaths and emigration. For a population in equilibrium, these factors balance out and there is no net change in the population abundance. When losses exceed gains, the population declines. When gains exceed losses, the population increases.

Births, deaths, immigrations (movements into the population) and emigrations (movements out of the population) are events that determine the numbers of individuals in a population. Population growth depends on the number of individuals added to the population from births and immigration, minus the number lost through deaths and emigration. This is expressed as:

Population growth =

Births – Deaths + Immigration – Emigration
(B)   (D)   (I)   (E)

The difference between immigration and emigration gives net migration. Ecologists usually measure the **rate** of these events. These rates are influenced by environmental factors and by the characteristics of the organisms themselves. Rates in population studies are commonly expressed in one of two ways:

- Numbers per unit time, e.g. 20,150 live births per year.
- Per capita rate (number per head of population), e.g. 122 live births per 1000 individuals per year (12.2%).

Limiting factors

Population size is also affected by limiting factors, i.e. factors or resources that control a process such as organism growth, or population growth or distribution. Examples include availability of food, predation pressure, or available habitat.

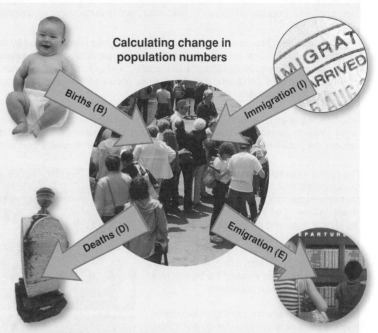

Calculating change in population numbers

Births (B) · Immigration (I) · Deaths (D) · Emigration (E)

Human populations often appear exempt from limiting factors as technology and efficiency solve many food and shelter problems. However, as the last arable land is used and agriculture reaches its limits of efficiency, it is estimated that the human population may peak at around 10 billion by 2050.

---

1. Define the following terms used to describe changes in population numbers:

   (a) Death rate (mortality): _____

   (b) Birth rate (natality): _____

   (c) Net migration rate: _____

2. Explain how the concept of limiting factors applies to population biology: _____

   _____

3. Using the terms, B, D, I, and E (above), construct equations to express the following (the first is completed for you):

   (a) A population in equilibrium: $B + I = D + E$

   (b) A declining population: _____

   (c) An increasing population: _____

4. A population started with a total number of 100 individuals. Over the following year, population data were collected. Calculate birth rates, death rates, net migration rate, and rate of population change for the data below (as percentages):

   (a) Births = 14: Birth rate = _____ (b) Net migration = +2: Net migration rate = _____

   (c) Deaths = 20: Death rate = _____ (d) Rate of population change = _____

   (e) State whether the population is increasing or declining: _____

5. The human population is now more than 7.5 billion. Describe two limiting factors for population growth in humans:

   _____

   _____

©2017 **BIOZONE** International
ISBN: 978-1-927309-65-0
Photocopying Prohibited

PRACTICES   WEB

207

DATA

208 Patterns of Population Growth

Key Idea: Populations typically show either exponential or logistic growth. The maximum sustainable population size is limited by the environment's carrying capacity.

Population growth is the change in a population's numbers over time (dN/dt or ΔN/Δt). It is regulated by the **carrying capacity (K)**, which is the maximum number the environment can sustain. Population growth falls into two main types:

exponential or logistic. Both can be defined mathematically. In these mathematical models, the per capita (or intrinsic) growth rate is denoted by a lower case *r*, determined by the per capita births minus deaths, i.e. (B-D)/N. **Exponential growth** occurs when resources are essentially unlimited. **Logistic growth** begins exponentially, but slows as the population approaches environmental carrying capacity.

Exponential growth occurs when the population growth rate is not affected by the population size, N. In this case, the population growth rate is simply r (the maximum per capita rate of increase) multiplied by N so that dN/dt = rN. On a graph, exponential growth is characterized by a J shaped curve. A lag phase occurs early in population growth due to low population numbers.

In nature, exponential growth is observed in two circumstances: (1) a few individuals begin a new population in a new habitat with plenty of resources, (2) a natural disaster reduces the population to a few survivors, and the population recovers from a low base.

The human population is currently in an exponential phase of growth. In ancient times, the human population remained relatively stable, but low. It was not until the end of the Middle Ages and the beginning of the Renaissance that the population began to grow. The Industrial Revolution increased living standards and population with it. Antibiotics and the Green Revolution sparked the current rapid increase in the human population.

Gray wolves were hunted almost to extinction in many parts of North America. By the 1930s they had mostly disappeared from the Northern Rockies. Repopulation began from packs that migrated from Canada in the 1980s. The table below shows the wolf population in Montana since 1979 (numbers are approximate). The wolves were taken off the endangered species list in 2011. Since then numbers have declined as hunting has once more been permitted.

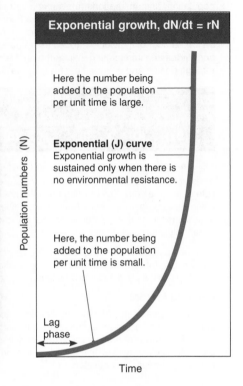

Exponential growth, dN/dt = rN

Here the number being added to the population per unit time is large.

Exponential (J) curve
Exponential growth is sustained only when there is no environmental resistance.

Here, the number being added to the population per unit time is small.

Lag phase

Population numbers (N)

Time

Gray wolf population in Montana			
Year	Population	Year	Population
1979	5	1997	52
1981	5	1999	75
1983	9	2001	125
1985	20	2003	190
1987	15	2005	252
1989	16	2007	420
1991	38	2009	525
1993	52	2011	651
1995	60	2013	620
		2014	552

1. Produce a line graph of the gray wolf population on the grid above: _____

2. Around which year(s) did explosive exponential growth begin in the wolf population? _____

3. Use the data to calculate the approximate doubling time of the wolf population: _____

4. Why did the human population only begin its rapid increase after the Renaissance? _____

_____

©2017 **BIOZONE** International
ISBN: 978-1-927309-65-0
Photocopying Prohibited

In nature, the population growth of most organisms follows a logistic growth curve. When entering a new environment a founding population will enter a period of exponential growth. The maximum size of population that the environment can support is called the carrying capacity (K). As the population nears K and the resources become limiting, population growth slows.

Under the logistic growth model, $dN/dt = rN$ is multiplied by the proportion of K that is left unfilled or unused. As the population increases, the proportion of K available decreases and individuals find it difficult to find or utilize space and resources. The rate of population increase therefore slows as population size approaches carrying capacity.

Occasionally a population's growth rate may not slow as it approaches K. This usually occurs in rapidly breeding organisms when there is a time lag between the depression in resources and the population response. In this case, the population overshoots K and then declines again as it responds to low resource availability. In time, populations usually stabilize around K.

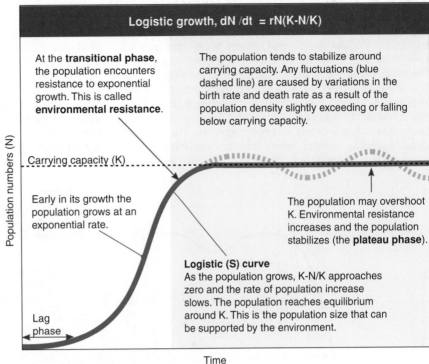

Logistic growth, $dN/dt = rN(K-N/K)$

At the **transitional phase**, the population encounters resistance to exponential growth. This is called **environmental resistance**.

The population tends to stabilize around carrying capacity. Any fluctuations (blue dashed line) are caused by variations in the birth rate and death rate as a result of population density slightly exceeding or falling below carrying capacity.

Carrying capacity (K)

Early in its growth the population grows at an exponential rate.

The population may overshoot K. Environmental resistance increases and the population stabilizes (the **plateau phase**).

Logistic (S) curve
As the population grows, K-N/K approaches zero and the rate of population increase slows. The population reaches equilibrium around K. This is the population size that can be supported by the environment.

Lag phase

Population numbers (N) — vertical axis

Time — horizontal axis

Logistic growth curve on a spreadsheet.

Plotting a logistic growth curve on a spreadsheet can help in understanding the effect of population size on the growth rate and how the logistic equation applies. For a hypothetical population of 2, r is 0.15 and K 100. The following formulae can be entered into the spreadsheet:

	A	B	C	D	E	F	G
1	r	t (period)	N	K	K-N/K	dN/dt	
2	0.15	0	2	100	=($D$2-C2)/$D$2	=$A$2*C2*E2	
3		=B2+1	=C2+F2				
4							
5							
6							
7							
8							

Population at t_1 = population at t_0 + dN/dt (the amount of population change over 1 time period)

The cells can then be filled down. The first three steps have been filled here. Fill the cells down to about 60 time periods, then plot t vs N.

	A	B	C	D	E	F	G
1	r	t (period)	N	K	K-N/K	dN/dt	
2	0.15	0	2.00	100	0.98	0.29	
3		1	2.29		0.98	0.34	
4		2	2.63		0.97	0.38	
5		3	3.01				
6							
7							

5. Why don't populations continue to increase exponentially in an environment? _____

_____

6. What is meant by **carrying capacity**? _____

_____

7. Describe and explain the phases of the logistic growth curve: _____

_____

_____

_____

_____

_____

8. (a) Around which time period does the curve on the spreadsheet above begin to flatten out? _____

(b) Describe how dN/dt changes over time: _____

_____

(c) What is the general shape of the logistic curve: _____

©2017 **BIOZONE** International
ISBN: 978-1-927309-65-0
Photocopying Prohibited

209 *r* and K Selection

Key Idea: *r*-selected species have high biotic potentials (*r*) and typically show exponential growth. K-selected species have lower biotic potentials and exist near carrying capacity.

The maximum rate at which a population can grow (its intrinsic rate of increase or *r*) is also called its biotic potential. It is a measure of reproductive capacity and is assigned a set value that is specific to the organism involved. Species with a high biotic potential are called *r*-selected species. They include algae, bacteria, rodents, many insects, and most annual plants. These species show life history features associated with rapid growth in disturbed environments and they usually exist well below the carrying capacity of the environment (K). Species with lower biotic potential tend to exist at or near K. These species, which include most large mammals, birds of prey, and large, long-lived plants are forced, through their interactions with other species, to use resources more efficiently. These species have fewer offspring and longer lives, and put their energy into nurturing their young to reproductive age. Most organisms have reproductive patterns between these two extremes.

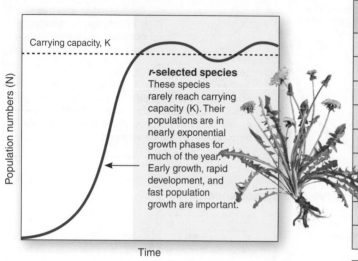

Population numbers (N) vs Time

Carrying capacity, K

r-selected species
These species rarely reach carrying capacity (K). Their populations are in nearly exponential growth phases for much of the year. Early growth, rapid development, and fast population growth are important.

Correlates of *r*-selected species	
Climate	Variable and/or unpredictable
Mortality	Density independent
Survivorship	Often Type III (early loss)
Population size	Fluctuates wildly. Often below K
Competition	Variable, often lax. Generalist niche.
Selection favors	Rapid development, high biotic potential, early reproduction, small body size, single reproduction (annual).
Length of life	Short (usually less than one year)
Leads to:	Productivity

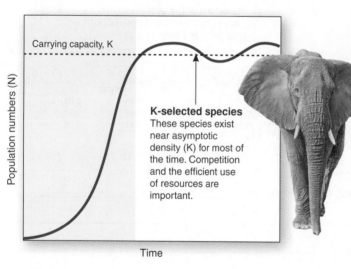

Population numbers (N) vs Time

Carrying capacity, K

K-selected species
These species exist near asymptotic density (K) for most of the time. Competition and the efficient use of resources are important.

Correlates of K-selected species	
Climate	Fairly constant and/or predictable
Mortality	Density dependent
Survivorship	Types I or II (late or constant loss)
Population size	Fairly constant in time. Near equilibrium with the environment.
Competition	Usually keen. Specialist niche.
Selection favors	Slower development, larger body size, greater competitive ability, delayed reproduction, repeated reproduction.
Length of life	Longer (greater than one year)
Leads to:	Efficiency

1. Explain why *r*-selected species tend to predominate in unstable, disturbed, or early successional communities:

_____

_____

2. Explain why many K-selected species tend to predominate in stable, climax communities: _____

_____

_____

3. Describe factors that might cause a change in the predominance of K-selected species in a climax community:

_____

_____

_____

©2017 **BIOZONE** International
ISBN: 978-1-927309-65-0
Photocopying Prohibited

210 Modeling Population Growth

Key Idea: Computer programs can be used to model population growth. This activity uses Populus 5.5.

Population growth can be simulated using spreadsheets or computer programs. This activity uses Populus 5.5, a Javascript program, which will run on Mac or Windows platforms. It models continuous and discrete population growth as well as the effects of competition. In this activity you will model continuous density-independent (exponential)

and density-dependent (logistic) growth. Using Populus, you can also model discrete growth, which uses λ instead of *r*, where λ is the discrete-time per capita growth rate. Discrete models are used for organisms with a discrete breeding season (e.g. annual plants and insects that breed once a year) because population growth occurs in 'steps' only within a discrete time period (not continuously) and there is no population growth outside those times.

Populus is shareware. Download it free from
https://cbs.umn.edu/populus/overview

(you can also download via Weblinks)

The opening screen looks like this.
- ▶ **Model** allows you to choose which type of simulation you want to run.
- ▶ **Preferences** lets you to load saved files and save new ones.
- ▶ **Help** loads a comprehensive PDF file covering all aspects of the program.

If it fills the entire screen grab the lower corner and resize it with the mouse.

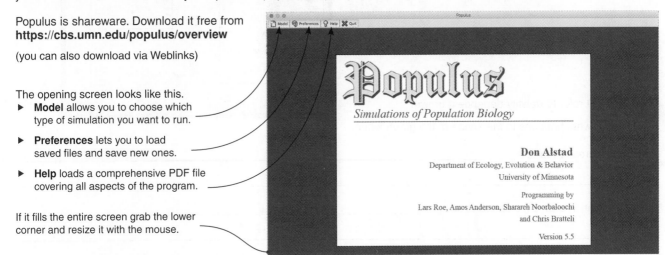

Density independent growth
- ▶ Click on the Model in the menu bar
- ▶ Select Single-Species Dynamics
- ▶ Then choose **Density-Independent** Growth

Set the model type to continuous (as in continuous growth). This produces a single line in the output window. Discrete produces a series of points (as in discrete bursts of growth).

Set plot type to N vs t. This models population vs time.

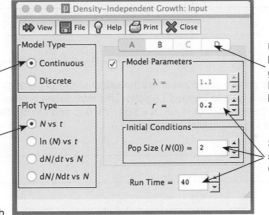

Up to four populations can be displayed on the one graph, using A, B, C, and D. Make sure the check box is ticked.

Set r to 0.2 and population size N to 2. Set run time 40. Click View to see the graph.

Questions 1-4 refer to density independent growth

1. What is the shape of the graph produced? _____

2. Describe what happens to the shape of the graph when:

   (a) *r* is increased to 0.4: _____

   (b) Population size is increased to 20: _____

   (c) Population size is increased to 20 but *r* is reduced to -0.2: _____

3. Set the parameters back to N = 2 and r = 0.2. Set the plot type to dN/dt vs N and view the plot. Describe the shape of the graph and explain what it means: _____

4. What is the value of *r* if the population doubles over one time period? _____

SAVE AND PRINT ALL YOUR SIMULATIONS AND ATTACH THEM TO THIS PAGE

PRACTICES PRACTICES WEB 210

KNOW

Density dependent growth

- ▶ Click on the Model in the menu bar
- ▶ Then select Single-Species Dynamics
- ▶ Then select **Density-Dependent Growth**

- ▶ As before set the model type to continuous.
- ▶ Produce a plot for N = 5, K = 500, r = 0.2, and t to 50.

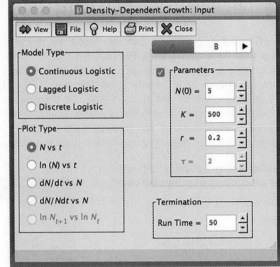

Questions 5-9 refer to density dependent growth

5. Describe what happens to the shape of the graph when:

(a) r is increased to 0.4: _____

_____

(b) Population size is increased to 50: _____

_____

(c) Reset the parameters and plot a graph of dN/dt vs N. Describe the shape of this graph and explain what it means:

_____

_____

_____

6. The standard logistic growth curve assumes the effect of the population size immediately affects the population growth rate. Now set the graph type to Lagged Logistic. This introduces a time lag between the population size and its effect on growth rate. Set the parameters to N = 5, K = 500, r = 0.2, and t to 50. Set the time lag T to 4 and view the graph. What is the effect of the time lag on population growth?

_____

_____

_____

7. (a) Now set r to 0.5 and t to 150. Describe the shape of the graph: _____

_____

(b) What kind of species (r-selected or K-selected) would show this type of growth? _____

8. (a) Keep T at 4 and set r to 0.2 view the graph. Describe the shape of the graph now: _____

_____

(b) What kind of species (r-selected or K-selected) would show this type of growth? _____

9. Keeping r at 0.2, vary T between 1 and 10. How does increasing the lag affect how the population oscillates around K?

_____

_____

_____

_____

SAVE AND PRINT ALL YOUR SIMULATIONS AND ATTACH THEM TO THIS PAGE

©2017 **BIOZONE** International
ISBN: 978-1-927309-65-0
Photocopying Prohibited

211 Human Demography

Key Idea: As human populations move from pre-industrial to industrial societies the structure of the population and birth and death rates also change.

Human populations through time have undergone demographic shifts related to societal changes and economic development. The demographic transition model (DTM) was developed in 1929 to explain the transformation of countries from high birth rates and high death rates to low birth rates and low death rates as part of their economic development from a pre-industrial to an industrialized economy. The transition involves four stages, or possibly five (with some nations recognized as moving beyond stage four). Each stage of the transition reflects the changes in birth and death rates observed in human societies over the last 200 years. Most developed countries are beyond stage three of the model; the majority of developing countries are in stage two or stage three. The model was based on the changes seen in Europe, so these countries follow the DTM relatively well. Many developing countries have moved into stage three. The exceptions include some poor countries, mainly in sub-Saharan Africa and some Middle Eastern countries, which are poor or affected by government policy or civil strife.

Stage one: Birth and death rates balanced but high as a result of starvation and disease.

Stage two: Improvement in food supplies and public health result in reduced death rates.

Stage three moves the population towards stability through a decline in the birth rate.

Stage four: Birth and death rates are both low and the total population is high and stable. New Zealand (above) is entering stage 5.

The demographic transition model

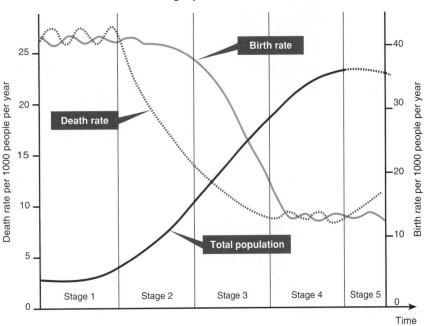

Stage 1 | Stage 2 | Stage 3 | Stage 4 | Stage 5

Stage one (high stationary): A balance between birth and death rates as was true of all populations until the late 18th Century. Children are important contributors to the household economy. Losses as a result of starvation and disease are high. Population growth is typically very slow in this stage, because it is constrained by the available food supply. Unless the society develops new technologies to increase food production, any fluctuations in birth rates are quickly matched by death rates.

Stage two (early expanding): Rapid population expansion as a result of high birth rates and a decline in death rates. The changes leading to this stage in Europe were initiated in the Agricultural Revolution of the 18th century but have been more rapid in developing countries since then. Stage two is associated with more reliable food supplies and improvements in public health.

Stage three (late expanding): The population moves towards stability through a decline in the birth rate. This stage is associated with increasing urbanization and a decreased reliance on children as a source of family wealth. Family planning in nations such as Malaysia (left) has been instrumental in their move to stage three.

Stage four (low stationary): Birth and death rates are both low and the total population is high and stable. The population ages and in some cases the fertility rate falls below replacement.

Stage five (declining): Proposed by some theorists as representing countries that have undergone the economic transition from manufacturing based industries into service and information based industries and the population reproduces well below replacement levels. Countries in stage five include the United Kingdom (the earliest nation recognized as reaching Stage Five), Germany, and Japan. Stage 5 models are largely theoretical because population outcomes can be altered by immigration.

©2017 **BIOZONE** International
ISBN: 978-1-927309-65-0
Photocopying Prohibited

PRACTICES CONNECT **254** AP2 WEB **211**

KNOW

1. Each of the first four stages of the DTM is associated with a particular age structure. The diagrams below show structures for three of the five stages. They are not in order.

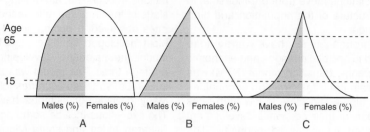

Age 65

15

Males (%) Females (%)   Males (%) Females (%)   Males (%) Females (%)

        A   B   C

(a) Identify which of the diagrams corresponds to stage one of the DTM and explain your choice:

_____

_____

_____

(b) Identify which of the diagrams corresponds to stage three of the DTM and explain your choice:

_____

_____

_____

2. Suggest why it might become less important to have a large number of children in more economically developed nations:

_____

_____

_____

_____

3. Explain why birth and death rates are high in a stage one society: _____

_____

_____

4. Among births there is a slightly higher male birth rate that female birth rate (a ratio of about 1.1: 1). What effect does this have on the replacement birth rate (i.e. the amount of births per female required to maintain a stable population).

_____

_____

_____

5. In many countries, the government pays a benefit (superannuation) to retired citizens. In others, there is no government help for people who are elderly and cannot work, or are retired. Explain why both of these systems require a large population of younger workers and cannot be supported by declining populations where young people are the minority.

_____

_____

_____

6. It might be expected that wealthy countries would have high birth rates because wealthy families can support more children. However this is not the pattern seen in most developed countries. Suggest some reasons why this might be so:

_____

_____

_____

_____

212 Modeling Human Survivorship

Key Idea: Changes in human survivorship can be studied using historical data of births and deaths.
Cemeteries are an excellent place to study changes in human demographics. Data collected from headstones can be used to calculate death rates and produce survivorship curves. It is also possible to compare survivorship curves over different periods and see how certain factors (e.g. war, medical advances) have altered survivorship.

Death data males and females

The data (right) represents age of death data for males and females collected over two different time periods; pre-1950 and post 1950. The pre-1950s was characterized by two world wars, and the prevalence of diseases such as polio and tuberculosis. The post 1950s have also seen global conflict, but to a lesser degree than the pre-1950 period. Many advances in medicine (e.g. vaccines) and technology have been made during this time.

The data used in this exercise has been collected from the online records of several cemeteries across five different states in the United States to provide representative data.

Pre-1950

Males age at death			Females age at death		
81	89	71	9	43	1
40	31	27	76	64	84
54	10	64	0	67	68
70	42	0	78	42	58
75	1	41	69	39	19
64	5	77	6	4	24
45	0	21	46	18	62
22	24	1	60	71	52
71	70	75	84	2	29
62	39	50	75	63	8

Post 1950

Males age at death			Females age at death		
80	60	64	92	87	87
81	71	41	46	76	82
79	62	76	44	63	80
81	63	17	70	33	90
8	83	40	80	99	85
30	31	79	71	76	63
88	78	74	88	92	58
90	56	46	65	96	89
84	86	71	51	65	56
64	80	90	80	54	86

Data source: http://www.interment.net/us/index.htm

1. Complete the following table using the cemetery data provided. The males Pre-1950 data have been completed for you.

(a) In the number of deaths column, record the number of deaths for each age category.

(b) **Calculate the survivorship** for each age category. For each column, enter the total number of individuals in the study (30) in the 0-9 age survivorship cell. This is the survivorship for the 0-9 age group. Subtract the number of deaths at age 0-9 from the survivorship value at age 0-9. This is the survivorship at the 10-19 age category. To calculate the survivorship for age 20-29, subtract the number of deaths at the age 10-19 age category from the survivorship value for age 10-19. Continue until you have completed the column.

Age	Males pre-1950 No. of deaths	Survivorship	Females pre-1950 No. of deaths	Survivorship	Males post 1950 No. of deaths	Survivorship	Females post 1950 No. of deaths	Survivorship
0-9	5	30						
10-19	1	25						
20-29	4	24						
30-39	2	20						
40-49	4	18						
50-59	2	14						
60-69	3	12						
70-79	7	9						
80-89	2	2						
90-99	0	0						
Total	30							

2. (a) On a separate piece of graph paper, construct a graph to compare the survivorship curves for each category. Staple the graph into this workbook once you have completed the activity.

(b) What conclusions can you make about survivorship before 1950 and after 1950? _____

_____

_____

(c) What factors might cause these differences? _____

_____

PRACTICES   PRACTICES   PRACTICES

DATA

213 World Population Growth

Key Idea: The human population has grown substantially since the 1950s, although growth rates are slowly declining. For most of human history, humans have not been very numerous compared to other species. It took all of human history to reach a population of 1 billion in 1804, but little more than 150 years to reach 3 billion in 1960. The world's population, now at 7.6 billion, is growing at the rate of about 80 million per year. World population increase carries important environmental consequences, particularly when it is associated with increasing urbanization. The global average fertility rate is 2.36 and has been reducing since World War II. Continued declines may give human populations time to address some to the major problems posed by the increasing scope and intensity of human activities.

World population density

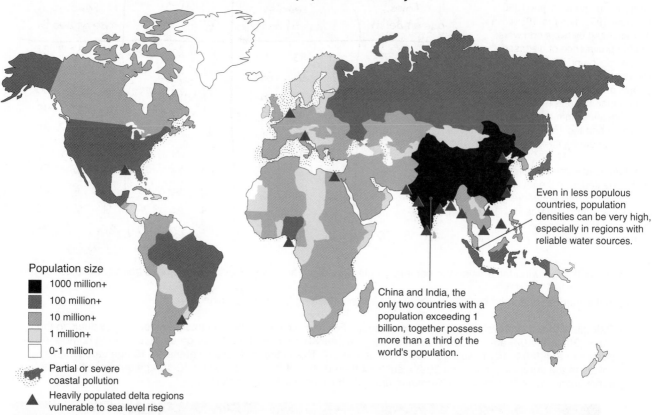

Population size
- 1000 million+
- 100 million+
- 10 million+
- 1 million+
- 0-1 million

Partial or severe coastal pollution

Heavily populated delta regions vulnerable to sea level rise

Even in less populous countries, population densities can be very high, especially in regions with reliable water sources.

China and India, the only two countries with a population exceeding 1 billion, together possess more than a third of the world's population.

World population trends

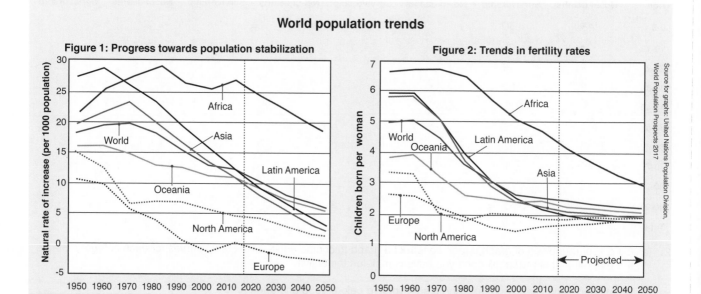

Figure 1: Progress towards population stabilization

Natural rate of increase (per 1000 population) — Africa, World, Asia, Latin America, Oceania, North America, Europe

Figure 2: Trends in fertility rates

Children born per woman — Africa, World, Oceania, Latin America, Asia, Europe, North America, ← Projected →

Source for graphs: United Nations Population Division, World Population Prospects 2017

The graphs above show two important trends in the growth of the human population and their projections forward to 2050. Estimates are always uncertain and political, social, and environmental changes affect population growth. Fertility rates are trending downwards, but more slowly than once thought. Twenty years ago fertility rates were projected to merge by 2040, this may now not happen until 2100. Even though the rate of natural increase and fertility rate are trending down, the human population is large enough that exponential growth is still continuing.

World population growth

Year	Global population (billions)	Developed nations (billions)	Developing nations (billions)
1950	2.5	0.8	1.7
1960	3.0	0.9	2.1
1970	3.7	1.0	2.7
1980	4.5	1.1	3.4
1990	5.3	1.1	4.2
2000	6.1	1.2	4.9
2010	6.9	1.2	5.7
2020	7.8	1.3	6.5
2030	8.6	1.3	7.3
2040	9.2	1.3	7.9
2050	9.7	1.3	8.4

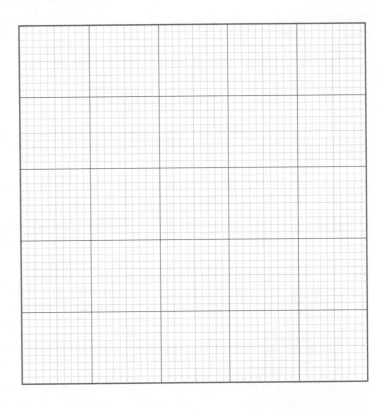

The world population has grown rapidly since the discovery of antibiotics and advances in food production. Projections into the future are difficult and depend on predicted fertility rate. The numbers shown above are based on a moderate fertility rate. High fertility rates could produce a population of 11 billion by 2050 or low fertility rates could produce a population of 8.1 billion.

1. Fertility rates of populations for all geographic regions are predicted to decline over the next 30 years.

    (a) Which region has the highest fertility rate? _____

    (b) Which region currently has the lowest fertility rate? _____

    (c) Suggest reasons why these regions are so different in their fertility rates: _____

    _____

    _____

    _____

2. Which direction would you expect Europe's population to trend over the next 30 years, assuming no migration?

    _____

3. Which region(s) has the highest population of people? _____

    _____

4. Calculate the population densities of the following places:

    (a) Monaco. Population: 37,550. Land area (km^2): 2.02: _____

    (b) United States. Population 325,737,272. Land area (km^2): 9,833,517: _____

    (c) World: Population: 7,566,417,000. Land area (km^2): 148,940,000: _____

5. (a) Plot a graph of the global population together with numbers for developed and developing nations:

    (b) Comment on the shape of the lines. Is the population stabilizing beyond 2020? _____

    _____

    _____

    _____

    _____

214 The Rise and Fall of Human Populations

Key Idea: Human populations are subject to rapid increases and declines in the same way as natural animal populations.

Throughout history, there have been a number of peaks of human civilization followed by collapse. The rise and collapse of civilizations is best studied in isolated areas or cities. The reasons for collapse are complex and triggered by various events but can generally be attributed either to the spread of disease, the collapse of a food source (normally agriculture), or pressure from foreign populations.

The human population remained small for most of history. The first very large cities of hundreds of thousands to millions appeared around 700 CE (AD). One such city was the Mayan city of Tikal.

TIKAL: At its peak around 800 AD, Tikal and its surrounds were inhabited by over 400,000 people. Extensive fields were used to grow crops and the total area of the city and its satellite towns and fields may have reached over 250 km². Eventually, the carrying capacity of the tropical, nutrient-poor land was overextended and people began to starve. By 900 AD the city had been deserted and the surrounding area abandoned.

EASTER ISLAND: Easter Island is located 3,000 km from South America. It has a mild climate and fertile volcanic soils. When Europeans arrived in the early 1700s, about 2000 people lived on the island. Initial archeological studies claimed a much larger population had existed and having run out of resources collapsed in war and civil unrest. However the latest evidence shows this may not be accurate. It seems the true collapse of the Easter Island population occurred after its discovery by Europeans. It appears that within 50 years of the first European visit, a once relatively stable society became unsettled. Many of the large statues were toppled by 1774. Slave raiding devastated the population in the 1860s and epidemics of introduced diseases almost wiped out the remaining inhabitants. By around 1880 the population was barely over 100. Although the fall of the Easter Island population may not have been ecological as once thought, it is still a powerful reminder of what can happen when new and unforeseen conditions are encountered by a society.

EUROPE: Europe has had its share of deadly pandemics. The Plague of Justinian (the first bubonic plague) wiped out half of Europe's population between 550 and 700 CE. In 1350, the bubonic plague again swept through Europe and again reduced its population by almost half, reducing the world's population to from 450 million in 1350 to 350 million in 1450 CE.

The 1918 flu pandemic involved the H1N1 influenza virus. The virus spread around the world as soldiers returned home from World War I. The epidemic lasted until 1920 and killed up to 100 million people, about 5% of the then world's population. Despite this and other epidemics, the world population has continued to grow. However, signs may be appearing that the human population is approaching maximum sustainable levels. Annual crop yields have ceased increasing and many common illnesses are becoming more difficult to treat. The rapid spread of modern pandemics, e.g. Ebola, illustrates the vulnerability of modern human populations. Could it be, perhaps, that another great reduction in the human population is imminent?

1. What are the general triggers for the collapse of a civilization? _____

_____

_____

2. What seems to have caused the collapse of Tikal? _____

_____

3. What effect did the two epidemics of bubonic plague have on Europe's population? _____

_____

4. How can studying the collapse of civilizations and disease epidemics help us plan for the future? _____

_____

_____

_____

_____

_____

PRACTICES

COMP

©2017 **BIOZONE** International
ISBN: 978-1-927309-65-0
Photocopying Prohibited

215 Population Regulation

Key Idea: A population's size is regulated by factors that affect population growth by changing birth or death rates. Population size is regulated by factors that limit population growth. The diagram below illustrates how population size can be regulated by environmental factors. **Density independent** factors are those that act independently of population density and affect all individuals more or less equally. **Density dependent** factors have a greater effect when the population density is higher. They become less important when the population density is low.

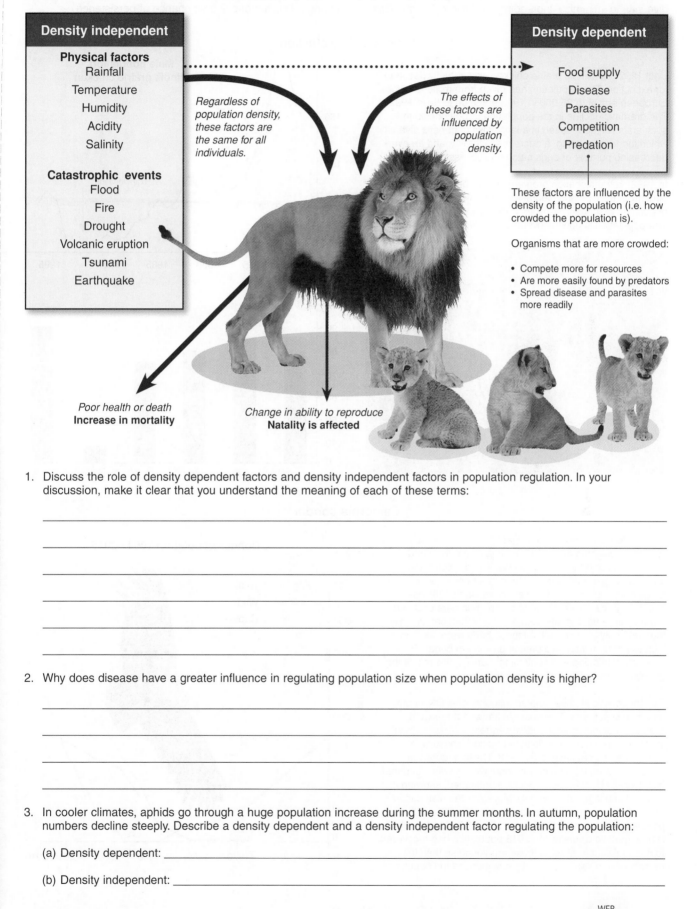

Density independent

Physical factors
Rainfall
Temperature
Humidity
Acidity
Salinity

Catastrophic events
Flood
Fire
Drought
Volcanic eruption
Tsunami
Earthquake

Regardless of population density, these factors are the same for all individuals.

The effects of these factors are influenced by population density.

Density dependent

Food supply
Disease
Parasites
Competition
Predation

These factors are influenced by the density of the population (i.e. how crowded the population is).

Organisms that are more crowded:

- Compete more for resources
- Are more easily found by predators
- Spread disease and parasites more readily

Poor health or death
Increase in mortality

Change in ability to reproduce
Natality is affected

1. Discuss the role of density dependent factors and density independent factors in population regulation. In your discussion, make it clear that you understand the meaning of each of these terms:

_____

_____

_____

_____

_____

_____

2. Why does disease have a greater influence in regulating population size when population density is higher?

_____

_____

_____

_____

3. In cooler climates, aphids go through a huge population increase during the summer months. In autumn, population numbers decline steeply. Describe a density dependent and a density independent factor regulating the population:

(a) Density dependent: _____

(b) Density independent: _____

216 Genetic Diversity and Population Viability

Key Idea: Genetic diversity refers to the variety of alleles and genotypes present in a population. Genetic diversity is important to the survival and adaptability of a species. For a population to be viable it needs to be large enough to replace deaths with births and it needs enough genetic diversity to minimize inbreeding, reducing the chances of genetic defects caused by inbreeding depression. Populations with low genetic diversity may not be able to respond to environmental change and are at risk of becoming extinct. In contrast, species with more genetic diversity have an improved ability to be able to adapt and respond to environmental change. This increases their chance of persistence.

Illinois prairie chicken

Until 1992, the Illinois prairie chicken was virtually destined for extinction. The population had fallen from millions before European arrival to 25,000 in 1933 and then to 50 in 1992. The dramatic decline in the population (a bottleneck) in such a short time resulted in a large loss of genetic diversity, which led to inbreeding, a decrease in fertility, and an ever-decreasing number of eggs hatching successfully.

In 1992, a translocation program began, bringing in 271 birds from Kansas and Nebraska. There was a rapid population response, as fertility and egg viability increased. The population is now recovering.

GregTheBusker CC2.0

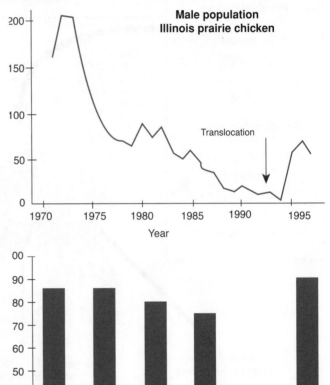

Male population Illinois prairie chicken

Translocation

California condor

The California condor is the largest North American land bird, weighing around 12 kg. It is a long lived bird, living for about 60 years. It is a New World vulture and as a scavenger eats carrion. The condor once had a range covering much of the west coast and south west of North America. However its range and numbers declined dramatically over the late 19th and early 20th century. By the 1987 only 27 were left. All these birds were taken into captivity and an intensive breeding program began. By carefully cataloging and controlling matings, the remaining genetic diversity was able to be retained.

Recent genetic studies isolated mitochondrial DNA from museum specimens. The analysis showed there were around 18 unique mitochondrial sequences (haplotypes) in the original North American population. These were simply labeled haplotypes (H) 1-18. These sequences were compared to the modern population. It was found that haplotypes H1, H4, and H7 were represented, showing a large loss of genetic diversity. Moreover the distribution of the modern haplotypes was uneven, with 62% of the population being H1, 15% H4, and 23% H7. Despite this lack of genetic diversity, the 2016 population had recovered to 446 individuals. Most of these are young so their full reproductive potential has not yet influenced the population.

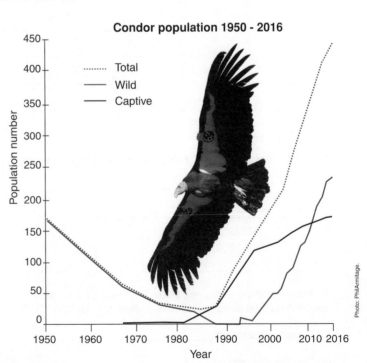

Condor population 1950 - 2016

.... Total
——— Wild
——— Captive

Photo: PhilArmitage.

WEB **216** KNOW

CONNECT **195** AP2

CONNECT **222** AP1

CONNECT **227** AP1

PRACTICES

PRACTICES

©2017 **BIOZONE** International
ISBN: 978-1-927309-65-0
Photocopying Prohibited

The great potato famine

Potatoes were the staple food for most rural Irish families during the 17th century. At the time, farmers favored one variety (the Irish Lumper) and, because potatoes are cultivated by vegetative propagation, genetic diversity between potato crops was very low. The effect of this low diversity was clearly demonstrated when the environment changed due to the arrival of the oomycete (water mold) *Phytophthora infestans* in the mid 1840s. *Phytophthora infestans* reached Ireland from Mexico via Europe, causing a disease in potatoes called late blight. It swept through the country, infecting all non-resistant potato plants. Records show that up to three quarters of potato crops were wiped out in 1846. Studies have identified the strain of *Phytophthora infestans* that caused the disease as HERB-1 (now extinct).

The loss of the potato crops, which were a staple diet for many Irish, contributed to a famine in which over one million people died (political decisions played a part also). Another 1 million emigrated to America, Canada, and elsewhere. If the farmers had planted a wider variety of potato crops (with greater genetic diversity) the chances of some being disease-resistant would have been higher, and crop losses would have been lower.

Infected potatoes are shrunken and rotted.

Diversity allows a range of responses

Prey species must avoid predators while maximizing the amount of time spent foraging. Differences in avoidance behavior is the result of many different influences, both environmental and internal. Some of the difference is genetic. Under controlled conditions, different individuals will emerge from hiding from a predator after different lengths of time. Those that remain hidden too long have a lower chance of being attacked, but lose time feeding, whereas the opposite is true for those that emerge sooner.

The graph right shows the hiding times for fiddler crabs after a simulated predator attack. Male hiding duration varied according to the size of the crab's carapace, although each category had its own range of times. The extremes of the range (outliers) are not shown, with some crabs taken up to 500 seconds to reemerge.

Duration of hiding in burrows after predator attack

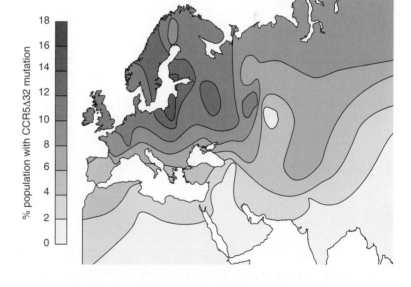

Genetic diversity and HIV-1 resistance

Before the advent of large scale global travel (effectively only 150 years old) human populations were largely isolated, and populations developed regional variations in alleles. Some of this allelic diversity affects our immune system and causes some of the variation in immunity that is seen in humans. This diversity may allow some people to be naturally resistant to some diseases.

A modern example of genetic diversity leading to differential responses to selective pressures is the HIV/AIDS epidemic. HIV/AIDS has killed more than 25 million people globally and infected another 33 million since it was first widely recognized in 1981.

In the mid 1990s, it was found that the HIV-1 virus entered T-cells of the immune system by docking with the receptor encoded by the CCR5 gene. Soon after this, it was discovered that a mutation in the gene (called CCR5Δ32) caused resistance to HIV-1.

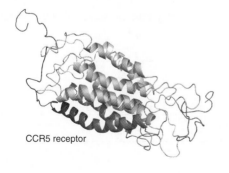

CCR5 receptor

Geographical studies have found that the CCR5Δ32 mutation is found in Caucasian populations in some areas of northern Europe where it is carried by up to 18% of the population. The mutation is virtually absent in Asian, Middle Eastern, and American Indian populations.

The CCR5Δ32 mutation produces a premature stop codon in the mRNA. People with this mutation in one allele produce T-cells with a reduced number of CCR5 receptors. HIV-1 infects these cells only slowly, taking 2-3 years longer than normal to progress to AIDS. People with mutations in both alleles produce T-cells with no CCR5 receptors. HIV-1 is effectively unable to enter these cells.

1. Explain what is meant by **genetic diversity**: _____
   _____
   _____

2. (a) What was the effect of a declining population of Illinois prairie chickens on the percentage of eggs hatching and what might have caused this?
   _____
   _____
   _____

   (b) Why did the translocation of 271 birds from outside Illinois into the Illinois population halt the population decline?
   _____
   _____
   _____

3. What was the effect of the population decline of the genetic diversity of the Californian condor?
   _____
   _____
   _____

4. Explain how low genetic diversity contributed to the great potato famine: _____
   _____
   _____
   _____

5. What is the effect of genetic diversity on the hiding behavior of fiddler crabs and how might this affect the survival of the fiddler crab population?
   _____
   _____
   _____
   _____
   _____

6. (a) Explain the effect of the CCR5Δ32 mutation on the entry of HIV-1 into T-cells: _____
   _____
   _____
   _____

   (b) In which populations in this mutation most common? _____

   (c) How could this genetic diversity affect the spread of HIV-1 (and conversely, how might the presence of HIV-1 affect the spread of the CCR5Δ32)?
   _____
   _____
   _____
   _____

©2017 **BIOZONE** International
ISBN: 978-1-927309-65-0
Photocopying Prohibited

217 The Ecological Niche

Key Idea: An organism's niche describes its functional role within its environment.

The **ecological niche** describes the functional position of a species in its ecosystem. It includes how the species responds to the distribution of resources and how it alters those resources for other species. The full range of environmental conditions under which an organism can exist describes its fundamental niche. As a result of interactions with other organisms, species usually occupy a realized niche that is narrower than this. Central to the niche concept is the idea that two species with exactly the same niche cannot coexist, because they would compete for the same resources and one would exclude the other. This is known as **Gause's competitive exclusion principle**. More often, species compete for some of the same resources. Competition will be intense where their resource use curves overlap.

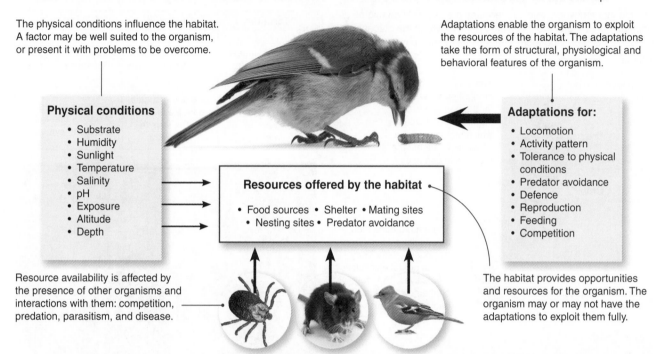

The physical conditions influence the habitat. A factor may be well suited to the organism, or present it with problems to be overcome.

Adaptations enable the organism to exploit the resources of the habitat. The adaptations take the form of structural, physiological and behavioral features of the organism.

Physical conditions
- Substrate
- Humidity
- Sunlight
- Temperature
- Salinity
- pH
- Exposure
- Altitude
- Depth

Resources offered by the habitat
- Food sources • Shelter • Mating sites
- Nesting sites • Predator avoidance

Adaptations for:
- Locomotion
- Activity pattern
- Tolerance to physical conditions
- Predator avoidance
- Defence
- Reproduction
- Feeding
- Competition

Resource availability is affected by the presence of other organisms and interactions with them: competition, predation, parasitism, and disease.

The habitat provides opportunities and resources for the organism. The organism may or may not have the adaptations to exploit them fully.

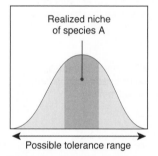

The realized niche

Realized niche of species A

Possible tolerance range

The tolerance range represents the **fundamental niche** a species could exploit. The actual or **realized niche** of a species is narrower than this because of competition with other species.

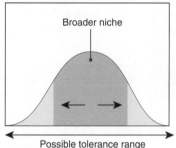

Intraspecific competition

Broader niche

Possible tolerance range

Competition is intense between individuals of the same species because they exploit the same resources. Individuals must exploit resources at the extremes of their tolerance range and the realized niche expands.

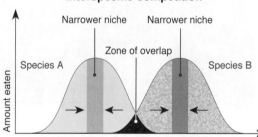

Interspecific competition

Narrower niche   Narrower niche

Zone of overlap

Species A   Species B

Amount eaten

Resource use as measured by food item size

If two (or more) species compete for some of the same resources, their resource use curves will overlap. Within the zone of overlap, resource competition will be intense and selection will favour niche specialization so that one or both species occupy a narrower niche.

1. (a) In what way could the realized niche be regarded as flexible? _____

_____

_____

(b) What factors might further constrain the extent of the realized niche? _____

_____

_____

2. Contrast the effects of interspecific competition and intraspecific competition on niche breadth: _____

_____

218 Adaptations for Cellulose Digestion

Key Idea: Cellulose digestion in insects is achieved through a complex symbiosis with both protozoa and bacteria.

Many insects are herbivores. Many of these feed on leaves where nutriment can be gained from starch and other molecules in the leaf. Few feed directly on woody material, which is made up mostly of cellulose. Insects that feed on woody material (e.g. termites) usually do so with the help of bacterial or protozoan symbionts harbored in the gut. Some insects, including stick insects, are able to produce cellulase enzymes that break down cellulose, but still rely on bacteria for the most part. Termites are a successful insect group that use gut protozoans and bacteria to break down cellulose.

It's complicated! A symbiont within a symbiont

Insects that eat wood, such as termites (below), rely on microbial populations to break the chemical bonds between the glucose molecules and provide them with molecules they can use.

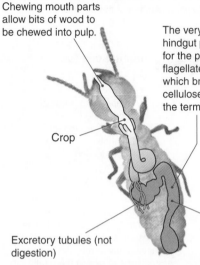

Chewing mouth parts allow bits of wood to be chewed into pulp.

The very large pouch-like hindgut provides room for the population of flagellated protozoans, which break down the cellulose to sugars that the termite can use.

Crop

Excretory tubules (not digestion)

The gut protozoa feed by engulfing wood fibers by phagocytosis. They do not produce cellulase themselves. They rely on their own mutualistic bacteria to do this.

Termites	Protozoa	Bacteria
Chew and ingest woody material	Engulf wood fibers using phagocytosis	Digest cellulose using cellulase enzymes
Absorb sugars released from protozoan symbionts	Use some sugars and release some to the termite symbiont	Use some sugars and release some to the protozoan symbiont

Althepal CC 2.5

The flagellated protozoans in the termite gut are very mobile. They are more easily kept as gut residents than free bacteria, which would be lost.

Termite damage behind a cupboard wall. In many countries, termites cause significant damage to homes each year by forming nests in wall cavities and chewing the wood framing.

1. Describe the features of a termite that allow it to feed on woody material: _____

_____

_____

_____

2. Describe how termites derive glucose from wood: _____

_____

_____

3. Use brief notes to compare and contrast the digestion of cellulose in termites and ruminants (Activity 83):

_____

_____

_____

_____

_____

_____

©2017 **BIOZONE** International
ISBN: 978-1-927309-65-0
Photocopying Prohibited

219 Adaptations of the Snow Bunting

Key Idea: Snow buntings have a number of adaptations to survive in the Arctic environment.

The adaptive features that evolve in species are the result of selection pressures on them through the course of their evolution. These features enable an organism to function most effectively in its niche, enhancing its exploitation of its environment and therefore its survival. The example below of the snow bunting below illustrates adaptations of a migratory Arctic bird. Note that adaptations may be associated with an animal's morphology, its internal physiology, or its behavior.

Snow bunting
(*Plectrophenax nivalis*)

The snow bunting is a small ground feeding bird with a circumpolar Arctic breeding range throughout the northern hemisphere. Although migratory, snow buntings do not move to traditional winter homes but prefer winter habitats that resemble their Arctic breeding grounds, such as bleak shores or open fields of northern Britain and the eastern United States. During the Arctic summer breeding season, the male develops black and white plumage. During the winter months both male and female are a brown color.

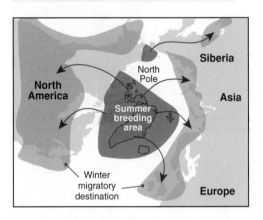

Very few small birds breed in the Arctic, because most small birds lose more heat than larger ones. In addition, birds that breed in the brief Arctic summer must migrate before the onset of winter, often traveling over large expanses of water. Large, long winged birds are better able to do this. However, the snow bunting is superbly adapted to survive in the extreme cold of the Arctic region.

Less heat is lost from white plumage compared to dark plumage.

White feathers are hollow and filled with air, which acts as an insulator. In the dark colored feathers the internal spaces are filled with pigmented cells.

Snow buntings, on average, lay one or two more eggs than equivalent species further south. They are able to rear more young because the continuous daylight and the abundance of insects at high latitudes enables them to feed their chicks around the clock.

During snow storms or periods of high wind, snow buntings will burrow into snowdrifts for shelter.

Habitat and ecology: Widespread throughout Arctic and sub-Arctic Islands. Active throughout the day and night, resting for only 2-3 hours in any 24 hour period. Snow buntings may migrate up to 6000 km but are always found at high latitudes.

Reproduction and behavior: The nest, which is concealed amongst stones, is made from dead grass, moss, and lichen. The male bird feeds his mate during the incubation period and helps to feed the young.

1. Describe a structural, physiological, and behavioral adaptation of the snow bunting, explaining how each aids survival:

(a) Structural adaptation: _____

_____

_____

(b) Physiological adaptation: _____

_____

_____

(c) Behavioral adaptation: _____

_____

_____

2. Snow buntings feed on a wide variety of food during summer and winter. The summer diet is quite different to the winter diet. How does this range of diet help its survival?

_____

_____

_____

PRACTICES   CONNECT

207
AP1

KNOW

220 Adaptations to Low Nutrient Environments

Key Idea: Insectivorous plants capture and digest small invertebrates to meet their nitrogen requirements.

Plants that live in acid bogs or in nutrient-poor soils often obtain extra nutrients (particularly nitrogen) by capturing and digesting small invertebrates. These plants are called **insectivorous** (or sometimes carnivorous). They photosynthesize to make their own sugars, but obtain some of their nitrogen and minerals from animal tissues. The traps are leaf modifications and usually contain special glands that secrete digestive enzymes. Insectivorous plants are usually small because their nutrient poor environment does not support the growth of large plants.

Long hairs tipped with sticky droplets

The leaf curls to form a temporary 'stomach' in which digestion occurs

Sundew

Sundews (*Drosera*) capture their prey in the sticky secretions on their leaf hairs. These make a trap like flypaper. The hairs bend over the prey and restrain it and the entire leaf cups around to enclose the prey. Sundews are widespread globally, with centers of diversity in South America and Australia.

Insects are attracted to the pitcher's colorful and prominent lip region by sweet secretions just over the rim.

Insects climb over the lip to find themselves on a nearly vertical surface made slippery by waxy secretions. This causes them to fall into the digestive fluid below.

Each leaf has a spring-like hinge of thin-walled cells down its midrib. When triggered, these cells rapidly lose water causing the two halves of the leaf to close together.

Spines line the edge of the leaf, creating a cage when the leaf folds together.

Gland cells line the lower part of the inside of the pitcher. They secrete digestive enzymes and are sometimes involved in the absorption of food.

Pitcher plant

Venus flytrap

Insects touch these trigger hairs on the leaf surface.

The digestive fluid that fills the lower part of the pitcher contains at least two potent, protein splitting enzymes. One is similar to pepsin, the enzyme found in vertebrate stomachs.

Venus fly trap: The Venus fly trap consists of two, lobed modified leaves that can rapidly close together to trap prey (usually small insects). The trigger for closing is a touch on the sensory hairs of the leaves.

Pitcher plant: This plant is a passive trap. Prey fall into the water collected at the base of the pitcher where they drown and their tissues are slowly digested by the plant's digestive enzymes. Pitcher plants are found in bogs and marshes throughout the Americas.

1. (a) What does an insectivorous plant gain from digesting insects? _____

   (b) How this is an advantage to the plant in its habitat? _____

   _____

   _____

2. (a) Describe one cost (disadvantage) to the plant of producing insect-trapping modified leaves: _____

   _____

   (b) Why are insectivorous plants not usually found in nutrient rich soils where non-insectivorous plants are present?

   _____

   _____

3. For one of the named plants above, describe the modifications to the leaf structure that enable capture of insects:

   _____

   _____

©2017 **BIOZONE** International
ISBN: 978-1-927309-65-0
Photocopying Prohibited

221 Adaptations to Saline Environments

Key Idea: Mangroves are salt tolerant (halophytes) and specifically adapted to the high salt, water-logged environments of estuaries, tidal flats, and salt marshes. Mangroves are **halophytes**, a group of plants with adaptations for growth in seawater or salty, water-logged soil. They grow in the upper part of the intertidal zone, but also extend further inland to form salt marshes and other coastal wetland communities throughout the tropics and subtropics.

Mangroves grow from the upper part of the intertidal zone to the high water mark, forming some of the most complex and productive ecosystems on Earth. The high salt environment would kill most other kinds of plants as high salt levels cause water to flow out of the cells. Mangroves overcome this by storing salt in their cell vacuoles and maintaining a high concentration of solutes in the cytoplasm of their cells. This reverses the osmotic gradient and maintains the transpiration stream.

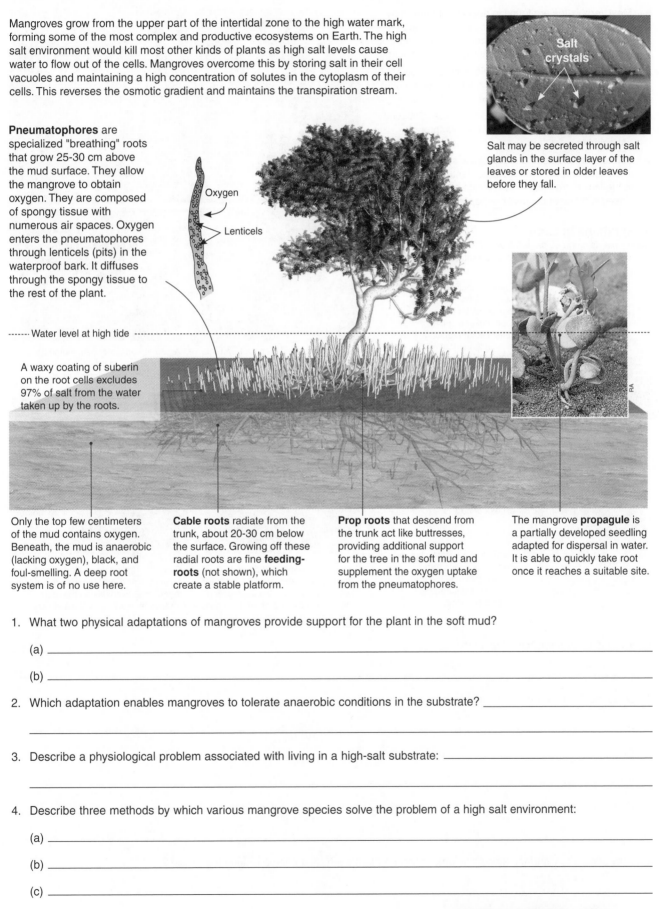

Salt crystals

Salt may be secreted through salt glands in the surface layer of the leaves or stored in older leaves before they fall.

Pneumatophores are specialized "breathing" roots that grow 25-30 cm above the mud surface. They allow the mangrove to obtain oxygen. They are composed of spongy tissue with numerous air spaces. Oxygen enters the pneumatophores through lenticels (pits) in the waterproof bark. It diffuses through the spongy tissue to the rest of the plant.

Oxygen

Lenticels

-------- Water level at high tide --------

A waxy coating of suberin on the root cells excludes 97% of salt from the water taken up by the roots.

Only the top few centimeters of the mud contains oxygen. Beneath, the mud is anaerobic (lacking oxygen), black, and foul-smelling. A deep root system is of no use here.

Cable roots radiate from the trunk, about 20-30 cm below the surface. Growing off these radial roots are fine **feeding-roots** (not shown), which create a stable platform.

Prop roots that descend from the trunk act like buttresses, providing additional support for the tree in the soft mud and supplement the oxygen uptake from the pneumatophores.

The mangrove **propagule** is a partially developed seedling adapted for dispersal in water. It is able to quickly take root once it reaches a suitable site.

1. What two physical adaptations of mangroves provide support for the plant in the soft mud?

   (a) _____

   (b) _____

2. Which adaptation enables mangroves to tolerate anaerobic conditions in the substrate? _____

   _____

3. Describe a physiological problem associated with living in a high-salt substrate: _____

   _____

4. Describe three methods by which various mangrove species solve the problem of a high salt environment:

   (a) _____

   (b) _____

   (c) _____

PRACTICES   CONNECT   WEB

207 AP1   **221**

KNOW

222 Adaptations to Arid Environments

Key Idea: Xerophytes are plants with adaptations that allow them to conserve water and survive in dry environments. Plants adapted to dry conditions are called **xerophytes**. Xerophytes are found in a number of environments globally, but all show similar adaptations (through convergent evolution) to conserve water. These adaptations include small, hard leaves, an epidermis with a thick cuticle, sunken stomata, succulence, and permanent or temporary absence of leaves.

▶ Most xerophytes are found in deserts, but they may be found in humid environments, provided that their roots are in dry micro-environments (e.g. the roots of epiphytic plants that grow on tree trunks or branches).

▶ Many xerophytes have a succulent morphology. Their stems are often thickened and retain a large amount of water in the tissues, e.g. *Aloe*.

▶ Many xerophytes have a low surface area to volume ratio, reducing the amount of water lost through transpiration.

▶ Salt tolerant plants and many alpine species may show xeromorphic (arid-adapted) features in response to the lack of free water and high transpirational losses in these often windy, exposed environments.

Desert xerophytes in Arizona have a succulent photosynthetic stem

Adaptations in cacti
Desert plants, such as cacti (below), must cope with low or sporadic rainfall and high transpiration rates.

Leaves modified into spines or hairs to reduce water loss. Light colored spines reflect solar radiation.

Rounded shape reduces surface area.

Stem becomes the major photosynthetic organ, plus a reservoir for water storage.

The surface tissues of many cacti are tolerant of temperatures in excess of 50°C.

Cacti have a shallow, but extensive fibrous root system. When in the ground the roots are spread out around the plant.

Acacia trees have deep root systems, allowing them to draw water from sources deep underground.

An outer surface coated in fine hairs traps air close to the surface and reduces the transpiration rate.

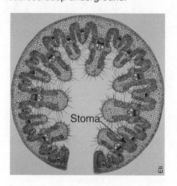
Grasses on coastal sand dunes (e.g. marram grass, above) curl their leaves. Stomata are sunken in pits, creating a moist microclimate around the pore, which reduces transpiration rate.

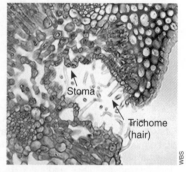
Oleander has a thick multi-layered epidermis and the stomata are sunken in trichome-filled pits on the leaf underside which restrict water loss.

1. What is a xeromorphic adaptation? _____

_____

2. Describe three xeromorphic adaptations of plants that reduce water loss:

(a) _____

(b) _____

(c) _____

3. How does creating a moist microclimate around the areas of water loss reduce the transpiration rate? _____

_____

4. How does a low surface area to volume ratio in a plant such as a cactus reduce water loss? _____

_____

©2017 **BIOZONE** International
ISBN: 978-1-927309-65-0
Photocopying Prohibited

223 Species Interactions

Key Idea: All species interact with their own and other species. These interactions carry costs and/or benefits to the parties involved.

No organism exists in isolation. Each interacts with other organisms of its own and other species. Species interactions (those between different species) involve benefit to at least one party. If one party benefits at the expense of another, the relationship is an **exploitation**. If one species benefits and one is unaffected, the relationship is said to be **commensal**. Some species interactions involve a close association or **symbiosis** (living together) between the parties involved. Symbioses include parasitism and mutualism. Many species interactions involve coevolution, in which there is a reciprocal evolution of adaptations in both parties.

Type of interaction between species

Mutualism	Exploitation			Competition
	Predation	**Herbivory**	**Parasitism**	
A ⇌ B Benefits Benefits	A → B Benefits Harmed	A → B Benefits Harmed	A → B Benefits Harmed	A ⇌ B Harmed Harmed
Both species benefit from the association. **Examples**: Tick bird on zebra removes parasites and alerts zebra to danger, while tick bird gains access to food. Flowering plants and their insect pollinators have a mutualistic relationship. Flowers are pollinated and the insect gains food (below).	Predator kills the prey outright and eats it. **Examples**: Lion preying on wildebeest or praying mantis (below) consuming insect prey. The adaptations of predators and prey are the result of coevolution: predators have adaptations to capture prey and prey have adaptations to avoid capture.	Herbivore eats parts of a plant and usually does not kill it. Plants often have defences to limit the impact of herbivory. **Example**: Giraffes browsing acacia trees. Browsing stimulates the acacia to produce toxic alkaloids, which cause the giraffe to move on to another plant. Acacia thorns also deter many browsers.	The parasite lives in or on the host, taking (usually all) its nutrition from it. The host is harmed but usually not killed. **Examples**: Pork tapeworm (below) in a pig's gut. Some plants (e.g. mistletoes) are semi-parasitic (hemi-parasites). They photosynthesize but rob the host plant of nutrients and water.	Individuals of the same or different species compete for the same resources, with both parties suffering, especially when resources are limited. **Examples**: Monarch caterpillars compete for access to milkweed. Those hatched later in the season may starve. Plants growing close to each other compete for light and soil nutrients.

1. For the purposes of this exercise, assume that species A in the diagram represents humans. Briefly describe an example of our interaction with another species (B in the diagram above) that matches each of the following interaction types:

   (a) Mutualism: _____

   _____

   (b) Exploitation: _____

   _____

   (c) Competition: _____

   _____

2. Plants are not defenceless against herbivores. They have evolved physical and chemical defences to deter herbivores. In some cases (as in grasses) grazing stimulates growth in the plant.

   (a) What is the acacia's response to giraffe browsing? _____

   (b) How might this response prevent over-browsing?_____

   _____

©2017 **BIOZONE** International
ISBN: 978-1-927309-65-0
Photocopying Prohibited

PRACTICES   CONNECT **163** AP2   CONNECT **162** AP2   CONNECT **161** AP2   WEB **223**

KNOW

Examples of interactions between different species are illustrated below. For each example, identify the type of interaction, and explain how each species in the relationship is affected.

3. The European honey bee *Apis mellifera* collects pollen in pollen baskets on its back legs and in doing so spreads pollen from one flower to the next, pollinating the flowers as it visits to feed on nectar.

   (a) Identify this type of interaction: _____

   (b) Describe how each species is affected (benefits/harmed/no effect):

   _____

   _____

4. The squat anemone shrimp (or sexy shrimp), lives among the tentacles of sea anemones, where it gains protection and scavenges scraps of food from the anemone. The anemone is apparently neither harmed nor benefitted by the shrimp's presence although there is some evidence that ammonium released by the shrimp may benefit the anemone indirectly by supplying nutrients to the mutualistic photosynthetic algae that reside in the anemone's tissues.

   (a) Identify this type of interaction: _____

   (b) Describe how each species is affected (benefits/harmed/no effect):

   _____

   _____

   _____

5. Hyenas will kill and scavenge a range of species. They form large groups and attack and kill large animals, such as wildebeest, but will also scavenge carrion or drive other animals off their kills.

   (a) Identify this type of interaction: _____

   (b) Describe how each species is affected (benefits/harmed/no effect):

   _____

   _____

6. Ticks are obligate hemtaophages and must obtain blood to pass from one life stage to the next. Ticks attach to the outside of hosts where they suck blood and fluids and cause irritation.

   (a) Identify this type of interaction: _____

   (b) Describe how each species is affected (benefits/harmed/no effect):

   _____

   _____

7. Large herbivores expose insects in the vegetation as they graze. The cattle egret, which is widespread in tropical and subtropical regions, follows the herbivores as they graze, feeding on the insects disturbed by the herbivore.

   (a) Identify this type of interaction: _____

   (b) Describe how each species is affected (benefits/harmed/no effect):

   _____

   _____

8. Explain the similarities and differences between a predator and a parasite: _____

_____

_____

©2017 **BIOZONE** International
ISBN: 978-1-927309-65-0
Photocopying Prohibited

224 Interspecific Competition

Key Idea: Interspecific competition occurs between individuals of different species for resources. It can affect the size and distribution of populations sharing the same environment.

Interspecific competition (competition between different species) is usually less intense than intraspecific (same species) competition because coexisting species have evolved slight differences in their realized niches. However, when two species with very similar niche requirements are

brought into direct competition through the introduction of a foreign species, one usually benefits at the expense of the other, which is excluded (the **competitive exclusion principle**). The introduction of alien species is implicated in the competitive displacement and decline of many native species. Displacement of native species by introduced ones is more likely if the introduced competitor is adaptable and hardy, with high fertility.

Competition in *Paramecium*

In 1934, Georgii Gause, a Russian biologist, carried out a series of experiments on *Paramecium*. The results led him to propose the **competitive exclusion principle**, a fundamental idea in ecology. In the first stage of the experiments, he grew three species of *Paramecium* in isolation in a nutritive medium containing their essential resource (bacterial food). Their growth curves are shown below:

Paramecium grown in isolation

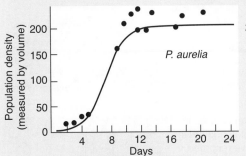

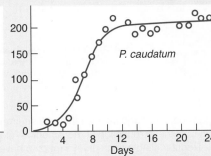

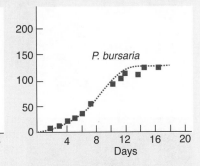

In the second stage of the experiment Gause grew *P. aurelia* and *P. caudatum* together. He found that *P. caudatum* was always out-competed and became extinct from the culture. Gause then grew *P. caudatum* with *P. bursaria*. He found they were able to exist together (but at lower numbers). Investigation found that *P. caudatum* occupied the oxygen rich top half of the culture tube, whereas *P. bursaria* retreated to the lower, poorly oxygenated region. *P. bursaria* contains symbiotic algae, which release oxygen in photosynthesis. This allows *P. bursaria* to remain in the anoxic zone.

Paramecium grown in competition

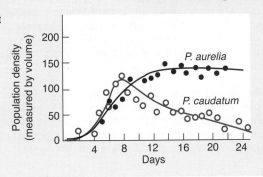

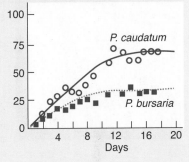

1. What is the competitive exclusion principle? _____

_____

2. (a) What type of growth curve do the *Paramecium* show when grown in isolation? _____

_____

(b) What was the fundamental niche of all the *Paramecium* species? _____

_____

(c) What were the realized niches of *P. caudatum* and *P. bursaria*? _____

_____

_____

(d) How do the experiments support Gause's competitive exclusion principle? In what way do they not? _____

_____

_____

_____

PRACTICES   CONNECT   WEB

164
AP2

224

KNOW

Change in distribution of red and gray squirrels in the UK

In Britain, introduction of the larger, more aggressive, gray squirrel in 1876 has contributed to a contraction in range of the native red squirrel

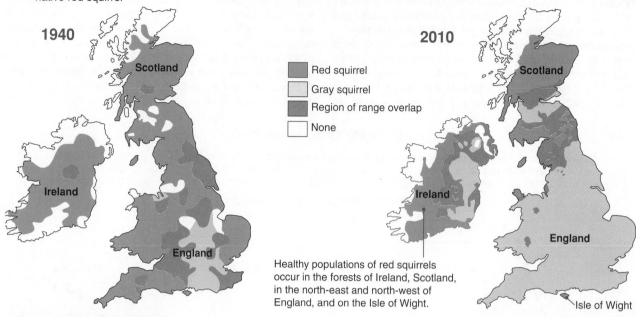

1940

2010

Scotland

Ireland

England

Scotland

Ireland

England

Isle of Wight

- ☐ Red squirrel
- ☐ Gray squirrel
- ☐ Region of range overlap
- ☐ None

Healthy populations of red squirrels occur in the forests of Ireland, Scotland, in the north-east and north-west of England, and on the Isle of Wight.

Paul Whippey cc 3.0

Red squirrel, *Sciurus vulgaris*

The European red squirrel was the only squirrel species in Britain until the introduction of the American gray squirrel in 1876. Regular distribution surveys (above) have recorded the range contraction of the reds, with the larger, more aggressive gray squirrel displacing populations of reds over much of England. Gray squirrels can exploit tannin-rich foods, which are unpalatable to reds. In mixed woodland and in competition with grays, reds may not gain enough food to survive the winter and breed. Reds are also very susceptible to several viral diseases, including squirrelpox, which is transmitted by grays.

Whereas red squirrels once occupied a range of forest types, they are now almost solely restricted to coniferous forest. The data suggest that the gray squirrel is probably responsible for the red squirrel decline, but other factors, such as habitat loss, are also likely to be important.

BirdPhotos.com cc 3.0

Gray squirrel, *Sciurus carolinesis*

3. Outline the evidence to support the view that the red-gray squirrel distributions in Britain are an example of the competitive exclusion principle:

_____

_____

_____

_____

4. The ability of red and gray squirrels to coexist appears to depend on the diversity of habitat type and availability of food sources (reds appear to be more successful in regions of coniferous forest). Suggest why careful habitat management is thought to offer the best hope for the long term survival of red squirrel populations in Britain:

_____

_____

_____

_____

_____

_____

_____

©2017 **BIOZONE** International
ISBN: 978-1-927309-65-0

225 Competition in Barnacles

Key Idea: Evidence from studies of natural populations shows that competition can reduce niche breadth.
Seashores provide a wide range of habitats and opportunities. Not surprisingly, this results in a high diversity and abundance of organisms competing for the benefits of living there.

Competition in species of barnacle, a common crustacean on rocky shores, was studied by J.H. Connell in Scotland. By removing one barnacle species and observing the effect on another, it was possible to determine the extent of the fundamental niches and compare them to the realized niches.

Competitive exclusion in barnacles

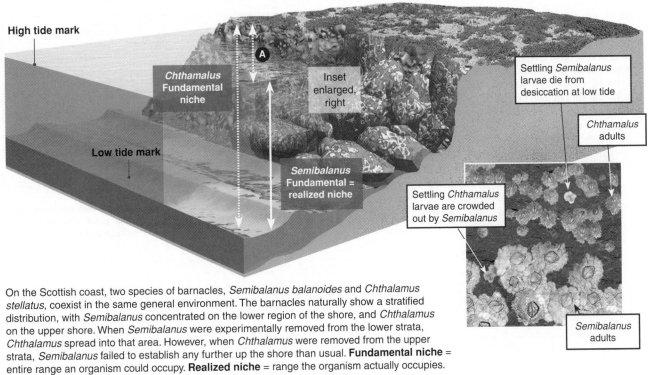

High tide mark

A

Chthamalus Fundamental niche

Inset enlarged, right

Settling *Semibalanus* larvae die from desiccation at low tide

Chthamalus adults

Low tide mark

Semibalanus Fundamental = realized niche

Settling *Chthamalus* larvae are crowded out by *Semibalanus*

Semibalanus adults

On the Scottish coast, two species of barnacles, *Semibalanus balanoides* and *Chthalamus stellatus*, coexist in the same general environment. The barnacles naturally show a stratified distribution, with *Semibalanus* concentrated on the lower region of the shore, and *Chthalamus* on the upper shore. When *Semibalanus* were experimentally removed from the lower strata, *Chthalamus* spread into that area. However, when *Chthalamus* were removed from the upper strata, *Semibalanus* failed to establish any further up the shore than usual. **Fundamental niche** = entire range an organism could occupy. **Realized niche** = range the organism actually occupies.

1. (a) In the example of the barnacles above, describe what is represented by the zone labeled with the arrow A:

_____

_____

_____

(b) Outline the evidence for the barnacle distribution being the result of competitive exclusion: _____

_____

_____

_____

2. (a) What keeps *Semibalanus* larvae from establishing at higher shore levels? _____

_____

(b) What is the consequence of this to the realized niche compared to the fundamental niche of *Semibalanus*?

_____

_____

3. There are many studies underway about the effect global warming and rising sea levels might have on marine communities. What effect might a rise in sea level have on the *Chthamalus/Semibalanus* community?

_____

_____

_____

PRACTICES PRACTICES CONNECT CONNECT WEB

?

217 AP2

164 AP2

225

KNOW

226 Modeling Interspecific Competition

Key Idea: Competition between two species can be modeled by accounting for the effect of each species on the other.

Populations of species living in the same area usually exploit different resources. This avoids competition that would affect the survival of the competing populations. When a new species (species 2) moves into an area and its resource use overlaps with an existing species (species 1), there will be resource competition. Species 2 may be better at acquiring the resource and so affect the population growth and viability of species 1. This effect of species 2 on species 1 can be quantified by the competition coefficient α. The greater the value of α, the greater the effect of species 2 on species 1. Similarly species 1 must in some way affect species 2 and this is termed β. By adding these terms into the logistic equation we can simulate what will happen when two competing species are placed in the same environment.

Here we are again using the Populus program (see Activity 210).

From the menu bar select Model, then Multi-species Dynamics, then Lotka-Volterra Competition.
To start set the parameters for both species to the following:
▶ N = 10, r = 0.4, K = 100, α and β= 0.5.
▶ Set the Termination Condition to Run until steady state.
▶ View the graph as N vs t and note its shape.

Competition coefficients in the Lotka-Volterra model:
• α quantifies the per capita reduction in the population growth of species 1 caused by species 2
• β quantifies the per capita reduction in the population growth of species 2 caused by species 1

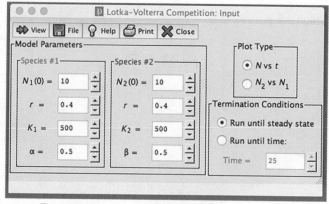

The Lotka-Volterra equations are a simple mathematical model of the population dynamics of competing species.

1. (a) Describe the shape of the graph: _____

_____

(b) What is the effect of competition on the populations? Do they reach K? _____

_____

_____

_____

2. Now set α to 0.9 and view the graph again. Describe the effect of increasing α on the population growth of each species:

_____

_____

_____

3. Now set α back to 0.5 and set *r* for species 1 to 0.6. View the graph and describe the effect on the population growth each of species:

_____

_____

_____

4. Using this model we can simulate what would happen when an aggressively competing species is introduced (either deliberately or by accident) into an area where a native species utilizes the same resource but is not an aggressive competitor. Set the parameters for species 1 to N = 1000, *r* = 0.15, K = 1000 and α = 0.9. Set the parameters for species 2 to N = 4, *r* = 0.85, K = 1000, and β = 0.1.

(a) Before you run the simulation, predict what will happen: _____

_____

(b) Were you correct?_____

(c) What level of α allows the populations reach a similar size? _____

(d) What level of β allows the populations reach a similar size? b = _____

SAVE AND PRINTb ALL YOUR SIMULATIONS AND ATTACH THEM TO THIS PAGE

©2017 **BIOZONE** International
ISBN: 978-1-927309-65-0
Photocopying Prohibited

227 Intraspecific Competition

Key Idea: Individuals of the same species exploit the same resources, so competition between them is usually intense and will act to limit population growth.

As a population grows, the resources available to each individual become fewer and **intraspecific competition** (competition between members of the same species) increases. When the demand for a resource (e.g. food or light) exceeds supply, that resource becomes a limiting factor to the number of individuals the environment can

support (the **carrying capacity**). Populations respond to resource limitation by reducing growth rate (e.g. lower birth rates or higher mortality). The response of individuals to limited resources varies. In many invertebrates and some vertebrates, individuals reduce their growth rate and mature at a smaller size. In many vertebrates, territories space individuals apart according to resource availability and only those individuals able to secure a territory will have sufficient resources to breed.

Scramble competition in caterpillars

Contest competition in wolves

Display of a male anole

Direct competition for available food between members of the same species is called **scramble competition.** In some situations where scramble competition is intense, none of the competitors gets enough food to survive.

In some cases, direct competition is limited by hierarchies existing within a social group. Dominant individuals receive adequate food, but individuals low in the hierarchy must **contest** the remaining resources and may miss out.

Competition within species is often associated with breeding. In anole lizards (above), males have a bright red throat pouch and use much of their energy displaying to compete with other males for available mates.

Competition between tadpoles of *Rana tigrina*

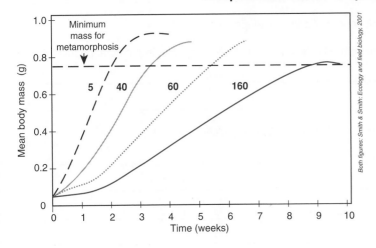

Food shortage reduces both individual growth rate and survival, and population growth. In some organisms, where there is a metamorphosis or a series of molts before adulthood (e.g. frogs, crustacean zooplankton, and butterflies), individuals may die before they mature. The graph (left) shows how the growth rate of tadpoles (*Rana tigrina*) declines as the density increases from 5 to 160 individuals (in the same sized space).

▶ At high densities, tadpoles grow more slowly, take longer to reach the minimum size for metamorphosis (0.75 g), and have less chance of metamorphosing into frogs.

▶ Tadpoles held at lower densities grow faster to a larger size, metamorphosing at an average size of 0.889 g.

▶ In some species, such as frogs and butterflies, the adults and juveniles reduce the intensity of intraspecific competition by exploiting different food resources.

1. Using an example, predict the likely effects of intraspecific competition on each of the following:

(a) Individual growth rate: _____

(b) Population growth rate: _____

(c) Final population size: _____

334

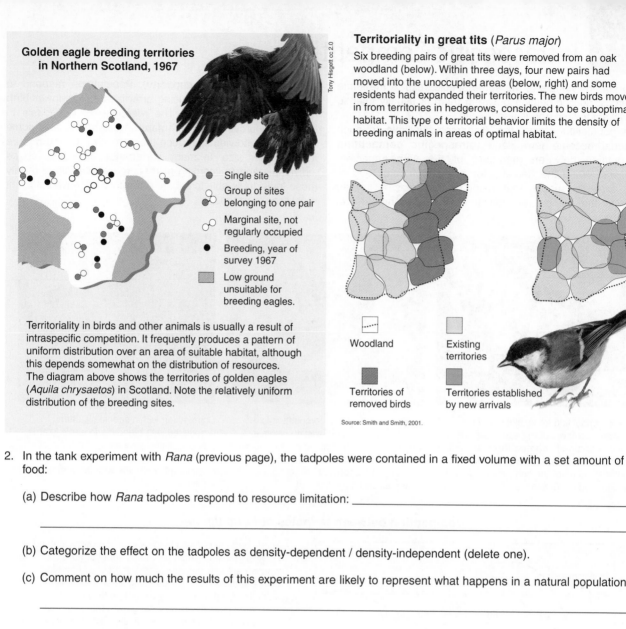

Golden eagle breeding territories in Northern Scotland, 1967

- Single site
- Group of sites belonging to one pair
- Marginal site, not regularly occupied
- Breeding, year of survey 1967
- Low ground unsuitable for breeding eagles.

Territoriality in birds and other animals is usually a result of intraspecific competition. It frequently produces a pattern of uniform distribution over an area of suitable habitat, although this depends somewhat on the distribution of resources. The diagram above shows the territories of golden eagles (*Aquila chrysaetos*) in Scotland. Note the relatively uniform distribution of the breeding sites.

Territoriality in great tits (*Parus major*)

Six breeding pairs of great tits were removed from an oak woodland (below). Within three days, four new pairs had moved into the unoccupied areas (below, right) and some residents had expanded their territories. The new birds moved in from territories in hedgerows, considered to be suboptimal habitat. This type of territorial behavior limits the density of breeding animals in areas of optimal habitat.

- Woodland
- Existing territories
- Territories of removed birds
- Territories established by new arrivals

Source: Smith and Smith, 2001.

2. In the tank experiment with *Rana* (previous page), the tadpoles were contained in a fixed volume with a set amount of food:

(a) Describe how *Rana* tadpoles respond to resource limitation: _____

(b) Categorize the effect on the tadpoles as density-dependent / density-independent (delete one).

(c) Comment on how much the results of this experiment are likely to represent what happens in a natural population: _____

3. Identify two ways in which animals can reduce the intensity of intraspecific competition:

(a) _____

(b) _____

4. (a) Suggest why carrying capacity of an ecosystem might decline: _____

(b) Predict how a decline in carrying capacity might affect final population size: _____

5. Using appropriate examples, discuss the role of territoriality in reducing intraspecific competition: _____

228 Predator-Prey Interactions

Key Idea: Predator and prey populations frequently show regular population cycles. The predator cycle is usually based on the intrinsic population cycle of the prey species.

It was once thought that predators regulated the population numbers of their prey. However, we now know that this is not usually the case. Prey species are more likely to be regulated by other factors such as the availability of food. However, predator population cycles are often regulated by the availability of prey, especially when there is little opportunity for switching to alternative prey species.

Fluctuations in hypothetical prey and predator populations

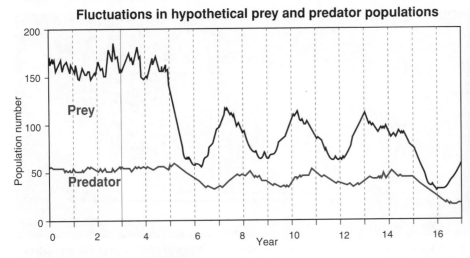

Predators capture and eat their prey, but do they control their population numbers?

Crown-of-thorns starfish

Population cycles in crown-of-thorns starfish

The crown-of-thorns starfish (right) is a voracious predator of coral on the Great Barrier Reef. A single starfish can destroy 5 m² of coral in a year. There have been two major starfish population explosions in recent times. One lasted from the early 1960s until the mid 1970s. The other began in the late 1970s and receded in the early 1990s. The reason for these cyclic outbreaks is not yet known. A favorite explanation is that shell collecting has led to a decline in the numbers of triton shells; one of the crown-of-thorn's few predators. However, tritons have a varied diet and feed on other organisms and they have never been common.

Coral

1. The graph above shows cycles in hypothetical predator and prey populations over 17 years. Answer the following:

   (a) State which population shows the greatest variation in population numbers: _____

   (b) Give the year at which a change occurred in the stability of the two population levels: _____

   (c) Describe the nature of the pattern of population numbers after the change: _____

   (d) Determine the period (duration) of the cycle exhibited by the two populations: _____

   (e) Determine how much the peaks of the two populations are out of phase (time lag between peaks): _____

2. The crown-of-thorns starfish on the Great Barrier Reef have undergone huge population increases in recent times.

   (a) Describe a favorite suggested cause for this population increase: _____

   _____

   _____

   (b) Why is this unlikely to be the major reason for the increase in starfish numbers? _____

   _____

   _____

3. At various times, a number of exotic species (such as rabbits and prickly pear cactus) were released into Australia's natural ecosystems. Why did many of these introduced species become uncontrollable pests?

   _____

   _____

©2017 **BIOZONE** International
ISBN: 978-1-927309-65-0
Photocopying Prohibited

229 Population Cycles

Key Idea: Predator-prey cycles can be maintained in a laboratory if the natural environment's complexity is created. Mathematical models predict that predator and prey populations will form stable cycles of population increase and decrease. Early ecologists set out to verify these population oscillations in small model ecosystems. Two researchers, Gause and Huffaker, each worked on this question. Their results gave great insight into the nature of predator-prey interactions and the factors that control population size and enable stable oscillations.

Gause's experiments

We have already examined several of Gause's experiments on competition in an earlier activity. Gause's also studied predator-prey interactions in two protists, *Paramecium* and its predator *Didinium,* in simple test tube 'microcosms'. When *Didinium* was added to a culture of *Paramecium*, it quickly ate all the *Paramecium* and then died out. When sediment was placed in the microcosm, *Paramecium* could hide, *Didinium* died out, and the *Paramecium* population recovered. Adding immigrants into the microcosm produced a predator-prey like cycle.

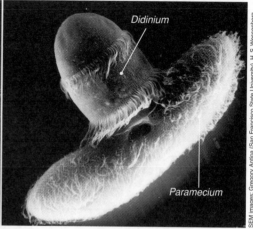
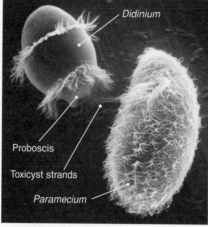

Didinium, a unicellular ciliate, feeds almost exclusively on *Paramecium*. It captures *Paramecium* by shooting toxicysts into the *Paramecium* (left). It then reels the *Paramecium* in to its proboscis.

SEM images: Gregory Antipa (San Francisco State University), H. S. Wessenberg

1 **Without sediment**

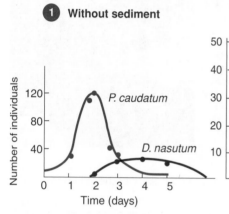

Five *Paramecium* were added to the medium and three *Didinium* added after two days.

2 **With sediment**

Medium included sediment in which the *Paramecium* could hide and not be encountered by the predator.

3 **Without sediment**

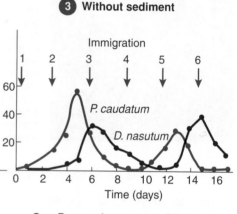

One *Paramecium* and one *Didinium* were introduced into the microcosm at day 0 and then every third day (immigration).

1. Why did the *Paramecium* die out in the first experiment? _____

_____

2. Why did the *Paramecium* survive in the second experiment? _____

_____

3. Why is there a lag in the predator population compared to the prey population in the third experiment?

_____

_____

_____

4. What did Gause's simple microcosm experiments tell us about the role of predation in limiting prey populations?

_____

_____

_____

_____

©2017 **BIOZONE** International
ISBN: 978-1-927309-65-0
Photocopying Prohibited

Huffaker's experiments

▶ Huffaker built on Gause's findings and attempted to design artificial systems that would better model a real world system. He worked on two mite species, the six spotted mite and its predator. Oranges provided both the habitat and the food for the prey.

▶ In a simple system, such as a small number of oranges grouped together, predators quickly ate all the prey and then died out.

▶ Huffaker then created a more complex system with arrays of 120 oranges (below). The amount of available food on each orange was controlled by sealing off parts of each orange with wax. Patchiness in the environment was created using balls (representing unsuitable habitat). Sticks aided dispersal of prey mites and petroleum jelly was used to form barriers for predatory mite dispersal. In this system, the predator and prey coexisted for three full cycles (> year). In the diagram below, the arrays depict the distribution and density of the populations at the arrowed points. The circles represent oranges or balls and the dots the predatory mites.

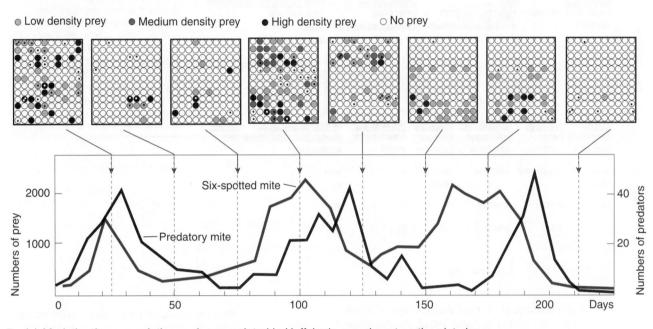

5. (a) Mark the three population cycles completed in Huffaker's experiment on the plot above.

   (b) In a different color, mark the lag in the predator population response to change in prey numbers.

   (c) What does the lag represent? _____

   _____

   _____

6. How well do you think Huffaker's model system approximated a real ecosystem? Use evidence from the arrays to discuss how variation in habitat makes it possible for populations to persist despite periodic declines in their numbers.

   _____

   _____

   _____

   _____

7. In 1960 the Alaska Department of Fish and Game released two wolves onto Coronation Island (116 km^2) to control the deer population. Indications were by 1961 the wolves had begun breeding and the population was increasing. By 1964 there were at least 13 wolves on the island. In 1965 wolves were seen on all the beaches, but there were few signs of deer. In 1968 only one wolf was sighted on the island. In 1983, there were no wolves, but plentiful deer once more.

   (a) From the information above, which of Gause's experiments does this data most closely follow?

   _____

   (b) Outline why the predator-prey populations behaved as they did: _____

   _____

   _____

   _____

230 Vertical Distribution in a Lake Community

Key Idea: Lake communities often show seasonal and daily changes in their patterns of distribution in the water column. A lake's water column is not uniform. Light intensity, light quality, temperature, and oxygen level may all show considerable variation with depth and these changes affect the movements and distribution of the plankton community. The distribution of animal plankton (zooplankton) in winter is often very different to their distribution in summer. Even on a daily basis, zooplankton populations will migrate vertically in the water column.

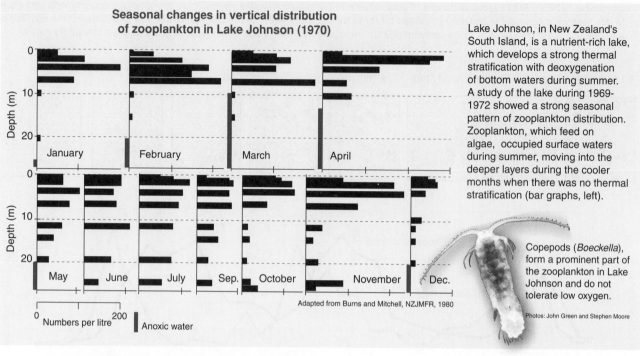

Seasonal changes in vertical distribution of zooplankton in Lake Johnson (1970)

Adapted from Burns and Mitchell, NZJMFR, 1980

0 Numbers per litre 200   ▮ Anoxic water

Lake Johnson, in New Zealand's South Island, is a nutrient-rich lake, which develops a strong thermal stratification with deoxygenation of bottom waters during summer. A study of the lake during 1969-1972 showed a strong seasonal pattern of zooplankton distribution. Zooplankton, which feed on algae, occupied surface waters during summer, moving into the deeper layers during the cooler months when there was no thermal stratification (bar graphs, left).

Copepods (*Boeckella*), form a prominent part of the zooplankton in Lake Johnson and do not tolerate low oxygen.

Photos: John Green and Stephen Moore

Daily pattern of vertical distribution of adult *Daphnia galeata* in Fuller Pond, Connecticut

Daphnia adults are prey for golden shiners, a small carnivorous fish. Golden shiners are visual predators and eat larger, adult *Daphnia*. They require enough light to see and capture their prey (graph, right). The kite graphs below show how *Daphnia* in Fuller Pond migrate in the water column over a 24 hour period.

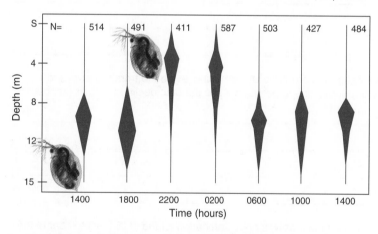

Predation rate at different light intensities

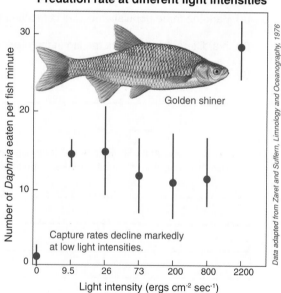

Golden shiner

Capture rates decline markedly at low light intensities.

Data adapted from Zaret and Suffern, Limnology and Oceanography, 1976

1. For the data relating to the Lake Johnson zooplankton community:

   (a) Describe the seasonal pattern of zooplankton distribution: _____

   (b) Identify the factor that appears to govern this pattern: _____

   (c) Is it biotic or abiotic? _____

2. What evidence is there, from the data presented, to indicate that *Daphnia* migrate in the water column to avoid predation?

   _____

   _____

   _____

PRACTICES

DATA

©2017 **BIOZONE** International
ISBN: 978-1-927309-65-0
Photocopying Prohibited

231 KEY TERMS AND IDEAS: Did You Get It?

1. Study the graph of population growth for a hypothetical population below and answer the following questions:

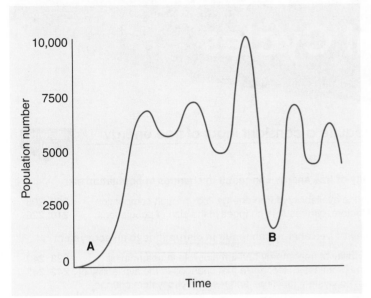

(a) Estimate the carrying capacity of the environment:

_____

(b) What happened at point **A** on the diagram?

_____

_____

_____

_____

(c) What happened at point **B** on the diagram?

_____

_____

_____

(d) What factors might have caused this? _____

_____

_____

2. Test your vocabulary by matching each term to its definition, as identified by its preceding letter code:

abiotic factor ....

biotic factor ....

community ....

competition ....

distribution ....

mortality ....

natality ....

population ....

predation ....

A The total number of individuals of a species within a set habitat or area.

B Exploitation involving the capture and consumption of an individual (the prey). The prey is killed outright.

C The term used in ecology for any contribution to the environment by a living organism.

D The birth rate of a population (usually expressed as the number of births per 1000 individuals).

E A term for any non-living part of the environment, e.g. rainfall or temperature.

F The location of individuals of a population within an area.

G The death rate of a population (usually expressed as the number of deaths per 1000 individuals).

H An interaction between organisms exploiting the same resource.

I A naturally occurring group of different species living within the same environment and interacting together.

3. The diagram right shows an area of lawn invaded in dandelions. Quadrats have been used to sample the dandelion population:

(a) From the quadrat based sample, what is the population density of dandelions?

_____

(b) What is the actual population density? _____

(c) Does this sample over- or underestimate the extent of the dandelion invasion? Can you suggest why?

_____

_____

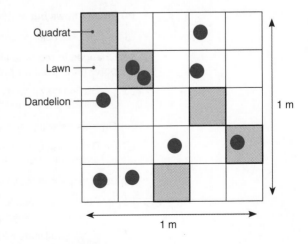

©2017 **BIOZONE** International
ISBN: 978-1-927309-65-0
Photocopying Prohibited

TEST

Energy Flow and Nutrient Cycles

2.A.1 All living systems require a constant input of free energy

Activity
number

Essential knowledge

(e) Changes in the availability of free energy can result in changes in population size

☐ 1 Recall how changes in the availability of free energy, e.g. through competition   207   208
or change in physical factors, can result in changes in the size of populations.   210   226

(f) Changes in the availability of free energy can result in disruptions to an ecosystem

☐ 1 Describe how the availability of free energy has a major role in determining   238   240
ecosystem structure and processes. Recognize that changes to the availability of   242   243
free energy may disrupt ecosystem functions and result in ecosystem change.

2.A.2 Organisms capture and store free energy for use in biological processes

Activity
number

Essential knowledge

(a) Autotrophs capture free energy from physical sources in the environment

☐ 1 Explain how photosynthetic organisms capture free energy in sunlight.   232   233
Recognize that photosynthetic is the basis for most food chains on Earth.   **recall 2**

☐ 2 Explain how chemosynthetic organisms capture free energy from small   233   237
inorganic molecules, often in the absence of oxygen.   **recall 2**

(b) Heterotrophs capture free energy in carbon compounds produced by other organisms

☐ 1 Describe how heterotrophs metabolize carbohydrates, proteins, and lipids as   233
sources of free energy. Identify the sources of these for heterotrophs.   **recall 2**

4.A.6 Interactions among living systems and with their physical environment result in the movement of matter and energy

Activity
number

Essential knowledge

(a) Energy flows but matter is recycled

☐ 1 Use examples to show how energy flows but matter is recycled in ecosystems.   241-247

(b) Changes in regional and global climates and in atmospheric composition influence patterns of primary productivity

☐ 1 Explain how primary productivity varies with changes in regional and global   238   239
climate and in atmospheric composition, e.g. as atmospheric CO_2 levels increase.

(c) Organisms within food webs and food chains interact

☐ 1 Distinguish between food chains and webs. Define trophic level and explain how   234-237
organisms interact through their feeding relationships.

(d) Food webs and food chains are dependent on primary productivity

☐ 1 Recall how producer biomass supports all consumers on Earth. Show this   232   237
relationship in energy flow diagrams and ecological pyramids.   241   242

☐   **PR-10** Estimate NPP and secondary productivity in the lab and determine   243
efficiencies of energy transfers from producer to consumer.

232 Plants as Producers

Key Idea: Photosynthesis by producers (photoautotrophs) directly and indirectly sustains almost all heterotrophic life. Plants, algae, and some bacteria are producers (they make their own food). They capture light energy and convert it into sugars through a process called **photosynthesis**.

The chemical energy stored in this food fuels the reactions that sustain life. Heterotrophs rely on producers directly or indirectly for their energy. The photosynthesis that occurs in the oceans is vital to the Earth's functioning, providing oxygen and absorbing carbon dioxide.

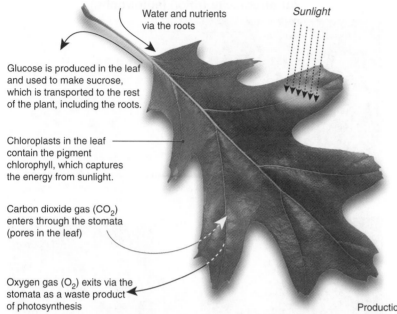

Water and nutrients via the roots

Sunlight

Glucose is produced in the leaf and used to make sucrose, which is transported to the rest of the plant, including the roots.

Chloroplasts in the leaf contain the pigment chlorophyll, which captures the energy from sunlight.

Carbon dioxide gas (CO_2) enters through the stomata (pores in the leaf)

Oxygen gas (O_2) exits via the stomata as a waste product of photosynthesis

Requirements for photosynthesis

Plants need only a few raw materials to make their own food:

- Light energy from the sun
- Chlorophyll absorbs light energy
- CO_2 gas is reduced to carbohydrate
- Water is split to provide the electrons for the fixation of carbon as carbohydrate

Photosynthesis is not a single process but two complex processes (the light dependent and light independent reactions) each with multiple steps.

Production of carbohydrate (light independent reactions) occur in the fluid stroma of chloroplast. This is commonly called carbon 'fixation'.

Energy capture (light dependent reactions) occur in the inner membranes of the chloroplast.

Photosynthesis equation	$6CO_2 + 12H_2O \xrightarrow[\text{Chlorophyll}]{\text{Light}} C_6H_{12}O_6 + 6O_2 + 6H_2O$

The photosynthesis of marine algae supplies a substantial portion of the world's oxygen. The oceans also act as sinks for absorbing large amounts of carbon dioxide.

Macroalgae, like this giant kelp, are important marine producers. Algae living near the ocean surface get access to light used in photosynthesis (the red wavelength).

On land, vascular plants (such as trees with transport vessels) are the main producers of food. Plants at different levels in a forest receive different intensity and quality of light.

1. Why is photosynthesis important for life on Earth? _____

_____

_____

_____

_____

2. (a) Identify the raw materials (inputs) for photosynthesis: _____

   (b) Identify the products (outputs) of photosynthesis: _____

CONNECT
14
AP2

WEB
232

KNOW

233 Autotrophs and Heterotrophs

Key Idea: Heterotrophs feed on other things to gain energy and carbon. Autotrophs (self feeders) use light or chemical energy to make their own food.

The nutritional mode of an organism describes how it obtains its energy and carbon. **Autotrophs** make their food from simple inorganic substances using the free energy in sunlight or chemical energy. **Heterotrophs** feed on other organisms to obtain their energy. They depend either directly on other organisms (dead or alive) or their by-products (e.g. feces, cell walls, or food stores).

1. Distinguish between photoautotrophs, chemoautotrophs, and chemoheterotrophs in terms of their sources of energy and carbon:

   (a) Photoautotroph

   Source of energy: _____

   _____

   Source of carbon: _____

   _____

   (b) Chemoautotrophs

   Source of energy: _____

   _____

   Source of carbon: _____

   _____

   (c) Chemoheterotroph

   Source of energy: _____

   _____

   Source of carbon: _____

   _____

2. Describe the three main nutritional modes of chemoheterotrophs:

   _____

   _____

   _____

   _____

   _____

   _____

   _____

3. What is the main difference between parasites and saprotrophs in the way in which obtain their nutrition?

   _____

   _____

   _____

   _____

   _____

Autotrophic nutrition (autotrophs)

Green plant

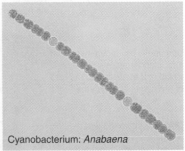
Cyanobacterium: *Anabaena*

Photoautotrophs (photosynthetic organisms) use light as their energy source, and carbon dioxide as a source of carbon to make their own food. They include bacteria, cyanobacteria, algae (photosynthetic protists), and green plants.

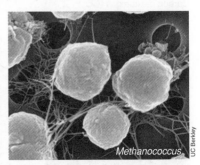
Methanococcus

Chemoautotrophs (chemosynthetic organisms) use inorganic compounds (e.g. elemental hydrogen) as a source of energy, and CO_2 as a source of carbon. Most are bacteria or archaea that live in hostile environments, such as geothermal and deep sea vents, e.g. *Methanococcus* (above) uses hydrogen to reduce CO_2 to methane.

Heterotrophic nutrition (chemoheterotrophs)

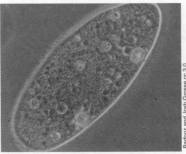

Heterotrophs need an organic source of carbon and energy (this is usually glucose). Many bacteria and many protists, e.g. *Paramecium* (above left) are heterotrophic. All fungi (above right) are chemoheterotrophs and most are decomposers (**saprotrophs**), obtaining nutrition from the extracellular digestion of dead organic material.

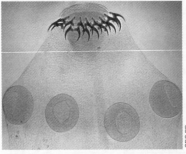

All animals are heterotrophs, relying on glucose (from plants, dead material, or other animals) for energy and carbon. **Parasites**, e.g. tapeworms (above left), live on or within their living host organism for part or all of their life. Bacteria, fungi, protists, and animals all have parasitic representatives. The ingestion of solid or liquid organic material from other organisms (**holozoic nutrition**) is the main feeding mode of animals.

©2017 **BIOZONE** International
ISBN: 978-1-927309-65-0
Photocopying Prohibited

234 Food Chains

Key Idea: A food chain is a model to illustrate the feeding relationships between organisms.

Organisms in ecosystems interact by way of their feeding (trophic) relationships. These interactions can be shown in a **food chain**, which illustrates how energy, in the form of food, passes from one organism to the next. The levels of a food chain are called **trophic levels**. Energy flows through trophic levels rather inefficiently, with only 5-20% of the energy at one level being transferred to the next (the rest is lost as heat). For this reason, food chains usually have fewer than six links. An organism is assigned to a trophic level based on its position in the food chain, but they may occupy different trophic levels in different food chains or during different stages of their life. Arrows link the organisms in a food chain. The direction of the arrow shows the flow of energy through the trophic levels. Most food chains begin with a producer, which is eaten by a primary consumer (herbivore). Higher level consumers (carnivores and omnivores) eat other consumers.

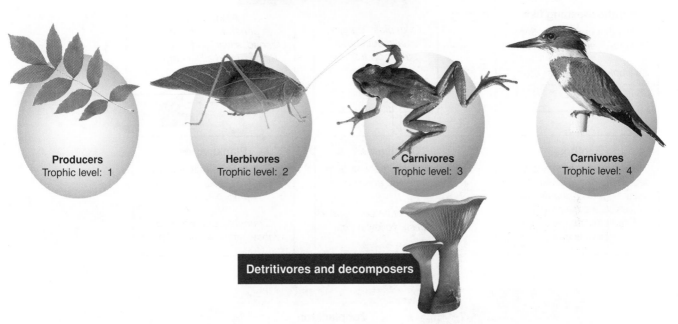

Respiration

Producers
Trophic level: 1

Herbivores
Trophic level: 2

Carnivores
Trophic level: 3

Carnivores
Trophic level: 4

Detritivores and decomposers

The diagram above represents the basic elements of a food chain. In the questions below, you are asked to add to the diagram the features that indicate the flow of energy through the community of organisms.

1. (a) State the original energy source for this food chain: _____

   (b) Draw arrows on the diagram above to show how the energy flows through the organisms in the food chain.

   (c) Label each of the arrows with the process that carries out this transfer of energy.

   (d) Draw arrows on the diagram to show how the energy is lost as heat by way of respiration.

2. (a) What happens to the amount of energy available to each successive trophic level in a food chain? _____

   _____

   (b) How does this limit the number of links in a food chain? _____

   _____

3. Discuss the trophic structure of ecosystems, including reference to food chains and trophic levels: _____

   _____

   _____

   _____

   _____

   _____

4. What could you infer about the trophic level(s) of the kingfisher if it was found to eat both katydids and frogs?

   _____

CONNECT
223
AP2

WEB
234

KNOW

235 Constructing a Food Web

Key Idea: Food chains intersect to form food webs. The complexity of a food web depends on the number of food chains and trophic levels involved.

Although feeding relationships can be shown as simple food chains, a more complex representation, called a food web, more accurately represents the complexity of trophic relationships in a community. Food webs recognize the different foods taken by organisms and the different trophic levels at which they feed. Species are assigned to trophic levels on the basis of their nutrition, with the first trophic level (the producers), ultimately supporting all other (consumer) levels. Consumers are ranked according to the trophic level they occupy, although some may feed at several levels. A simplified food web for Lake Erie is shown below.

A simplified Great Lake food web: Lake Erie (open water)

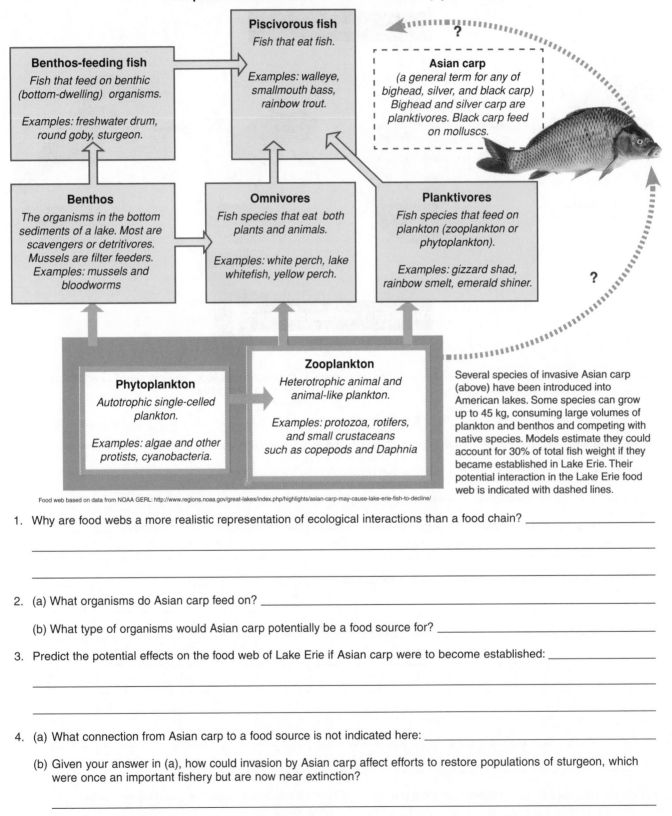

Piscivorous fish
Fish that eat fish.

Examples: walleye, smallmouth bass, rainbow trout.

Benthos-feeding fish
Fish that feed on benthic (bottom-dwelling) organisms.

Examples: freshwater drum, round goby, sturgeon.

Asian carp
(a general term for any of bighead, silver, and black carp)
Bighead and silver carp are planktivores. Black carp feed on molluscs.

Benthos
The organisms in the bottom sediments of a lake. Most are scavengers or detritivores. Mussels are filter feeders. Examples: mussels and bloodworms

Omnivores
Fish species that eat both plants and animals.

Examples: white perch, lake whitefish, yellow perch.

Planktivores
Fish species that feed on plankton (zooplankton or phytoplankton).

Examples: gizzard shad, rainbow smelt, emerald shiner.

Phytoplankton
Autotrophic single-celled plankton.

Examples: algae and other protists, cyanobacteria.

Zooplankton
Heterotrophic animal and animal-like plankton.

Examples: protozoa, rotifers, and small crustaceans such as copepods and Daphnia

Several species of invasive Asian carp (above) have been introduced into American lakes. Some species can grow up to 45 kg, consuming large volumes of plankton and benthos and competing with native species. Models estimate they could account for 30% of total fish weight if they became established in Lake Erie. Their potential interaction in the Lake Erie food web is indicated with dashed lines.

Food web based on data from NOAA GERL: http://www.regions.noaa.gov/great-lakes/index.php/highlights/asian-carp-may-cause-lake-erie-fish-to-decline/

1. Why are food webs a more realistic representation of ecological interactions than a food chain? _____

_____

_____

2. (a) What organisms do Asian carp feed on? _____

(b) What type of organisms would Asian carp potentially be a food source for? _____

3. Predict the potential effects on the food web of Lake Erie if Asian carp were to become established: _____

_____

_____

4. (a) What connection from Asian carp to a food source is not indicated here: _____

(b) Given your answer in (a), how could invasion by Asian carp affect efforts to restore populations of sturgeon, which were once an important fishery but are now near extinction?

_____

_____

©2017 **BIOZONE** International
ISBN: 978-1-927309-65-0
Photocopying Prohibited

5. From the information provided for the Lake Erie food web components on the previous page, construct four generalized food chains to show the feeding relationships between the organisms. Exclude Asian carp from your answer.

(a) _____

(b) _____

(c) _____

(d) _____

6. Construct a food chain to represent the inclusion of Asian carp into Lake Erie:

_____

7. Classify the organisms in the Lake Erie food web according to their trophic level:

(a) Producers (trophic level 1): _____

(b) Primary consumers (trophic level 2): _____

(c) Secondary consumers (trophic level 3): _____

(d) Tertiary consumers (trophic level 4): _____

(e) What trophic levels would Asian carp occupy in this food web? _____

8. The diagram below shows a simple grassland food web. Decide whether the following statements are true or false:

Sparrow

Goshawk

Owl

Snake

Mouse

Grasshopper

Grass

(a) The grass is the only producer in this food web: _____

(b) The grasshopper is a herbivore: _____

(c) The diagram shows two trophic levels: _____

(d) All the animals are consumers: _____

(e) The snake and the goshawk are at the same trophic level: _____

(f) The snake and the sparrow are at the same trophic level : _____

(g) The owl is both a secondary and a tertiary consumer: _____

236 Cave Food Webs

Key Idea: The food webs of cave ecosystems are fragile and based on only a few resources.

Cave environments lack the light that sustains most ecosystems. Despite this, a wide range of animals are adapted to live there. Some animals, such as bats, are not permanent cave-dwellers, but rest and breed there. Around the entrance of the cave, the owl (1) preys on the mouse (2) which itself feeds on the vegetation outside the cave. The owl and the mouse leave droppings that support the cave dung beetle (3) and millipede (4). The cave cricket (5) scavenges dead birds and mammals near the entrance. The harvestman

(6) is a predator of the dung beetle, the millipede, and the cricket. Inside the cave, the horseshoe bat (7) roosts and breeds, leaving the cave to feed on flying insects. The bats produce vast quantities of guano (droppings). The guano is eaten by the blind cave beetle (8), millipede (4) and springtail (9). These invertebrates are hunted by the predatory cave spider (10). In underground pools, the bat guano supports the growth of bacteria (11). Flatworms (12) and isopods (13) feed on the bacteria and are eaten by the blind cave shrimp (14). The blind cave fish (15) is the top predator in this system, feeding on isopods and the blind cave shrimps.

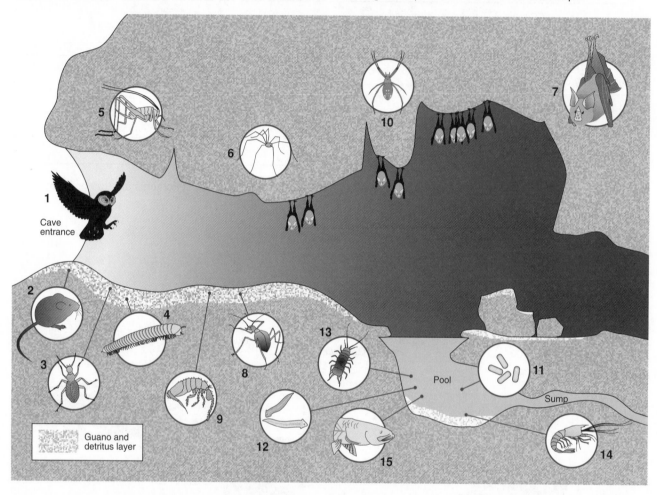

1. Using the lake food web activity as a guide, construct a food web for the cave ecosystem on a separate sheet of paper. For animals that feed outside the cave, do not include this outside source of food. As in the lake food web, label each species with the following codes to indicate its diet type (e.g. producer, detritivore, herbivore, carnivore, omnivore) and its position in the food chain if it is a consumer (1st, 2nd, 3rd, 4th order consumer). Staple your finished web to this page.

2. Which major trophic level is missing from the cave food web? _____

3. How is energy imported into the cave's food web? _____

_____

4. How might energy be removed from the cave ecosystem? _____

_____

_____

5. In many parts of the world, cave-dwelling bat species are endangered, often taken as food by humans or killed as pests. Explain how the cave food web would be affected if bat numbers were to fall substantially:

_____

_____

©2017 **BIOZONE** International
ISBN: 978-1-927309-65-0
Photocopying Prohibited

237 The Darkest Depths

Key Idea: Deep sea hydrothermal vents are the site of unique communities supported by chemosynthetic bacteria.

Deep sea hydrothermal vents occur at around 2000 m depth and where tectonic plates meet. Buckling in the plates causes fault lines to form where water can move down into the crust before being heated and ejected at temperatures of up to 350°C. Temperatures this high are possible because the pressure of the ocean prevents the water from boiling into steam. The high water temperature dissolves minerals from crustal rocks and, when they reach the surface, they precipitate into formations called **black smokers** (mineral chimneys up to 60 m high). At this depth, no light penetrates and the amount of organic debris falling from above is minimal because much of it has been used up by the time it reaches the bottom. The organisms here are restricted in their movement; only a few tens of meters from the mouth of the vent, the water temperature plummets to barely above freezing. The isolation has resulted in the evolution of a unique fauna.

The water spewing from the hydrothermal vents is rich in minerals and the chemosynthetic bacteria living there use oxygen and hydrogen sulfide (highly toxic to most organisms) to build organic molecules. They are the producers on which the vent community is based. They form thick mats around the vents or float in aggregations resembling snow storms.

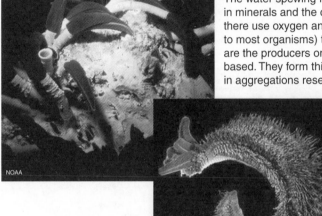

NOAA

Photo: US Federal Govt

Tube worms, one of the larger organisms in these communities, provide shelter for the bacteria and benefit from the products of bacterial chemosynthesis. Vent mussels also have bacteria living within them and have abandoned a filter-feeding lifestyle to form a mutualistic relationship with bacteria in their tissues. Blind shrimps and crabs scavenge on decaying material and the bacterial mats. Octopi and fish also make up part of the food web, preying on smaller animals. Of most interest to scientists is the Pompeii worm. It can withstand temperatures of 80°C; higher than any other complex organism. The hairy coat that covers it is, in fact, mats of bacteria on which the worm feeds.

1. Describe the environmental conditions found around deep sea hydrothermal vents: _____

_____

_____

_____

2. Explain reasons for the uniqueness of the vent communities: _____

_____

_____

3. Discuss the relationships between the organisms of the vent community and use the information to construct a basic food web in the space provided:

_____

_____

_____

_____

_____

_____

_____

_____

_____

_____

©2017 **BIOZONE** International
ISBN: 978-1-927309-65-0
Photocopying Prohibited

PRACTICES

KNOW

238 Global Primary Productivity

Key Idea: Primary productivity varies between regions, and is influenced by atmospheric conditions, such as CO_2 levels. Primary production is the synthesis of organic compounds from carbon dioxide. Glucose and other carbohydrates are the primary organic compounds produced, mostly through the process of photosynthesis, but chemosynthetic processes are also involved. Changes to atmospheric conditions (e.g. CO_2 or precipitation levels) can influence patterns of primary productivity. Such variations can occur on both a global scale and locally, varying between regions and ecosystem types.

Earth's climate and biomes

▶ The Earth's biomes are the largest geographically-based biotic communities that can be conveniently recognized.

▶ **Biomes** are large areas where the vegetation type shares a particular suite of physical requirements. Biomes have characteristic features, but the boundaries between them are not distinct.

▶ The same biome may occur in widely separated regions of the world wherever the climatic and soil conditions are similar.

▶ Terrestrial biomes are recognized for all the major climatic regions of the world. They are classified by their predominant vegetation type. Biomes are closely related to the major air cells that circle the Earth and are reflected in the Northern and Southern Hemispheres.

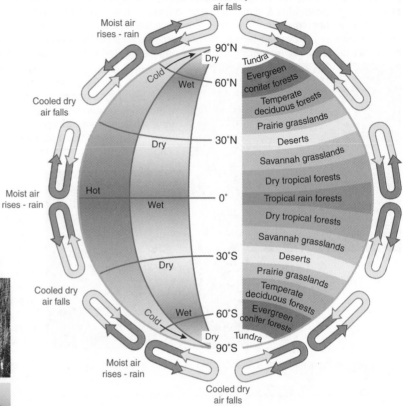

Prairie grassland

Evergreen conifer forest

Desert

Tropical rain forest

The productivity of ecosystems varies

The **gross primary productivity** of an ecosystem will depend on the capacity of the producers to capture and fix carbon in organic compounds. In most ecosystems, this is limited by constraints on photosynthesis (availability of light, nutrients, or water for example).

The **net primary productivity** (NPP) is then determined by how much of this goes into plant biomass per unit time after respiratory needs are met. This will be the amount available to the next trophic level.

Productive swamp ecosystem: Southern Louisiana

Jan Kronsell CC3.0 SA

Estuaries
Swamps and marshes
Tropical rainforest
Temperate forest
Boreal forest
Savanna
Agricultural land
Woodland and shrubland
Temperate grassland
Lakes and streams
Continental shelf
Tundra
Open ocean
Desert scrub
Extreme desert

5 10 15 20 25 30 35 40 45 50
Average net primary productivity (x 1000 kJ m^{-2}y^{-1})

Globally, the least productive ecosystems are those that are limited by heat energy and water. The most productive are those with high light and temperature, plenty of water, and non-limiting supplies of soil nitrogen. The primary productivity of oceans is lower overall than that of terrestrial ecosystems because the water reflects (or absorbs) much of the light energy before it reaches and is utilized by producers. Many regions of the open ocean are also low in nutrients.

©2017 **BIOZONE** International
ISBN: 978-1-927309-65-0

The effect of climate change on NPP

Maintaining NPP above a certain level is important for food security and ecosystem sustainability. Some regions, such as the African continent, are highly vulnerable to climate variability. More than 40% of Africa's population live in arid or semiarid regions, and changes in rainfall (e.g. drought) can severely reduce productivity (reduced NPP), including that of food crops.

Africa has undergone a number of climatic changes, all of which have affected NPP. These include a 5°C increase in air temperature over the last 100 years, and an increase in CO_2 from 280 ppm (preindustry) to 380 ppm in 2005. Spatial variations in rainfall across the continent were also a contributing factor. Some areas received more rainfall, others less, and some suffered from drought. Nitrogen deposition (in the form of added fertilizer) is also a major factor in determining NPP. These changes have a strong effect on NPP in Africa (right).

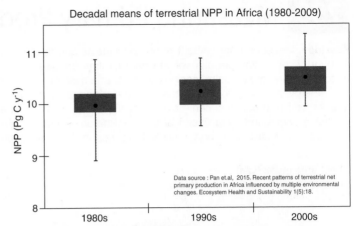

Decadal means of terrestrial NPP in Africa (1980-2009)

Data source : Pan et.al, 2015. Recent patterns of terrestrial net primary production in Africa influenced by multiple environmental changes. Ecosystem Health and Sustainability 1(5):18.

The plot above shows how terrestrial NPP has changed over three decades. These changes are primarily due to variation in precipitation (rainfall) levels, elevated CO_2 levels, and nitrogen deposition. Circles = mean NPP. Boxes = the first and third quartiles (medians of the lower and upper halves of the data set respectively). Whiskers = minimum and maximum values.

1. Define a biome: _____
_____

2. (a) What is net primary productivity (NPP)?_____
_____

(b) What process accounts for most NPP? _____

(c) Describe how NPP is calculated: _____
_____

3. Studying the ecosystem NPP graph on the previous page:

(a) Identify the three most productive ecosystems: _____

(b) What factors are likely to contribute to this high productivity? _____
_____

(c) Why do deserts have low NPP? _____
_____

(d) Why is primary productivity lower in the oceans than on land? _____
_____
_____

4. Study the graph of NPP in Africa at the top of this page.

(a) Describe the trend in NPP between 1980 - 2009: _____
_____
_____

(b) Identify the factors contributing to this trend: _____
_____
_____
_____

(c) Predict the effect a sustained drought would have on NPP: _____
_____

©2017 **BIOZONE** International
ISBN: 978-1-927309-65-0
Photocopying Prohibited

239 Measuring Primary Productivity

Key Idea: Measuring the amount of photosynthesis and the amount of respiration per unit volume per unit time enables us to determine the gross primary productivity of a system. The energy available to ecosystem is determined by the rate at which producers can convert sunlight energy or inorganic compounds into chemical energy. Photosynthesis accounts for most of the energy available to most of Earth's ecosystems. The total energy fixed by photosynthesis per unit area or volume per unit time is the **gross primary productivity** (GPP) and it is usually expressed as J m^{-2} (or kJ m^{-2}), or as g m^{-2}. However, some of this energy is required for respiration. Subtracting respiration from GPP gives the **net primary productivity** (NPP). This represents the energy or biomass (mass of biological material) available to consumers.

Leaf area index (LAI)
Leaf area index is a measure of the total leaf area of a given plant.

Relative growth rate (R)
Relative growth rate is the gain in mass of plant tissue per unit time.

$$\frac{\text{Increase in dry mass in unit time}}{\text{Original dry mass of the plant}}$$

Harvestable dry biomass
Used for commercial purposes, it is the dry mass of crop available for sale or use.

Net assimilation rate (NAR)
NAR is the increase in plant weight per unit of leaf area per unit time. Essentially it is the balance between carbon gain from photosynthesis and carbon loss from respiration.

$$NAR = \frac{\text{Increase in dry mass in unit time}}{\text{Leaf area}}$$

Measuring productivity

Measuring gross primary productivity (GPP) can be difficult due to the effect of on-going respiration, which uses up some of the organic material produced (glucose). One method for measuring GPP is to measure the difference in production between plants kept in the dark and those in the light. A simple method for measuring GPP in phytoplankton is illustrated below.

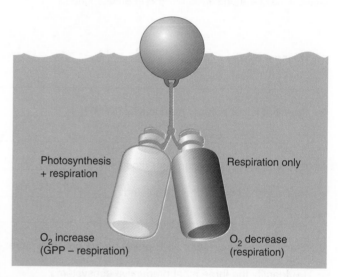

Photosynthesis + respiration

Respiration only

O_2 increase (GPP – respiration)

O_2 decrease (respiration)

Two bottles are lowered into the ocean or lake to a specified depth, filled with water, and stoppered. One bottle is transparent, the other is opaque. The O_2 concentration of the water around the bottles is measured and the bottles are left for a specified amount of time. The phytoplankton in the transparent bottle will photosynthesize, increasing the O_2 concentration, and respire, using some of that O_2. The phytoplankton in the opaque bottle will only respire. The final measured difference in O_2 between the bottles gives the amount of O_2 produced by the phytoplankton in the specified time (including that used for respiration). The amount of O_2 used allows us to determine the amount of glucose produced and therefore the GPP of the phytoplankton.

1. What would be the effect on subsequent trophic levels if gross primary productivity was to drop? _____

_____

_____

2. An experiment was carried out to measure the gross primary production of a lake system. The lake was initially measured to have 8 mg O_2 L^{-1}. A clear flask and an opaque flask were lowered into the lake filled and stoppered. When the flasks were retrieved it was found the clear flask contained 10 mg O_2 L^{-1} while the opaque contained 5 mg O_2 L^{-1}.

   (a) How much O_2 was used (respired) in the opaque flask? _____

   (b) What is the net O_2 production in the clear flask? _____

   (c) What is the gross O_2 production in the system? _____

   (d) For every 10 g of O_2 formed during photosynthesis, 9.4 grams of glucose is formed. How much glucose formed during the experiment?

   _____

©2017 **BIOZONE** International
ISBN: 978-1-927309-65-0
Photocopying Prohibited

240 Production and Trophic Efficiency

Key Idea: The net primary productivity of an ecosystem determines the amount of biomass available to primary consumers. It varies widely between different ecosystems. The energy entering ecosystems is fixed by producers at a rate that depends on limiting factors such as temperature and the availability of light and water. This energy is converted to biomass by anabolic reactions. The rate of biomass production (net primary productivity), is the biomass produced per area

per unit time. **Trophic (or ecological) efficiency** refers to the efficiency of energy transfer from one trophic level to the next. The trophic efficiencies of herbivores vary widely, depending on how much of the producer biomass is consumed and assimilated (incorporated into new biomass). In some natural ecosystems this can be surprisingly high. Humans intervene in natural energy flows by simplifying systems and reducing the number of transfers occurring between trophic levels.

The productivity of natural grassland ecosystems

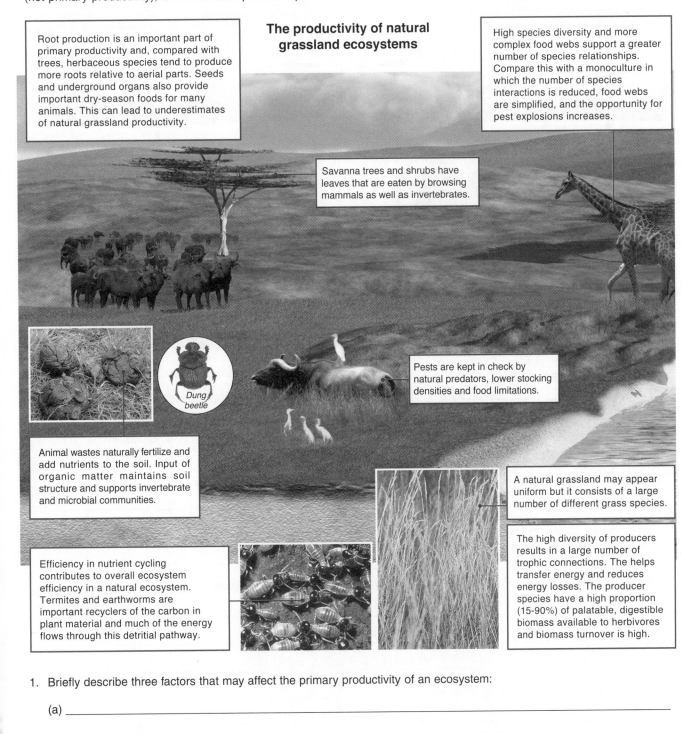

Root production is an important part of primary productivity and, compared with trees, herbaceous species tend to produce more roots relative to aerial parts. Seeds and underground organs also provide important dry-season foods for many animals. This can lead to underestimates of natural grassland productivity.

High species diversity and more complex food webs support a greater number of species relationships. Compare this with a monoculture in which the number of species interactions is reduced, food webs are simplified, and the opportunity for pest explosions increases.

Savanna trees and shrubs have leaves that are eaten by browsing mammals as well as invertebrates.

Dung beetle

Pests are kept in check by natural predators, lower stocking densities and food limitations.

Animal wastes naturally fertilize and add nutrients to the soil. Input of organic matter maintains soil structure and supports invertebrate and microbial communities.

A natural grassland may appear uniform but it consists of a large number of different grass species.

Efficiency in nutrient cycling contributes to overall ecosystem efficiency in a natural ecosystem. Termites and earthworms are important recyclers of the carbon in plant material and much of the energy flows through this detrital pathway.

The high diversity of producers results in a large number of trophic connections. The helps transfer energy and reduces energy losses. The producer species have a high proportion (15-90%) of palatable, digestible biomass available to herbivores and biomass turnover is high.

1. Briefly describe three factors that may affect the primary productivity of an ecosystem:

(a) _____

(b) _____

(c) _____

©2017 **BIOZONE** International
ISBN: 978-1-927309-65-0
Photocopying Prohibited

WEB
240
KNOW

Agriculture and productivity

Increasing net productivity in agriculture (increasing yield) is a matter of manipulating and maximizing energy flow through a reduced number of trophic levels. On a farm, the simplest way to increase the net primary productivity is to produce a monoculture (single crop). Monocultures reduce competition between the desirable crop and weed species, allowing crops to put more energy into biomass. Other agricultural practices designed to increase productivity in crops include fertilizer (nitrogen) application, pest (herbivore) control and spraying to reduce disease. Higher productivity in feed-crops also allows greater secondary productivity (e.g. in livestock). Here, similar agricultural practices make sure the energy from feed-crops is efficiently assimilated by livestock.

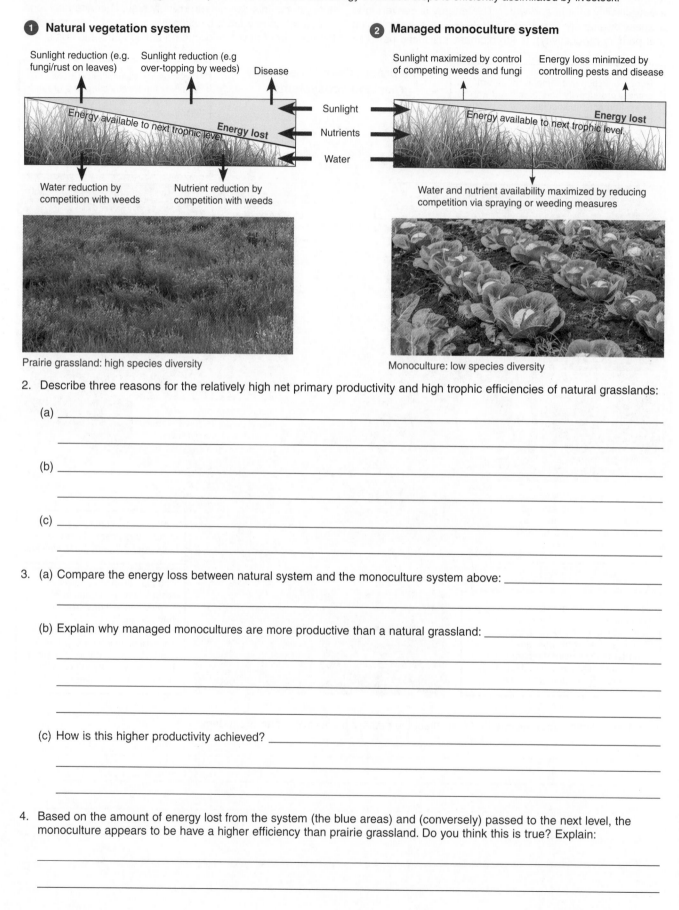

1 **Natural vegetation system**

Sunlight reduction (e.g. fungi/rust on leaves)

Sunlight reduction (e.g over-topping by weeds)

Disease

Energy available to next trophic level

Energy lost

Sunlight

Nutrients

Water

Water reduction by competition with weeds

Nutrient reduction by competition with weeds

Prairie grassland: high species diversity

2 **Managed monoculture system**

Sunlight maximized by control of competing weeds and fungi

Energy loss minimized by controlling pests and disease

Energy available to next trophic level

Energy lost

Water and nutrient availability maximized by reducing competition via spraying or weeding measures

Monoculture: low species diversity

2. Describe three reasons for the relatively high net primary productivity and high trophic efficiencies of natural grasslands:

(a) _____

_____

(b) _____

_____

(c) _____

_____

3. (a) Compare the energy loss between natural system and the monoculture system above: _____

_____

(b) Explain why managed monocultures are more productive than a natural grassland: _____

_____

_____

_____

(c) How is this higher productivity achieved? _____

_____

_____

4. Based on the amount of energy lost from the system (the blue areas) and (conversely) passed to the next level, the monoculture appears to be have a higher efficiency than prairie grassland. Do you think this is true? Explain:

_____

_____

©2017 **BIOZONE** International
ISBN: 978-1-927309-65-0
Photocopying Prohibited

241 Quantifying Energy Flow in an Ecosystem

Key Idea: Chemical energy in the bonds of molecules flows through an ecosystem between trophic levels. Only 5-20% of energy is transferred from one trophic level to the next.

Energy cannot be created or destroyed, only transformed from one form (e.g. light energy) to another (e.g. chemical energy in the bonds of molecules). This means that the flow of energy through an ecosystem can be measured. Each time energy is transferred from one trophic level to the next (by eating, defecation, etc.), some energy is given out as heat to the environment, usually during cellular respiration. Living organisms cannot convert heat to other forms of energy, so the amount of energy available to one trophic level is always less than the amount at the previous level. Potentially, we can account for the transfer of energy from its input (as solar radiation) to its release as heat from organisms, because energy is conserved. Recall that the percentage of energy transferred from one trophic level to the next is the **trophic efficiency**. It varies between 5% and 20% and measures the efficiency of energy transfer. An average figure of 10% trophic efficiency is often used. This is called the **ten percent rule**.

Energy flow through an ecosystem

NOTE
Numbers represent **kilojoules** of energy per square meter per year ($kJ\ m^{-2}\ yr^{-1}$)

Sunlight falling on plant surfaces
7,000,000

Light absorbed by plants
1,700,000

Ⓐ

Energy absorbed from the previous trophic level
100

Energy lost as heat ← 65 | **Trophic level** | 15 → Energy lost to detritus

20

Energy passed on to the next trophic level

The energy available to each trophic level will always equal the amount entering that trophic level, minus total losses to that level (due to metabolic activity, death, excretion etc). Energy lost as heat will be lost from the ecosystem. Other losses become part of the detritus and may be utilized by other organisms in the ecosystem

Producers
87,400

50,450

22,950

(a)

Heat loss in metabolic activity

7,800

Primary consumers

Ⓑ

1600

4,600

Ⓖ

(b)

Secondary consumers

1,330

2,000

Detritus

10,465

Ⓓ

90

(c)

19,300

(d)

55

Tertiary consumers

Ⓒ

Ⓕ

Ⓔ

Decomposers and detritivores

19,200

1. Study the diagram above illustrating energy flow through a hypothetical ecosystem. Use the example at the top of the page as a guide to calculate the missing values (a)–(d) in the diagram. Note that the sum of the energy inputs always equals the sum of the energy outputs. Place your answers in the spaces provided on the diagram.

PRACTICES | PRACTICES | CONNECT | WEB

1
AP2

241

DATA

2. What is the original source of energy for this ecosystem? _____

3. Identify the processes occurring at the points labelled **A – G** on the diagram:

   A. _____   E. _____

   B. _____   F. _____

   C. _____   G. _____

   D. _____

4. (a) Calculate the percentage of light energy falling on the plants that is absorbed at point **A**:

   Light absorbed by plants ÷ sunlight falling on plant surfaces x 100 = _____

   (b) What happens to the light energy that is not absorbed? _____

   _____

5. (a) Calculate the percentage of light energy absorbed that is actually converted (fixed) into producer energy:

   Producers ÷ light absorbed by plants x 100 = _____

   (b) How much light energy is absorbed but not fixed: _____

   (c) Account for the difference between the amount of energy absorbed and the amount actually fixed by producers:

   _____

   _____

6. Of the total amount of energy fixed by producers in this ecosystem (at point **A**) calculate:

   (a) The total amount that ended up as metabolic waste heat (in kJ): _____

   (b) The percentage of the energy fixed that ended up as waste heat: _____

7. (a) State the groups for which detritus is an energy source: _____

   (b) How could detritus be removed or added to an ecosystem? _____

   _____

8. Under certain conditions, decomposition rates can be very low or even zero, allowing detritus to accumulate:

   (a) From your knowledge of biological processes, what conditions might slow decomposition rates?

   _____

   (b) What are the consequences of this lack of decomposer activity to the energy flow? _____

   _____

   (c) Add an additional arrow to the diagram on the previous page to illustrate your answer.

   (d) Describe three examples of materials that have resulted from a lack of decomposer activity on detrital material:

   _____

   _____

9. The ten percent rule states that the total energy content of a trophic level in an ecosystem is only about one-tenth (or 10%) that of the preceding level. For each of the trophic levels in the diagram on the preceding page, determine the amount of energy passed on to the next trophic level as a percentage:

   (a) Producer to primary consumer: _____

   (b) Primary consumer to secondary consumer: _____

   (c) Secondary consumer to tertiary consumer: _____

   (d) Which of these transfers is the most efficient? _____

©2017 **BIOZONE** International
ISBN: 978-1-927309-65-0
Photocopying Prohibited

242 Ecological Pyramids

Key Idea: Ecological pyramids are used to illustrate the number of organisms, amount of energy, or amount of biomass at each trophic level in an ecosystem.

The energy, biomass, or numbers of organisms at each trophic level in any ecosystem can be represented by an ecological pyramid. The first trophic level is placed at the bottom of the pyramid and subsequent trophic levels are stacked on top in their 'feeding sequence'. Ecological pyramids provide a convenient model to illustrate the relationship between different trophic levels in an ecosystem.

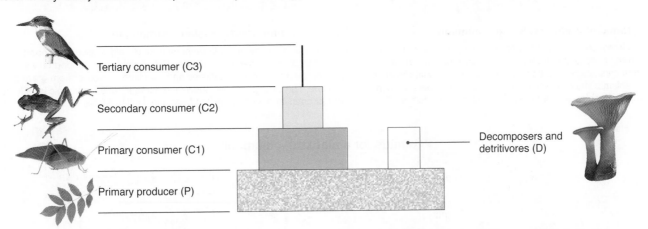

Tertiary consumer (C3)

Secondary consumer (C2)

Primary consumer (C1)

Primary producer (P)

Decomposers and detritivores (D)

The generalized ecological pyramid pictured above shows a conventional pyramid shape, with a large number (or biomass) of producers forming the base for an increasingly small number (or biomass) of consumers. Decomposers are placed at the level of the primary consumers and off to the side. They may obtain energy from many different trophic levels and so do not fit into the conventional pyramid structure. For any particular ecosystem at any one time (e.g. the forest ecosystem below), the shape of this typical pyramid can vary greatly depending on whether the trophic relationships are expressed as numbers, biomass or energy

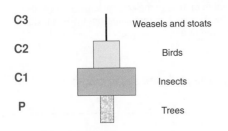

C3 — Weasels and stoats

C2 — Birds

C1 — Insects

P — Trees

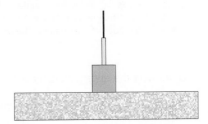

Numbers in a forest community

Pyramids of numbers display the number of individual organisms at each trophic level. The pyramid above has few producers, but they may be of a very large size (e.g. trees). This gives an 'inverted pyramid', although not all pyramids of numbers are like this.

Biomass in a forest community

Biomass pyramids measure the 'weight' of biological material at each trophic level. Water content of organisms varies, so 'dry weight' is often used. Organism size is taken into account, allowing meaningful comparisons of different trophic levels.

Energy in a forest community

Pyramids of energy are often very similar to biomass pyramids. The energy content at each trophic level is generally comparable to the biomass (i.e. similar amounts of dry biomass tend to have about the same energy content).

1. What do each of the following types of ecological pyramids measure?

(a) Number pyramid: _____

(b) Biomass pyramid: _____

(c) Energy pyramid: _____

2. What is the advantage of using a biomass or energy pyramid rather than a pyramid of numbers to express the relationship between different trophic levels?

_____

_____

_____

_____

3. How can a forest community with relatively few producers (see next page) support a large number of consumers?

_____

_____

©2017 **BIOZONE** International
ISBN: 978-1-927309-65-0
Photocopying Prohibited

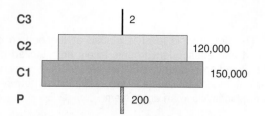

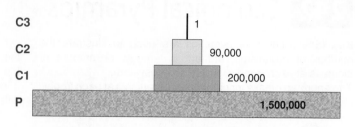

Pyramid of numbers: forest community

In a forest community, a few producers may support a large number of consumers. This is due to the large size of the producers; large trees can support many individual consumer organisms. The example above shows the numbers at each trophic level for an oak forest in England, in an area of 10 m².

Pyramid of numbers: grassland community

In a grassland community, a large number of (small) producers support a much smaller number of consumers. Grass plants can support only a few individual consumer organisms and take time to recover from grazing pressure. The example above shows the numbers at each trophic level for a derelict grassland area (10 m²) in Michigan, United States.

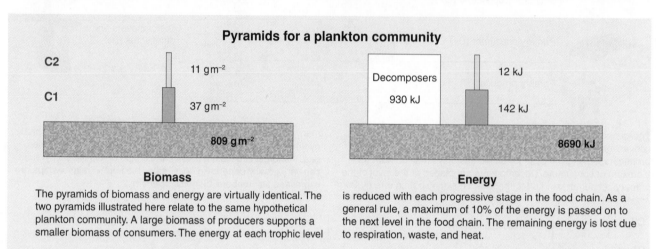

Pyramids for a plankton community

Biomass

The pyramids of biomass and energy are virtually identical. The two pyramids illustrated here relate to the same hypothetical plankton community. A large biomass of producers supports a smaller biomass of consumers. The energy at each trophic level

Energy

is reduced with each progressive stage in the food chain. As a general rule, a maximum of 10% of the energy is passed on to the next level in the food chain. The remaining energy is lost due to respiration, waste, and heat.

4. Determine the energy transfer between trophic levels in the plankton community example in the above diagram:

(a) Between producers and the primary consumers: _____

(b) Between the primary consumers and the secondary consumers: _____

(c) Why is the amount of energy transferred from the producer level to primary consumers considerably less than the expected 10% that occurs in many other communities?

_____

_____

(d) After the producers, which trophic group has the greatest energy content? _____

(e) Give a likely explanation why this is the case: _____

_____

_____

_____

An unusual biomass pyramid

The biomass pyramids of some ecosystems appear rather unusual with an inverted shape. The first trophic level has a lower biomass than the second level. What this pyramid does not show is the rate at which the producers (algae) are reproducing in order to support the larger biomass of consumers.

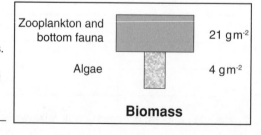

5. Give a possible explanation of how a small biomass of producers (algae) can support a larger biomass of consumers (zooplankton):

_____

_____

_____

©2017 **BIOZONE** International
ISBN: 978-1-927309-65-0
Photocopying Prohibited

243 Investigating Trophic Efficiencies

Key Idea: The efficiency of energy transfers in ecosystems can be quantified if we know the amount of energy entering and leaving the different trophic levels.

The gross primary production of any ecosystem will be determined by the efficiency with which solar energy is captured by photosynthesis. The efficiency of subsequent energy transfers will determine the amount of energy available to consumers. These energy transfers can be quantified using measurements of dry mass. In this activity, you will calculate energy and biomass transfers in real and experimental systems. This analysis will help you to more easily plan and carry out your own investigation (AP #10).

Production vs productivity: What's the difference?

Strictly speaking, the primary production of an ecosystem is distinct from its productivity, which is the amount of production per unit time (a rate). However because values for production (accumulated biomass) are usually given for a certain period of time in order to be meaningful, the two terms are often used interchangeably.

Corn field

Mature pasture

1. The energy budgets of two agricultural systems (4000 m² area) were measured over a growing season of 100 days. The results are tabulated right.

   (a) For each system, calculate the percentage efficiency of energy utilization (how much incident solar radiation is captured by photosynthesis):

   Corn: _____

   Mature pasture: _____

   (b) For each system, calculate the percentage losses to respiration:

   Corn: _____

   Mature pasture: _____

   (c) For each system, calculate the percentage efficiency of NPP:

   Corn: _____

   Mature pasture: _____

   (d) Which system has the greatest efficiency of energy transfer to biomass? _____

	Corn field	Mature pasture
	kJ x 10⁶	kJ x 10⁶
Incident solar radiation	8548	1971
Plant utilization		
Net primary production (NPP)	105.8	20.7
Respiration (R)	32.2	3.7
Gross primary production (GPP)	138.0	24.4

Estimating NPP in *Brassica rapa*

Background

Brassica rapa (right) is a fast growing brassica species, which can complete its life cycle in as little as 40 days if growth conditions are favorable. A class of students wished to estimate the gross and net primary productivity of a crop of these plants using wet and dry mass measurements made at three intervals over 21 days.

The method

▸ Seven groups of three students each grew 60 *B. rapa* plants in plant trays under controlled conditions. On day 7, each group made a random selection of 10 plants and removed them, with roots intact. The 10 plants were washed, blotted dry, and then weighed collectively (giving wet mass).

▸ The 10 plants were placed in a ceramic drying bowl and placed in a drying oven at 200°C for 24 hours, then weighed (giving dry mass).

▸ On day 14 and again on day 21, the procedure was repeated with a further 10 plants (randomly selected).

▸ The full results for group 1 are presented in Table 1 on the next page. You will complete the calculation columns.

©2017 **BIOZONE** International
ISBN: 978-1-927309-65-0
Photocopying Prohibited

PRACTICES PRACTICES PRACTICES PRACTICES CONNECT

14
AP2

DATA

Table 1: Group 1's results for growth of 10 *B. rapa* **plants over 21 days**

Age in days	Wet mass of 10 plants (g)	Dry mass of 10 plants (g)	Percent biomass	Energy in 10 plants (kJ)	Energy per plant (kJ)	NPP (kJ plant⁻¹ d⁻¹)
7	19.6	4.2				
14	38.4	9.3				
21	55.2	15.5				

2. Calculate percent biomass using the equation: % biomass = dry mass ÷ wet mass x 100. Enter the values in Table 1.

3. Each gram of dry biomass is equivalent to 18.2 kJ of energy. Calculate the amount of energy per 10 plants and per plant for plants at 7, 14, and 21 days. Enter the values in Table 1.

4. Calculate the Net Primary Productivity per plant, i.e. the amount of energy stored as biomass per day (kJ plant⁻¹ d⁻¹). Enter the values in Table 1. We are using per plant in this exercise as we do not have a unit area of harvest.

5. The other 6 groups of students completed the same procedure and, at the end of the 21 days, the groups compared their results for NPP. The results are presented in Table 2, below.

    Transfer group 1's NPP results from Table 1 to complete the table of results and calculate the mean NPP for *B. rapa*.

Table 2: Class results for NPP of *B. rapa* **over 21 days**

Time in days (d)	Group NPP (kJ plant⁻¹ d⁻¹)							Mean NPP
	1	2	3	4	5	6	7	
7		1.05	1.05	1.13	1.09	1.13	1.09	
14		1.17	1.21	1.25	1.21	1.25	1.17	
21		1.30	1.34	1.30	1.34	1.38	1.34	

6. On the grid (right), plot the class mean NPP vs time.

7. (a) What is happening to the NPP over time?

    _____

    _____

    (b) Explain why this is happening: _____

    _____

    _____

    _____

    _____

    _____

    _____

8. What would you need to know to determine the gross primary productivity of *B. rapa*?

    _____

    _____

9. Net production in consumers (N), or secondary production, can be expressed as N = I - (F+R). Red meat contains approximately 700 kJ per 100 grams. If N = 20% of the energy gain (I), how much energy is lost as F and R?

    _____

©2017 **BIOZONE** International
ISBN: 978-1-927309-65-0
Photocopying Prohibited

Calculating energy flow from producers to primary consumers

Secondary production is the generation of primary consumer (heterotrophic) biomass in a system. In this experiment, students determined the net secondary production and respiratory losses using 12 day old cabbage white larvae feeding on Brussels sprouts. Of the NPP from the Brussels sprouts that is consumed by the larvae, some will be used in cellular respiration, some will be available to secondary consumers (the **net secondary production**) and some will be lost as egested waste products (**frass**).

The method

▶ The wet mass of ten, 12 day old larvae, and approximately 30 g Brussels sprouts was accurately measured and recorded.

▶ The larvae and Brussel sprouts were placed into an aerated container. After three days the container was disassembled and the wet mass of the Brussels sprouts, larvae, and frass was individually measured and recorded.

▶ The Brussels sprouts, larvae and frass were placed in separate containers and placed in a drying oven and their dry mass was recorded.

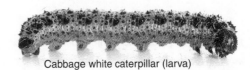

Cabbage white caterpillar (larva)

Note: We assume the % biomass of Brussels sprouts and caterpillars on day 1 is the same as the calculated value from day 3.

Table 3: Brussels sprouts

	Day 1	Day 3	
Wet mass of Brussels sprouts	30 g	11 g	g consumed =
Dry mass of Brussels sprouts	–	2.2 g	
Plant proportion biomass (dry/wet)			
Plant energy consumed (wet mass x proportion biomass x 18.2 kJ)			kJ consumed per 10 larvae =
Plant energy consumed ÷ no. of larvae			kJ consumed per larva (E) =

Table 5: Frass

	Day 3
Dry mass frass from 10 larvae	0.5 g
Frass energy (waste) = frass dry mass x 19.87 kJ	
Dry mass frass from 1 larva (W)	

Table 4: Caterpillars (larvae)

	Day 1	Day 3	
Wet mass of 10 larvae	0.3 g	1.8 g	g gained =
Wet mass per larva			g gained per larva =
Dry mass of 10 larvae	–	0.27 g	
Larva proportion biomass (dry/wet)			
Energy production per larva (wet mass x proportion biomass x 23.0 kJ)			kJ gained per larva (S) =

10. Complete the calculations in tables 3-5 above.

11. (a) Write the net secondary production per larva value here: _____

(b) Write the equation to calculate the percentage efficiency of energy transfer from producers to consumers (use the notation provided) and calculate the value here:

_____

(c) Is this value roughly what you would expect? Explain: _____

_____

_____

12. (a) Write the equation to calculate respiratory losses per larva (use the notation provided): _____

(b) Calculate the respiratory losses per larva here: _____

13. Why can't we measure the actual dry biomass of Brussels sprouts and larvae on day 1? _____

_____

244 Nutrient Cycles

Key Idea: Matter cycles through the biotic and abiotic compartments of Earth's ecosystems in nutrient cycles. Nutrient cycles move and transfer chemical elements (e.g. carbon, hydrogen, nitrogen, and oxygen) through the abiotic and biotic components of an ecosystem. Commonly, nutrients must be in an ionic (rather than elemental) form in order for plants and animals to have access to them. The supply of nutrients in an ecosystem is finite and limited.

Essential nutrients

Macronutrient	Common form	Function
Carbon (C)	CO_2	Organic molecules
Oxygen (O)	O_2	Respiration
Hydrogen (H)	H_2O	Cellular hydration
Nitrogen (N)	N_2, NO_3^-, NH_4^+	Proteins, nucleic acids
Potassium (K)	K^+	Principal ion in cells
Phosphorus (P)	$H_2PO_4^-, HPO_4^{2-}$	Nucleic acids, lipids
Calcium (Ca)	Ca^{2+}	Membrane permeability
Magnesium (Mg)	Mg^{2+}	Chlorophyll
Sulfur (S)	SO_4^{2-}	Proteins
Micronutrient	**Common form**	**Function**
Iron (Fe)	Fe^{2+}, Fe^{3+}	Chlorophyll, blood
Manganese (Mn)	Mn^{2+}	Enzyme activation
Molybdenum (Mo)	MoO_4^-	Nitrogen metabolism
Copper (Cu)	Cu^{2+}	Enzyme activation
Sodium (Na)	Na^+	Ion in cells
Silicon (Si)	$Si(OH)_4$	Support tissues

Tropical rainforest

Nutrients in plants and animals

High speed return of nutrients to plants and animals leaving soil nutrient deficient

Nutrients in soil

Temperate woodland

Nutrients in plants and animals

Slow return of nutrients allowing greater build up of nutrients in the soil

Nutrients in soil

The speed of nutrient cycling can vary, some nutrients are cycled slowly, others quickly. The environment and diversity of an ecosystem can also have a large effect on the speed at which nutrients are recycled.

The role of organisms in nutrient cycling

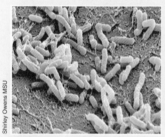

Shirley Owens MSU

Bacteria
Bacteria play an essential role in nutrient cycles. They act as decomposers, but can also convert nutrients into forms accessible to plants and animals.

Fungi
Fungi are saprophytes and are important decomposers, returning nutrients to the soil or converting them into forms accessible to plants and animals.

Plants
Plants have a role in absorbing nutrients from the soil and making them directly available to browsing animals. They also add their own decaying matter to soils.

Animals
Animals utilize and break down materials from bacteria, plants and fungi and return the nutrients to soils and water via their wastes and when they die.

1. (a) Briefly outline how matter (e.g. nutrients) moves through an ecosystem: _____

_____

 (b) How does the rate of this movement vary between ecosystems: _____

_____

_____

2. Briefly describe the role of each of the following in nutrient cycling:

 (a) Bacteria: _____

 (b) Fungi: _____

 (c) Plants: _____

 (d) Animals: _____

©2017 **BIOZONE** International
ISBN: 978-1-927309-65-0
Photocopying Prohibited

245 The Hydrologic Cycle

Key Idea: The hydrologic cycle results from the cycling of water from the oceans to the land and back.

The **hydrologic cycle** (water cycle), collects, purifies, and distributes the Earth's fixed supply of water. Besides replenishing inland water supplies, rainwater causes erosion and is a major medium for transporting dissolved nutrients within and among ecosystems. On a global scale, evaporation (conversion of water to gaseous water vapor) exceeds precipitation (rain, snow etc.) over the oceans. This results in a net movement of water vapor (carried by winds) over the land. On land, precipitation exceeds evaporation. Some of this precipitation becomes locked up in snow and ice but most forms surface and groundwater systems that flow back to the sea, completing the major part of the cycle. Over the sea, most of the water vapor is due to evaporation alone. However on land, about 90% of the vapor results from plant transpiration. Animals (particularly humans) intervene in the cycle by utilizing the resource for their own needs.

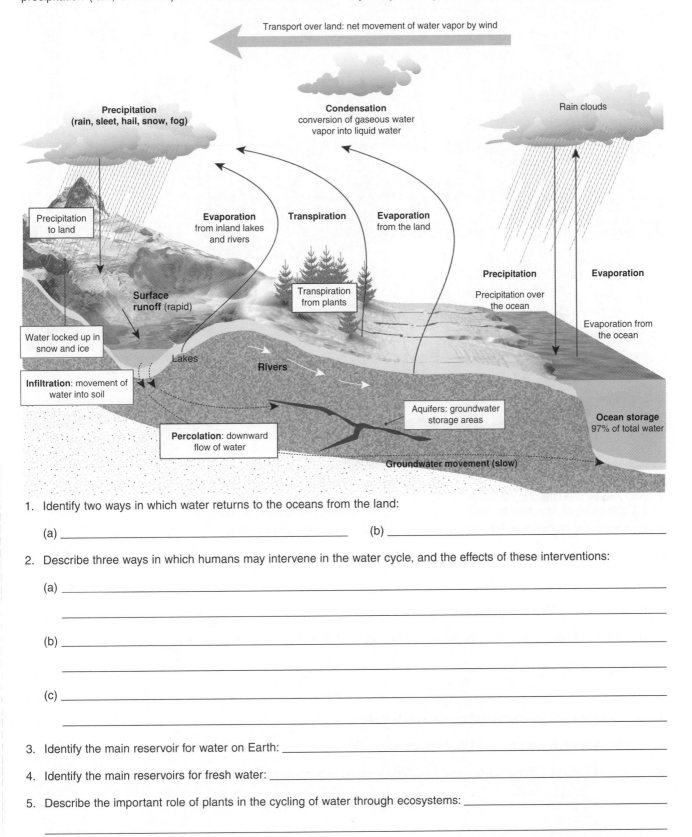

Transport over land: net movement of water vapor by wind

Precipitation
(rain, sleet, hail, snow, fog)

Condensation
conversion of gaseous water vapor into liquid water

Rain clouds

Precipitation to land

Evaporation
from inland lakes and rivers

Transpiration

Evaporation
from the land

Precipitation

Evaporation

Precipitation over the ocean

Transpiration from plants

Evaporation from the ocean

Surface runoff (rapid)

Water locked up in snow and ice

Lakes

Rivers

Infiltration: movement of water into soil

Percolation: downward flow of water

Aquifers: groundwater storage areas

Ocean storage
97% of total water

Groundwater movement (slow)

1. Identify two ways in which water returns to the oceans from the land:

   (a) _____   (b) _____

2. Describe three ways in which humans may intervene in the water cycle, and the effects of these interventions:

   (a) _____

   _____

   (b) _____

   _____

   (c) _____

   _____

3. Identify the main reservoir for water on Earth: _____

4. Identify the main reservoirs for fresh water: _____

5. Describe the important role of plants in the cycling of water through ecosystems: _____

   _____

PRACTICES   WEB

245

KNOW

246 The Carbon Cycle

Key Idea: Carbon is an essential element of life. Its continued availability depends on carbon cycling through the abiotic and biotic components of interconnected ecosystems.

Carbon is an essential element of life and is incorporated into the organic molecules that make up living organisms. Large quantities of carbon are stored in **sinks**, which include the atmosphere as carbon dioxide gas (CO_2), the ocean

as carbonate and bicarbonate, and rocks such as coal and limestone. Carbon moves between the biotic and abiotic environment. Autotrophs convert CO_2 into carbohydrates via photosynthesis and CO_2 is returned to the atmosphere through respiration. Some of the sinks and processes involved in the carbon cycle, together with the movement of carbon (carbon fluxes), are shown below.

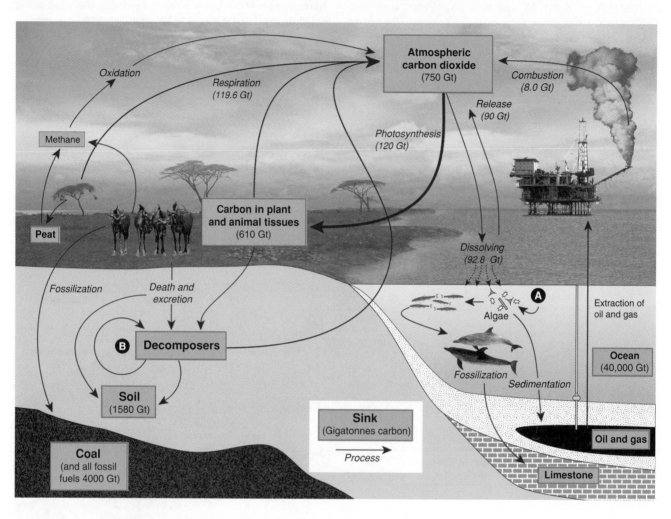

1. Add arrows and labels to the diagram above to show:
   (a) Dissolving of limestone by acid rain
   (b) Release of carbon from the marine food chain
   (c) Mining and burning of coal
   (d) Burning of plant material.

2. (a) Name the processes that release carbon into the atmosphere: _____

   _____

   (b) In what form is the carbon released? _____

3. Name the four geological reservoirs (sinks), in the diagram above, that can act as a source of carbon:

   (a) _____ (c) _____

   (b) _____ (d) _____

4. (a) Identify the process carried out by algae at point [**A**]: _____

   (b) Identify the process carried out by decomposers at [**B**]: _____

5. What would be the effect on carbon cycling if there were no decomposers present in an ecosystem? _____

   _____

   _____

©2017 **BIOZONE** International
ISBN: 978-1-927309-65-0
Photocopying Prohibited

Bracket fungus on tree trunk

Coal mine in Wyoming

Carbon may be locked up in biotic or abiotic systems for long periods of time, e.g. in the wood of trees or in fossil fuels such as coal or oil. Human activity, e.g. extraction and combustion of fossil fuels, has disturbed the balance of the carbon cycle.

Organisms break down organic material to release carbon. Fungi and decomposing bacteria break down dead plant matter in the leaf litter of forests. Termites, with the aid of symbiotic protozoans and bacteria in their guts, digest the cellulose of woody tissue.

Coal is formed from the remains of terrestrial plant material buried in shallow swamps and subsequently compacted under sediments to form a hard black material. Coal is composed primarily of carbon and is a widely used fuel source.

Oil and natural gas formed in the past when dead algae and zooplankton settled to the bottom of shallow seas and lakes. These remains were buried and compressed under layers of non-porous sediment.

Limestone is a type of sedimentary rock composed mostly of calcium carbonate. It forms when the shells of molluscs and other marine organisms with calcium carbonate ($CaCO_3$) skeletons become fossilized.

Peat (partly decayed organic material) forms when plant material is not fully decomposed due to acidic or anaerobic conditions. Peatlands are a very efficient carbon sink but are easily lost through oxidation when land is drained.

6. Describe the biological origin of the following geological deposits:

   (a) Coal: _____

   (b) Oil: _____

   (c) Limestone: _____

   (d) Peat: _____

7. Describe the role of living organisms in the carbon cycle: _____

   _____

   _____

   _____

   _____

   _____

   _____

8. In natural circumstances, accumulated reserves of carbon such as peat, coal and oil represent a sink or natural diversion from the cycle. Eventually, the carbon in these sinks returns to the cycle through the action of geological processes which return deposits to the surface for oxidation.

   (a) What is the effect of human activity on the amount of carbon stored in sinks? _____

   _____

   (b) Describe the effect of human activity on atmospheric CO_2 levels: _____

   _____

247 The Nitrogen Cycle

Key Idea: The nitrogen cycle describes how nitrogen is converted between its various chemical forms. Nitrogen gas is converted to nitrates which are taken up by plants. Heterotrophs obtain their nitrogen by eating other organisms. Nitrogen is an essential component of proteins and nucleic acids and required by all living things. The Earth's atmosphere is about 78% nitrogen gas (N_2), but this is so stable that it is unavailable to most organisms and nitrogen is often limiting in ecosystems. Lightning oxidizes nitrogen gas to nitrogen oxides, which form nitrate with water, but most transformations are biological. Bacteria transfer nitrogen between the biotic and abiotic environments. Some fix atmospheric nitrogen, while others convert ammonia to nitrate, making it available to plants. Nitrogen-fixing bacteria are found free in the soil (*Azotobacter*) and in symbioses with some plants in root nodules (*Rhizobium*). Denitrifying bacteria return fixed nitrogen to the atmosphere. Humans intervene in the nitrogen cycle by applying nitrogen fertilizers to the land.

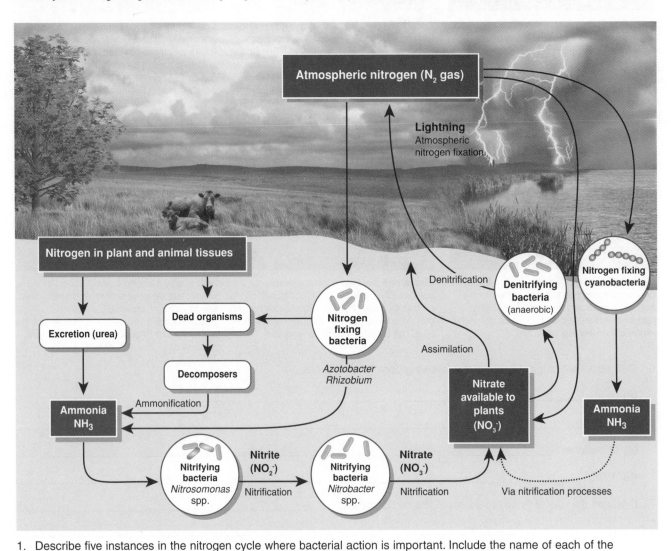

1. Describe five instances in the nitrogen cycle where bacterial action is important. Include the name of each of the processes and the changes to the form of nitrogen involved:

(a) _____

_____

(b) _____

_____

(c) _____

_____

(d) _____

_____

(e) _____

_____

©2017 **BIOZONE** International
ISBN: 978-1-927309-65-0
Photocopying Prohibited

Nitrogen fixation in root nodules

Root nodules are a root symbiosis between a higher plant and a bacterium. The bacteria fix atmospheric nitrogen and are extremely important to the nutrition of many plants, including the economically important legume family. Root nodules are extensions of the root tissue caused by entry of a bacterium. In legumes, this bacterium is *Rhizobium*. Other bacterial genera are involved in the root nodule symbioses in non-legumes.

The bacteria in these symbioses live in the nodule where they fix atmospheric nitrogen and provide the plant with most, or all, of its nitrogen requirements. In return, they have access to a rich supply of carbohydrate. The fixation of atmospheric nitrogen to ammonia occurs within the nodule, using the enzyme nitrogenase. Nitrogenase is inhibited by oxygen and the nodule provides a low O_2 environment in which fixation can occur.

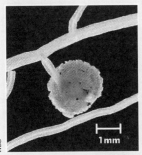

1mm

WBS

Two examples of legume nodules caused by *Rhizobium*. The images above show the size of a single nodule (left), and the nodules forming clusters around the roots of *Acacia* (right).

Human intervention in the nitrogen cycle

The largest interventions in the nitrogen cycle by humans occur through farming and effluent discharges. Other interventions include burning, which releases nitrogen oxides into the atmosphere, and irrigation and land clearance, which leach nitrate ions from the soil.

Farmers apply organic nitrogen fertilizers to their land in the form of green crops and manures, replacing the nitrogen lost through cropping and harvest. Until the 1950s, atmospheric nitrogen could not be made available to plants except through microbial nitrogen fixation (left). However, during World War II, Fritz Haber developed the Haber process, combining nitrogen and hydrogen gas to form gaseous ammonia. The ammonia is converted into ammonium salts and sold as inorganic fertilizer. This process, although energy expensive, made inorganic nitrogen fertilizers readily available and revolutionized farming practices and crop yields.

Two examples of human intervention in the nitrogen cycle. The photographs above show the aerial application of a commercial fertilizer (left), and the harvesting of an agricultural crop (right).

2. Identify three processes that fix atmospheric nitrogen:

(a) _____   (b) _____   (c) _____

3. What process releases nitrogen gas into the atmosphere? _____

4. What is the primary reservoir for nitrogen? _____

5. What form of nitrogen is most readily available to most plants? _____

6. Name one essential organic compound that plants need nitrogen for: _____

7. How do animals acquire the nitrogen they need? _____

8. Why might farmers plow a crop of legumes into the ground rather than harvest it? _____

9. Describe five ways in which humans may intervene in the nitrogen cycle and the effects of these interventions:

(a) _____

(b) _____

(c) _____

(d) _____

(e) _____

248 KEY TERMS AND IDEAS: Did You Get It?

1. (a) The graph (right) shows primary production in the oceans. Explain the shape of the curves:

_____

_____

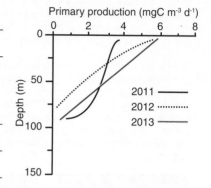

(b) About 90% of all marine life lives in the photic zone (the depth to which light penetrates). Suggest why this is so:

_____

_____

_____

2. The schematic below shows the movement of energy and minerals from producers to consumers.

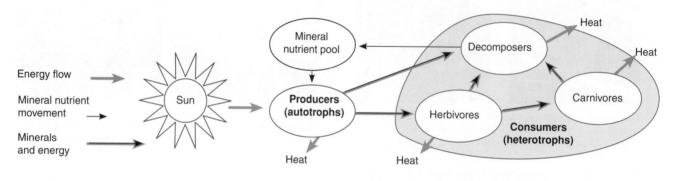

(a) How are the movements of minerals and energy different? _____

_____

_____

(b) What process is responsible for losses of energy from the system? _____

3. Test your vocabulary by matching each term to its correct definition, as identified by its preceding letter code.

carbon cycle ....

chemoautotroph ....

consumer ....

ecological pyramid ....

food chain ....

food web ....

gross primary productivity ....

net primary productivity ....

nutrient cycle ....

photosynthesis ....

producer ....

trophic level ....

A A graphical representation of the numbers, energy, or biomass at each trophic level in an ecosystem. Pyramidal in shape, but sometimes inversely so.

B An organism that obtains its carbon and energy from other organisms.

C An organism capable of manufacturing its own food.

D A sequence of steps describing how an organism derives energy from those before it.

E Cycle in which inorganic nutrients move between the biotic and abiotic components of the environment.

F The biochemical process that fixes atmospheric carbon as carbohydrate.

G The total amount of energy fixed by producers per unit area.

H An organism that uses inorganic energy sources, such as hydrogen sulfide or elemental sulfur to synthesize its own organic compounds.

I The position an organism occupies on the food chain.

J The total amount of energy fixed by producers per unit area, less the costs of respiration. It is effectively the amount of biomass available to consumer levels.

K Biogeochemical cycle by which carbon is exchanged among the biotic and abiotic components of the Earth.

L A complex series of interactions showing the feeding relationships between organisms in an ecosystem.

PRACTICES

TEST

©2017 **BIOZONE** International
ISBN: 978-1-927309-65-0
Photocopying Prohibited

The Diversity and Stability of Ecosystems

Key terms

biodiversity

climate change

climate model

community

continental drift

ecosystem

ecosystem resilience

ecosystem resistance

ecosystem stability

El Niño

extinction

global warming

greenhouse effect

greenhouse gas

infrastructure

slash and burn agriculture

species diversity

urbanization

2.D.3 Biological systems are affected by disruptions to their dynamic homeostasis

Essential knowledge

(b) Disruptions to ecosystems impact dynamic homeostasis or balance of the ecosystem

☐ 1 Use examples to describe and explain the impact of disruptions on the dynamic homeostasis or balance of ecosystems. Examples could include:
  • The impact of invasive species, e.g. kudzu, zebra mussels.
  • The impact of human activity, e.g. as a result of climate change, pollution, urbanization, or deforestation
  • The effect of natural hazards such as hurricanes, fires, floods, earthquakes, or volcanic eruption.
  • The effect of water limitation or salination.

249-252
254-258
261-266
270

4.A.5 Communities are composed of populations of organisms that interact in complex ways

Essential knowledge

(a) Community structure is described in terms of species composition and diversity

☐ 1 Recall how the structure of communities can be quantified and described by the species composition and the species diversity. Changes to either of these measures can be indicative of environmental change.

191 194
250-252

(b) Models are used to investigate and illustrate population interactions within, and environmental impacts on, a community

☐ 1 Use the example of climate change modeling to explain how mathematical or computer models can be used to illustrate and investigate the effect of environmental impact on a community.

226 229
258 259

4.A.6 Interactions among living systems and with their physical environment result in the movement of matter and energy

Essential knowledge [4.A.6.a-c & 4.A.6. e, g are covered in previous chapters]

(f) Human activities have an impact on ecosystems on local, regional, and global scales

☐ 1 Explain how the increase in the numbers of human populations has resulted in a magnification of their impact on the habitats of other species on Earth.

254 255

☐ 2 Using illustrative examples, explain how this, in turn, has reduced the population size of affected species and resulted in habitat destruction and, in some cases, extinction.

257 263
266 267

4.B.3 Interactions between and within populations influence patterns of species distribution and abundance

Activity number

Essential knowledge [4.B.3. a-b are covered in 'Populations & Communities]

(c) Species-specific and environmental catastrophes, geological events, and sudden changes in resources or increased human activity affect species distribution and abundance

☐ 1 Use examples to explain how species distribution and abundance can be affected by:

253-256
261 267
268 270

- species-specific catastrophes, e.g. loss of keystone species, Dutch elm disease, kauri dieback [*also 4.C.4.b*]
- environmental disasters, e.g. flood, fire, drought
- geological events, e.g. earthquakes, El Niño, continental drift [*also 4.B.4.b*]
- sudden influxes or depletion of abiotic, resources, e.g. nitrogen pollution
- increased human activity, e.g. urbanization, spread of Kudzu

4.B.4 Distribution of ecosystems changes over time

Activity number

Essential knowledge

(a) Human impact accelerates change at local and global levels

☐ 1 Use examples to illustrate how the impact of human activities accelerates change in natural ecosystems both locally and globally. Examples include:

254
255-258
261-266
269-272

- Logging, slash and burn agriculture, urbanization, intensive agriculture (monocultures), infrastructure development (dams, roads, transmission lines), and global climate change and their threat to ecosystems and life on Earth.
- Introduced species, which can exploit niches free of competition or predation and so exploit new resources , e.g. Kudzu, red imported fire ants.
- New diseases, which an devastate native populations when introduced, e.g. Dutch elm disease (caused by Ascomycete microfungi), kauri dieback and potato blight (both caused by species of the water mold *Phytophthora*), smallpox epidemics in Native American populations (including in Mexico). [*also 1.A.2.d: Humans have an impact on the diversity in other species*].

(b) Geological and meteorological events have an impact on ecosystem distribution

☐ 1 Explain how biogeographical studies can document changes in the distribution of ecosystems as a result of geological or meteorological events. Examples include: 267 268

- El Niño (the warm phase of the naturally occurring El Niño Southern Oscillation).
- Continental drift (how has continental drift affected geographical distribution)?
- Meteor impact on the dinosaurs (mass extinction and rise of the mammals).

4.C.4 The diversity of species within an ecosystem may influence the stability of the ecosystem

Activity number

Essential knowledge

(a) Natural and artificial ecosystems with fewer component parts and with little diversity among the parts are often less resilient to changes in the environment

☐ 1 Recognize the components of ecosystem stability: resistance (the ability to remain unchanged when subjected to a disturbance) and resilience (the ability to recover from disturbance. Use examples from natural or experimental systems that illustrate the relationship between ecosystem diversity and stability and suggest probable explanations for this relationship. [*also 1.C.1. b: Species extinction rates are rapid at times of ecological stress*]. 249-252

(b) Keystone species, producers, and essential biotic and abiotic factors contribute to maintaining the diversity of an ecosystem

☐ 1 Explain what is meant by a keystone species and explain why their effects on ecosystem structure and function are disproportionate relative to their abundance. Use examples to illustrate the consequences to ecosystem structure and function when keystone species are removed or depleted (or conversely, returned after a period of absence). Examples include: 253

- Grey wolves in Yellowstone National Park.
- Sea otters and the californian kelp forests.
- Elephants in savanna grasslands.
- Termites in the savannas of Australia and Africa

249 The Stability of Ecosystems

Key Idea: The components of ecosystem stability are resistance and resilience. High biodiversity is correlated with higher ecosystem stability.

Ecological theory suggests that all species in an ecosystem contribute in some way to ecosystem function. Therefore, species loss past a certain point is likely to have a detrimental effect on the functioning of the ecosystem and on its ability to resist change over time (its **stability**). Although many species still await discovery, we know that the rate of species extinction is increasing. Scientists estimate that human destruction of natural habitats is implicated in the extinction of up to 100,000 species every year. This substantial loss of biodiversity has serious implications for the long term stability of many ecosystems.

Ecosystem stability has various components, including **resistance** (the ability to resist disturbance) and **resilience** (ability to recover from external disturbances). Ecosystem stability is closely linked to the biodiversity of the system, although it is difficult to predict which factors will stress an ecosystem beyond its range of tolerance. It was once thought that the most stable ecosystems were those with the most species, because they had the greatest number of biotic interactions operating to buffer them against change. This assumption is supported by experimental evidence but there is uncertainty over what level of biodiversity provides an insurance against catastrophe.

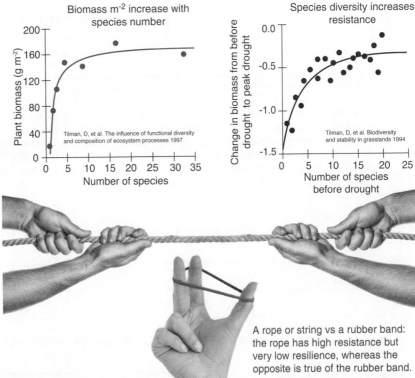

Biomass m⁻² increase with species number

Tilman, D, et al. The influence of functional diversity and composition of ecosystem processes 1997

Species diversity increases resistance

Tilman, D, et al. Biodiversity and stability in grasslands 1994

Resilience and resistance can be modeled by a rubber band and a piece of string.

The rubber band is highly resilient. It can be stretched out of shape by an external force. Once the force is removed, the rubber band quickly returns to its original shape. However it is not very resistant, even a small external force can stretch the band out of shape.

A piece of string is highly resistant, a large force does not change its shape. It is not resilient though as a large enough force will break the string so it can never return to its original shape.

A rope or string vs a rubber band: the rope has high resistance but very low resilience, whereas the opposite is true of the rubber band.

Monoculture

Brian Dell Public Domain

Natural grassland (savanna)

Lubasi cc 2.0

Amazon rainforest (aerial)

Deforestation

Single species crops (monocultures), such as the soy bean crop (above, left), represent low diversity systems that can be vulnerable to disease, pests, and disturbance. In contrast, natural grasslands (above, right) may appear homogeneous, but contain many species which vary in their predominance seasonally. Although they may be easily disturbed (e.g. by burning) they are very resilient and usually recover quickly.

Tropical rainforests (above, left) represent the highest diversity systems on Earth. Whilst these ecosystems are generally resistant to disturbance, once degraded, (above, right) they have little ability to recover. The biodiversity of ecosystems at low latitudes is generally higher than that at high latitudes, where climates are harsher, niches are broader, and systems may be dependent on a small number of key species.

1. (a) Define ecosystem resilience: _____

_____

(b) Define ecosystem resistance: _____

_____

2. What is the effect of species number on plant biomass production? _____

_____

3. What is the effect of latitude on biodiversity? _____

_____

PRACTICES

CONNECT
240 AP2

WEB
249

KNOW

Community response to environmental change

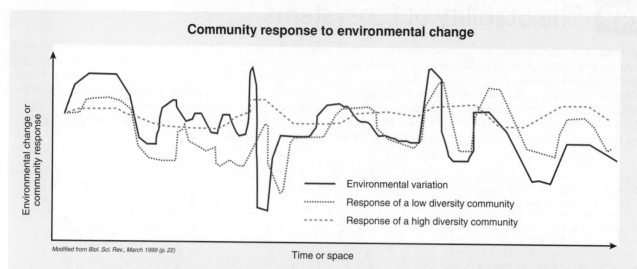

Environmental change or community response

Modified from Biol. Sci. Rev., March 1999 (p. 22)

Time or space

— Environmental variation

·········· Response of a low diversity community

- - - - - Response of a high diversity community

In models of ecosystem function, a higher species diversity increases the stability of ecosystem functions such as productivity and nutrient cycling. In the graph above, note how the low diversity system varies more consistently with the environmental variation, whereas the high diversity system is buffered against major fluctuations. In any one ecosystem, some species may be more influential than others in the stability of the system. Such **keystone (key) species** have a disproportionate effect on ecosystem function due to their pivotal role in some ecosystem function such as nutrient recycling or production of plant biomass.

4. Suggest one probable reason why high biodiversity promotes greater ecosystem stability: _____

_____

_____

_____

5. Explain why monoculture crops require an input of energy (from the farmer) to maintain their structure: _____

_____

_____

_____

6. A student used blocks to crudely model ecosystem stability. She placed once set of blocks as in figure 1 and another set of blocks as in figure 2. Figure 1 she said modeled an low diversity ecosystem, figure 2 she said modeled a high diversity ecosystem.

   Explain how well each model represents the system described, referring to stability, resilience, and resistance.

Figure 1   Figure 2

250 Primary Succession

Key Idea: Ecological succession is the process by which communities change over time.

Succession occurs as a result of the interactions between biotic and abiotic factors. Early communities modify the physical environment. This in turn alters the biotic community, which further alters the physical environment and so on. Each successive community makes the environment more favorable for the establishment of a different assemblage of species. An idealized succession proceeds in stages to a climax community, which is generally stable until further disturbance. Early successional communities have low species diversity and a simple structure. Climax communities are more complex, with a greater number of species interactions and higher species diversity.

USGS

Primary succession is the biological colonization and development of a region with no preexisting vegetation or soil. A primary succession may take hundreds of years, but the time scale depends strongly on factors such as distance from vegetated areas and availability of pollinating species such as birds and insects.

Examples include the emergence of new volcanic islands, new coral atolls, or islands where the previous community has been destroyed by a volcanic eruption.

Many texts show a sequence of colonization beginning with lichens, mosses, and liverworts, then progressing to ferns, grasses, shrubs, and finally a climax community of mature forest. In reality, this sequence is very rare. The sequence below shows the recolonization of Mount St Helens (left) after it erupted in 1980.

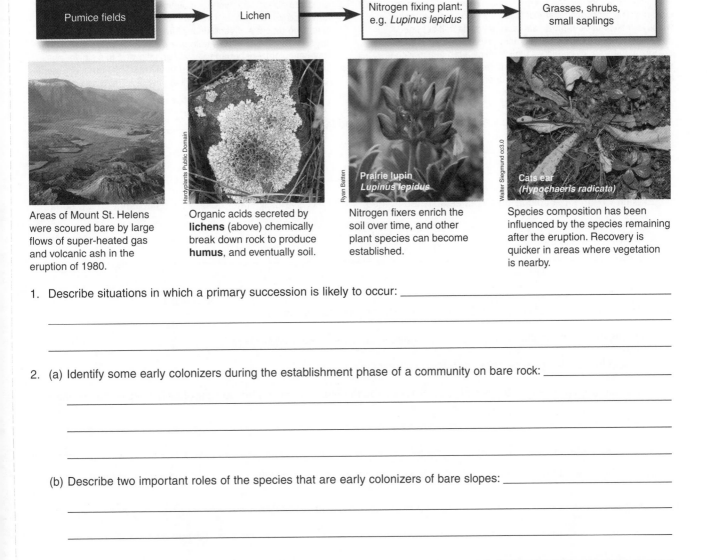

| Pumice fields | → | Lichen | → | Nitrogen fixing plant: e.g. *Lupinus lepidus* | → | Grasses, shrubs, small saplings |

Prairie lupin
Lupinus lepidus
Ryan Batten

Cats ear
(Hypochaeris radicata)
Walter Siegmund cc3.0

Hardyplants Public Domain

Areas of Mount St. Helens were scoured bare by large flows of super-heated gas and volcanic ash in the eruption of 1980.

Organic acids secreted by **lichens** (above) chemically break down rock to produce **humus**, and eventually soil.

Nitrogen fixers enrich the soil over time, and other plant species can become established.

Species composition has been influenced by the species remaining after the eruption. Recovery is quicker in areas where vegetation is nearby.

1. Describe situations in which a primary succession is likely to occur: _____

_____

_____

2. (a) Identify some early colonizers during the establishment phase of a community on bare rock: _____

_____

_____

_____

(b) Describe two important roles of the species that are early colonizers of bare slopes: _____

_____

_____

_____

©2017 **BIOZONE** International
ISBN: 978-1-927309-65-0
Photocopying Prohibited

PRACTICES   PRACTICES   WEB

250

KNOW

Surtsey: A case study in primary succession

Surtsey Island is a volcanic island lying 33 km off the southern coast of Iceland. The island was formed over four years from 1963 to 1967 when a submarine volcano 130 m below the ocean surface built up an island that initially reached 174 m above sea level and covered 2.7 km^2. Erosion has since reduced the island to around 150 m above sea level and 1.4 km^2.

As an entirely new island, Surtsey was able to provide researchers with an ideal environment to study primary succession in detail. The colonization of the island by plants and animals has been recorded since the island's formation. The first vascular plant there (sea rocket) was discovered in 1965, two years before the eruptions on the island ended. Since then, 69 plant species have colonized the island and there are a number of established seabird colonies.

Number of vascular plant species found on Surtsey

Sea rocket

H. peploides

P. annua

S. phylicifolia

The first stage of colonization on Surtsey was dominated by shore plants colonizing the northern shores, brought by ocean currents. The most successful of these was *Honckenya peploides*, which established on tephra sand and gravel flats. It set seed in 1971 and subsequently spread across the island. This initial colonization by shore plants was followed by a lag phase with few new colonizers. A number of new plant species arrived after a gull colony became established at the southern end of the island.

Populations of plants within or near the gull colony expanded rapidly to about 3 ha, while populations outside the colony remained low but stable. Grasses such as *Poa annua* formed extensive patches of vegetation. After this rapid increase in plant species, the arrival of new colonizers again slowed. A second wave of colonizers began to establish following this slower phase and soil organic matter increased markedly. The first bushy plants established in 1998, with the arrival of willow *Salix phylicifolia*.

3. Explain why Surtsey provided ideal conditions for studying primary succession: _____

_____

_____

4. Explain why the first colonizing plants established in the north of the island, but later colonizers established in the south.

_____

_____

_____

5. There are three distinct phases on Surtsey where species increased rapidly.

(a) Label on the graph the three phases of increase in species number on Surtsey:

(b) Label the two lag phases where species number increased slowly:

6. A gull colony established on the island in 1985. What was the effect on this on the number of plant species on the island?

_____

_____

7. Why is the living number of plant species on the island less than the cumulative number of species reaching the island?

_____

_____

_____

©2017 **BIOZONE** International
ISBN: 978-1-927309-65-0
Photocopying Prohibited

251 Secondary Succession

Key Idea: Secondary succession occurs in cleared land where soil and seeds have not been lost.

Secondary succession occurs when land is cleared of vegetation (e.g. after a fire or landslide). These events do not involve the loss of soil or seed, and root stocks are often undamaged. As a result, secondary succession tends to be more rapid than primary succession, although the time scale depends on the species involved, soil composition, and climate. Secondary succession may occur over a wide area or in smaller areas where single trees have fallen or abandoned farmland has been left to regenerate. Tree falls result in a type of secondary succession called **gap regeneration** (below).

Redwoods (*Sequoia* and *Sequoiadendron*) are very long-lived species (500-1000 years on average). They are a keystone species in the ecosystem, supporting many different species, some of which are endangered.

The light open spaces in the canopy allow smaller understory plants to become established as a brushy growth. Examples include tan oak, huckleberry (left), blackberry, and rhododendrons.

Catastrophic events such as large wind-storms storms, forest fires, or landslides can destroy large areas of redwood forest (left).

Many trees may fall and their broken logs can remain on the forest floor for decades. Fallen trees open up the canopy, creating gaps and allowing light to reach the forest floor, encouraging the growth of seedlings.

Fallen trees play an important role in the ecosystem diversity of a forest. They support a variety of invertebrates, bacteria, and fungi.

Redwood seedlings begin to grow through the understory. As they grow taller, they begin to out-compete the smaller plants. While the saplings grow up, the crowns of the remaining canopy trees will close some of the gap. This may take decades.

1. Describe how secondary succession differs from primary succession: _____

_____

_____

2. Explain why secondary succession usually takes place more rapidly than primary succession: _____

_____

_____

_____

_____

3. Using an example, explain how the outcome of a secondary succession may depend on an interplay of unpredictable biotic and abiotic factors:

_____

_____

_____

_____

_____

_____

WEB
251

KNOW

252 Wetland Succession

Key Idea: Wetlands can slowly fill in with sediment and transition to dry land over hundreds or thousands of years. Wetland areas present a special case of ecological succession. They are constantly changing as plant invasion of open water leads to siltation and infilling. This process is accelerated by eutrophication (nutrient enrichment). In well drained areas, pasture or **heath** may develop as a result of succession from freshwater to dry land. When the soil conditions remain non-acid and poorly drained, a swamp will eventually develop into a seasonally dry **fen**. In special circumstances, an acid **peat bog** may develop. The domes of peat produce a hummocky landscape with a unique biota. Wetland peat ecosystems may take more than 5000 years to form but are easily destroyed by excavation and lowering of the water table.

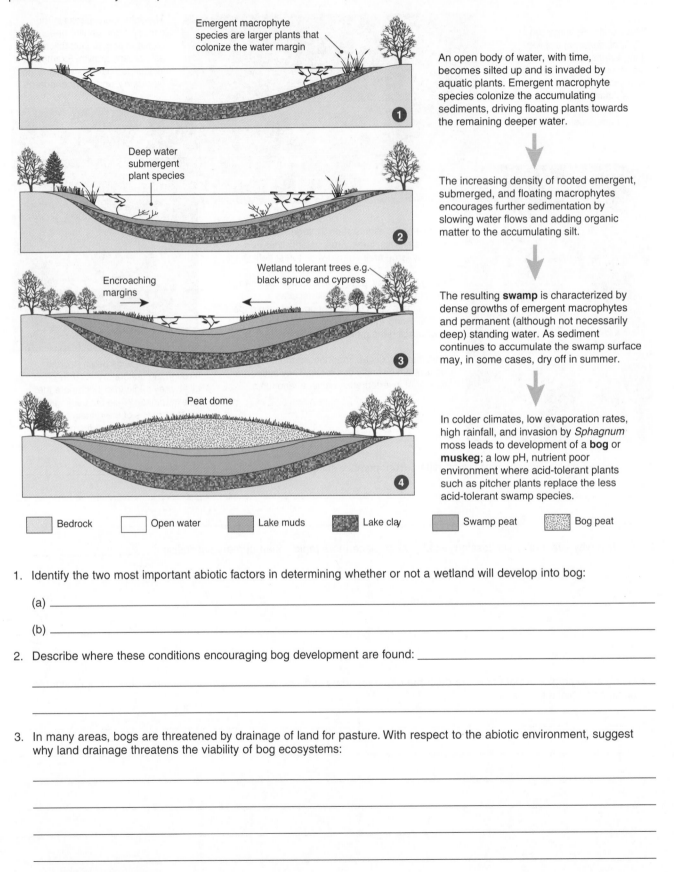

Emergent macrophyte species are larger plants that colonize the water margin

1 An open body of water, with time, becomes silted up and is invaded by aquatic plants. Emergent macrophyte species colonize the accumulating sediments, driving floating plants towards the remaining deeper water.

Deep water submergent plant species

2 The increasing density of rooted emergent, submerged, and floating macrophytes encourages further sedimentation by slowing water flows and adding organic matter to the accumulating silt.

Encroaching margins

Wetland tolerant trees e.g. black spruce and cypress

3 The resulting **swamp** is characterized by dense growths of emergent macrophytes and permanent (although not necessarily deep) standing water. As sediment continues to accumulate the swamp surface may, in some cases, dry off in summer.

Peat dome

4 In colder climates, low evaporation rates, high rainfall, and invasion by *Sphagnum* moss leads to development of a **bog** or **muskeg**; a low pH, nutrient poor environment where acid-tolerant plants such as pitcher plants replace the less acid-tolerant swamp species.

Bedrock | Open water | Lake muds | Lake clay | Swamp peat | Bog peat

1. Identify the two most important abiotic factors in determining whether or not a wetland will develop into bog:

   (a) _____

   (b) _____

2. Describe where these conditions encouraging bog development are found: _____

   _____

   _____

3. In many areas, bogs are threatened by drainage of land for pasture. With respect to the abiotic environment, suggest why land drainage threatens the viability of bog ecosystems:

   _____

   _____

   _____

   _____

©2017 **BIOZONE** International
ISBN: 978-1-927309-65-0
Photocopying Prohibited

253 The Importance of Keystone Species

Key Idea: Keystone species play a disproportionately important role in ecosystem function.

Some species, called keystone species, have a disproportionate effect on the stability of an ecosystem. The term keystone species comes from the analogy of the keystone in a true arch. If the keystone is removed, the arch collapses. Similarly if the keystone species in removed from the environment there is usually a rapid change and often collapse of the ecosystem from high diversity to low diversity. The pivotal role of keystone species is a result of their influence in some aspect of ecosystem functioning, e.g. as predators, prey, or processors of biological material.

Examples of keystone species

The starfish *Pisaster* is found along the coasts of North America where it feeds on mussels. In an experiment, scientist Robert Paine removed it from an area. As a result, the mussels dominated the area, crowding out most of the algae and leading to a decrease in the number of herbivore species.

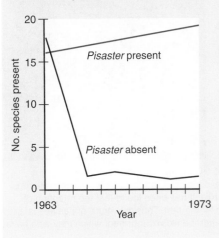

Graph: No. species present (y-axis, 0 to 20) vs Year (x-axis, 1963 to 1973). Lines labelled "Pisaster present" and "Pisaster absent".

Beaver, *Castor canadensis*

Sea otter, *Enhydra lutris*

Two smaller mammals are important keystone species in North America. **Beavers** (top) play a crucial role in biodiversity and many species, including 43% of North America's endangered species, depend partly or entirely on beaver ponds. **Sea otters** are also critical to ecosystem function. When their numbers were decimated by the fur trade, sea urchin populations exploded and the kelp forests, on which many species depend, were destroyed.

Quaking aspen

Quaking aspen (*Populus tremuloides*) is one of the most widely distributed tree species in North America, and aspen communities are among the most biologically diverse in the region, with a rich understorey flora supporting an abundance of wildlife. Moose, elk, deer, black bear, and snowshoe hare browse its bark, and aspen groves support up to 34 species of birds, including ruffed grouse, which depends heavily on aspen for its winter survival.

1. Explain why keystone species are so important to ecosystem function: _____

_____

_____

_____

2. For each of the following species, discuss features of their biology that contribute to their position as keystone species:

(a) *Pisaster*: _____

_____

(b) Beaver: _____

_____

(c) Sea otter: _____

_____

(d) Quaking aspen: _____

_____

PRACTICES   CONNECT   WEB

240 AP2   253

KNOW

Case study: The wolves of Yellowstone Park

Gray or **timber wolves** (*Canis lupus*) are a keystone predator and were once widespread in North American ecosystems. Wolves were eliminated from Yellowstone National Park by 1926 because of their perceived threat to humans and livestock. As a result, elk populations increased to the point that they adversely affected other flora and fauna. 31 wolves were reintroduced to Yellowstone in 1995. Since then the wolf population has increased and the elk population has decreased. Parts of the park have regenerated and various animal populations that had decreased since the wolves extermination has begun to increase again (for example the beaver population has increased). Recently there has been debate over how much of the restoration of Yellowstone's ecosystem can be credited to the return of the gray wolf. There is agreement, however, that ecosystems are far more complex than we think.

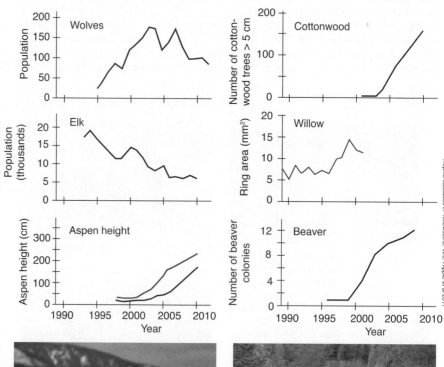

Trophic cascades in Yellowstone. W.J. Ripple et al 2011

Gray wolves

Elk

Beaver dam

Weigmund cc 2.5

Since the introduction of wolves to Yellowstone, the elk population is about a third of its peak of 18,000 and the wolf population has stabilized at about 100. Studies have found that although trees such as aspen and willow are expanding in range since the wolves retuned, the reduction in elk numbers may not have been entirely responsible, although it has played a role. It has been found that the retention of moisture in the ground caused by the production of lakes behind beaver dams (above) is just as important. Thus the return of beavers is as important in the park's restoration. Yellowstone Park emerged from a multi-year drought soon after the wolves returned, which may also have helped tree growth.

3. What effect did introducing wolves to Yellowstone have on the elk population? _____

_____

4. Suggest why the wolf population has declined from an earlier peak: _____

_____

_____

5. (a) The regrowth of various plant types in the park is correlated with the reduction in elk. What else is it correlated with?

_____

(b) Why might this make scientists question how much direct effect the wolves have had on the park ecosystem?

_____

_____

_____

_____

_____

©2017 **BIOZONE** International
ISBN: 978-1-927309-65-0

254 The Impact of Humans on Ecosystems

Key Idea: The exploitation of environmental resources by humans has had a negative impact on the environment.
Human activity can have very negative effects on surrounding ecosystems. Unrestricted mining, logging, or harvesting of natural populations can alter the balance of an ecosystem.

There are many examples documenting how human overexploitation of resources adversely affects the diversity, functioning, and stability of natural ecosystems. There are many other examples too of humans simply exterminating an entire species e.g. the dodo and passenger pigeon.

How humans affect the environment

Projections of current human population growth predict the human population will reach between 9 and 11 billion by 2050. This growth will create an increasing demand on natural resources, particularly water, arable land, and minerals.

As the human population increases, cities expand, fragmenting or destroying the natural ecosystems surrounding them. Careful management of urban development and resource use will be needed to prevent local extinctions and loss of biodiversity.

Human exploitation of natural populations has driven some species to the brink of extinction, or significantly reduced their populations. Examples include the hunting of whales and overfishing of many fish populations such as North Sea cod.

Most industry depends on the combustion of fossil fuels and this contributes to **global warming**, the continuing rise in the average temperature of the Earth's surface. Global warming and associated shifts in climate will affect species distributions, breeding cycles, and patterns of migration. Species unable to adapt are at risk of extinction.

Eighty percent of Earth's documented species are found in tropical rainforests. These rainforests are being destroyed rapidly as land is cleared for agriculture or cattle ranching, to supply logs, for mining, or to build dams. Deforestation places a majority of the Earth's biodiversity at risk and threatens the stability of important ecosystems.

Demand for food increases as the population grows. Modern farming techniques favour **monocultures** to maximize yield and profit. However monocultures, in which a single crop type is grown year after year, are a low diversity system and food supplies are vulnerable if the crop fails. The UN estimates that 12 plant species provide 75% of our total food supply.

1. The human population is growing at a rapid rate and as a result the demand for resources in increasing. Discuss how human activities are affecting local and global ecosystems:

_____

_____

_____

_____

_____

_____

_____

_____

_____

©2017 **BIOZONE** International
ISBN: 978-1-927309-65-0
Photocopying Prohibited

PRACTICES	CONNECT	CONNECT	CONNECT	CONNECT	WEB
	247 AP2	246 AP2	245 AP2	240 AP2	254

KNOW

Plastic in the Pacific

Human activity is global. There is no part of the planet that is not affected in some way by human activity. Even remote parts of the world are affected by activities thousands of kilometers away.

It is estimated around 8 million tonnes of plastic enters the oceans each year. Much of this comes from waste thrown away in cities. It finds its way into rivers and is washed into the seas and oceans. The circulation of the oceans tends to concentrate the plastic into certain areas of the ocean. In the South Pacific, plastic is concentrated by the South Pacific Gyre, which a huge system of ocean currents that rotates in an anticlockwise direction. In the North Pacific, plastic is concentrated by the North Pacific Gyre, which rotates in a clockwise direction. The concentration of plastic in the North Pacific is sometimes called the Great Pacific Garbage Patch.

The effect of plastic in the marine environment can be dramatically seen on islands near the center of these gyres. Henderson Island and Midway Atoll are two of these islands.

Henderson Island

Henderson Island sits in the South Pacific Ocean. It is part of the Pitcairn group and is about 5000 km from the nearest significant land mass. The island is just 37.3 km^2 and uninhabited. A study in 2017 led by Jennifer L. Lavers measured the amount of plastic on the island's beaches. Her team measured the amount of plastic already on the beach, and then cleared a control area, to measure the rate at which plastic accumulated on the beach (below).

Site	Mean density on beach (items m^{-2})	Rate of accumulation (items km^{-1} d^{-1})	Estimated total debris on beach (items)		Estimated island total including buried items and back beach (items)	
			Number	Mass (kg)	Number	Mass (kg)
North Beach	30.3	13,316	812,116	2985	7,634,052	4,744
East beach	239.4	–	3,053,901	12,611	30,027, 343	12,857
Total			3,866,017	15,597	37,661,395	17,601

Jennifer L. Lavers. Exceptional and rapid accumulation of anthropogenic debris on one of the world's most remote and pristine islands, 2017

Midway Atoll

Midway Atoll (land area 6.2 km^2) is in the North Pacific Ocean, 2,400 km west of Hawaii (the nearest significant land) and near the center of the North Pacific Gyre. As with Henderson Island the circulation of the ocean washes vast quantities of plastic on to the beaches. NOAA regularly removes plastic debris from the beaches. Since 1999 they have removed 125 tonnes of plastic.

On both these islands, and many others, plastic is mistaken by seabirds for food and they eat it, or feed it to their young. Every year on Midway Island, thousands of albatross young die from ingesting plastic products. It is estimated that at the current rate of plastic accumulation in the oceans, by 2050 there will be more plastic by mass in the ocean than fish. Microplastic and nanoplastic, tiny pieces of plastic including microbeads, enters the food chain and has been demonstrated to concentrate toxins, and so increase toxins in the fish that eat them. Nanoplastic can enter cells and affect their functioning.

Plastic in albatross chick, Midway Atoll

Plastic debris on beach, Henderson Island

2. Explain how plastic thrown away in a city on the west coast of North America can end up in a Laysan albatross chick on Midway Atoll:

_____

_____

_____

3. What effect do micro and nanoplastics have on the animals that ingest them: _____

_____

_____

255 The Impact of Urbanization

Key Idea: Urbanization describes the concentration of human populations in cities. It can be associated with environmental problems, including air and water pollution.

Urbanization refers to the movement of people into urban areas. One might think that this would have positive effects on the wider environment as a concentration of people in cities should leave fewer people to directly affect outlying unpopulated regions. However, large cities tend to concentrate problems, such as water runoff and pollution, which have wide-ranging impacts.

Urbanization affects the wider environment

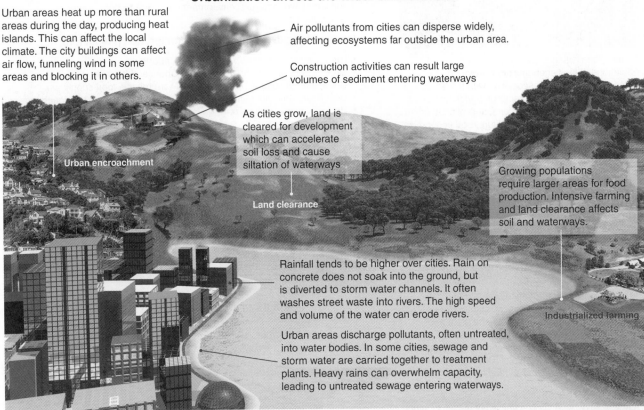

Urban areas heat up more than rural areas during the day, producing heat islands. This can affect the local climate. The city buildings can affect air flow, funneling wind in some areas and blocking it in others.

Air pollutants from cities can disperse widely, affecting ecosystems far outside the urban area.

Construction activities can result large volumes of sediment entering waterways

As cities grow, land is cleared for development which can accelerate soil loss and cause siltation of waterways

Urban encroachment

Land clearance

Growing populations require larger areas for food production. Intensive farming and land clearance affects soil and waterways.

Rainfall tends to be higher over cities. Rain on concrete does not soak into the ground, but is diverted to storm water channels. It often washes street waste into rivers. The high speed and volume of the water can erode rivers.

Industrialized farming

Urban areas discharge pollutants, often untreated, into water bodies. In some cities, sewage and storm water are carried together to treatment plants. Heavy rains can overwhelm capacity, leading to untreated sewage entering waterways.

Cities generally are not producers of food. As the urban population grows, more pressure is put on the rural areas to produce enough food to feed both them and the city population.

Cities tend to concentrate the negative aspects of human life. People crowded into poorly designed and built housing results in higher disease and pollution. Pollutants wash into waterways, polluting water needed for drinking and irrigation.

Fidel Gonzalez CC 3.0

Urban areas tend to concentrate pollutants and wastes. Smog from vehicle exhausts and industry emissions affect respiratory health. Heavy metals tend to be concentrated in wastes and can lead to serious defects in wildlife once in the environment.

1. Give three reasons why urbanization affects the environment more widely than just the urban area itself:

(a) _____

_____

_____

(b) _____

_____

(c) _____

_____

©2017 **BIOZONE** International
ISBN: 978-1-927309-65-0
Photocopying Prohibited

CONNECT **247** AP2   CONNECT **246** AP2   CONNECT **245** AP2   WEB **255**

KNOW

The shift to urban living

The traditional villages characteristic of the rural populations of less economically developed countries have a close association with the land. The households depend directly on agriculture or harvesting natural resources for their livelihood and are linked through family ties, culture, and economics.

The computerization and mechanization of agriculture has affected rural populations. More powerful machinery can plow land and plant and harvest crops more efficiently. Computerization of farming mechanisms allows fewer people to do more work, and from remote locations. This reduces jobs and pushes people away from rural areas.

Cities are differentiated communities, where the majority of the population does not depend directly on natural resource-based occupations. While cities are centers of commerce, education, and communication, they are also centers of crowding, pollution, and disease.

The redistribution of people from rural to urban environments, or urbanization, has been an important characteristic of human societies. About half of the people in the world already live in urban areas and this figure is predicted to increase to 64% by 2050. Urban populations can grow through natural increase (i.e more births than deaths) or by immigration. Immigration is driven by **push factors** that encourage people to leave their rural environment and **pull factors** that draw them to the cities.

Immigration push factors

- Rural overpopulation
- Lack of work or food
- Changing agricultural practices
- Desire for better education
- Racial or religious conflict
- Political stability

Immigration pull factors

- Opportunity
- Chance of better housing
- More reliable food supply
- Opportunity for greater wealth
- Freedom from village traditions
- Government policy

Rural and urban population in the USA

— % of US population (urban)
···· % of US population (rural)

■ Total urban population
□ Total rural population

The United States underwent a dramatic rural to urban shift in the 19th and early 20th centuries. Many developing countries are now experiencing similar shifts. Graph compiled from UN data

2. Describe the kinds of changes in agricultural practices that could contribute to urbanization: _____

_____

_____

3. (a) Describe some of the positive effects of urbanization: _____

_____

_____

(b) Describe some negative effects of urbanization: _____

_____

_____

©2017 **BIOZONE** International
ISBN: 978-1-927309-65-0
Photocopying Prohibited

256 Nitrogen Pollution

Key Idea: Excessive nitrogen in the environment can cause groundwater contamination and eutrophication of waterways. Excess nitrogen in the environment (**nitrogen pollution**) has several effects depending on the compound formed. Nitrogen gas (N_2) makes up almost 80% of the atmosphere but is unreactive at normal pressure and temperature. However, at the high pressures and temperatures of factory processes and combustion engines, N_2 forms nitrogen oxides, which contribute to air pollution. Nitrates in fertilizers are washed into groundwater and slowly make their way to lakes and rivers and eventually out to sea. This process can take time to become noticeable because groundwater may take decades to reach a waterway. Likewise, once nitrate loads are reduced, it may take years for a recovery to occur.

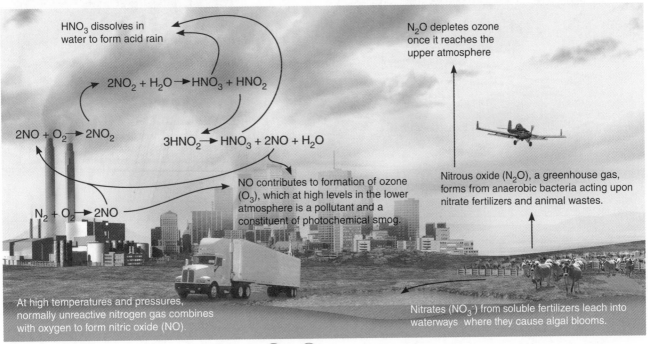

HNO_3 dissolves in water to form acid rain

$2NO_2 + H_2O \rightarrow HNO_3 + HNO_2$

$2NO + O_2 \rightarrow 2NO_2$

$3HNO_2 \rightarrow HNO_3 + 2NO + H_2O$

$N_2 + O_2 \rightarrow 2NO$

NO contributes to formation of ozone (O_3), which at high levels in the lower atmosphere is a pollutant and a constituent of photochemical smog.

N_2O depletes ozone once it reaches the upper atmosphere

Nitrous oxide (N_2O), a greenhouse gas, forms from anaerobic bacteria acting upon nitrate fertilizers and animal wastes.

At high temperatures and pressures, normally unreactive nitrogen gas combines with oxygen to form nitric oxide (NO).

Nitrates (NO_3^-) from soluble fertilizers leach into waterways where they cause algal blooms.

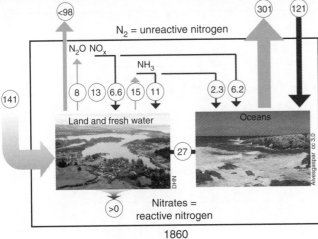

N_2 = unreactive nitrogen

<98 | 301 | 121 | 141

N_2O NO_x

NH_3

8 | 13 | 6.6 | 15 | 11 | 2.3 | 6.2

Land and fresh water | Oceans

27

Nitrates = reactive nitrogen

>0

1860

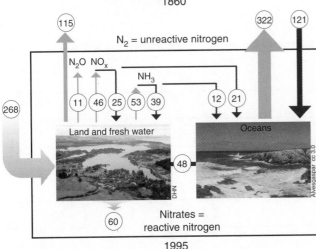

N_2 = unreactive nitrogen

115 | 322 | 121 | 268

N_2O NO_x

NH_3

11 | 46 | 25 | 53 | 39 | 12 | 21

Land and fresh water | Oceans

48

Nitrates = reactive nitrogen

60

1995

Changes in nitrogen inputs and outputs between 1860 and 1995 in million Tonne (modified from Galloway et al 2004)

Early last century, the Haber-Bosch process made nitrate fertilizers readily available for the first time. Since then, the use of nitrogen fertilizers has increased at an almost exponential rate. Importantly, this has led to an increase in the levels of nitrogen in land and water by up to 60 times those of 100 years ago. This extra nitrogen load is one of the causes of human-induced accelerated enrichment (**cultural eutrophication**) of lakes and coastal waters. An increase in algal production also results in higher decomposer activity and, consequently, oxygen depletion, fish deaths, and depletion of aquatic biodiversity. Many aquatic microorganisms also produce toxins, which may accumulate in the water, fish, and shellfish. The diagrams (left) show the increase in nitrates in water sources from 1860 to 1995. The rate at which nitrates are added has increased faster than the rate at which nitrates are returned to the atmosphere as unreactive N_2 gas. This has led to the widespread accumulation of nitrogen.

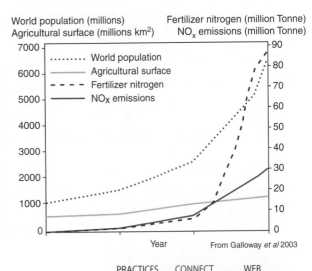

World population (millions)
Agricultural surface (millions km^2)

Fertilizer nitrogen (million Tonne)
NO_x emissions (million Tonne)

········· World population
——— Agricultural surface
- - - - Fertilizer nitrogen
——— NO_x emissions

Year

From Galloway et al 2003

PRACTICES

CONNECT **247** AP2

WEB **256**

KNOW

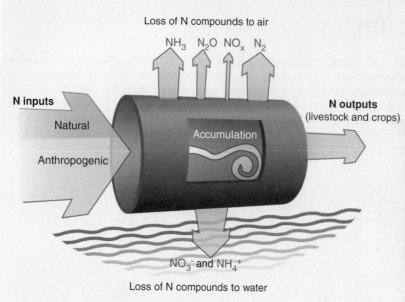

Loss of N compounds to air

NH$_3$ N$_2$O NO$_x$ N$_2$

N inputs

Natural

Anthropogenic

Accumulation

N outputs
(livestock and crops)

NO$_3^-$ and NH$_4^+$

Loss of N compounds to water

From O. Oenema *et al* 2007

The "hole in the pipe" model (left) demonstrates inefficiencies in nitrogen fertilizer use. Nitrogen that is added to the soil and not immediately taken up by plants is washed into waterways or released into the air by bacterial action. These losses can be minimized to an extent by using slow release fertilizers during periods of wet weather and by careful irrigation practices.

Algal blooms

Satellite photo of algal blooms around Florida. Excessive nitrogen contributes to algal blooms in both coastal and inlands waters. *Image: NASA*

1. Describe the effect each of the following nitrogen compounds have on air and water quality:

   (a) NO: _____

   _____

   (b) N$_2$O: _____

   _____

   (c) NO$_2$: _____

   _____

   (d) NO$_3^-$: _____

   _____

2. Explain why the formation of NO can cause large scale and long term environmental problems: _____

   _____

   _____

3. Why would an immediate halt in the use of nitrogen fertilizers not cause an immediate stop in their effects?

   _____

   _____

   _____

4. (a) Calculate the increase in nitrogen deposition in the oceans from 1860 to 1995 and compare this to the increase in release of nitrogen from the oceans.

   _____

   (b) What is the effect of this increase on the oceans? _____

   _____

5. (a) Why do nitrogen inputs tend to be so much more than outputs in crops and from livestock? _____

   _____

   (b) Suggest how the nitrogen losses could be minimized: _____

   _____

   _____

©2017 **BIOZONE** International
ISBN: 978-1-927309-65-0
Photocopying Prohibited

257 Deforestation

Key Idea: Deforestation is the permanent removal of forest from an area. It has many negative environmental and social or cultural effects.

Tropical rainforests prevail in places where the climate is very moist throughout the year (annual rainfall of 200 to 450 cm). Almost half of the world's rainforests are in just three countries: **Brazil** in South America, **Zaire** in Africa, and **Indonesia** in Southeast Asia. Much of the world's biodiversity resides in rainforests. Destruction of the forests will contribute towards global warming through a large reduction in photosynthesis. In the Amazon, 75% of deforestation has occurred within 50 km of Brazil's roads. Many potential drugs could still be discovered in rainforest plants, and loss of species through deforestation may mean they will never be found. Rainforests can provide economically sustainable crops (rubber, coffee, nuts, fruits, and oils) for local people.

Deforestation

At the end of the last glacial period, about 10,000 years ago, forests covered an estimated 6 billion hectares, about 45% of the Earth's land surface. Forests currently cover about 4 billion hectares of land (31% of Earth's surface). They include the cooler temperate forests of North and South America, Europe, China and Australasia, and the tropical forests of equatorial regions. Over the last 5000 years, the loss of forest cover is estimated at 1.8 billion hectares. 5.2 million hectares has been lost in the last 10 years alone. Temperate regions where human civilizations have historically existed the longest (e.g. Europe) have suffered the most but now the vast majority of deforestation is occurring in the tropics. Intensive clearance of forests during settlement of the most recently discovered lands has extensively altered their landscapes (e.g. in New Zealand, 75% of the original forest was lost in a few hundred years).

Causes of deforestation

Deforestation is the end result of many interrelated causes, which often center around socioeconomic drivers. In many tropical regions, the vast majority of deforestation is the result of subsistence farming. Poverty and a lack of secure land can be partially solved by clearing small areas of forest and producing family plots. However huge areas of forests have been cleared for agriculture, including ranching and plantations for the production of palm oil. These produce revenue for governments through taxes and permits, producing an incentive to clear more forest. Just 14% of deforestation is attributable to commercial logging (although combined with illegal logging it may be higher).

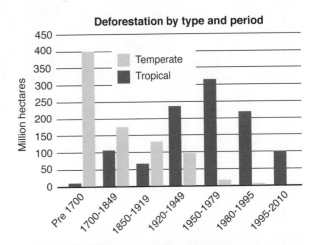

Deforestation by type and period

Causes of deforestation	%
Fuel	5
Logging	14
Commercial agriculture	32
Subsistence farming	48

Tropical deforestation globally

Percentage loss of primary tropical forest

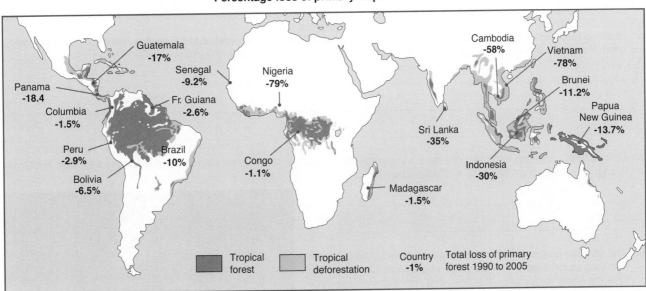

Guatemala -17%
Senegal -9.2%
Panama -18.4
Columbia -1.5%
Fr. Guiana -2.6%
Nigeria -79%
Peru -2.9%
Brazil -10%
Bolivia -6.5%
Congo -1.1%
Madagascar -1.5%
Sri Lanka -35%
Cambodia -58%
Vietnam -78%
Brunei -11.2%
Papua New Guinea -13.7%
Indonesia -30%

Tropical forest | Tropical deforestation | Country -1% | Total loss of primary forest 1990 to 2005

▶ It is important to distinguish between deforestation involving primary (old growth) forest and deforestation in plantation forests. Plantations are regularly cut down and replaced and can artificially inflate a country's apparent forest cover or rate of deforestation. The loss of primary forests is far more important as these are refuges of high biodiversity, including for rare species, many of which are endemic to relatively small geographical regions (i.e. they are found nowhere else).

▶ Although temperate deforestation is still a concern, it is in equatorial regions that the pace of deforestation is accelerating (above). This is of global concern as species diversity is highest in the tropics and habitat loss puts a greater number of species at risk.

©2017 **BIOZONE** International
ISBN: 978-1-927309-65-0
Photocopying Prohibited

PRACTICES   PRACTICES   CONNECT   WEB

274 AP2   **257**

KNOW

The building of new road networks into regions with tropical rainforests causes considerable environmental damage. In areas with very high rainfall there is an increased risk of erosion and loss of topsoil.

Up to 80% of Earth's terrestrial species are found in tropical rainforests. Many tropical species are endemic to specific parts of the forest or even tree species. Loss of these areas could exterminate hundreds of species.

The soil in tropical rainforests in typically poor. Once the trees are removed and crops planted the soil quickly loses its nutrients and productivity decreases within a few seasons.

1. Identify the three main human activities that cause tropical deforestation and briefly describe their detrimental effects:

    (a) _____

    (b _____

    (c) _____

2. Describe the trend in temperate and tropical deforestation over the last 300 years: _____

    _____

    _____

3. Deforestation in temperate regions has largely stabilized and there has been substantial forest regrowth. However, these second growth forests differ in structure and composition to the forests that were lost. Why might this be of concern?

    _____

    _____

    _____

4. Building roads in primary forests can have serious negative effects on the forest, apart from the damage of building the road. Suggest why roads through primary forests might be damaging to the forest in the long term?

    _____

    _____

    _____

5. Calculate the area (in hectares) of forest lost per minute globally in the last ten years (to visualize the size, one hectare is slightly smaller than two NFL football pitches).

    _____

    _____

6. The satellite image shown right shows an area of the Amazon that was deforested over the course of one year (2012 - 2013). Use the scale to calculate the area that was lost over this time:

    _____

    _____

    _____

amshiyacu

deforested area

2 km

NASA

258 Climate Change

Key Idea: Global warming refers to the continuing rise in the average temperature of the Earth's surface. It is associated with global climate change.

Since the mid 20th century, the Earth's surface temperature has been steadily increasing. The consensus scientific view (97% of publishing climate scientists) is that this phenomenon, called **global warming**, is attributable to the increase in atmospheric levels of CO_2 and other greenhouse gases emitted as a result of human activity. The increased warming is associated with changes in the Earth's climate.

Changes in near-surface temperature

This graph right shows how the mean temperature for each year from 1860-2010 (bars) compares with the average temperature between 1961 and 1990. The blue line represents the fitted curve and shows the general trend indicated by the annual data.

Most anomalies since 1977 have been above normal and warmer than the long term mean, indicating that global temperatures are tracking upwards. The last ten years have been some of the warmest on record.

Source: Hadley Centre for Prediction and Research

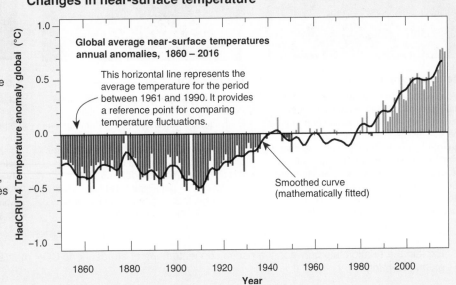

HadCRUT4 Temperature anomaly global (°C)

Global average near-surface temperatures annual anomalies, 1860 – 2016

This horizontal line represents the average temperature for the period between 1961 and 1990. It provides a reference point for comparing temperature fluctuations.

Smoothed curve (mathematically fitted)

Year

Changes in atmospheric CO_2

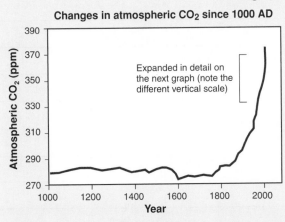

Changes in atmospheric CO_2 since 1000 AD

Atmospheric CO_2 (ppm)

Expanded in detail on the next graph (note the different vertical scale)

Year

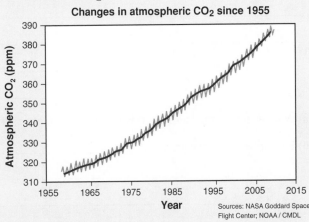

Changes in atmospheric CO_2 since 1955

Atmospheric CO_2 (ppm)

Year

Sources: NASA Goddard Space Flight Center; NOAA / CMDL

Atmospheric CO_2 has been rapidly increasing since the 1800s. In 2012, the world emitted a record (till then) 34.5 billion tonnes of CO_2 from fossil fuels. In total, humans have emitted 545 billion tonnes of CO_2. CO_2 levels fluctuate seasonally, especially in the northern hemisphere because of its much larger landmass and forests.

During the Industrial Revolution (1760-1840), coal was burned in huge quantities to power machinery. The increase in CO_2 released is attributed to an increase in average global temperatures. The combustion of fossil fuels (coal, oil, and natural gas) continues to pump CO_2 into the atmosphere and contribute to the current global warming.

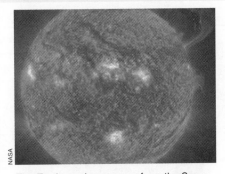

The Earth receives energy from the Sun (above) as UV, visible, and near-infrared radiation. Some is absorbed by the Earth's surface and the rest is reflected away as long-wavelength thermal radiation (heat). Much of this is trapped by the greenhouse gases and directed back to Earth, further increasing the mean surface temperature.

The oceans act as a carbon sink, absorbing the CO_2 produced from burning fossil fuels. The CO_2 reacts in the water, forming carbonic acid, lowering ocean pH, and reducing the availability of carbonate ions. This makes it harder for corals (above) to build their calcium carbonate exoskeletons and is causing significant coral reef damage.

PRACTICES   PRACTICES   PRACTICES   WEB

258

KNOW

Confusing the debate

Media coverage

Global warming and associated climate changes are complex issues and most people obtain their information from popular media. Despite the scientific consensus on the role of human activity in climate change, some media still provide biased or inaccurate information. In order to make an informed decision, people must read or listen to a wide range of balanced media, or read scientific documents and make their own decision.

Lobby groups

Lobby groups with specific interests strive constantly to influence policy makers. Reducing CO_2 emissions by restricting coal and oil use will help reduce global warming. However, fossil fuel consumption generates billions of dollars of revenue for coal and oil companies, so they lobby against legislation that penalizes fossil fuel use. If successful, lobbying could result in less effective climate change policies.

Controversy

All scientific bodies of international standing agree that human activity has contributed disproportionately to global warming. However, there are still some in the political, scientific, and commercial community who claim that global warming is not occurring. These people often command media attention and engage a poorly informed public audience, who are often suspicious of the scientific community.

1. Explain the relationship between the rise in concentrations of atmospheric CO_2, methane, and oxides of nitrogen, and global warming:

_____

_____

_____

_____

_____

2. (a) What effect did the Industrial Revolution have on atmospheric CO_2 levels? _____

(b) Explain why this occurred: _____

_____

_____

3. Divide your class into small groups and discuss and evaluate claims by climate change sceptics that human activities are not causing climate change. Summarize your arguments below and present them to the class in a short presentation:

_____

_____

_____

_____

_____

_____

_____

_____

_____

_____

_____

©2017 **BIOZONE** International
ISBN: 978-1-927309-65-0

259 The History of Climate Modeling

Key Idea: Climate models have become more complicated and sophisticated over time, allowing scientists to better predict climate change.

Climate models have been in use since the 1950s, but these very early versions really only modeled the weather in a particular region. The sophistication and accuracy of climate models has increased over time (below). This is because our knowledge about factors contributing to climate has increased and also because developments in computing and mathematics have allowed the more accurate prediction of more complicated scenarios. In 1988, the Intergovernmental Panel on Climate Change (IPCC) was established. Its role is to analyze published climate data and inform the international community about their findings.

Climate models have become more sophisticated over time

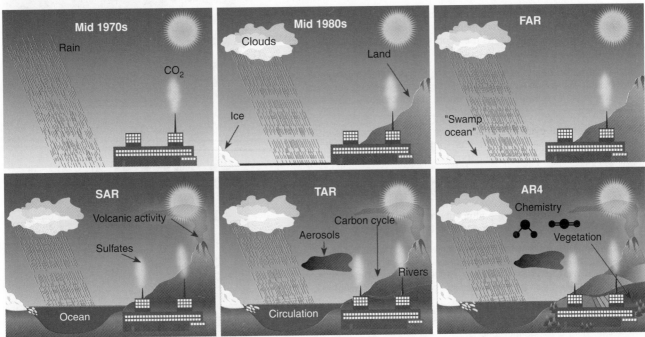

▶ The diagrams above show how the sophistication of climate models has changed over time. Note how the complexity has increased as more elements are incorporated into the models. Early models in the 1970s were very simple and factored in only a few components (incoming sunlight, rainfall, and CO_2 concentration).

▶ By the 1980s, the models were becoming more complex and other features were added such as clouds, land surface features, and ice. After the establishment of the IPCC, several climate models were developed in relatively quick succession.

▶ The First Assessment Report (FAR) in the early 1990s, the Second Assessment Report (SAR) in 1995, the Third Assessment Report (TAR) in 2001, and the Fourth Assessment Report (AR4) in 2007. FAR included the ocean's effect for the first time, and subsequent models became more sophisticated, including adding the effect of atmospheric constituents such as sulfates and aerosols, the role of the carbon cycle, atmospheric chemistry, and vegetation.

How are climate models tested?

To see how well models work, scientists enter past data and see how accurately they predict the climate changes that have already occurred. If the models recreate historical trends accurately, we can have confidence that they will also accurately predict future trends in climate change.

The graph on the right shows an example of how climate models are tested. The grey band represents data from 14 models and 58 different simulations. The black line represents the average of all 58 simulations. The grey line represents the average actual (observed) data for the same period. The gray vertical lines represent large volcanic eruptions during the period.

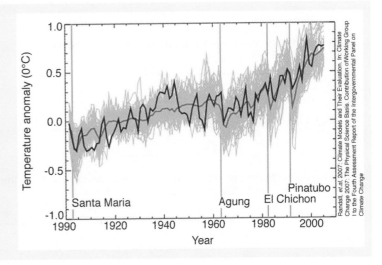

Randall, et.al. 2007. Climate Models and Their Evaluation. In: Climate Change 2007: The Physical Science Basis. Contribution of Working Group I to the Fourth Assessment Report of the Intergovernmental Panel on Climate Change

1. When were the effects of the ocean first included in climate models? _____

2. What effect do eruptions have on the observed temperatures? _____

_____

 PRACTICES PRACTICES PRACTICES WEB **259**

KNOW

What should a climate model include?

Climate models predict climate change more accurately when the model incorporates all the factors contributing to climate change. Some components influencing climate (e.g. the ocean and atmosphere) have their own models to better understand how the individual components can be influenced. Data from these separate models can provide more detailed information about the climate model as a whole. As we have already seen, climate models have become more complicated over time. Most now incorporate the following components:

Sea ice

- **Atmosphere**: This includes cloud cover, air pressure, water vapor, and gas concentrations.

- **Oceans**: Oceans have a key role in climate regulation. They help to buffer (neutralize) the effects of increasing levels of greenhouse gases in the atmosphere by acting as a carbon sink. They also act as a heat store, preventing rapid rises in global atmospheric temperature.

- **Ice sheets and sea ice (the cryosphere)**: These factors influence how much of the Sun's heat is reflected or retained. Increased ice levels reflect more heat away from Earth. Less ice allows more heat to be retained.

Carbon emissions

- **Biogeochemical cycles**: Levels of some atmospheric compounds can greatly influence climate change. Carbon is the most significant, but others such as nitrogen, phosphorus, and sulfur can also influence climate.

- **Biosphere**: The level of plant cover on Earth has a significant impact on the amount of carbon in the atmosphere. During photosynthesis, plants utilize carbon dioxide from the atmosphere to produce carbohydrates, effectively removing a major greenhouse gas from the atmosphere.

- **Human activity**: Human activity has increased the rate of global warming, especially through the actions of deforestation and carbon emissions into the atmosphere. The addition of greenhouse gases into the atmosphere through human activity is driving current climate change.

Deforestation

- **External influences**: These include energy variations from the Sun (e.g. through sunspot cycles) and levels of carbon dioxide and other aerosols released during volcanic eruptions.

3. (a) How has the complexity of climate models changed over time? _____

_____

_____

(b) What has been the significance of this? _____

_____

_____

4. (a) How do scientists check the accuracy of their models? _____

_____

_____

(b) Why is it important that they do this? _____

_____

(c) Study the testing results on the previous page. Do you think the average data from the models accurately reflects the historical data? Why or why not?

_____

_____

_____

5. (a) Working in pairs or small groups, select one component of a climate model and research its significance to climate change. Summarize your findings and report back to the class.

(b) Once all the presentations have been made, determine if any factor(s) has a larger influence than another.

©2017 **BIOZONE** International
ISBN: 978-1-927309-65-0
Photocopying Prohibited

260 Models of Climate Change

Key Idea: Climate change models provide best-case and worst-case predictions for increase in global temperatures. There are elements of uncertainty even in well tested models. The major source of uncertainty is human activity and, in particular, how much the consumption of fossil fuels will change in the future. The level of greenhouse gases in the atmosphere will have a significant impact on future climate change. The IPCC often run a number of different scenarios to predict climate change. Between them, the results provide best-case and worst-case scenarios.

Using climate models to predict change

▶ The major scenarios are presented below, but there are subcategories (e.g. A1B) to help make them more accurate:

- **A1** assumes rapid economic and technological growth, a low rate of population growth, and a very high level of energy use. Differences between "rich" and "poor" countries narrow.

- **A2** assumes high population growth, slower technological change and economic growth, and a larger difference between countries and regions than in other scenarios. Energy use is high.

- **B1** assumes a high level of environmental and social consciousness and sustainable development. There is low population growth, high economic and technological advancement, and low energy use. The area devoted to agriculture decreases and reforestation increases.

- **B2** has similar assumptions to B1. However, there are more disparities between industrialized and developing nations, technological and economic growth is slower than in B1, and population growth is greater (but less than A2). Energy use is midway between B1 and A2. Changes in land use are less dramatic than in B1.

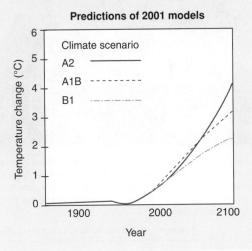

Predictions of 2001 models

Climate scenario: A2, A1B, B1

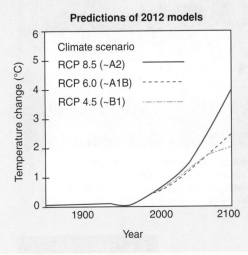

Predictions of 2012 models

Climate scenario: RCP 8.5 (~A2), RCP 6.0 (~A1B), RCP 4.5 (~B1)

1. Why do scientists simulate a number of different scenarios when they run a climate model? _____

_____

_____

2. Study the 2001 and 2012 models of climate change predictions (above).

(a) In the 2001 model, identify which scenario was predicted to produce the highest temperature change by 2100:

_____

(b) What factors are likely to contribute to this? _____

_____

(c) Why would scenario B1 produce the lowest temperature increase? _____

_____

_____

(d) How do the predictions between the 2001 and 2012 models differ? _____

_____

_____

PRACTICES   PRACTICES   PRACTICES   WEB

  260

KNOW

261 Ice Sheet Melting

Key Idea: Higher average temperatures melt sea-ice. Less heat is reflected back to space, warming sea temperature and promoting further melting of the ice.

The surface temperature of the Earth is in part regulated by the amount of ice on its surface, which reflects a large amount of heat into space. However, the area and thickness of the polar sea-ice has almost halved since 1980. This melting of sea-ice can trigger a cycle where less heat is reflected into space during summer, warming seawater and reducing the area and thickness of ice forming in the winter.

Arctic sea-ice summer minimum
1980: 7.8 million km²

Arctic sea-ice summer minimum 2012:
Record low, 4.1 million km²

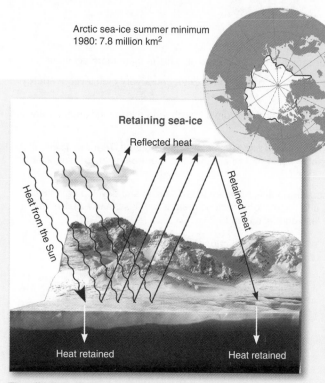

Retaining sea-ice

The **albedo** (reflectivity of sea-ice) helps to maintain its presence. Thin sea-ice has a lower albedo than thick sea-ice. More heat is reflected when sea-ice is thick and covers a greater area. This helps to regulate the temperature of the sea, keeping it cool.

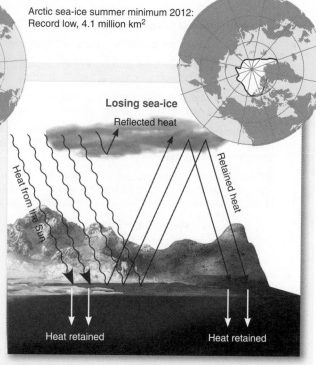

Losing sea-ice

As sea-ice retreats, more non-reflective surface is exposed. Heat is retained instead of being reflected, warming both the air and water and causing sea-ice to form later in the autumn than usual. Thinner and less reflective ice forms and perpetuates the cycle.

The temperature in the Arctic has been above average every year since 1988. Coupled with the reduction in summer sea-ice, this is having dire effects on Arctic wildlife such as polar bears, which hunt out on the ice. The reduction in sea-ice reduces their hunting range and forces them to swim longer distances to firm ice. Studies have already shown an increase in drowning deaths of polar bears.

Average* Arctic air temperature fluctuations

Data source: National Geographic

| | 1900-1919 | 1920-1939 | 1940-1959 | 1960-1979 | 1980-1999 | 2000-2008 |

+2.0ºC
+1.5ºC
+1.0ºC
+0.5ºC
-6.8ºC
-0.5ºC
-1.0ºC
-1.5ºC
-1.7ºC

*Figure shows deviation from the average annual surface air temperature over land. Average calculated on the years 1961-2000.

1. Explain how low sea-ice albedo and volume affects the next year's sea-ice cover: _____

_____

_____

2. Discuss the effects of decreasing summer sea-ice on polar wildlife: _____

_____

_____

_____

©2017 **BIOZONE** International
ISBN: 978-1-927309-65-0
Photocopying Prohibited

262 The Modern Atlantis?

Key Idea: Many low lying nations are at risk of flooding as sea levels rise.

It is estimated that sea levels may rise by between 100 and 900 mm over the next century. This is mainly due to the thermal expansion of the oceans as they increase in temperature, but also includes the melting of large ice sheets. This rise in sea level will have a significant effect on low lying islands and coral atolls as many are presently only a few meters above sea level. The following news article focuses on the island nation of Kiribati in the Pacific.

Vanishing Lands

The newspaper, and author name of the following article are fictitious, but the text is based on real events and information.

The Tribune
By Michael Anton: Saturday 7 June 2008

After many years of unanswered appeals for action on climate change, the tiny South Pacific nation of Kiribati has concluded that it is doomed. On Thursday its President, Anote Tong, used World Environment Day to request international help to evacuate his country before it disappears.

Kiribati consists of just 33 coral atolls scattered across 5 million square kilometers of the Pacific Ocean and it has limited scope for coping with impending global climate changes with most of the land being barely 2 metres above sea level.

Speaking from New Zealand, Mr Tong said his fellow countrymen, i-Kiribati, as they are known, had no alternative but to leave. "We may be beyond redemption," he said. "We may be at the point of no return, where the emissions in the atmosphere will carry on contributing to climate change, to produce a sea level change. So in time our small, low-lying islands will be submerged."

A London economics graduate, Mr Tong said the emigration of his people needed to start immediately. "We don't want to believe this, and our people don't want to believe this. It gives us a deep sense of frustration. What do we do?"

Kiribati is home to 97,000 people most of them living on the main atoll of Tarawa, a ring of islets surrounding a central lagoon. It is regarded as one of the places most vulnerable to climate change along with Vanauta, Tuvalu and the Marshall Islands.

Currently, the most serious problem Kiribati faces is erosion caused by flooding and storms. "We have to find the next highest spot," said Mr Tong. "At the moment there's only the coconut trees." But even the coconut trees are dying, caused by drought and a rising level of salt in ground water, which is also not being replaced due to the fact there has little rain at all for the last three years.

Mr Tong was in New Zealand – which, after committing itself to becoming carbon neutral, was chosen to host the UN's World Environment Day – for talks with Prime Minister, Helen Clark, whom he hopes to persuade to help resettle his people. But he also appealed to other countries for help relocating the i-Kiribati.

However, New Zealand, already with a large population of Pacific Islanders, would have immense trouble absorbing the 97,000 immigrants which would strain its generosity to the limit and total almost 2.5% of its current national population.

And that is just Kiribati. Talks have not yet begun with many of the other island nations which soon may also be submerged. In 2006, the Australian government issued a warning of a flood of environmental refugees across the Asian-Pacific region.

President Tong said he had heard many national leaders argue that measures to combat climate change could negatively affect their economic development, but pointed out that for the i-Kiribati it was not a matter of economics, it was a matter of survival. He said that while international scientists argue about the causes of climate change, the effects were already beginning to show on his nation. "I am not a scientist, but what I know is that things are happening we did not experience in the past... Every second week, when we get the high tides, there are always reports of erosion."

Villages, after occupying the same site for centuries, are having to be relocated due to the encroaching water. "We're doing it now... it's that urgent," he said. "Where they have been living over the past few decades is no longer there. It is being eroded."

Worse case scenarios suggest the i-Kiribati could be uninhabitable within 60 years, Mr Tong said. "I've appealed to the international community that we need to address this challenge. It's a challenge for the whole global community."

Leading industrialized nations last month pledged to cut their carbon emissions in half by 2050. But they stopped short of setting firm targets for 2020, which many scientists argue is crucial if the planet is to be saved. But for Kiribati, its saviour may have come too late.

---

1. From the information provided in the article, explain why Kiribati is vulnerable to the effects of global warming:

2. The article states that global warming is already adversely affecting the island nation of Kiribati. What are the physical effects it refers to and what will be their impact on the inhabitants of Kiribati in the short term?

3. The inhabitants of Kiribati may be forced to relocate to other countries. What could this mean for the identity of the i-Kiribati and their culture and language?

4. Using other research tools, such as the library or internet, find and read other articles relating to sea level change in the Pacific. Do they all agree that there will be a rise in sea level if carbon emissions are not reduced? On a separate sheet explain why you think some articles take different viewpoints. Staple your explanation to this page:

PRACTICES

COMP

263 Biodiversity and Global Warming

Key Idea: Global warming has the potential to change many habitats. Many organisms may contract or expand their range. Accelerated global warming is changing the habitats of plants and animals and this could have significant effects on the biodiversity of specific regions as well as on the planet overall. As temperatures rise, organisms may have to migrate to new areas to find the conditions to which they are adapted. Those that cannot move face extinction, as temperatures move outside their limits of tolerance. Changes in precipitation as a result of climate change also affect where organisms can live. Long term changes in climate could see the contraction or expansion of many habitats. Habitat migration (the shift in habitats to new latitudes or altitudes) will also become more frequent. Already there are a number of recorded cases showing the direct effects of climate change on the distribution, behavior, and genetics of a range of organisms.

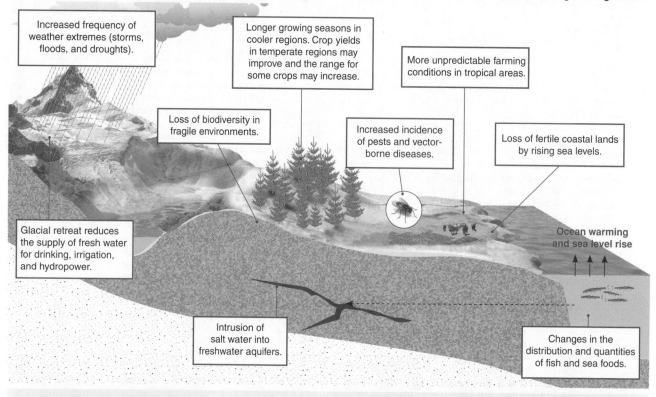

Increased frequency of weather extremes (storms, floods, and droughts).

Longer growing seasons in cooler regions. Crop yields in temperate regions may improve and the range for some crops may increase.

More unpredictable farming conditions in tropical areas.

Loss of biodiversity in fragile environments.

Increased incidence of pests and vector-borne diseases.

Loss of fertile coastal lands by rising sea levels.

Glacial retreat reduces the supply of fresh water for drinking, irrigation, and hydropower.

Ocean warming and sea level rise

Intrusion of salt water into freshwater aquifers.

Changes in the distribution and quantities of fish and sea foods.

Studies of the distributions of butterfly species in many countries show their populations are shifting. Surveys of Edith's checkerspot butterfly (*Euphydryas editha*) in western North America have shown it to be moving north and to higher altitudes.

Studies of sea life along the Californian coast have shown that between 1931 and 1996, shoreline ocean temperatures increased by 0.79°C and populations of invertebrates including sea stars, limpets and snails moved northward in their distributions.

Studies of forests in the United States have shown that although there will be increases and decreases in the distribution ranges of various tree species, overall there will be an 11% decrease in forest cover, with an increase in savanna and arid woodland. Communities of oak/pine and oak/hickory are predicted to increase in range while spruce/fir and maple/beech/birch communities will decrease.

An Australian study in 2004 found the centre of distribution for the AdhS gene in *Drosophila*, which helps survival in hot and dry conditions, had shifted 400 kilometers south in twenty years.

A 2009 study of 200 million year old plant fossils from Greenland has provided evidence of a sudden collapse in biodiversity that is correlated with, and appears to be caused by, a very slight rise in CO_2 levels.

©2017 **BIOZONE** International
ISBN: 978-1-927309-65-0
Photocopying Prohibited

Effects of increases in temperature on animal populations

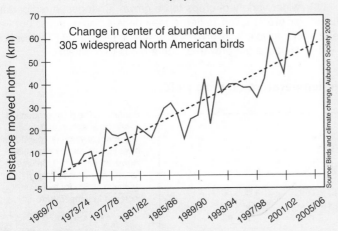

Change in center of abundance in 305 widespread North American birds

Distance moved north (km)

70
60
50
40
30
20
10
0
-5

1969/70 1973/74 1977/78 1981/82 1985/86 1989/90 1993/94 1997/98 2001/02 2005/06

Source: Birds and climate change, Aububon Society 2009

A number of studies indicate that animals are beginning to be affected by increases in global temperatures. Data sets from around the world show that birds are migrating up to two weeks earlier to summer feeding grounds and are often not migrating as far south in winter.

Animals living at altitude are also affected by warming climates and are being forced to shift their normal range. As temperatures increase, the snow line increases in altitude pushing alpine animals to higher altitudes. In some areas of North America this has resulting the local extinction of the North American pika (*Ochotona princeps*).

Wiki Commons, Public Domain

1. Describe some of the likely effects of global warming on physical aspects of the environment:

_____

_____

_____

_____

2. Describe the possible effects of global warming on plants: _____

_____

_____

3. Discuss the evidence that insect populations are affected by global temperature: _____

_____

_____

_____

4. (a) Describe how increases in global temperatures have affected some migratory birds: _____

_____

_____

_____

   (b) Explain how these changes in migratory patterns might affect food availability for these populations:

_____

_____

_____

5. Explain how global warming could lead to the local extinction of some alpine species: _____

_____

_____

_____

264 Climate Change and Agriculture

Key Idea: The range and variety of crops may change due to the climate changes associated with global warming.

The impacts of climate change on agriculture and horticulture in North America will vary because of the size and range of its geography. In some regions, temperature changes will increase the growing season for existing crops, or enable a wider variety of crops to be grown. Changes in temperature or precipitation patterns may benefit some crops, but have negative effects on others. Higher atmospheric CO_2 levels will enhance the growth of some crops (e.g. wheat).

Effects of increases in temperature on crop yields

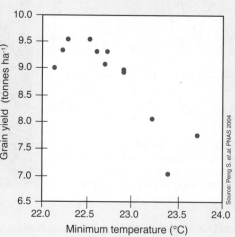

Studies on the grain production of rice have shown that maximum daytime temperatures have little effect on crop yield. However minimum night time temperatures lower crop yield by as much as 5% for every 0.5°C increase in temperature.

Source: Peng S. *et.al.* PNAS 2004

Possible effects of increases in temperature on crop damage

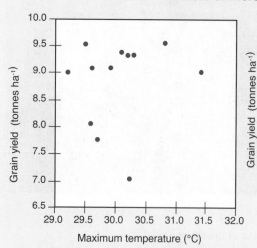

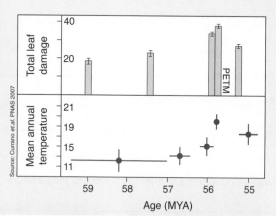

Source: Currano *et.al.* PNAS 2007

The fossil record shows that global temperatures rose sharply around 56 million years ago. Studies of fossil leaves with insect browse damage indicate that leaf damage peaked at the same time as the Paleocene Eocene Thermal Maximum (PETM). This gives some historical evidence that as temperatures increase, plant damage caused by insects also rises. This could have implications for agricultural crops.

Ellen Levy Finch cc3.0

Citrus production will shift slightly north with reduced yields in Texas and Florida.

Grain crops (such as wheat, above) are at higher risk of crop failures if precipitation decreases.

Brocken Inaglory cc3.0

Californian wine grapes

Crops grown near to their climate threshold may suffer reductions in yield, quality, or both.

USDA

Corn rootworm beetle

Milder winters and longer growing seasons may see the distribution of agricultural pest species spread.

1. Why will a warming climate benefit some agricultural crops, while disadvantaging others? _____

_____

_____

2. Explain how a warming climate can influence the distribution of pest species, and in turn affect agriculture: _____

_____

_____

©2017 **BIOZONE** International
ISBN: 978-1-927309-65-0
Photocopying Prohibited

265 Temperature and Enzyme Activity

Key Idea: A rise in global temperatures may affect the metabolic rates and enzyme activity of many organisms.

All enzymes, including those involved in metabolic processes, have a limited range of conditions over which they can function. Each enzyme has a set of optimal conditions within which they work most effectively. Outside these conditions, enzyme activity rapidly declines. As global temperatures rise, ectotherms that do not thermoregulate (so their body temperature varies with the environmental temperature) may find that higher temperatures reduce their metabolic efficiency. Those that are able to move, may shift to more suitable climates. Those with limited or no mobility could become extinct if they are unable to tolerate the rising temperature. Endotherms will have fewer problems because body temperature is regulated, but they may still have to move to stay within their preferred temperature zone.

Enzymes and temperature

Temperature affects the rate of enzyme activity. All enzymes have a temperature range they function within. At temperatures below this range, enzyme activity is very slow. As the temperature increases, so too does the enzyme activity, until the temperature is high enough to damage the enzyme's functional structure. At this point, the enzyme is denatured and ceases to function.

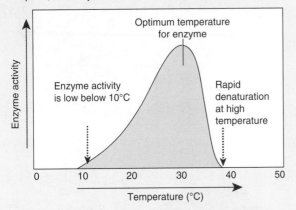

Enzyme activity in rice

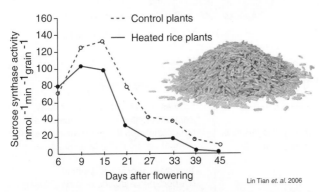

Rice contains starch and it is one of the world's most important food plants. To make starch, sugars (including sucrose) are first hydrolyzed into monomers and then the components are linked together. The degradation of sucrose by the enzyme sucrose synthase is the first step in the synthesis of starch. Recent studies (above) have shown that a temperature increase of 0.3 - 2°C reduces sucrose synthase activity. This would inhibit or slow the formation of starch within a rice grain and change its nutritional value.

Coral reef bleaching

An increase in sea temperatures could mean the death of coral reefs. Healthy coral reefs depend upon the symbiotic relationship between a coral polyp that builds the reef, and photosynthetic protozoa called zooxanthellae. Zooxanthellae live within the polyp tissues and provides it with most of its energy. A 1-2°C temperature increase is sufficient to disrupt the photosynthetic enzymes. The zooxanthellae either dies, or is expelled from the coral due to stress. The result is coral bleaching

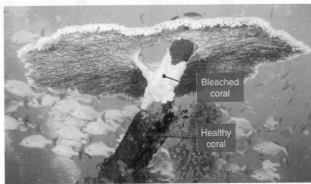

Photosynthesis

The photosynthetic enzyme **RuBisCO** (below) is activated by a companion enzyme, **rubisco activase**. Rubisco activase is inhibited at temperatures above 35°C and also by high CO_2 levels. In many countries, a warming climate could push the temperature above 35°C. Rubisco activase would become less active, reducing the activity of the RuBisCo enzyme. The result would be a reduction in net photosynthesis and productivity.

The RuBisCO enzyme (above) is the most abundant protein found in plant leaves. It catalyzes the first major step in the carbon fixation process.

1. Describe why enzymes could be affected by an increase in global temperature: _____

2. Explain why it can be difficult to predict the effect of global warming on specific enzymes: _____

PRACTICES PRACTICES CONNECT 34 AP2 CONNECT 23 AP2 WEB 265 **KNOW**

266 Temperature and the Distribution of Species

Key Idea: Global warming will have important implications for the development, distribution, and survival of many species. Many similar species avoid competition by breeding at different times and under different conditions. Temperature shifts may force similar species into competition within an overlapping range. Many habitats will change in a warmer climate. Climatic zones will shift and some species will need to relocate or adapt to the new conditions in order to survive.

Distribution and breeding of leopard frogs in North America

Water temperature during breeding and embryo development of *Rana* spp.

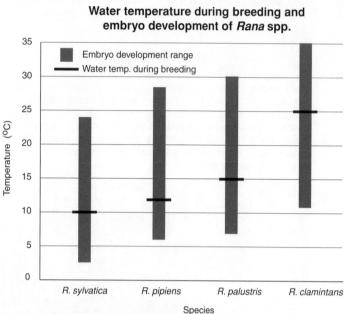

Legend:
- Embryo development range
- Water temp. during breeding

Species: R. sylvatica, R. pipiens, R. palustris, R. clamintans

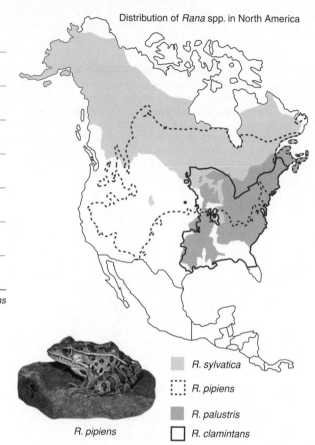

Distribution of *Rana* spp. in North America

- R. sylvatica
- R. pipiens
- R. palustris
- R. clamintans

R. pipiens

The frog genus *Rana* is relatively common and widely distributed in North America (top right). The graph above shows the preferred water temperature for breeding in four common species of *Rana*, as well as the temperature tolerance range for embryonic development. Outside these ranges, embryonic development rate decreases or the embryos die. Increases in temperature could reduce the available breeding habitat for some species (e.g. *Rana sylvatica* requires low temperature to breed). Another likely outcome would be a change to the timing of breeding or a shift in the distribution patterns of populations.

Effects of temperature on life history parameters in *Ceriodaphnia*

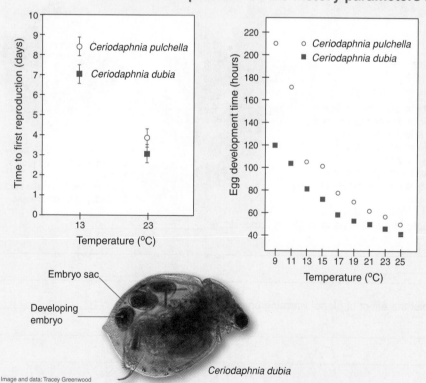

Time to first reproduction (days) vs Temperature (°C):
- Ceriodaphnia pulchella
- Ceriodaphnia dubia

Egg development time (hours) vs Temperature (°C):
- Ceriodaphnia pulchella
- Ceriodaphnia dubia

Embryo sac
Developing embryo

Ceriodaphnia dubia

Image and data: Tracey Greenwood

Studies of lake ecosystems in New Zealand highlight the significance of temperature on the generation times and competitive outcome between organisms.

Two small species of daphnid (*Ceriodaphnia pulchella* and *C. dubia*) show marked temperature-dependent differences in important aspects of their life histories, such as **egg development time** and time to maturity. At lower temperatures, *C. dubia* has a significant breeding advantage over *C. pulchella* because its eggs develop faster and it matures earlier. This means that fall populations can establish quickly without competition from *C. pulchella,* which matures one and a half days later. At higher temperatures *C. dubia* loses this competitive advantage (far left).

The competitive advantage enjoyed by *C. dubia* during cooler seasons could be lost if lake temperatures were to rise by even 1 or 2°C. The warming could provide a subtle but measurable change in selection pressures, shift the species balance, and so influence the ecology of the lake.

WEB 266

CONNECT 192 AP2

CONNECT 196 AP2

PRACTICES

PRACTICES

PRACTICES

KNOW

©2017 **BIOZONE** International
ISBN: 978-1-927309-65-0
Photocopying Prohibited

Disappearing sea ice

Temperatures have risen in the Arctic every year for the last 27 years. At the same time, the average summer sea ice minimum has decreased by almost half. Polar bears mainly hunt seals which surface at breathing holes in the ice. The reduced sea ice levels have changed the distribution patterns of seals and many polar bears have been forced to swim for miles to hunt. Between 1979 and 1991, 87% of bears observed were on sea ice during summer. However in the period between 1992 to 2004, this reduced to just 33%. In addition, the thinner sea ice cannot hold the bear's weight, and they are forced to the mainland before they have built up their winter fat stores. The loss of condition is affecting reproduction rates, and juvenile survival rates are lower as a consequence. Pregnant females must also swim for longer distances to reach their dens, and so lose more condition in the process. A 2007 study (right) shows a decrease in the number of bears denning in pack ice over a 20 year period.

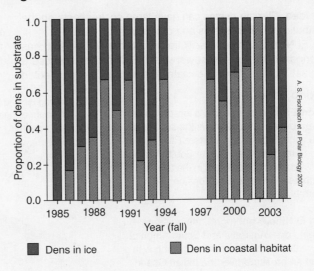

A. S. Fischbach et al Polar Biology 2007

■ Dens in ice   ■ Dens in coastal habitat

1. Discuss the potential effects of a rise in global temperature on the North American distribution of the frog genus *Rana*:

_____

_____

_____

_____

_____

2. (a) Describe the general effect of temperature on egg development rates of *C. pulchella* and *C. dubia*: _____

_____

   (b) Describe the implication of this to competition between the two species: _____

_____

3. (a) Describe the effect of temperature on differences in time to first reproduction between *C. pulchella* and *C. dubia*:

_____

_____

   (b) Describe the effect of increasing temperature on egg development for both *C. pulchella* and *C. dubia:* _____

_____

_____

   (c) Predict the effect of an increase in temperature on competition between the two species: _____

_____

_____

_____

4. Discuss the impact of the reduction of the Arctic ice sheet on the polar bear population: _____

_____

_____

_____

_____

_____

267 The Impact of El Niño

Key Idea: Climate cycles can have large scale effects on the distribution of ecosystems and organisms.

The El Niño-Southern Oscillation is a climate cycle that has a periodicity of three to seven years. El Niño years cause a reversal of the ordinary climate regime and are connected to such economically disastrous events as the collapse of fisheries stocks (e.g the Peruvian anchovy stock), severe flooding in the Mississippi Valley, drought-induced crop failures, and forest fires in Australia and Indonesia. It is also responsible for greening of the Chilean deserts.

El Niño effect

In **non-El Niño** conditions, a low pressure system over Australia draws the southeast trade winds across the eastern Pacific from a high pressure system over South America. This system produces rain in the area of Australasia and dry conditions on the coast of South America.

In an **El Niño** event (right), the pressure systems over Australia and South America are weakened or reversed, beginning with a rise in air pressure over the Indian Ocean, Indonesia, and Australia. Warm waters block the nutrient upwelling along the west coast of the Americas. El Niño brings drought to Indonesia and northeastern South America, while heavy rain over Peru and Chile causes the deserts to bloom.

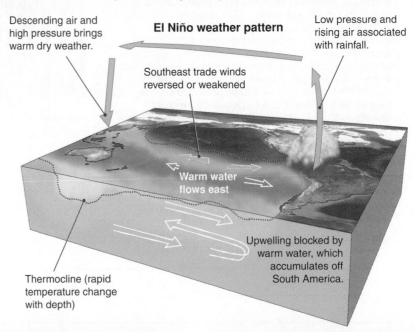

El Niño weather pattern

Descending air and high pressure brings warm dry weather.

Low pressure and rising air associated with rainfall.

Southeast trade winds reversed or weakened

Warm water flows east

Upwelling blocked by warm water, which accumulates off South America.

Thermocline (rapid temperature change with depth)

During non-El Niño years, cool nutrient-rich waters along the South American coast sustain huge populations of fish such as anchovy. During El Niño events, warm waters reduce nutrients, and fish populations either crash or move to feeding grounds elsewhere.

El Niño events bring more rain to deserts in parts of South America and parts of Baja California. On the islands of the Gulf of California, plant cover increases from 0-4% during non El Niño years to 54-89% during El Niño years. In Northern Chile plant cover increased over five times during El Niño.

El Niño events affect terrestrial animal communities too. On the islands of the Gulf of California, spider densities doubled in response to high levels of insect prey resulting from increases in plant abundance. The population later crashed as parasitoid wasps proliferated.

1. Describe the events that cause El Niño conditions and its effects on ocean circulation: _____

_____

_____

_____

_____

2. Describe the effects of El Niño on the ecosystems of:

(a) Chilean deserts: _____

_____

(b) Islands of the Gulf of California: _____

_____

_____

©2017 **BIOZONE** International
ISBN: 978-1-927309-65-0
Photocopying Prohibited

268 Continental Drift and Species Distribution

Key Idea: The distribution of many groups of organisms can be explained by continental drift.

Continental drift is a theory explaining the past and current positions of the continents. It has been incorporated into the theory of plate tectonics, but is still useful when talking about the continents themselves, rather than the mechanisms that move them. About 200 million years ago, the continents were joined in one large land mass called Pangea. By 135 million years ago this had broken into two large subcontinents, Laurasia in the north and Gondwana in the south. Gondwana consisted of what are now the continents of Africa, Antarctica, Australia, South America, and India. These continents contain organisms, both extinct and living, with distributions that are best explained by the continents having all once been joined.

Cycads

Cycads has a scattered distribution, being found in mainly tropical parts of Australia, southern Africa, Malaysia and the Americas. Today these places are separated by large areas of ocean. Because the seeds of most cycads sink in water and disperse only short distances on land (they don't tend to be dispersed by flying animals) this varied distribution is difficult to explain. However, the cycad fossil record dates back to around 300- 280 million years ago. This was when the all the continents were still joined as Pangea. This scattered distribution then makes sense if it is explained by the cycad being present on various continents as they moved apart.

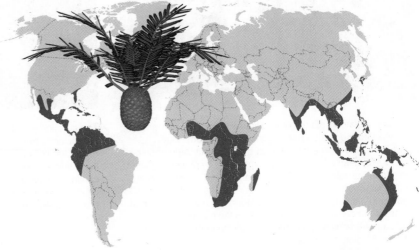

Glossopteris

Glossopteris is a genus of extinct seed fern that existed during the Permian period 300 million years ago. Fossils of it are found in all the continents that made up the ancient supercontinent of Gondwana. These continents are now widely separated, but the distribution of fossils can be explained by the separation of the continents after the plant evolved in Gondwana.

Glossopteris

Africa

India

South America

Antarctica

Australia

Gondwana

Marsupials

About 70% of marsupials (pouched mammals) are found in Australia, the rest are found in South and Central America. One species is found in North America. Australia is about 12,000 km from South America. This is far further than any marsupial could swim or fly (there are no truly flying or marine marsupials anyway!). There are no known land bridges between Australia and South America, indeed the ocean depth between the two continents is an average of at least 3000 m. Again, the answer to the question of their distribution is explained by continental drift.

The first true marsupials appeared in North America about 80 million years ago (ancestral marsupials appeared in China about 120 million years ago). They then spread to South America before Gondwana split from Laurasia. The fossil record shows marsupials appeared in Antarctica about 40 million years ago and in Australia between 40 and 30 million years ago. Because Australia then drifted away from the rest of the continents before being reached by placental mammals, marsupials became the dominant mammalian group. However, when South America rejoined North America via the Isthmus of Panama about 3-4 million years ago, placental mammals spread south and replaced many South American marsupials. (Placentals tend to out-compete marsupials in resource-rich environments that support high energy use).

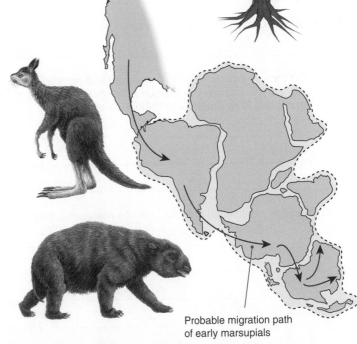

Probable migration path of early marsupials

©2017 **BIOZONE** International
ISBN: 978-1-927309-65-0
Photocopying Prohibited

PRACTICES

PRACTICES

CONNECT **248** AP1

CONNECT **246** AP1

CONNECT **192** AP2

WEB **268**

KNOW

Continental drift creates new ecosystems

Continental drift can have effects on ecosystems that may last for extended periods of time and cover large areas. Examples include the production of the Himalayas and the isolation of Antarctica.

Antarctica is isolated from the other continents by the Southern Ocean which formed when Antarctica separated from South America. This allowed the formation of the Antarctic Circumpolar Current, which helped to rapidly cool Antarctica by preventing warm currents from the Atlantic and Pacific reaching it. This isolation and Antarctica's current position at the South Pole has produced a polar ecosystem that has no large land predators. The only large animal that lives on the continent through winter is the emperor penguin. The predator free environment means the emperor penguin can form huge colonies than nest directly on the ice.

The collision of India with Europe raised up the Himalayas, producing numerous climatic changes and new ecosystems. The rise of the Himalayas produced the high Tibetan plateau and affects the seasonal rains in India by causing warm, moist air to rise. The cooling air then loses its moisture as precipitation as it moves south, causing the monsoons. The rain shadow effect of the Himalayas produces the large cold deserts of the Gobi and the Taklamakan.

1. (a) What features of cycad seeds make it unlikely that cycads would spread across oceans? _____

    _____

    (b) How then did cycads achieve a global tropical distribution? _____

    _____

2. Describe the current distribution of *Glossopteris* fossils and explain why this distribution can both be explained by continental drift and is evidence for continental drift:

    _____

    _____

    _____

    _____

3. (a) Why can the distribution of marsupials not be explained by a recent migration? _____

    _____

    (b) The theory of continental drift predicted that marsupial fossils would be found in Antarctica long before they were. How did the discovery of marsupial fossils in Antarctica in 1982 give support for continental drift and the migration of marsupials from South America to Australia?

    _____

    _____

    _____

4. Explain how continental drift can produce new ecosystems: _____

    _____

    _____

    _____

    _____

269 Ocean Acidification

Key Idea: Carbon dioxide reacts with water to reduce its pH. The oceans act as a carbon sink, absorbing much of the CO_2 produced from burning fossil fuels. When CO_2 reacts with water it forms carbonic acid, which decreases the pH of the oceans. This could have major effects on marine life, especially shell making organisms. Ocean acidification is relative term, referring to the oceans becoming less basic as the pH decreases. Ocean pH is still above pH 7.0

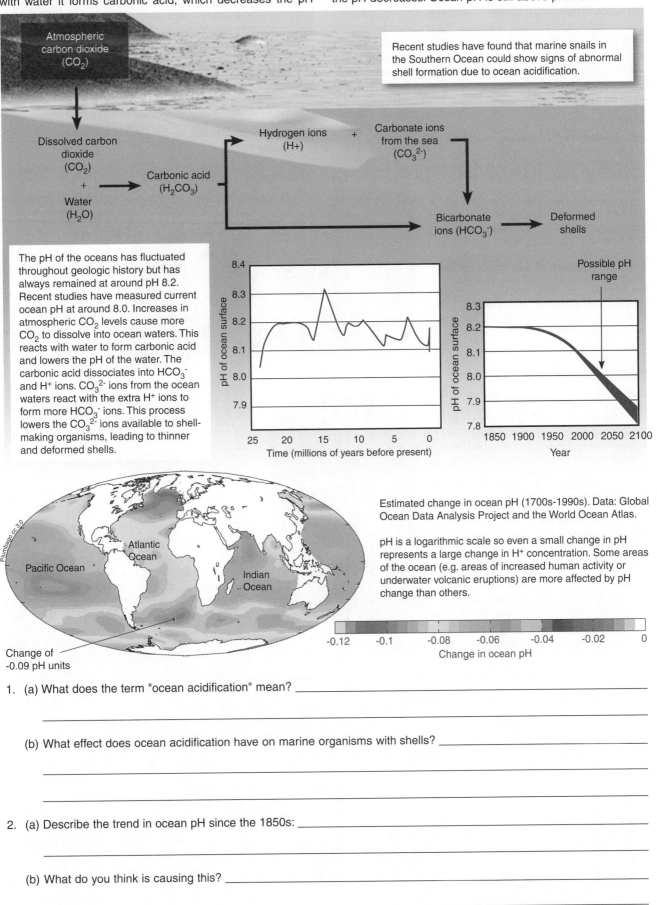

Recent studies have found that marine snails in the Southern Ocean could show signs of abnormal shell formation due to ocean acidification.

The pH of the oceans has fluctuated throughout geologic history but has always remained at around pH 8.2. Recent studies have measured current ocean pH at around 8.0. Increases in atmospheric CO_2 levels cause more CO_2 to dissolve into ocean waters. This reacts with water to form carbonic acid and lowers the pH of the water. The carbonic acid dissociates into HCO_3^- and H^+ ions. CO_3^{2-} ions from the ocean waters react with the extra H^+ ions to form more HCO_3^- ions. This process lowers the CO_3^{2-} ions available to shell-making organisms, leading to thinner and deformed shells.

Estimated change in ocean pH (1700s-1990s). Data: Global Ocean Data Analysis Project and the World Ocean Atlas.

pH is a logarithmic scale so even a small change in pH represents a large change in H^+ concentration. Some areas of the ocean (e.g. areas of increased human activity or underwater volcanic eruptions) are more affected by pH change than others.

1. (a) What does the term "ocean acidification" mean? _____

(b) What effect does ocean acidification have on marine organisms with shells? _____

2. (a) Describe the trend in ocean pH since the 1850s: _____

(b) What do you think is causing this? _____

270 The Impact of Alien Species

Key Idea: Species introduced from one ecosystem to another are often highly competitive in their new environment and detrimentally affect the dynamics of the native community.

Introduced (or **alien**) **species** are those that have evolved at one place in the world and have been transported by humans, either intentionally or in advertently, to another region. Some of these introductions are beneficial, e.g. introduced agricultural plants and animals and Japanese clams and oysters (the mainstays of global shellfish industries). **Invasive species** are those alien species that have a detrimental effect on the ecosystems into which they have been imported.

There are hundreds of these species with varying degrees of undesirability to humans. Humans have deliberately introduced many species into new environments whereas others have been imported with cargo shipments or in the ballast of ships. Some have been deliberately introduced to control another pest species and have themselves become a problem. Some of the most destructive of all alien species are fast growing plants, e.g. mile-a-minute weed (a perennial vine from Central and South America), velvet tree (a South American tree invading Hawaii and Tahiti), and *Caulerpa* seaweed, the aquarium strain now found in the Mediterranean.

Kudzu

Kudzu (*Pueraria lobata*) is a climbing vine native to south-east Asia. It spreads aggressively by vegetative reproduction and is a serious invasive pest in the southern US, where it has been spreading at a rate of 61,000 ha per annum. Kudzu was first introduced to the US in the 1800s as an ornamental plant for shade porches, and was subsequently widely distributed as a high-protein cattle fodder and as a cover plant to prevent soil erosion. It grew virtually unchecked in the climate of the Southeastern US and was finally listed as a weed in 1970, more than a decade after it was removed from a list of suggested cover plants. Today, kudzu is estimated to cover 3 million ha of land in the southeastern US.

Kudzu is now widely distributed (dark blue, above). The photo left was taken in Atlanta, Georgia.

Zebra mussels

Zebra mussels (*Dreissena polymorpha*) are freshwater mussels native to the Black, Caspian, and Azov Seas. They have become a highly invasive species in many parts of the world. Zebra mussels were first discovered in North America in 1988. They reproduce prolifically, crowding out native species and severely affecting food chains where native mussels and clams are food for other species.

Zebra mussels have also affected human activities. They often grow in the intake pipes of hydroelectric turbines and the bottom of boats and anchor chains, and must be regularly removed. Anti-fouling cupronickel alloy coatings prevent the mussels establishing but these measures are costly. It is estimated zebra mussels have cost North American communities around $5 billion since their introduction.

Zebra mussels were discovered in Lake St. Clair in 1988. It is thought they arrived in ballast water of ships traversing the St. Lawrence Seaway. The mussels now inhabit hundreds of North America lakes, including all the Great Lakes.

1. How do alien species arrive in a new environment? _____

_____

2. Why did Kudzu first spread through the USA? _____

_____

3. What makes zebra mussels an economically costly invasive species? _____

_____

_____

4. Investigate an introduced species in your region. When did it arrive? What area does it cover? What effects is it having on the environment? Are there any plans in place for its removal?

_____

_____

©2017 **BIOZONE** International
ISBN: 978-1-927309-65-0
Photocopying Prohibited

271 The Impact of New Diseases

Key Idea: Diseases introduced to new areas can have serious effects on the organisms there.

The organisms in an ecosystem have evolved to survive with the suite of other species in their ecosystem, including competitors, predators, and pathogens. Therefore, they are often poorly equipped (e.g. in behaviors or physiology) to survive the challenges posed by the introduction of a new organism. When the new organism is a microscopic pathogen, the original inhabitants of the ecosystem are especially vulnerable, as they can not actively avoid it. Populations in which the disease is endemic are usually the descendents of earlier generations that survived an epidemic and so have some immunity to the disease, gained by natural selection. When the disease is introduced to a new population that has never encountered it, no such immunity exists and the effects of the disease are more severe. Occasionally a new strain of a disease appears and this has a similar effect to the introduction of a new disease, where no immunity exists.

Dutch elm disease

Dutch elm disease (DED) is caused by the fungus *Ophiostoma ulmi* and affects trees in the elm family (*Ulmus* and *Zelkova*) causing wilt. The disease is spread by the elm bark beetle. It originated in Asia and was accidently introduced to Europe, North America, and much later, New Zealand. Where the disease has been introduced it has quickly caused the loss of large numbers of elm trees by clogging their vascular tissue.

DED in North America

DED arrived in North America in 1928. Efforts to contain the disease were reduced during WWII and by 1989 North America has lost an estimated 75% of its elms.

DED in Minnesota

As in other U.S. states, elms were commonly used in Minnesota as shade trees in municipal areas, such as in parks and along streets. By 1950s, Minnesota had around 140 million elms in parks and lining streets and streams. DED did not reach Minnesota until about 1960, although it was known to be in many surrounding states. Measures that could have been taken to stop its introduction into Minnesota, such as restricting elm planting and controlling the import of logs, could have reduced the impact when the disease arrived. The city of Minneapolis was estimated to have an elm population of over 200,000 in 1970. Losses began slowly, but rapidly increased until an aggressive sanitation campaign began in 1977. However, by around 1990 the elm population had reduced to about 100,000.

This campaign, including trimming, stump removal, and replanting, cost Minneapolis around $60 million between 1978 and 1988.

Elm trees lost in Minneapolis	
Year	Losses per year
1972	222
1973	225
1974	937
1975	1688
1976	7239
1977	31,475
1978	20,813
1979	6751
1980	4184
1981	5068
1982	3389
1983	2144
1984	4965
1985	4087
1986	2896
1987	2280

Elm afflicted by DED

DED in the United Kingdom

DED appeared in Europe in 1910 and a much deadlier strain appeared in 1967. In the UK, these strains have killed 60 million trees. France has lost nearly 90% of its elm trees.

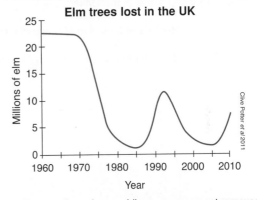

Elm trees lost in the UK

(Clive Potter *et al* 2011)

Tree numbers drop rapidly as management removes infected trees. New trees grow, but new infections, outbreaks, and repeated management reduces numbers.

Nests in elm trees for kestrel and stock dove		
	Kestrel (*Falco tinnunculus*)	Stock dove (*Columba oenas*)
1970	15	20
1971	12	15
1972	11	11
1973	10	26
1974	11	5
1975	8	10
1976	8	8
1977	7	7

The number of nests of various British birds recorded in elm trees reduced significantly over the 1970s as elm trees succumbed to DED. Interestingly, the kestrel actually increased its population numbers five fold over this time.

1. Using an example, explain why a disease entering a new population is likely to cause a severe epidemic: _____

_____

_____

_____

_____

©2017 **BIOZONE** International
ISBN: 978-1-927309-65-0
Photocopying Prohibited

PRACTICES   PRACTICES   CONNECT   CONNECT   WEB

   274 AP1   **122** AP2   **271**

KNOW

Smallpox in North America

Smallpox caused by *Variola major* is a deadly disease with a mortality of around 35%. It did not reach the New World until European settlers arrived in the 16th century. Native Americans both in North and South America had no immunity to the disease and it caused enormous losses of life, wiping out half the population of some areas.

The first epidemic among native Americans occurred in 1617 in Massachusetts. The epidemic spread across the continent, causing many deaths. Smallpox eliminated half the people of Huron and Iroquois in the Great Lakes region. In 1738, it destroyed half the Cherokee, and in 1759 almost half the Catawbas. During the time of the American Revolution, the Piegan lost half its people to smallpox and around 1803, two thirds of the Omahas died from the disease. In the 1820s, smallpox killed 80% of the people of the Columbia River. In 1837, smallpox returned to the Great Plains region, killing about half the people there.

Public Domain

The Three Chiefs - Piegan. Photograph c1900 by Edward S. Curtis

2. (a) What causes Dutch elm disease? _____

   (b) How is it spread? _____

   _____

   (c) When did it first arrive in the United States? _____

   (d) Plot a graph of elm tree losses in Minneapolis from the data on the previous page:

   (e) Calculate the total number of trees lost in Minneapolis from 1972 to 1987: _____

3. Explain why the number of elm trees lost in the UK peaked again around 1990: _____

   _____

   _____

   _____

4. What was the effect of Dutch elm disease on the bird population of the UK? _____

   _____

   _____

©2017 **BIOZONE** International
ISBN: 978-1-927309-65-0
Photocopying Prohibited

272 Extinction or Evolution?

Key Idea: Phenotypic plasticity may allow organisms to shift its phenotype in response to environmental change. Species with limited phenotypic plasticity may be unable to survive environmental change.

Environmental change puts stress on organisms and if they cannot adapt to the new environmental conditions, their populations may decline or become extinct. Recent studies have investigated the phenotypic response of organisms to climate change. These include changes to life history patterns (e.g. changing breeding times) and morphological changes (e.g. becoming smaller). However, most of these variations are the result of **phenotypic plasticity** rather than adaptation through evolution. Phenotypic plasticity involves changes to the phenotype without a change in genotype and may provide species with a way to track environmental changes giving them the time required to adapt genetically.

Phenotypic plasticity allows an organism to shift its phenotype in response to environmental changes within certain, genetically constrained, extremes. It is more important for immobile organisms because if they cannot adjust to the new environment they will die.

The buckeye butterfly (*Junonia coenia*) changes color as the breeding season progresses. Those hatched in early summer are light colored (almost yellow). Buckeyes hatched in early autumn are darker, so they absorb heat and warm up faster during the cooler days.

Such phenotypic changes allow a species to quickly adapt to changes in its environment, but there is no overall change to the gene pool.

Early season buckeye

Late season buckeye

Photos: Megan McCarty

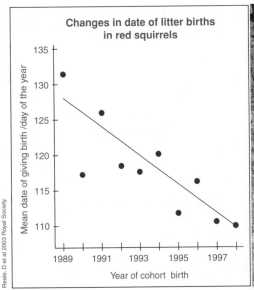

Gilles Gonthier cc 2.0

Reale, D et al 2003 Royal Society.

Changes in date of litter births in red squirrels

Mean date of giving birth /day of the year

Year of cohort birth

North American red squirrels (*Tamiasciurus hudsonicus*) in Canada have adapted to a 2°C increase in spring temperature by breeding earlier in the year. Records were kept of cohorts of female squirrels to determine the day of the year they gave birth. Over a period of ten years, the squirrel breeding time shifted to be earlier in the year (by 18 days). The change was linked to an increase in the abundance of spruce cones, an important food source for the red squirrels.

Some organisms have limited phenotypic plasticity and will be unable to respond immediately to shifts in climate. For these organisms, relocation may be their only chance for survival. For species with an already limited range (e.g. polar bears) there may be no where to go and, as food supplies dwindle, they may be faced with extinction.

1. (a) Explain the difference between phenotypic plasticity and adaptation through evolution: _____

_____

_____

(b) Explain why adapting to the environment takes longer than a physiological adjustment: _____

_____

_____

2. Explain why some organisms are more likely to become extinct as a result of climate change than others:

_____

_____

PRACTICES

CONNECT
274
AP1

WEB
272

KNOW

273 KEY TERMS AND IDEAS: Did You Get It?

1. Test your vocabulary by matching each term to its definition, as identified by its preceding letter code.

climate change

climate model

continental drift

ecosystem resilience

ecosystem resistance

El Niño

extinction

global warming

urbanization

A The complete dying out a taxon, usually a species.

B The process of the Earth's average surface temperature steadily increasing in temperature (and its projection continuation).

C The change in distribution of weather patterns over a long period of time.

D The gradual increase in the proportion of people living in urban areas.

E Mathematical or computation model that predicts future changes in the climate.

F The movement of the Earth's continents relative to each other.

G The capacity of an ecosystem to recover from disturbance or withstand ongoing pressures.

H The property of communities to remain essentially unchanged when subjected to disturbance.

I One phase of a naturally occurring global climate cycle, which disrupts normal weather patterns and can lead to storms in some places and droughts in others.

2. The plot right show the results of the experimental removal of the starfish *Pisaster* from a region of an ecosystem:

(a) What does the graph show? _____

_____

(b) What do the results suggest about *Pisaster*'s role in this ecosystem?

_____

_____

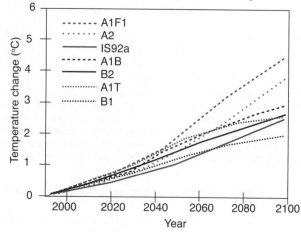

3. Use the graphs below to answer the following questions:

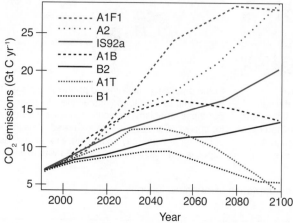

(a) Which scenario produces the highest CO_2 emissions by 2100? _____

(b) Which scenario produces the lowest CO_2 emissions by 2100? _____

(c) Which scenario produces the greatest temperature change by 2100? _____

(d) Which scenario produces the smallest temperature change by 2100? _____

(e) Explain why even in the scenarios where CO_2 emissions are falling, temperature still rises:

_____

_____

TEST

274 Synoptic Questions

Questions 1-2 relate to the plot below. Identify the correct answer

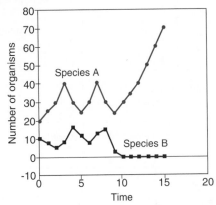

1. Species A and B are found in the same ecosystem. What could be inferred from the data presented:

   (a) Species B is the prey of species A

   (b) Species B is the predator of species A

   (c) Species A is a keystone species

   (d) Species A is a parasite of species B

2. Which of the following statements would explain why a species would show cyclic population changes:

   (a) A keystone species in the ecosystem is removed.

   (b) The vegetation on which it depends varies seasonally.

   (c) Climate shifts alter the prevalence of parasites that affect the species.

   (d) A predator of the species is preyed on by a species that is introduced to the ecosystem.

3. Pumas prey on a species of deer in a region. The deer feed on vegetation. What is the most likely outcome if the puma population declines markedly as a result of hunting by humans:

   (a) The deer population will cycle between high and low numbers until the predator returns.

   (b) The deer population will increase and strip the region of vegetation.

   (c) The deer population will decline and vegetation cover will increase.

   (d) The deer population will be stable.

4. Which of the following might be expected if a keystone species is eliminated from an ecosystem:

   (a) Interspecific competition increases.

   (b) There is one less niche available.

   (c) The ecosystem diversity declines.

   (d) There is no significant change.

5. When two species have a close relationship in which both obtain benefits, the relationship is called:

   (a) Symbiosis

   (b) Mutualism

   (c) Commensalism

   (d) Competition

6. The plot below shows the mean net productivity of various biomes. The productivity varies between them because:

   (a) Productivity increases from the poles to the equator.

   (b) The oceans have fewer photosynthetic organisms than terrestrial biomes.

   (c) Productivity is determined largely by available water, light, and nutrients.

   (d) Grasses have lower productivity than trees.

7. The photograph below depicts:

   (a) Intraspecific competition and predation

   (b) Cooperation

   (c) Intra- and interspecific competition and predation

   (d) Interspecific competition and predation

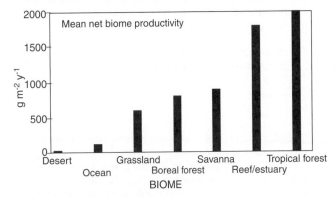

8. The cartoon sequence depicted below illustrates?

   (a) Primary succession and ecosystem resilience

   (b) Secondary succession and ecosystem resilience

   (c) Secondary succession and ecosystem resistance

   (d) Primary succession and ecosystem resistance

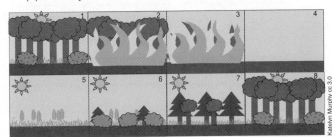

9. Which of the following could accelerate current rates of global warming:

   (a) Increased rate of fossil fuel use

   (b) Deforestation without replanting

   (c) Decrease in ice-albedo

   (d) All of the above

PRACTICES

TEST

10. (a) Use the following labels to complete the diagram of the carbon cycle shown right: *atmosphere, geosphere, oceans, respiration (R), photosynthesis (PS)*.

(b) Add arrows to show deforestation (D) and combustion (C) of fossil fuels:

(c) Combustion and deforestation result in another 9 petagrams of carbon being added to the carbon cycle. Add this value to the diagram.

(d) About 3 petagrams is taken up by photosynthesis and 2 petagrams is taken up by the oceans. Add these values to the appropriate labels on the diagram.

(e) How much extra carbon is actually added to the atmosphere by deforestation and combustion?

_____

Exchanges

Land, plants, animals, soil

Burial   Weathering

Coal, oil, gas

11. The plot below right shows the results of a 60 day experiment to investigate the interactions between rocky shore organisms; **whelks** (*Nucella* spp.), the starfish *Pisaster*, and a species of mussel (*Mytilus*) which is preyed on by both *Nucella* and *Pisaster*. The experiment involved replicate groups of mussels in experimental areas of the shore. From some areas, the researchers removed both *Pisaster* and *Nucella*, from some areas only *Pisaster* or only *Nucella*, and from some areas neither predator was removed.

(a) Which line represents the control? _____

(b) What was the effect of removing both predators?

_____

_____

(c) What was the effect of removing only *Pisaster*?

_____

_____

(d) What was the effect of removing only *Nucella*?

_____

_____

Sample data from wave-exposed shore, Boiler Bay, Oregon

Proportion of live mussels

Number of days

Legend

1. Both predators removed.
2. *Pisaster* only removed.
3. *Nucella* only removed.
4. Neither predator removed.

(e) Do these results support the statement that *Pisaster* is a keystone species in rocky shore ecosystems in Oregon? Explain your reasoning, using the data above to justify your statement:

_____

_____

_____

_____

_____

_____

_____

_____

_____

©2017 **BIOZONE** International
ISBN: 978-1-927309-65-0
Photocopying Prohibited

12. The diagram right depicts a food web for a meadow habitat with an area of 25.6 km² . The producer biomass is evenly distributed throughout the area and totals 1500 kg km⁻². Proposed development of the area will reduce the producer biomass by 50% and remove all the rabbits and deer. Predict the likely outcome of this development:

_____

_____

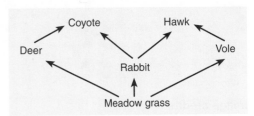

13. The diagram right depicts carbon flow in a grassland ecosystem. Based on the values given, calculate the amount of carbon (in g m⁻²) released into the atmosphere as a result of the metabolic activity of predators:

_____

_____

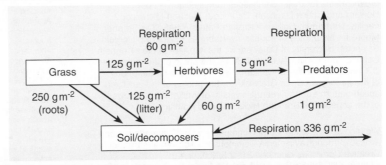

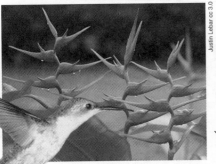

Heliconia:
Bright red or orange, tubular flowers with copious nectar but no scent. *Heliconia* flowers are pollinated virtually exclusively by hummingbirds (birds have a poor sense of smell), which can reach inside the long bracts housing the true flowers.

Northern catalpa:
White trumpet-shaped flowers with yellow stripes and purple spots inside. The floral markings are a UV nectar guide for pollinating insects, which include bumblebees, the large carpenter bee, various moths, and honeybees.

California pipevine:
A vine endemic to California. Flowers have a musty unpleasant odor, which is attractive to their pollinators, carrion-feeding flies and fungus gnats. The insects crawl into the convoluted flowers where they receive a small amount of nectar while temporarily trapped.

14. The photographs above depict three different plants, each typical of their genus and pollinated by a specific animal pollinator. Read the captions and answer the following questions:

(a) What type of ecological relationship is represented by these three examples? _____

(b) What does each participant in the relationship (animal and plant) gain from the interaction? _____

_____

_____

(c) What evolutionary mechanism is responsible for the specificity of the relationships depicted here? Explain:

_____

_____

_____

(d) *Heliconia* flowers are pollinated almost exclusively by hummingbirds. Predict the outcome for *Heliconia* species if hummingbird populations were to decline.

_____

_____

(e) Explain why the Northern catalpa and the California pipevine are less vulnerable in this respect:

_____

_____

Image Credits and Index

Image credits

The writing team would like to thank the following people and organizations for their contributions to this edition: • Louisa Howard and Chuck Daghlian, Dartmouth Electron Microscopy Facility, Dartmouth College • Kristian Peters • PASCO • Stephen Moore for his photos of aquatic invertebrates • Dept of Biological Sciences, University of Delaware for the capillary photo • Kenneth Catania, Vanderbilt University NSF for the photo of the mole • Dr D. Cooper: University of California San Francisco for the photo of the podocyte • Brian Gratwicke for the photos of the gull fixed action pattern • John Green (University of Waikato) for the image of the copepod • Brendan Hicks • David Hamilton

We also acknowledge the photographers who have made images available through Wikimedia Commons Public Domain or under Creative Commons Licences 2.0, 2.5, 3.0, or 4.0:
• Masur • HG6996 • Bob Blaylock • Professor Dr. Habil • Uwe Kills • Trance Gemini • Magne Flåten • Sakaori • flowergarden.noaa.gov • Nestor Galina • Charlesjsharp • Ragesoss • Ed Uthman • Tom Adriaenssens • RM Hunt • Ildar Sagdejev • Mckdanday • Capkuckokos • Dave Powell USDA • Tony Wills • diveofficer • Nhobgood •Bcexp • Saleem Hameed • Michael.Dodge • Jessie Eastland • Florian Prisch • Emmanueim • Pöllö • Volker Brinkmann PLOS • Kristof A. & Klussmann-Kolb A • Obli • Lusb • Eyewire • Ianaré Sév • Eric Wittman • Joseph Berger; Bugwood.org • Paula M Wolter • Fins • CSIRO • Bjørn Christian Tørrissen • Alvesgaspar • P Barden • Uwe Schmidt • Coolstock • Thergothon • Jeremy Kemp • kafka4prez • Tangopaso • Greg Miles • Jim Bendon • JJ Harrison • Ron Knight • Lip Kee • Bernard Dupont • Malene Thyssen • bauer • Opzwartbeek • artfarmer • Sini Merikallio • Beentree • Onno Zweers • Charlesjsharp • Snake3yes • Todd Huffman Lattice • Greg Hume at en.wikipedia • onathunder • Alex Wild • Christin Khan • maciejbledowski • ATamari • Caelio • Adrian A. Smith • Kalyan Varma • Roadnottaken • UC Regents David campus • Eric Houdas •Butterfly austral • Jud McCranie • USAID Bangladesh • Komencanto • GregTheBusker • PhilArmitage • Althepal • Luc Viatour www.Lucnix.be • Steve Garvie • Paul Whippey • BirdPhotos.com • Tony Hisgett • Gregory Antipa (San Francisco State University) • H. S. Wessenberg • UC Berkley • Barfooz and Josh Grosse •Brett Taylor • NJR ZA • US Federal Government • Jan Kronsell • Shirley Owens MSU • Wojsyl • Lubasi-Catedarl Verde-Floresta • Brian Dell • Hardyplants • Ryan Batten • Walter Siegmund • Janke • Mikrolit • Rasbak • Daderot • Victorgrigas • DUK • Wsiegmund • Piet Spaans • Duncan Wright • Fidel Gonzalez • Alvesgaspar • Brocken Inaglory • USDA • Javier Riubliar • Dan Harkless • Plumbago • Scott Ehhardt • Megan McCarty • Gilles Gonthier • Katelyn Murphy • Justin Lebar • Curtis Clark • Wissembourg

Contributors identified by coded credits are: **BH**: Brendan Hicks (University of Waikato), **CDC**: Centers for Diseases Control and prevention, Atlanta, USA, **EII**: Education Interactive Imaging, **KP**: Kent Pryor, **NASA**: National Aeronautics and Space Administration Earth Observatory **NCI**: National Cancer Institute, Rockville, Maryland U.SA, **NOAA**: National Oceanic and Atmospheric Administration www.photolib.noaa.gov, **RCN**: Ralph Cocklin, **USDA**: United States Department of Agriculture, **WBS**: Warrick B. Sylvester (University of Waikato), **Wintec**: Waikato Institute of Technology, **WMU**: Waikato Microscope Unit, **DHN**: David Hamilton, **JB-BU**: Jason Biggerstaff, Brandeis University,

We acknowledge our use of royalty-free images, purchased by BIOZONE International Ltd from the following sources: Adobe Stock, iStock images, Dollar Photo Club, Corel Corporation from various titles in their Professional Photos CD-ROM collection; ©1996 Digital Stock, Medicine and Health Care collection; ©Hemera Technologies Inc, 1997-2001; © 2005 JupiterImages Corporation www.clipart.com; ©1994., ©Digital Vision; Gazelle Technologies Inc.; ©1994-1996, Education Interactive Imaging (UK), PhotoDisc®, Inc. USA, www.photodisc.com. We also acknowledge the following clipart providers: TechPool Studios, for their clipart collection of human anatomy: Copyright ©1994, TechPool Studios Corp. USA (some of these images have been modified); Totem Graphics, for clipart; Corel Corporation, for vector art from the Corel MEGAGALLERY collection.

©2017 **BIOZONE** International
ISBN: 978-1-927309-65-0
Photocopying Prohibited

©2017 **BIOZONE** International
ISBN: 978-1-927309-65-0
Photocopying Prohibited